JACARANDA NATURE OF
BIOLOGY 1

VCE UNITS 1 AND 2 | SIXTH EDITION

JACARANDA NATURE OF
BIOLOGY 1

VCE UNITS 1 AND 2 | SIXTH EDITION

JUDITH KINNEAR

MARJORY MARTIN

SARAH BEAMISH

NICK HAMER

SACHA O'CONNOR-PRICE

CONTRIBUTING AUTHOR

Lucy Cassar

jacaranda
A Wiley Brand

Sixth edition published 2021 by
John Wiley & Sons Australia, Ltd
42 McDougall Street, Milton, Qld 4064

First edition published 1992
Second edition published 2000
Third edition published 2006
Fourth edition published 2013
Fifth edition published 2016

Typeset in 11/14 pt Times Ltd Std

© Judith Kinnear and Marjory Martin 1992, 2000, 2006, 2013, 2016, 2021

The moral rights of the authors have been asserted.

ISBN: 978-0-7303-8805-0

Reproduction and communication for educational purposes
The Australian *Copyright Act 1968* (the Act) allows a maximum of one chapter or 10% of the pages of this work, whichever is the greater, to be reproduced and/or communicated by any educational institution for its educational purposes provided that the educational institution (or the body that administers it) has given a remuneration notice to Copyright Agency Limited (CAL).

Reproduction and communication for other purposes
Except as permitted under the Act (for example, a fair dealing for the purposes of study, research, criticism or review), no part of this book may be reproduced, stored in a retrieval system, communicated or transmitted in any form or by any means without prior written permission. All inquiries should be made to the publisher.

Trademarks
Jacaranda, the JacPLUS logo, the learnON, assessON and studyON logos, Wiley and the Wiley logo, and any related trade dress are trademarks or registered trademarks of John Wiley & Sons Inc. and/or its affiliates in the United States, Australia and in other countries, and may not be used without written permission. All other trademarks are the property of their respective owners.

Cover image: © USAPD1/Alamy Stock Photo

Illustrated by various artists, diacriTech and Wiley Composition Services

Typeset in India by diacriTech

A catalogue record for this book is available from the National Library of Australia

Printed in Singapore
M WEP311139 210924

All activities have been written with the safety of both teacher and student in mind. Some, however, involve physical activity or the use of equipment or tools. **All due care should be taken when performing such activities.** Neither the publisher nor the authors can accept responsibility for any injury that may be sustained when completing activities described in this textbook.

This suite of print and digital resources may contain images of, or references to, members of Aboriginal and Torres Strait Islander communities who are, or may be, deceased. These images and references have been included to help Australian students from all cultural backgrounds develop a better understanding of Aboriginal and Torres Strait Islander peoples' history, culture and lived experience.

Contents

About this resource viii
Accessing online resources xi
Acknowledgements xiii

UNIT 1 HOW DO ORGANISMS REGULATE THEIR FUNCTIONS?   1

AREA OF STUDY 1 HOW DO CELLS FUNCTION?

1 Cellular structure and function   3
    1.1 Overview 4
    1.2 Cells — the basic unit of life 5
    1.3 Limitations to cell size 17
    1.4 Organelles 24
    1.5 The plasma membrane 53
    1.6 Review 78

2 The cell cycle and cell growth, death and differentiation   89
    2.1 Overview 90
    2.2 Cell replication — an asexual process 91
    2.3 Regulation of the cell cycle 103
    2.4 Cell differentiation 118
    2.5 Review 130

AREA OF STUDY 1 REVIEW
Practice examination 141
Practice school-assessed coursework 148

AREA OF STUDY 2 HOW DO PLANT AND ANIMAL SYSTEMS FUNCTION?

3 Functioning systems   151
    3.1 Overview 152
    3.2 Specialisation and organisation of plant cells 153
    3.3 Specialisation and organisation of animal cells 164
    3.4 Digestive system in animals 174
    3.5 Endocrine system in animals 205
    3.6 Excretory system in animals 227
    3.7 Review 243

4 Regulation of systems   251
    4.1 Overview 252
    4.2 Regulation of water balance in vascular plants 253
    4.3 Homeostatic mechanisms and stimulus–response models 263
    4.4 Regulation of body temperature in animals 278
    4.5 Regulation of blood glucose in animals 295
    4.6 Regulation of water balance in animals 305
    4.7 Malfunctions in homeostatic mechanisms 314
    4.8 Review 326

AREA OF STUDY 2 REVIEW
Practice examination 336
Practice school-assessed coursework 343

AREA OF STUDY 3 HOW DO SCIENTIFIC INVESTIGATIONS DEVELOP UNDERSTANDING OF HOW ORGANISMS REGUATE THEIR FUNCTIONS?

5 Scientific investigations *online only*
- 5.1 Overview
- 5.2 Key science skills and concepts in biology
- 5.3 Characteristics of scientific methodology and generation of primary data
- 5.4 Health, safety and ethical guidelines
- 5.5 Accuracy, precision, reproducibility, repeatability and validity of measurements
- 5.6 Ways of organising, analysing and evaluating primary data
- 5.7 Challenging scientific models and theories
- 5.8 The limitations of investigation methodology and conclusions
- 5.9 Presenting key findings and implications of scientific investigations
- 5.10 Conventions of scientific report writing
- 5.11 Review

UNIT 2 HOW DOES INHERITANCE IMPACT ON DIVERSITY? 349

AREA OF STUDY 1 HOW IS INHERITANCE EXPLAINED?

6 From chromosomes to genomes 351
- 6.1 Overview 352
- 6.2 BACKGROUND KNOWLEDGE The role of chromosomes as structures that package DNA 353
- 6.3 The distinction between genes, alleles and a genome 357
- 6.4 Homologous chromosomes, autosomes and sex chromosomes 369
- 6.5 Variability of chromosomes 375
- 6.6 Karyotypes 382
- 6.7 Production of haploid gametes 392
- 6.8 Review 402

7 Genotypes and phenotypes 411
- 7.1 Overview 412
- 7.2 Writing genotypes using symbols 413
- 7.3 Expression of dominant and recessive phenotypes 419
- 7.4 Influences on phenotypes of genetic material, and environmental and epigenetic factors 429
- 7.5 Review 438

8 Patterns of inheritance 445
- 8.1 Overview 446
- 8.2 Pedigree charts and patterns of inheritance 447
- 8.3 Predicting genetic outcomes: a monohybrid cross and a monohybrid test cross 460
- 8.4 Predicting genetic outcomes: linked genes and independently assorted genes 474
- 8.5 Review 488

AREA OF STUDY 1 REVIEW
Practice examination 497
Practice school-assessed coursework 505

AREA OF STUDY 2 HOW DO INHERITED ADAPTATIONS IMPACT ON DIVERSITY?

9 Reproductive strategies — 509
- **9.1** Overview — 510
- **9.2** Asexual reproduction — 511
- **9.3** Advantages of sexual reproduction — 529
- **9.4** Reproductive cloning technologies — 542
- **9.5** Review — 553

10 Adaptations and diversity — 561
- **10.1** Overview — 562
- **10.2** Importance of genetic diversity to survive change — 563
- **10.3** Adaptations for survival — 567
- **10.4** Survival through interdependence between species — 593
- **10.5** Australia's First Peoples — 627
- **10.6** Review — 634

AREA OF STUDY 2 REVIEW
Practice examination — 643
Practice school-assessed coursework — 651

AREA OF STUDY 3 HOW DO HUMANS USE SCIENCE TO EXPLORE AND COMMUNICATE CONTEMPORARY BIOETHICAL ISSUES?

online only

11 Exploring and communicating bioethical issues
- **11.1** Overview
- **11.2** Analysis and evaluation of bioethical issues
- **11.3** Scientific evidence
- **11.4** Characteristics of effective scientific communication
- **11.5** The use of data representations, models and theories
- **11.6** Conventions for referencing and acknowledging sources of information
- **11.7** Review

Glossary — 657
Index — 673

About this resource

Jacaranda Nature of Biology 1 Sixth Edition has been revised and reimagined to provide students and teachers with the most relevant and comprehensive resource on the market. This engaging and purposeful suite of resources is fully aligned to the VCE Biology Study Design (2022–2026).

Formats

Jacaranda Nature of Biology is now available in digital and print formats:

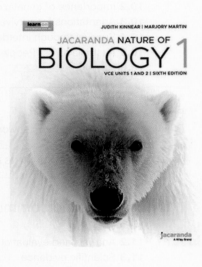

Fully aligned to the VCE Biology Study Design

Have confidence that you are covering the entire VCE Biology Study Design (2022–2026), with:
- key knowledge stated at the start of every topic and subtopic
- explicit support through dedicated topics for key science skills
- tailored exercise sets at the end of every subtopic, including past VCAA exam questions
- additional background information, case studies and extension easily distinguished from curriculum content
- comprehensive topic reviews including a summary flowchart, plus topic review exercises including exam questions
- practice SACs and exams for each Area of Study outcome
- glossary boxes to target key biological literacy
- key ideas at the end of each subtopic to help summarise important points
- suggested practical investigations to support VCAA requirements
- onResources boxes highlighting additional online resources

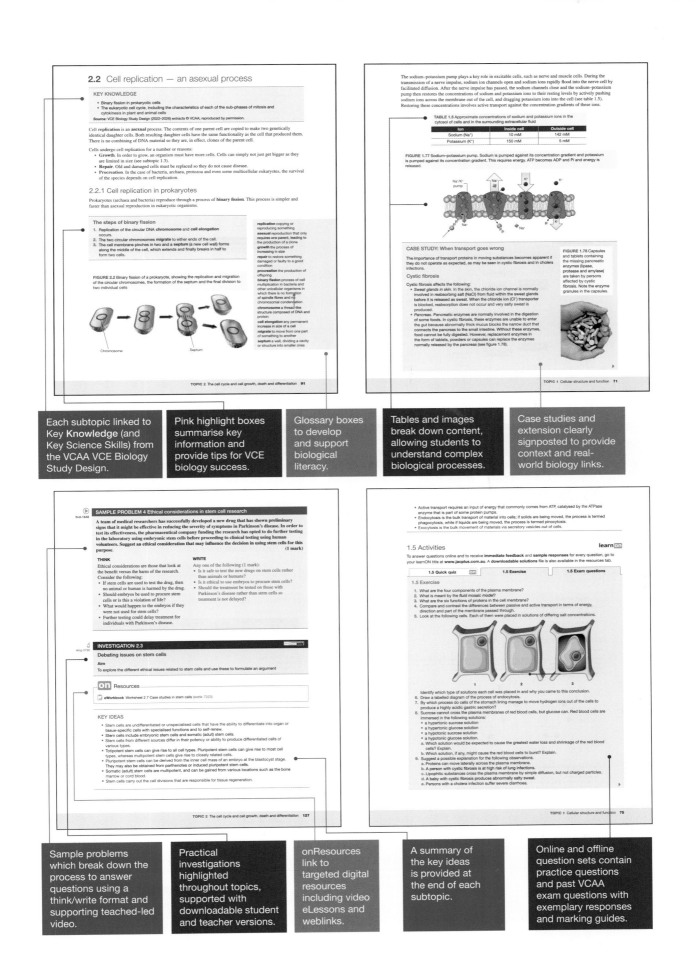

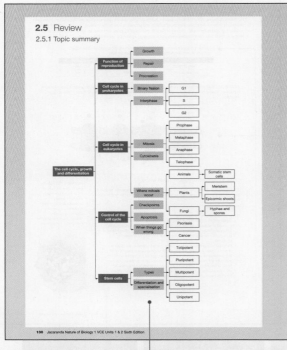

A summary flowchart that shows the interrelationship between the main ideas of the topic. This includes links to both Key Knowledge and Key Science Skills.

End of topic resources including a reflection, key terms glossary and key ideas summary.

Teacher-led videos for every VCAA exam questions through the topic.

End of topic exam questions, containing both multiple choice and short answer VCAA question questions that link together various concepts.

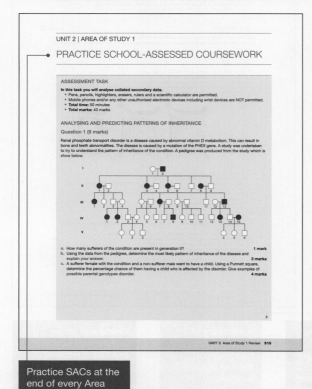

Practice SACs at the end of every Area of Study, providing exposure to various task types.

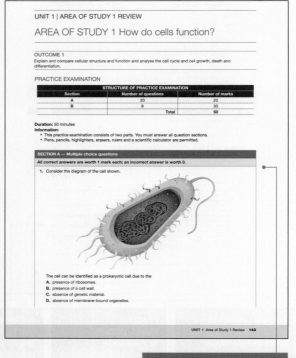

Practice exams at the end of every Area of Study to further support student learning across a broader range of concepts.

x ABOUT THIS RESOURCE

Accessing online resources

The power of learnON

The *Jacaranda Nature of Biology* series is now available on learnON, an immersive digital learning platform that provides teachers with valuable insights into their students' learning and engagement. It's so much more than a textbook! It empowers students to be independent learners and allows teachers to assign, mark and track student work.

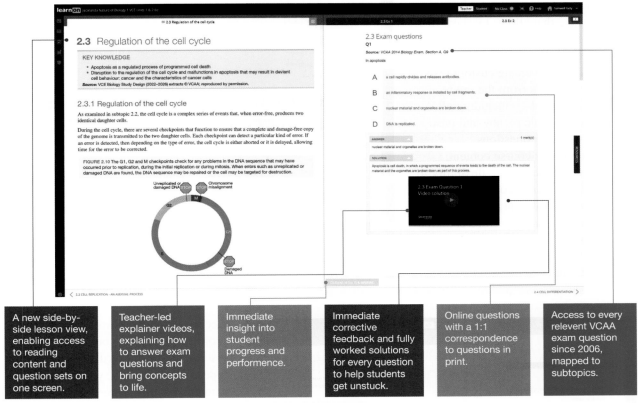

- A new side-by-side lesson view, enabling access to reading content and question sets on one screen.
- Teacher-led explainer videos, explaining how to answer exam questions and bring concepts to life.
- Immediate insight into student progress and performance.
- Immediate corrective feedback and fully worked solutions for every question to help students get unstuck.
- Online questions with a 1:1 correspondence to questions in print.
- Access to every relevent VCAA exam question since 2006, mapped to subtopics.

Alongside the powerful learnON platforms, a downloadable and customisable eWorkbook is available. This is further supplemented by a practical investigation eLogbook, complete with risk assessments, to support all aspects of VCE Biology.

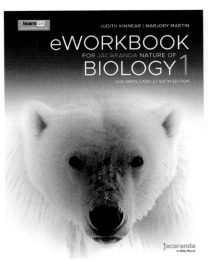
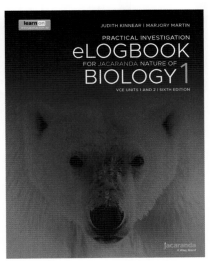

A wealth of teacher resources

Nature of Biology empowers teachers to teach their class their way, with the extensive range of teacher resources including:

- comprehensive support material, including work programs and teaching advice
- quarantined practice SACs and topic tests with answers and exemplary responses
- an easy to navigate marking interface that allows teachers to see student responses, add comments and mark their work
- the ability to create custom tests for your class from the entire question pool — including all subtopic, topic review and past VCAA exam questions
- customisable course content, giving teachers more flexibility to create their own course
- the ability to separate a class into subgroups, making differentiation simpler
- dashboards to track progress and insight to students' strengths and weaknesses.

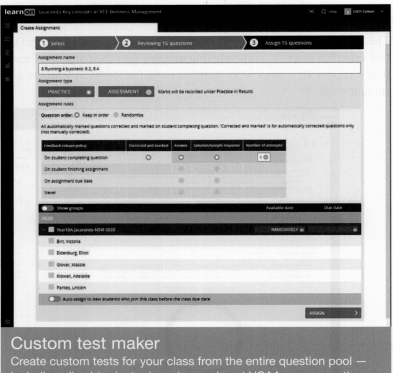

Custom test maker
Create custom tests for your class from the entire question pool — including all subtopic, topic review and past VCAA exam questions.

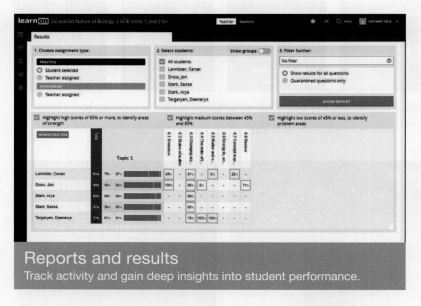

Reports and results
Track activity and gain deep insights into student performance.

Access detailed reports on student progress that allow you to filter results for specific skills or question types. With learnON, you can show students (or their parents or carers) their own assessment data in fine detail. You can filter their results to identify areas of strength or weakness. Results are also colour-coded to help students understand their strengths and weaknesses at a glance.

Acknowledgements

The authors would like to thank those people who have played a key role in the production of this text. Their families and friends were always patient and supportive, especially when deadlines were imminent. This project was greatly enhanced by the generous cooperation of many academic colleagues and friends.

The staff of Wiley are also deserving of the highest praise. Their professionalism and expertise are greatly admired and appreciated. We build on each other's work.

The authors and publisher would like to thank the following copyright holders, organisations and individuals for their permission to reproduce copyright material in this book.

Selected extracts from the VCE Biology Study Design (2022–2026) are copyright Victorian Curriculum and Assessment Authority (VCAA), reproduced by permission. VCE® is a registered trademark of the VCAA. The VCAA does not endorse this product and makes no warranties regarding the correctness or accuracy of its content. To the extent permitted by law, the VCAA excludes all liability for any loss or damage suffered or incurred as a result of accessing, using or relying on the content. Current VCE Study Designs and related content can be accessed directly at www.vcaa.vic.edu.au. Teachers are advised to check the VCAA Bulletin for updates.

Images

- © Marjory Martin: 71 • a. Africa Studio/Shutterstock; b. Ollyy/Shutterstock: 361 • a. Arina_B/Shutterstock; b. Viktoriia Kotliarchuk/Shutterstock; c. PixieMe/Shutterstock: 425 • a. martynowi.cz/Shutterstock; b. www.royaltystockphoto.com/Shutterstock; c. '© MichaelTaylor3d/Shutterstock': 13 • a. Natali Glado/Shutterstock; b. apiguide/Shutterstock; c. bluedogroom/Shutterstock; d. africa924/Shutterstock: 460
- a. Newspix/ Chris Crerar; b. © AAP / Joe Castro; c. Lee Davis/Alamy Stock Photo; d. Pat Sullivan/AAP Image; e. Handout/Getty Images: 548 • a. Samantzis/Shutterstock; b. © Dr Judith Kinnear; c. © Dr Judith Kinnear; d. © Dr Judith Kinnear: 600 • a. Science photo library/Alamy Stock Photo; b. Jose Luis Calvo/Shutterstock; c. Jose Luis Calvo/Shutterstock: 170 • a. © Dr. Torsten Wittmann; b. Dr. Cora-Ann Schoenenberger: 38 • a. © Marjory Martin; b. Seregraff/Shutterstock; c. © Marjory Martin: 362
- Achiichiii/Shutterstock: 11, 37642 • Alamy Stock Photo: 653 • Getty Images/iStockphoto: • Getty Images/Photo Researchers R: 33 • power and syred/science photo li: 375 • Carolina K. Smith/Shutterstock: 164 • © Legionella pneumophila multiplying inside a cultured human lung fibroblast. CDC/Dr. Edwin P. Ewing: Jr, 11 • © 2015 DigitalGlobe: Inc., 597 • © 2020 The Canadian Continence Foundation: 234
- © 3d_man/Shutterstock: 50 • © 3Dstock / Shutterstock: 4 • © AAP Image/Oscar Kornyei:
- © adison pangchai/Shutterstock: 31962 • © agefotostock/Alamy Australia Pty Ltd: 234 • © AJ Photo/Science photo library: 554 • © Akadet Guy/Shutterstock: 191 • © Akarat Phasura/Shutterstock: 185 • © Aldona Griskeviciene/Shutterstock: 26, 85, 255, 360, 414 • © Aldona Griskeviciene/Shutterstock: 9
- © Alessandro Cristiano/ Shutterstock: 446 • © Alila Medical Media/Shutterstock: 54, 61, 187
- © Alizada Studios/Shutterstock: • © Anasta/Shutterstock: 152 • © Andrea Danti/Alamy Stock Photo: 374, 37680 • © Andrea Danti/Shutterstock: 91, 182 • © Andrea Danti/Shutterstock: 186 • © Andrew Bajer: 98
- © Andrew Lock: • © Andrey_Popov/Shutterstock: 319 • © Andrii Vodolazhskyi/Shutterstock: 571
- © Anna Kucherova/Shutterstock: 90522 • © Anna LoFi/Shutterstock: 617 • © Anne Peters: MD/Wikimedia Commons, 319 • © ANTPhoto.com.au • © Ashley Cooper/Alamy Stock Photo: 606282579587 • © Ashley Whitworth/Shutterstock: • © Ashray Shah/Shutterstock: 138 • © Associate Professor Brian Cummings: 125
- © Astrid & Hanns-Frieder Michler/Science photo library: 599 • © Auscape International Pty Ltd/Alamy Stock Photo: 290 • © Barbol/Shutterstock: • © Auscape/Densey Clyne: • © Australian Institute of Health and Welfare: 316 • © Ben Schonewille/Shutterstock: 160338 • © Benedicte Desrus/Alamy Stock Photo: 433
- © blickwinkel/Alamy: 514 • © blickwinkel/Alamy Stock Photo: 193, 630 • © bluedogroom/Shutterstock: 533 • © BlueRingMedia/Shutterstock: 75, 141601 • © BMJ/Shutterstock: • © blvdone/Shutterstock: • © Bob Daemmrich/Alamy Live News: 472 • © Bohbeh/Shutterstock: 586354 • © Bonnie Taylor Barry/Shutterstock:

372 • © Bruce Wilson Photographer/Shutterstock: 588 • © Burger/Phanie/Science photo library: 382 • © c PhotoDisc: Inc, 596 • © Carolina K. Smith MD/Shutterstock: 340 • © Catzatsea/Shutterstock:575 • © Chamaiporn Naprom/Shutterstock: • © Chansom Pantip/Shutterstock: 309341 • © Charles V. Angelo/Science photo library: • © Chase Dekker/Shutterstock: 658618 • © Choksawatdikorn/Shutterstock: 39, 340, 353 • © Chompoo Suriyo/Shutterstock: • © CNRI/Science photo library: 320 • © Christell van der Vyver. Sourced from https://kids.frontiersin.org/: • © Christoph Burgstedt/Science photo library: 254515 • © chromatos/Shutterstock: 323 • © Chronicle/Alamy Stock Photo: 467 • © Claude Nuridsany & Marie Perennou/Science photo library: 561 • © clearviewstock/Shutterstock: 579 • © Cliparea l Custom media/Shutterstock: 215 • © Clusterx/Shutterstock: 5192, 297 • © cronodon.com:52622 • © CSIRO Sustainable Ecosystems Division. Image supplied by Science Image Online: • © dageldog / E+/ Getty Images: • © Dabarti CGI/Shutterstock: 342669 • © Dan Kitwood/Staff/Getty Images Australia: 219 • © DAntes Design/Shutterstock: 395, 396 • © David Cook/blueshiftstudios/Alamy Stock Photo: • © David Evison/Shutterstock: 623 • © David Fleetham/Alamy Stock Photo: 259531 • © David Havel/Shutterstock: 604 • © David Maitland/Getty Images: 606 • © David Massemin/Biosphoto/Alamy: • © David Mckee/Shutterstock: 613 • © Denis OByrne/ANTPhoto.com.au: 136615 • © Dennis Kunkel Microscopy/Science photo library: • © Designua/Shutterstock: 26, 42, 71, 110, 111, 147, 207, 267, 299, 315, 355, 432, 53319 • © Dimarion/Shutterstock: • © Designua/Shutterstock: 76 • © Dmitry Lobanov/Shutterstock: • Dew_gdragon/Shutterstock: 296258 • © Dr Alexia Ferrand: IMCF, University of Basel, 99, 101610 • © Dr Elizabeth Ng and Dr Andrew Elefanty/MCRI: 105 • © Dr Gopal Murti/Science photo library: • © Dr Daniel L. Nickrent/Science photo library: 89, 355396, 539126 • © Dr Jeremy Burgess/Science photo library: 366, 521, 565 • © Dr Jan West: 48, 107, 235, 254, 290, 323, 363, 518, 568, 578, 636 • © Dr Kari Lounatmaa/Science photo library: 523, 659 • © Dr Kari Lounatmaa/Science photo library: • © Dr Judith Kinnear: • © Dr Tony Brain/Science photo library: • © Dr. Igor Siwanowicz: Howard Hughes Medical Institute HHMI Janelia Research Campus Ashburn, Virginia, USA/Nikon Small World, 28 • © Dr. Morley Read/Shutterstock: 351 • © Dr. Norbert Lange/Shutterstock: 41557 • © Dr. Richard Kessel & Dr. Gene Shih: Visuals Unlimited/Science photo library, 109 • © Durden Images/Shutterstock: • © Ed Reschke/Getty Images Australia: 155, 157, 18847 • © ellepigrafica/Shutterstock: 328 • © Emre Terim/Shutterstock: 8, 160, 257, 431 • © EpicStockMedia/Shutterstock:62358, 112 • © Eric Isselee/Shutterstock: 240, 445 • © Eric Lafforgue/Alamy: 323 • © eveline himmelreich/Shutterstock: 423 • © Eye of Science/Science photo library: 44 • © FLPA/Alamy Stock Photo: • © Fotogrin/Shutterstock: 347255 • © Frankie Beggs: -265, S. Karger AG, Basel, 108383 • © From Genetic Analysis by Chromosome Sorting and Parting: Phylogenetic and Diagnostic Applications: by Prof. Malcolm Ferguson Smith, Euro J. Hum. Genet 1997; • © G.E. Schmida/ANTPhoto.com.au: 5: 253 • © GE Healthcare/Jane Stout: 98 • © GeoStock/Getty Images Australia: 598608604 • © gritsalak karalak/Shutterstock: • © Hamady / Shutterstock: • © Gerald Holmes: 577 • © Henk Bentlage/Shutterstock: 60759 • © Henk Vrieselaar/Shutterstock: • © Herbert Gehr / Contributor / Getty Images: 420 • © Ian Dyball/Shutterstock: 582 • © Ian Sedwell/Alamy Stock Photo: 196 • © ilusmedical/Shutterstock: 212, 234 • © imageBroker/Alamy Stock Photo: 284261 • © Images & Stories/Alamy Stock Photo: 520 • © Irina Strelnikova/Shutterstock: 296 • © ittipon/Shutterstock: 601 • © James King-Holmes/Science photo library: 545319 • © James Hager/Age Fotostock: 637 • © Jamilia Marini/Shutterstock: • © Jane Gould / Alamy Stock Photo: • © James Laurie/Shutterstock: • © JJ Gouin/Alamy Stock Photo: 56, 84 • © Jamilia Marini/Shutterstock: 77 • © Johann Schumacher/Alamy Stock Photo: 169284 • © John Bavosi/Science photo library: 257 • © John Devries/Science photo library: 639297524 • © John Radcliffe Hospital/Science photo library: • © John Panella/Shutterstock: • © John Wiley & Sons: Inc., • © Jose Calvo/Science photo library: 113159, 188, 210, 212, 217, 299 • © Jose Luis Calvo/Shutterstock: 4828, 168, 181, 182, 189, 191, 214, 228, 229 • © Jose Luis Calvo/Shutterstock: 123, 229 • © Juan Gaerthner/Shutterstock:15 • © Juan Gaertner/Shutterstock: • © Julian W/ Shutterstock: 13616 • © Judith Kinnear: • © karelnoppe/Shutterstock: 161385 • © Kateryna Kon/Shutterstock: 182, 218, 355, 378, 391, 392, 408216 • © Kay Hawkins/Alamy Stock Photo: • © Kateryna Kon/Shutterstock: • © KellyNelson/Shutterstock: 374 • © Ken Griffiths/ANTPhoto.com.au: 596 • © Ken Griffiths/Shutterstock: 607, 615546 • © Kittipong Somklang/Shutterstock: 312 • © Ken Griffiths/ANTPhoto.com.au: • © Kjersti Joergensen/Shutterstock:623574530 • © Kobkit Chamchod/Shutterstock: 254 • © Klaus Uhlenhut/ANTPhoto.com.au: • © Kym Smith/Newspix: • © LCleland/Shutterstock: • © Lebendkulturen.de/Shutterstock: 10615 • © Lebendkulturen.de/Shutterstock:

2041 • © Lena Lir/Shutterstock: 535 • © Leigh Ackland: • © Look at Sciences/Science photo library: 114379 • © Louise Murray/Science photo library: 570, 586536 • © lunatic67/Shutterstock: 607 • © Luke Shelley/Shutterstock: • © Madlen/Shutterstock: 423 • © magnetix/Shutterstock: 145 • © Manon van Os/Shutterstock: 520 • © MAPgraphics Pty Ltd: Brisbane,578 • © Marc Anderson/Alamy Australia Pty Ltd: 174195 • © Maridav/Shutterstock: 290 • © Marie Lochman/Lochman LT: 596107, 109, 131, 167, 362, 363, 385, 518, 519 • © Marjory Martin/Photo byTeresa Dibbayawan: 44 • © Marjory Martin: • © Mark Clarke/Science photo library: 484 • © Mark Sheridan-Johnson/Shutterstock: 195596 • © Mark Simmons: • © matteo_it/Shutterstock: • © Mary Malloy: • © Mazur Travel/Shutterstock: 610513 • © medicalstocks/Shutterstock: 416 • © Melinda Fawver/Shutterstock: 435 • © meyblume/Shutterstock: 193 • © Meter Group: 256 • © Micrograph by Dr. Conly L. Rieder: Wadsworth Centre, N.Y.S.Dept. of Health, Albany, New York 12201-0509, 96 • © Minden Pictures/Alamy Stock Photo: 537 • © Miriam Maslo/Science photo library: • © Monica Schroeder/Science Source/Science photo library: 564322 • © Monkey Business Images/Shutterstock: • © Nadiia Z/Shutterstock: • © My Litl Eye/Shutterstock: 52186 • © N.Vinoth Narasingam/Shutterstock: 286571, 5975 • © Nasky/Shutterstock: • © NASA: • © National Cancer Institute/Science photo library: 46257 • © National Institute of Diabetes and Digestive and Kidney Diseases: National Institutes of Health, 231, 309 • © Nature Picture Library/Alamy Stock Photo: 628 • © Neil Bromhall/Genesis Films/Science photo library: 206 • © Niday Picture Library/Alamy Stock Photo: 119466 • © no limit pictures/Getty Images: 524 • © Ocean Earth Images/Kevin Deacon: 530 • © Ody_Stocker/Shutterstock: 134 • © Olga Bolbot/Shutterstock: • © Ondrej Prosicky/Shutterstock: 176434 • © OpenStax College: 379 • © OpenStax Microbiology. Provided by: OpenStax CNX. Located at: http://cnx.org/contents/e42bd376-624b-4c0f-972f-e0c57998e765@4.2: • © P.Motta & T.Naguro/Science photo library: • © Paddy Ryan/ANTPhoto.com.au:36609 • © Patrick Guenette/Alamy Stock Vector: 258 • © Peter Hermes Furian/Shutterstock: 440 • © Peter Menzel/Science photo library: 629 • © Peter Parks/imagequestmarine.com: 612 • © Peter Scoones/Science photo library: 603 • © Peter Street: • © Phassa K/Shutterstock: 279411 • © Phillips D/Science photo library: 393 • © phloen/ Shutterstock: 320 • © Photo Researchers/Science History Images / Alamy: 105 • © PhotoEdit / Alamy Australia Pty Ltd: 214 • © photosgenius/Shutterstock: 423 • © Pics by Nick/Shutterstock: 627 • © piya Sukchit/Shutterstock: 159 • © Power and Syred/Science photo library: 96, 353, 359, 415 • © PR Philippe Vago: ISM/Science photo library, 389 • © Professor David Lambert: 595 • © Professor P. Motta & T. Naguro/Science photo library: 59933 • © Public Domain https://commons.wikimedia.org/wiki/Coprinellus_disseminatus\#/media/File:Coprinus_disseminatus.JPG: • © Published in Stem Cell Research: Trends and Perspectives on the Evolving International Landscape: page 21, published by Elsevier.http://www.elsevier.com/__data/assets/pdf_file/0005/53177/Stem-Cell-Report-Trends-and-Perspectives-on-the-Evolving-International-Landscape_Dec2013.pdf, 132 • © PureSolution / Shutterstock: 394 • © Quinn: Leonie & Yin, Da & Cranna, Nicola & Lee, Amanda & Mitchell, Naomi & Hannan, Ross. 2012. Steroid Hormones in Drosophila: How Ecdysone Coordinates Developmental Signalling with Cell Growth and Division. 10.5772/27927., 223 • © Quino Al/Unsplash: 252 • © Radboud University Nijmegen: Faculty of Science,www.vcbio.science.ru.nl Virtual Classroom Biology, 400 • © Ralph Hutchings/Getty Images: • © Randy Rimland/Shutterstock: 201 • © Reinhard Dirscherl/Science photo library: 230654 • © Reprinted by permission from Copyright Clearance Center: Nature Cell Research. Amphibian metamorphosis as a model for studying the developmental actions of thyroid hormone: Jamshed R Tata, [Copyright] 1998., 224531 • © Richard Bowen. Accessed from http://www.vivo.colostate.edu/hbooks/pathphys/endocrine/hypopit/overview.html: 215 • © Richard Whitcombe/Shutterstock: 509 • © Ron and Valerie Taylor/ANTPhoto.com.au: 606 • © Rostislav Stefanek / Shutterstock: 312 • © Roy Caldwell: • © royaltystockphoto.com/Shutterstock: 72 • © Sacha OConnor-Price: 538 • © Sakurra/Shutterstock: 316 • © Sari ONeal / Alamy Australia Pty Ltd: 221 • © Sasha Samardzija / Alamy Australia Pty Ltd: 195 • © Sashkin/Shutterstock: 505 • © Scenics & Science/Alamy Australia Pty Ltd: 212 • © Scenics & Science/Alamy Stock Photo: • © Science History Images/Alamy: 1569 • © Science photo library: 34, 351, 354 • © Science photo library - Steve Gschmeissner / Getty Images: 510 • © Science photo library/ Alamy Australia Pty Ltd: 180, 183, 211, 213 • © Science photo library/ Alamy Stock Photo: 151, 257, 267, 384, 562, 582, 625 • © Science Source/Science photo library: 352 • © Science VU: Visuals Unlimited/Science photo library, 178 • © sciencephotos/Alamy Stock Photo: 37 • © sciencepics/Shutterstock: 50

- © SciePro/Shutterstock: 216 • © Shebeko/Shutterstock: 420 • © sirtravelalot/Shutterstock: 267
- © skyfotostock/Shutterstock: 319 • © Soleil Nordic/Shutterstock: 58 • © Sovereign: ISM/Science photo library, 382 • © SpicyTruffel/Shutterstock: 386, 393, 461, 463, 475 • © Stacey Welu/Shutterstock: 237 • © Stephane Bidouze/Shutterstock: com, 237 • © Steve Cymro/Shutterstock: 50, 519 • © Steve Gschmeissner/Science photo library: 251 • © StevenRussellSmithPhotos/Shutterstock: 435
- © StMayQ/Shutterstock: 582 • © Studio BKK/Shutterstock: 423 • © StudioMolekuul/Shutterstock: 359 • © SunshineVector/Shutterstock: 468 • © Suthiporn Hanchana/Alamy Stock Photo: 289
- © Suzi Eszterhas/Nature Picture Library/Science photo library: 639 • © suzz/Shutterstock: 611
- © Tefi/Shutterstock: 209 • © The Atlas of Living Australia: 572 • © The Book Worm/Alamy Stock Photo: 221, 255 • © The Royal Botanic Gardens and Domain Trust/Jaime Plaza: 542 • © The Washington Post/Contributor/Getty Images: 433 • © ThermoSurvey: 279, 287 • © Timonina/Shutterstock: 317
- © Tinydevil/Shutterstock: 168 • © Tomasz Klejdysz/Shutterstock:223 • © udaix/Shutterstock: 153
- © USAPD1/Alamy Stock Photo: 431 • © VectorMine/Shutterstock:58, 264, 323 • © vetpathologist/Shutterstock: 214 • © Visual&Written SL/Alamy Stock Photo: 166 • © Volodymyr Dvornyk/Shutterstock: 657 • © W.Y. Sunshine/Shutterstock: 57, 59 • © wantanddo/Shutterstock: 412 • © Wikipedia/Public Domain https://commons.wikimedia.org/wiki/File\%3AIshihara_9.png: 465 • © Wildlife GmbH/Alamy Stock Photo: 533
- © Worachat Tokaew/Shutterstock: 611 • © Wuttichok Panichiwarapun/Shutterstock: 425
- © www.cronodon.com: 239, 240 • © Zivica Kerkez/Alamy Stock Photo: 285 • © Zuzanae/Shutterstock: 500, 508 • © Zvitaliy/Shutterstock: 358 • ©Woods Hole Oceanographic Institution.: 612

Every effort has been made to trace the ownership of copyright material. Information that will enable the publisher to rectify any error or omission in subsequent reprints will be welcome. In such cases, please contact the Permissions Section of John Wiley & Sons Australia, Ltd.

UNIT 1
How do organisms regulate their functions?

AREA OF STUDY 1

How do cells function?

OUTCOME 1

Explain and compare cellular structure and function and analyse the cell cycle and cell growth, death and differentiation.

1	Cellular structure and function	3
2	The cell cycle and cell growth, death and differentiation	89

AREA OF STUDY 2

How do plant and animal systems function?

OUTCOME 2

Explain and compare how cells are specialised and organised in plants and animals, and analyse how specific systems in plants and animals are regulated.

3	Functioning systems	151
4	Regulation of systems	251

AREA OF STUDY 3

How do scientific investigations develop understanding of how organisms regulate their functions?

OUTCOME 3

Adapt or design and then conduct a scientific investigation related to function and/or regulation of cells or systems, and draw a conclusion based on evidence from generated primary data.

5	Scientific investigations	347

Source: VCE Biology Study Design (2022–2026) extracts © VCAA; reproduced by permission.

AREA OF STUDY 1 HOW DO CELLS FUNCTION?

1 Cellular structure and function

KEY KNOWLEDGE

In this topic you will investigate:

Cellular structure and function
- cells as the basic structural feature of life on Earth, including the distinction between prokaryotic and eukaryotic cells
- surface area to volume ratio as an important factor in the limitations of cell size and the need for internal compartments (organelles) with specific cellular functions
- the structure and specialisation of plant and animal cell organelles for distinct functions, including chloroplasts and mitochondria
- the structure and function of the plasma membrane in the passage of water, hydrophilic and hydrophobic substances via osmosis, facilitated diffusion and active transport.

Source: VCE Biology Study Design (2022–2026) extracts © VCAA; reproduced by permission.

PRACTICAL WORK AND INVESTIGATIONS

Practical work is a central component of learning and assessment. Experiments and investigations, supported by a **practical investigation eLogbook** and **teacher-led videos**, are included in this topic to provide opportunities to undertake investigations and communicate findings.

1.1 Overview

Numerous **videos** and **interactivities** are available just where you need them, at the point of learning, in your digital formats, learnON and eBookPLUS at **www.jacplus.com.au**.

1.1.1 Introduction

Life is amazing. It exists in so many different forms. It is only known to exist on Earth and yet, is searched for throughout the cosmos. But what is it about life that makes it so extraordinary? Is it that all living things can move, grow and reproduce, passing attributes from parent to offspring? Or is it that life exists in so many different places, some more hospitable than others? Could it be that all life, no matter its shape, size or habitat, is made up of the same basic structural unit — the cell?

Cells are fascinating. They let some substances in but not others. They can have different compartments, which allow them to make all of the molecules that are required for an organism to function, communicate and defend itself from attack. It is this sophisticated functioning of each and every cell in an organism that is necessary for continued life on our planet.

FIGURE 1.1 Tardigrades, or water bears, demonstrate the resilience of life on Earth — found across our planet, from the deep sea, to mud volcanoes, to icy Antarctic waters, they have even survived exposure to outer space.

LEARNING SEQUENCE

1.1	Overview	4
1.2	Cells — the basic unit of life	5
1.3	Limitations to cell size	17
1.4	Organelles	24
1.5	The plasma membrane	53
1.6	Review	78

on Resources

- **eWorkbook** — eWorkbook — Topic 1 (ewbk-2290)
- **Practical investigation eLogbook** — Practical investigation eLogbook — Topic 1 (elog-0156)
- **Digital documents** — Key science skills — VCE Biology Units 1–4 (doc-34648)
  Key terms glossary — Topic 1 (doc-34649)
  Key ideas summary — Topic 1 (doc-34660)

1.2 Cells — the basic unit of life

> **KEY KNOWLEDGE**
>
> - Cells as the basic structural feature of life on Earth, including the distinction between prokaryotic and eukaryotic cells
>
> **Source:** VCE Biology Study Design (2022–2026) extracts © VCAA; reproduced by permission.

1.2.1 Cell theory

Living **organisms** can exist only where:

- an *energy source* is available that can be trapped and utilised by an organism for metabolic processes that maintain its living state
- *liquid water* is available to allow biochemical reactions to occur, and to dissolve chemicals and transport them both within cells and to and from cells
- the *chemical building blocks* required for life are available for use by an organism in cellular repair, growth and reproduction. These chemical building blocks include carbon, oxygen, nitrogen and hydrogen, and each is able to form chemical bonds with other elements. Carbon, in particular, is the most versatile chemical building block, as it can bond with many other elements, forming a variety of complex biomolecules, including long chains.
- *stable environmental conditions* exist within the range of tolerance of an organism, such as pressure, temperature, light intensity, pH and salinity.

FIGURE 1.2 *Dracula simia*, or monkey orchid, is an epiphytic orchid native to the mountains of South-eastern Ecuador, where it is found at altitudes of around 2000 m.

Where these conditions are met, living organisms can use energy to perform the complex set of chemical transformations (metabolic activities) within their cells that sustains their living state. These activities include not only capturing energy but also taking up nutrients and water, as well as removing wastes, so that their internal environment is kept within narrow limits.

All living things share the following attributes, remembered through the acronym MRS GREND.

M — movement: have some level of self-powered movement
R — respiration: the conversion of carbohydrates to a usable energy form (ATP)
S — sensitivity to stimuli: the response of an organism to its environment (e.g. plants responding to light, animals responding to external temperatures by sweating, shivering)
G — growth: an irreversible change in mass
R — reproduction: production of offspring, passing attributes from one generation to the next
E — excretion of wastes: produce wastes, such as dead cells or urine, that need to be removed
N — nutrition: intake of food or nutrients
D — DNA: the molecule that codes for the production of proteins

---

organism any living creature

Although life on Earth exists in many different forms, all life forms share one common feature — they are made up of **cells**. This is the idea that underpins **cell theory**, developed in 1839.

cell basic functional unit of all organisms

cell theory theory that all living things are made of cells

The three tenets of cell theory

1. All organisms are composed of cells.
2. The cell is the basic unit of structure and organisation in organisms.
3. All cells come from pre-existing cells.

BACKGROUND KNOWLEDGE: Development of cell theory — a joint effort!

Cell theory, like many other scientific theories, is derived from the work of many scientists and inventors. Organisms are built of one or more cells. Cells, with only a very few exceptions, are too small to be seen with an unaided eye. Their existence was not recognised until after the development of the first simple microscopes. This enabled the first observations of cells to be made in the 1660s. However, the recognition of cells as the basic unit of life did not occur until almost 200 years later.

Cell theory is credited to three scientists:
- Matthias Schleiden observed that all plants were composed of cells.
- Theodor Schwann observed that animal cells were also made up of cells.
- Rudolf Virchow observed that cells only come from other cells.

This work would not have been possible without the invention of the microscope. Many scientists have been credited with the development and improvement of the microscope: Zacharias Janssen, who was credited with inventing the first compound light microscope; Galileo Galilei, who improved the compound microscope; Robert Hooke, who was the first to describe a cell (figure 1.3a and b); and Anton van Leeuwenhoek, who first described bacteria as 'animalcules' (figure 1.3c and d). Leeuwenhoek built a simple microscope where the specimen was placed on top of the pin, the microscope was held up to the eye and the specimen was viewed through a quartz lens just 2 mm wide.

FIGURE 1.3 a. The microscope built by Robert Hooke that he used to make his first observations of 'little compartments', or cells **b.** The first drawing made by Hooke in 1665 of 'cells' from a thin piece of cork. Were these living cells? **c.** The simple microscope built by Leeuwenhoek **d.** Some of Leeuwenhoek's 'little animacules'

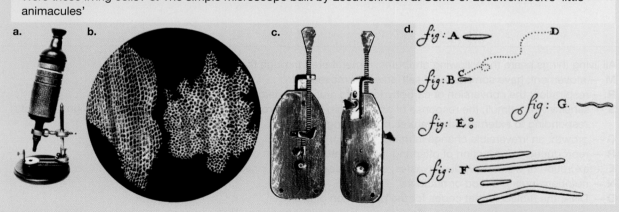

1.2.2 Basic structure of a cell

All cells consist of three components:
- a cell membrane — controls what enters and exits the cell
- nuclear material — the instructions for all processes and structures made by the genetic material, usually DNA
- cytosol — the fluid environment within the cell

1.2.3 Prokaryotic cells and eukaryotic cells

There are two main classes of cells: **prokaryotes** and **eukaryotes**. They are distinguished by the absence or presence of **membrane-bound organelles** (see subtopic 1.4).

A prokaryote is a single-celled organism that has no membrane-bound **organelles**. This means that they have no **nucleus**, so their **deoxyribonucleic acid (DNA)** is free-floating in the cell. Life is thought to have started with the prokaryote. Two different classification groups fit into this category: **bacteria** and **archaea**.

Eukaryotic cells encompass the four other classification groups: plants, animals, fungi and protists. They all have membrane-bound organelles.

Organelles are specialised structures within a cell. Often, but not always, they have a membrane (or two) around them that controls what moves into and out of the organelle. This membrane creates an internal environment in the organelle that can be different to that of the cell cytosol. This becomes important when we delve into the functions of each of these organelles (subtopic 1.4).

FIGURE 1.4 Classification of organisms is based on cell type. Note that in some classification schemes, an additional kingdom of Chromista is included with the eukaryotes, and protozoa is included instead of protista.

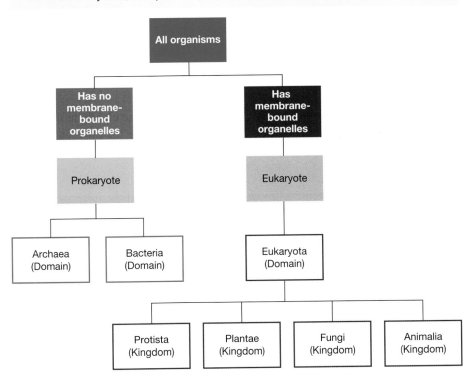

prokaryote any cell or organism without a membrane-bound nucleus

eukaryote any cell or organism with a membrane-bound nucleus

membrane-bound organelle an organelle that has a membrane surrounding it

organelle any specialised structure that performs a specific function

nucleus in eukaryotic cells, membrane-bound organelle containing the genetic material DNA

deoxyribonucleic acid (DNA) nucleic acid containing the four bases — adenine, guanine, cytosine and thymine — which forms the major component of chromosomes and contains coded genetic instructions

bacteria a group of prokaryotes that can reproduce by binary fission

archaea a group of prokaryotes that live in extreme environments; also known as extremophiles

Prokaryotic cells

On average, prokaryotic cells are about ten times smaller than plant and animal cells.

The general features of a prokaryotic cell (figure 1.5) are as follows:
- **Capsule** — made of polysaccharides
- **Cell wall** — made of peptidoglycan
- **Cell (plasma) membrane** — controls which substances move into and out of the cell
- Large, circular DNA — free-floating in the cell
- **Ribosomes** — synthesise proteins
- **Plasmids** — smaller pieces of DNA
- **Cytosol** — water environs that everything floats in.

The role of these features will be discussed further in subtopic 1.4. *Note:* Not all prokaryotic cells possess pili or flagella.

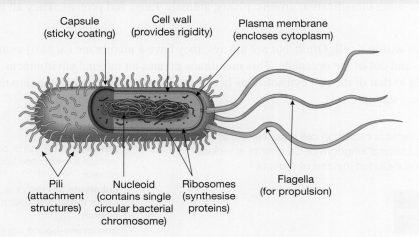

FIGURE 1.5 A generalised diagram of a prokaryotic cell. Prokaryotes lack membrane-bound organelles but do possess a cell membrane.

capsule polysaccharide layer outside the cell membrane for protection

cell wall semi-rigid structure located outside the plasma membrane in the cells of plants, algae, fungi and bacteria

cell (plasma) membrane partially permeable boundary of a cell separating it from its physical surroundings; boundary controlling entry to and exit of substances from a cell

ribosome organelle containing RNA that is the major site of protein production in cells

plasmid small ring of DNA found in prokaryotes

cytosol the aqueous part of the cell

Eukaryotic cells

In comparison to the simple architecture of the prokaryotic cell, eukaryotic cells have a more elaborate internal structure owing to the presence of many membrane-bound organelles. Table 1.1 summarises the difference between prokaryotic and eukaryotic cells. Figure 1.6 demonstrates this contrast.

TABLE 1.1 Comparison of prokaryotic and eukaryotic cells

Feature	Prokaryote	Eukaryote
Size	Small: typically ~1–2 μm diameter	Larger: typically in range 10–100 μm
Chromosomes	Present as single circular DNA molecule	Present as multiple linear DNA molecules
Ribosomes	Present: small size (70S)	Present: large size (80S)
Plasma membrane	Present	Present
Cell wall	Present and chemically complex	Present in plants, fungi and some protists, but chemically simple; absent in animal cells
Membrane-bound nucleus	Absent	Present
Membrane-bound cell organelles	Absent	Present; e.g. lysosomes, mitochondria
Cytoskeleton	Absent	Present

FIGURE 1.6 The organisational difference between eukaryotic and prokaryotic cells

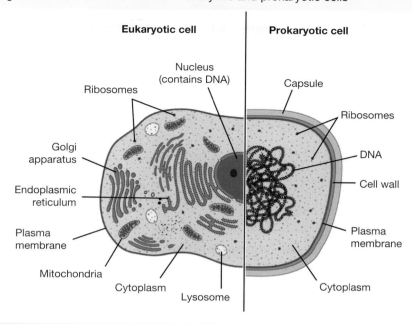

EXTENSION: The difference between archaea and bacteria

Until the 1970s there was only one recognised kingdom of prokaryotes — the kingdom Monera. However, due to the work of Carl Woese (1928–2012) and his colleagues at the University of Illinois in comparing genes at a molecular level (initially ribosomal RNA), it was found that there are significant differences between members of the archaea and the bacteria. These differences were enough to increase the number of domains recognised. Archaea are more similar to eukaryotes, having three RNA polymerases (the enzyme that joins RNA molecules together), whereas members of the bacteria domain only have one. Archaea that can survive in extreme environments are often termed **extremophiles**.

extremophile microbe that lives in extreme environmental conditions, such as high temperature and low pH

SAMPLE PROBLEM 1 Identifying bacterium cells

A student is given samples that contain different cells, as shown in the images.

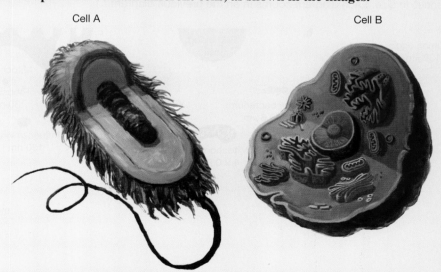

Which cell (A or B) corresponds to the bacterium? Explain your answer. (2 marks)

THINK

Consider the features of each cell to determine which is a bacterium (prokaryote).

Cell A:
- contains no visible compartments
- has free-floating DNA (not in a nucleus).

It must be a prokaryote.

Cell B:
- contains visible compartments
- therefore, has organelles.

It must be a eukaryote.

TIPS:

- When explaining your answer to questions, use the definition to support your answer.
- Keep answers concise by answering in dot point format. The number of marks can indicate how many dot points you should have.

WRITE

- Cell A is the bacterium (1 mark).
- Cell A has no membrane-bound organelles, which is a key feature of prokaryotes (1 mark).

1.2.4 Cell size

Cells are limited in size by their ability to exchange substances with their environment. Therefore, the size and shape of cells vary greatly depending on the organism and the cell's function.

Cells are typically **microscopic**. Only a few single cells are large enough to be seen with an unaided human eye — for example, human egg cells with diameters of about 0.1 mm and the common amoeba (*Amoeba proteus*), a unicellular organism with an average size ranging from 0.25 to 0.75 mm. (You would see an amoeba as about the size of a full stop on this page.) Contrast this with one of the smallest bacteria, *Pelagibacter ubique*, consisting of a cell of just 0.2 μm diameter. How many of these bacteria could fit across an amoeba that is 0.5 mm wide?

FIGURE 1.7 Diagrams, at increasing levels of magnification, showing cells, cell organelles and viruses. Note the extreme differences in size. Each diagram zooms in on the previous diagram.

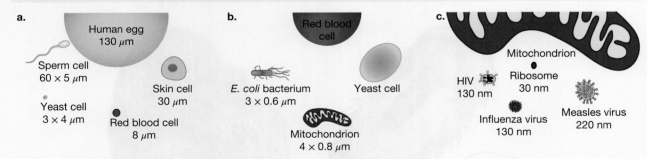

microscopic anything that is not visible to the naked eye and requires a microscope to view it

Table 1.2 shows some typical cell sizes. Cell size can be used to help you define what type of cell you have.

TABLE 1.2 Comparison of cell size based on cell type

Cell type	Cell class	Size range
Bacterial/microbial	Prokaryote	0.4 to 5 µm
Plant	Eukaryote	10 to 400 µm
Animal	Eukaryote	10 to 130 µm

CASE STUDY: Legionnaire's disease, the bacterial invasion of cells

Microbial cells are relatively much smaller than the cells of animals and plants, so some animal bacterial infections can involve the invasion of bacterial cells into the cells of the host, where they multiply.

Figure 1.8 shows a human lung fibroblast with numerous bacterial cells (small dark circular and ovoid shapes) in a single cell. The bacteria are *Legionella pneumophila*, the cause of several infections in people, including Legionnaires' disease.

This image highlights the size difference between microbial cells and the cells of animals (and plants).

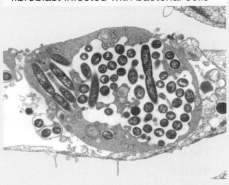

FIGURE 1.8 Transmission electron microscope (TEM) image of a lung fibroblast infected with bacterial cells

tlvd-1738

SAMPLE PROBLEM 2 Identifying types of cells

A laboratory technician has accidently mixed up her samples of cells before she has labelled them. The biology teacher takes this as an opportunity to see what the students have learnt so far by asking them to try to identify the samples. They know that they have a plant sample, an animal sample and a bacterium sample. The students take careful photos under ×400 magnification.

Cell A
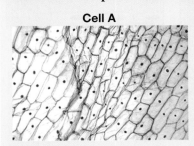

Cell B
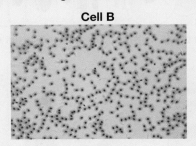

Cell C
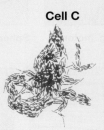

a. Which cell is the bacterial cell? Explain your response. (2 marks)

The teacher tells the students that that one of the eukaryotic cells is a human cheek cell, which averages 60 μm, and the other is an onion skin cell, which averages around 400 μm.

b. Using this information, which cell is likely to be the plant cell? (2 marks)

THINK

a. 1. Cells A and B have a nucleus. Therefore, they have membrane-bound organelles. They must be eukaryotes.
2. Cell C does not have a visible nucleus. It is the prokaryote.

b. Cell A is bigger than cell B.

The onion skin cell (400 μm) is significantly bigger than the human cheek cell (60 μm).

WRITE

As both A and B contain a nucleus, they must be eukaryotes (1 mark).

Cell C is the bacterial cell as it does not have a visible nucleus. It is a prokaryote (1 mark).

Cell A is the plant cell (1 mark).

At the same magnification it is significantly larger than cell B, making it more likely to be the plant cell (1 mark).

1.2.4 Cell shapes

Cells come in a variety of shapes and their shapes are reflective of their functions.

For example:
- Neurons (figure 1.9a) have a long axon, allowing electrical signals to be transferred over large distances quickly. This enables quick responses like reflex actions.
- The concave shape of a red blood cell (figure 1.9f) increases its surface area, allowing greater uptake of oxygen.

FIGURE 1.9 Examples of variations in cell shape: a. star-shaped b. spherical c. columnar d. flat e. elongated f. disc-shaped g. cuboidal

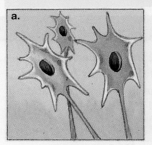
Star-shaped (e.g. motor neuron cells)

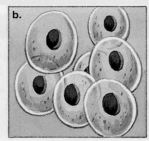
Spherical (e.g. egg cells)

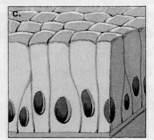
Columnar (e.g. gut cells)

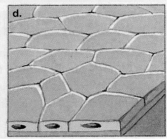
Flat (e.g. skin cells)

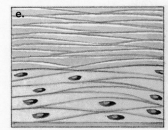
Elongated (e.g. human smooth muscle cells)

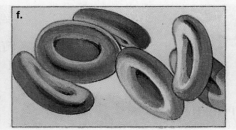
Disc-shaped (e.g. human red blood cells)

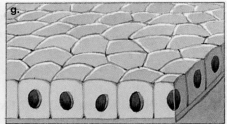
Cuboidal (e.g. human kidney cells)

Microbial cells also vary in shape (figure 1.10). Note that some bacteria are rod-shaped, such as the gut-dwelling bacterium *Escherichia coli*; some are corkscrew-shaped, such as *Borrelia burgdorferi*, the causative agent of Lyme disease; while others are more or less spherical, such as *Streptococcus pneumoniae*, the cause of many infections, including pneumonia.

FIGURE 1.10 Bacterial cells come in many shapes. Some are **a.** rod-shaped bacilli (singular = bacillus), **b.** spiral-shaped and **c.** spherical cocci (singular = coccus).

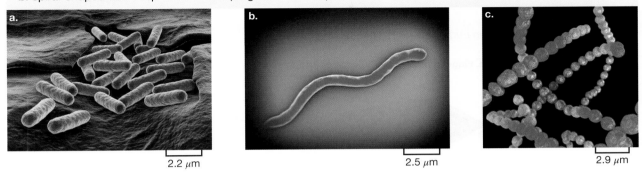

The shape of a cell enables it to have a specific function.

Changing shape

Not all cells have a fixed shape. For example, some cells are able to move actively, and these self-propelled cells do not have fixed shapes because their outer boundary is their flexible plasma membrane. So, as these cells move, their shapes change.

Examples of cells capable of active self-propelled movement include:
- *cancer cells* that migrate into capillaries and move around the body when a malignant tumour undergoes metastasis (see figure 1.11a). The thread-like protrusions (known as filopodia) that fold out from the plasma membrane of cancer cells make the cells self-mobile, so they are able to migrate from a primary tumour and invade other tissues.
- *white blood cells* that can squeeze from capillaries into the surrounding tissues where they travel to attack infectious microbes
- *amoebas* as they move across surfaces (see figure 1.11b).

FIGURE 1.11 a. Cancer cells. Note the many threadlike projections (filopodia) that enable these cells to be mobile, or self-propelling. The ability to move is an important factor in the spread of a malignant cancer. **b.** Outlines showing the changing shape of an amoeba as it moves

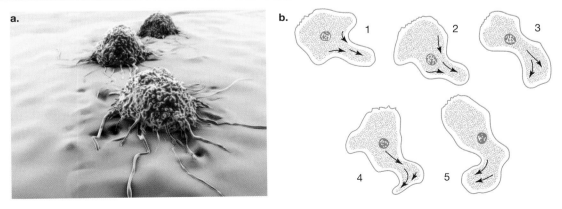

Some cells have a rigid cell wall, which means that changing shape is impossible, but this does not mean that they can't move.

Moving around

As discussed previously, some cells can change shape as they move. But how do cells with a fixed shape move? These cells have appendages such as **cilia** and **flagella** (see section 1.4.3).

cilia in eukaryote cells, whip-like structures formed by extensions of the plasma membrane involved in synchronised movement; singular = cilium

flagella whip-like cell organelles involved in movement; singular = flagellum

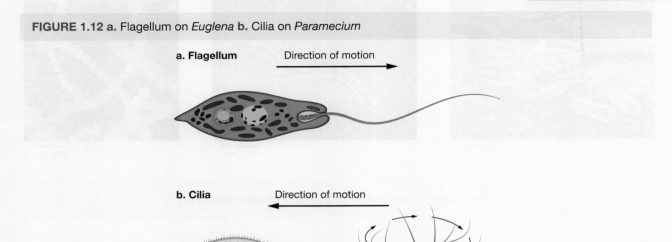

FIGURE 1.12 a. Flagellum on *Euglena* b. Cilia on *Paramecium*

elog-0774

INVESTIGATION 1.1

online only

Cells under the microscope

Aim

To observe differences between prokaryotes and eukaryotes under the microscope

 Resources

 eWorkbook Worksheet 1.1 Exploring cells (ewbk-7186)

 Video eLesson Living organisms are made of cells (eles-4165)

KEY IDEAS

- For life to survive in a place, there are a set of conditions that need to be met:
  - energy source
  - liquid water
  - building blocks to make necessary molecules
  - stable environment.
- Cells are the basic structural and functional units of life.
- Cells are typically too small to be seen by an unaided eye.
- The unit of measurement used for cell size is the micrometre (μm), one millionth of a metre.
- Microbial cells are much smaller than plant and animal cells.

- Prokaryotes (archaea and bacteria) differ from eukaryotes in that they have no internal compartments, called organelles.
- Archaea are also known as extremophiles as they live in extreme environments.
- Cells have different shapes that meet their function.
- Movement of the cell membrane enables some cells to change shape and to move through a watery environment.
- Cilia and flagella are appendages that allow movement of certain cell types.

1.2 Activities

To answer questions online and to receive **immediate feedback** and **sample responses** for every question, go to your learnON title at **www.jacplus.com.au**. A **downloadable solutions** file is also available in the resources tab.

| 1.2 Quick quiz | 1.2 Exercise | 1.2 Exam questions |

1.2 Exercise

1. Outline the three tenets of cell theory.
2. Several different types of cells have cell walls: bacteria, fungi and plants. How would the composition of the cell wall help to determine if an unknown cell is bacterial?
3. Skin epidermal cells have a flattened shape (see figure 1.9d). How does this shape enable their function as barrier for the body?
4. A student is given samples of different cells. These are shown in the images provided.

Sample A

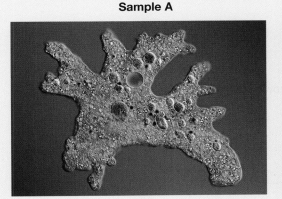

Sample B

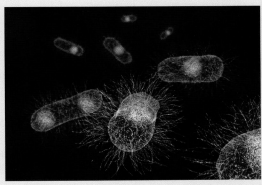

   a. Which of the cells, A or B, is the eukaryote? Explain your response.
   b. How would each of these cells move through their environment?

5. Students are asked to determine the cell types of three cell samples. They carefully measure each cell under the microscope and make note if there are internal compartments visible. They record the following results.

TABLE Cell sample sizes and observations

Cell sample	Cell size (μm)			Average	Compartments visible?
	1	2	3		
X	47	49	55	50	Yes
Y	100	150	70	110	Yes
Z	5	4	5	5	No

Based on this information, identify the likely cell types of X, Y and Z, giving reasons for your decision. Ensure you use the evidence from the table to justify your conclusions.

1.2 Exam questions

Question 1 (1 mark)
MC An example of a prokaryotic organism is
A. an amoeba.
B. a *Paramecium*.
C. a bacterium.
D. an *Euglena*.

Question 2 (1 mark)
MC Eukaryotic cells
A. all have cell walls.
B. can be found in fungi.
C. contain plasmids.
D. have large vacuoles.

Question 3 (2 marks)
Discuss two features of the cell shown that suggest that it is a prokaryotic cell.

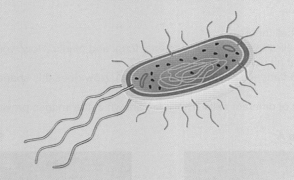

Question 4 (2 marks)
In our blood there are different cell types. The photograph shows a human blood smear.

Consider the cell labelled A.

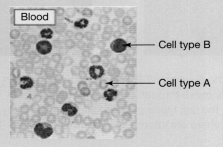

a. Describe what is unusual about this cell. **1 mark**
b. Explain why cell type B is a different size and shape to cell type A. **1 mark**

Question 5 (2 marks)
A multicellular organism may consist of many different types of cells rather than many cells of the same type. Explain how this may be an advantage for the multicellular organism.

More exam questions are available in your learnON title.

1.3 Limitations to cell size

KEY KNOWLEDGE

- Surface area to volume ratio as an important factor in the limitations of cell size and the need for internal compartments (organelles) with specific cellular functions

Source: VCE Biology Study Design (2022–2026) extracts © VCAA; reproduced by permission.

1.3.1 Substance movement limits cell size

Surface area to volume ratio

Why are cells microscopically small? Would it be more efficient to have a larger macroscopic unit to carry out cellular processes rather than many smaller units occupying the same space?

Every living cell must maintain its internal environment within a narrow range of conditions, such as pH and the concentrations of ions and chemical compounds. At the same time, a cell must carry out a variety of functions that are essential for life. These functions include trapping a source of energy; obtaining the chemical building blocks needed for cellular repair, growth and reproduction; taking up water and nutrients; and removing wastes.

The cell's essential functions rely on the plasma membrane to allow a constant exchange of material between the cell and the external environment.

The plasma membrane must:
- enable enough exchange between the external and internal environments to support these life functions
- ensure the exchange of materials occurs at rates sufficient to deliver substances fast enough into cells to meet their nutrient needs, and remove wastes fast enough from the cells to avoid their accumulation.

A critical issue in keeping a cell alive is the surface area of plasma membrane available to supply material to or remove wastes from the metabolically active **cytoplasm** of the cell.

The **surface area to volume ratio** (SA : V) is a measure of the critical balance between the surface area available for substance exchange and the volume of the cell.

The surface area of a cube is given by the equation:

$$SA = 6L^2$$

where L = the length of one side of the cube.

The volume of a cube is given by the equation:

$$V = L^3$$

cytoplasm formed by cell organelles, excluding the nucleus, and the cytosol

surface area to volume ratio a measure that identifies the number of units of surface area available to 'serve' each unit of internal volume of a cell, tissues or organism

Examine figure 1.13. Note that as the cubes increase in size, their volumes enlarge faster than their surface areas expand. As the side length doubles, the surface area increases by a factor of 4 but the volume increases by a factor of 8. This is reflected in a decrease in the SA : V ratio as the cube grows bigger.

This generalisation applies to other shapes; that is, the SA : V ratio of a smaller object is higher than that of a larger object with the same shape. The higher the SA : V ratio, the greater efficiency of two-way exchange of materials across the plasma membrane; that is, efficient uptake and output of dissolved material is favoured by a high SA : V ratio.

The same principle applies to cells:
- As cells increase in size through an increase in cytoplasm, both their surface areas and volumes increase, but *not* at the same rate.
- The internal volumes of cells expand at a greater rate than the areas of their plasma membrane. Therefore, the growth of an individual cell is accompanied by a relative decrease in the area of its plasma membrane.

FIGURE 1.13 The surface area (SA) and volume (V) of cubes with increasing side lengths (L). As a cube increases in size, the volume and surface area increase, but the surface area to volume ratio decreases.

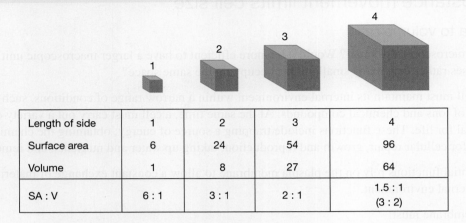

Length of side	1	2	3	4
Surface area	6	24	54	96
Volume	1	8	27	64
SA : V	6 : 1	3 : 1	2 : 1	1.5 : 1 (3 : 2)

How SA : V ratios limit cell size

- As the volume of a cell increases, the metabolic needs of the cell increase.
- These needs can only be met by an increase in the inputs and outputs of materials across the plasma membrane.
- This increase in materials can only increase in proportion to the cell's surface area. The increase in the surface area is always less than the increase in the volume.
- Therefore, as a cell increases in volume, its metabolic needs increase faster than the cell's ability to transport the materials into and out of the cell to meet those needs.

As cells increase in size, the continued decrease in SA : V ratio places an upper limit on cell size.

This is the one clue as to why metabolically active cells are so small.

Additionally, the rate at which nutrients enter and wastes leave a cell generally is inversely proportional to the cell size, as measured in metabolically active cytoplasm; in other words, the larger the cell, the slower the rate of movement of nutrients into and wastes out of a cell. Beyond a given cell size, the two-way exchange of materials across the plasma membrane cannot occur fast enough to sustain the volume of the cell's contents. If that cell is to carry out the functions necessary for living, it must divide into smaller cells or die.

Compensating for the decrease in SA : V ratio

Compare the surface areas of a clenched fist with that of an outspread hand in figure 1.14. Both the clenched fist and the hand with outstretched fingers have the same volume of tissue. However, the hand with outstretched fingers has a much greater surface area as compared with the clenched fist. If these were cell shapes, which cell would be more efficient in the exchange of materials into and out of the cell with the external environment?

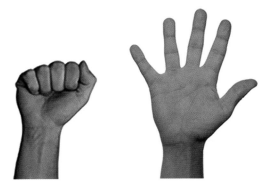

FIGURE 1.14 The presence of fingers greatly increases the surface area of the hand.

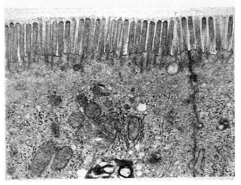

FIGURE 1.15 TEM image showing a section through part of two cells from the lining of the small intestine. Note the multiple folds of the plasma membranes.

The cells that function in the absorption of digested nutrients from your small intestine compensate for this decrease in SA : V ratio with increasing size in the same way as the hand in figure 1.14. They greatly increase their surface area with only a minimal increase in cell volume by extensive folding of the plasma membrane on the cell surface that faces into the gut lumen (figure 1.15). These folds are termed microvilli (singular = microvillus). A single cell lining the small intestine may have up to 10 000 microvilli on its apical surface facing into the gut lumen. Surfaces of other cells with either a major absorptive or secretory function also show microvilli.

tlvd-1739

SAMPLE PROBLEM 3 Surface area to volume ratios of cells

There are (broadly) two types of blood cells in the body. Red blood cells are responsible for the uptake of oxygen as they pass through the lungs and delivery of this oxygen to where it is needed around the body. A red blood cell is typically around 6 μm in size. A white blood cell is typically three times larger and has a role in the defence against diseases that infect the body.

a. Assuming that both these cells are spherical and the SA : V is given by $1 : \frac{3}{r}$, where r is the radius of the cell, calculate the surface area to volume ratio for each of these cells. **(2 marks)**
b. Explain the impact of this difference in surface area to volume ratio between red blood cells and white blood cells. **(2 marks)**

THINK

a. 1. A red blood cell (RBC) is 6 μm in diameter. So, it has a radius of 3 μm.
We can use the SA : V ratio formula given in the question.
Remember to show calculations to demonstrate understanding of process.

WRITE

The surface area to volume ratio (SA : V) of the red blood cell:

$$SA : V \text{ (RBC)} = 1 : \frac{3}{r}$$
$$= 1 : \frac{3}{3}$$
$$= 1 : 1 \text{ (1 mark)}$$

2. A white blood cell (WBC) is three times larger than the RBC.
Determine the radius of the WBC and use the SA : V ratio formula given in the question.
Remember to acknowledge abbreviations.

$$\text{Radius (WBC)} = 3 \times 3\,\mu m$$
$$= 9\,\mu m$$
$$\text{SA : V (WBC)} = 1 : \frac{3}{r}$$
$$= 1 : \frac{3}{9}$$
$$= 1 : 0.33 \text{ (1 mark)}$$

b. The SA : V is significantly larger in the RBC than the WBC.
This would allow faster uptake and removal of substances by the RBC than the WBC. Consider the roles of the RBCs and WBCs in the body.
- RBCs uptake oxygen for transport around the body.
- WBCs have a role in defence of the body against illness.

Remember to refer back to the stem of the question in your response.

The SA : V is significantly larger in the RBC than the WBC (1 mark).
This would allow faster uptake and removal of substances like oxygen by the RBC than the WBC, which would enable the RBCs function of transporting oxygen around the body (1 mark).

CASE STUDY:
Sea anemone — increasing the SA : V ratio in an organism

Surface area to volume ratio considerations apply not only to individual cells but also to entire organisms; for example, the sea anemone (figure 1.16) has many thin tentacles, each armed with stinging cells — these provide a greatly increased surface area for gaining nutrients for the whole animal (as compared with a single flat sheet of cells).

FIGURE 1.16 Anemones have a high SA : V ratio.

elog-0776

INVESTIGATION 1.2

online only

What limits the size of cells?

Aim

To examine the effects of the surface area to volume ratio by modelling the movement of material into cells

Resources

eWorkbook Worksheet 1.2 Surface area to volume ratio (ewbk-7188)

KEY IDEAS

- The metabolic needs of a cell are determined by its metabolically active cytoplasmic volume.
- The ability of a cell to meet its metabolic needs is determined by the surface area of the cell.
- As a cell increases in size, its internal volume expands at a greater rate than the area of its plasma membrane.
- The surface area to volume ratio (SA : V) of a smaller object is higher than that of a larger object with the same shape.
- The continued decrease in SA : V ratio as metabolically active cells increase in size places an upper limit on cell size.

1.3 Activities

learn on

To answer questions online and to receive **immediate feedback** and **sample responses** for every question, go to your learnON title at **www.jacplus.com.au**. A **downloadable solutions** file is also available in the resources tab.

| 1.3 Quick quiz | 1.3 Exercise | 1.3 Exam questions |

1.3 Exercise

1. Why is cell size limited?
2. Identify three ways in which surface area to volume ratio (SA : V) can be increased.
3. The following graph shows the SA : V ratio of cuboidal cells.

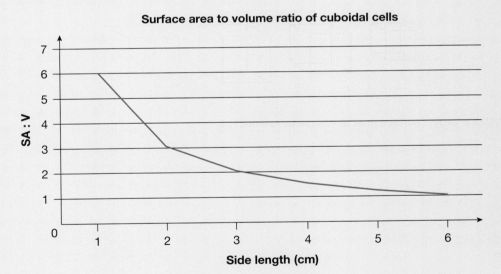

Surface area to volume ratio of cuboidal cells

a. Describe the relationship between side length and SA : V ratio.
b. What does an SA : V ratio of 3 tell you?

4. Two cells (P and Q) have the same volume, but the surface area of cell P is ten times greater than that of cell Q.
   a. Placed in the same environment, which cell would be expected to take up dissolved material at a greater rate? Why?
   b. What might reasonably be inferred about the shapes of these two cells?
   c. Which measure — surface area or volume — determines the rate at which essential materials can be supplied to a cell? Explain.
   d. Which measure — surface area or volume — determines the needs of a cell for essential materials? Explain.
   e. Briefly explain why the SA:V ratio provides a clue as to why cells are microscopically small.
5. The human skin cell is approximately 30 µm in diameter. Its primary role is to prevent infection. Motor neuron cells are responsible for carrying electrical messages from the brain to muscles to make them move. This process involves the fast movement of sodium and potassium ions along the length of the cell. They can be around 1 m (1 000 000 µm) in length and 4 µm in diameter.
   a. Assuming that the skin cells are cubes and the motor neuron cells are rectangular prisms, compare the SA:V ratios of these cells.
   b. How does the SA:V ratio help these cells meet their function?

1.3 Exam questions

Question 1 (1 mark)

MC The size of a cell is limited by its
A. surface area.
B. volume.
C. surface area to volume ratio.
D. mass.

Question 2 (1 mark)

MC Consider the following four shapes.

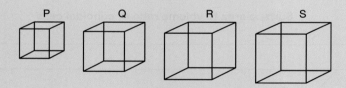

The shape with the largest surface area to volume ratio would be
A. P.
B. Q.
C. R.
D. S.

▶ Question 3 (1 mark)

MC The following cells all have the same volume.

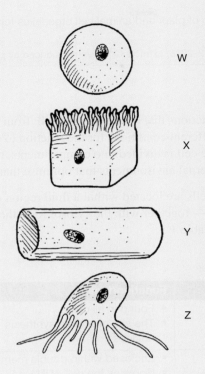

The cell with the smallest surface area to volume ratio is
A. W.
B. X.
C. Y.
D. Z.

▶ Question 4 (2 marks)

A diagram showing the shape of three different cells is shown.

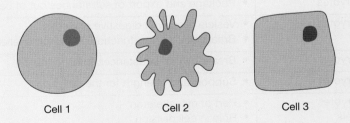

Choose which of the three cells would be most suitable for absorption of materials into the cell. Explain your choice.

▶ Question 5 (1 mark)

Cells are usually microscopic in size. Explain why cells remain very small in size.

More exam questions are available in your learnON title.

1.4 Organelles

> **KEY KNOWLEDGE**
>
> - The structure and specialisation of plant and animal cell organelles for distinct functions, including chloroplasts and mitochondria
>
> *Source:* VCE Biology Study Design (2022–2026) extracts © VCAA; reproduced by permission.

1.4.1 Inside the cell

There are a multitude of different reactions that occur in every cell, from DNA replication to cellular respiration. Each reaction can require slightly different conditions for the reaction to occur at its optimal rate. So how does a cell meet all these requirements? The cell is divided into compartments, called organelles, that have very specific functions and therefore will have internal environments that optimise that function.

The organelles in plant and animal cells are located within a fluid region of the cell called the cytosol. The organelles — excluding the nucleus — together with the cytosol form the cytoplasm of a cell. The organelles we will examine are summarised in table 1.3.

TABLE 1.3 Summary of cell organelles

Organelle	Found in	Function/feature
Ribosome	Prokaryotes Eukaryotes	• Found in all cells • The site of protein synthesis, using the genetic code from the nucleus
Nucleus	Eukaryotes	• Enclosed within double membrane (nuclear envelope) • Contains the cell's DNA • Controls DNA replication during cell division • Repairs genetic material • Initiates gene expression • Controls metabolic activities of a cell • Contains nucleolus
Mitochondria	Eukaryotes	• Main site of ATP production for energy (through cellular respiration)
Endoplasmic reticulum (ER)	Eukaryotes	• Transport system within cells • Protein modification (rough ER) • Lipid synthesis and storage, and detoxification (smooth ER)
Golgi apparatus	Eukaryotes	• Package and export of substances out of cell
Lysosome	Eukaryotes	• Vesicle filled with digestive enzymes • Breakdown of non-functioning cell organelles and substances
Peroxisome	Eukaryotes	• Breakdown of substances toxic to the cell
Cytoskeleton	Eukaryotes	• Support and strength for the cell
Centriole	Eukaryotes — animals	• Part of cytoskeleton • Role in cell division
Chloroplasts	Eukaryotes — photosynthetic species (such as plants)	• Convert energy from the Sun using photosynthesis

(continued)

TABLE 1.3 Summary of cell organelles *(continued)*

Organelle	Found in	Function/feature
Vacuole	Eukaryotes — plants	• Storage of nutrients and mineral salts • Waste disposal • Large and central in plant cells
Cell wall	Most prokaryotes Eukaryotes — plants, algae and fungi	• Provides protection, shape and support to the cell

1.4.2 Ribosomes: an organelle for all cells

Ribosomes are found in all cells. They lack a membrane and are extremely small — about 0.03 μm in diameter.

> The function of a ribosome is to convert a code from the nucleus (messenger RNA or mRNA) into a functional protein.

Ribosomes are made of **ribosomal RNA (rRNA)** and proteins. These form two sub-units that lock together around the **messenger RNA (mRNA)** (the code for the protein from the nucleus of the cell), as shown in figure 1.18, and convert it to a chain of amino acids to form a functional **protein** (figure 1.19).

Ribosomes are small, but are numerous in the cell. They are found either free-floating in the cytoplasm or, in eukaryotes, attached to the membrane of the **endoplasmic reticulum** (see section 1.4.3). Proteins made by ribosomes attached to the endoplasmic reticulum are generally for export from the cell, whereas those made by the free-floating ribosomes are generally proteins used within the cell.

ribosomal RNA (rRNA) ribosomal ribonucleic acid; synthesises proteins and is the primary component of ribosomes

messenger RNA (mRNA) single-stranded RNA formed by transcription of a DNA template strand in the nucleus; mRNA carries a copy of the genetic information into the cytoplasm

proteins macromolecules built of amino acid sub-units and linked by peptide bonds to form a polypeptide chain

endoplasmic reticulum cell organelle consisting of a system of membrane-bound channels that transport substances within the cell

FIGURE 1.17 Diagram of a eukaryotic animal cell; the ribosomes are highlighted in green

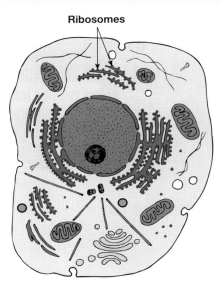

FIGURE 1.18 Ribosomes are made of two sub-units that lock together.

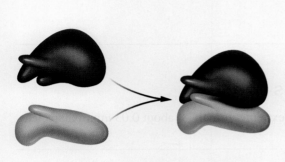

FIGURE 1.19 A ribosome locks around mRNA to convert it into a polypeptide chain of amino acids, which will later become a functional protein.

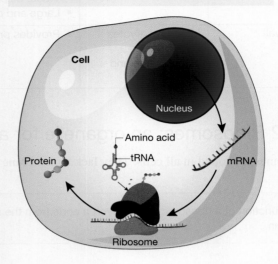

The importance of proteins

The polypeptide chain of amino acids assembled at the ribosomes is folded into a three-dimensional shape to make a functional protein.

Cells make a range of proteins for many purposes. For example, development of human red blood cells in the bone marrow manufacture of haemoglobin, an oxygen-transporting protein; manufacture of the contractile proteins, actin and myosin, by the muscle cells; and manufacture of the hormone insulin and digestive enzymes, including lipases by different cells of the pancreas.

EXTENSION: Protein production

The production of proteins is a particularly important process in cells. This process is split into two main stages — transcription and translation.

Protein production is an extension concept for the Units 1 & 2 course but is examinable in the Units 3 & 4 course.

To access more information on this extension concept please download the digital document.

FIGURE 1.20 Translation: mRNA is converted into a chain of amino acids.

on Resources

📄 **Digital document** Extension: Protein production (doc-35857)

1.4.3 Organelles for eukaryotes

Eukaryotes are defined as having membrane-bound organelles, compared to prokaryotes that have none. The membrane-bound organelles shown in figure 1.21a are found in all eukaryotic animal cells and the membrane-bound organelles shown in figure 1.21b are found in all eukaryotic plant cells.

FIGURE 1.21 Membrane-bound organelles found in all eukaryotic a. animal cells and b. plant cells

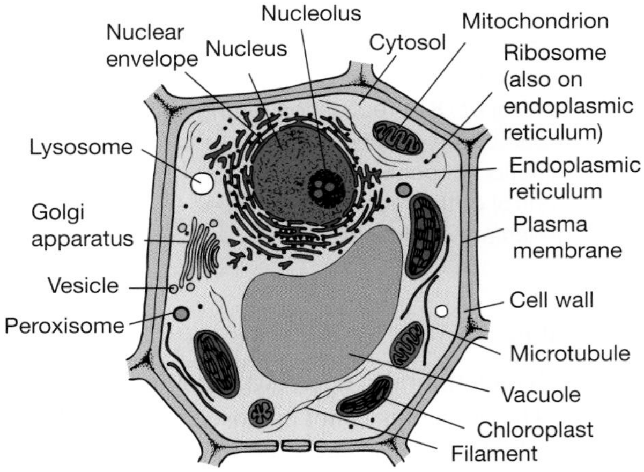

Nucleus: the control centre

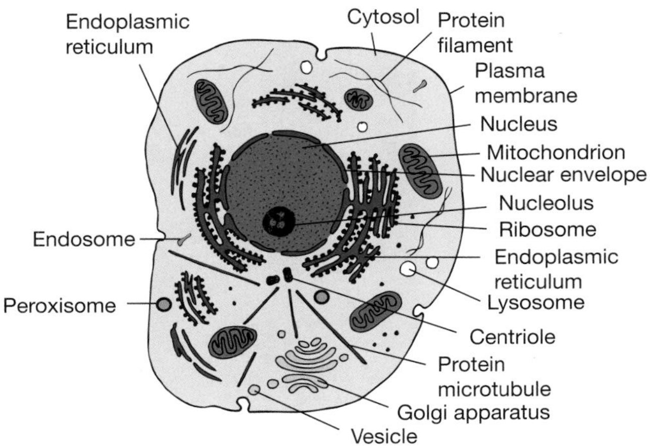

Cells have a complex internal organisation and are able to carry out many functions. The centre that controls these functions in the cells of animals, plants and fungi is the nucleus (see figure 1.22).

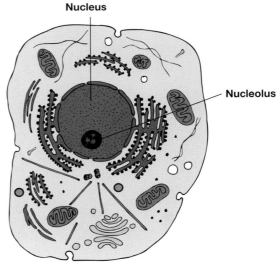

FIGURE 1.22 Diagram of a eukaryotic animal cell; the nucleus is highlighted in red

The nucleus is the defining feature of eukaryotic cells and it is a distinct spherical structure that is enclosed within a double membrane, known as the **nuclear envelope** (or nuclear membrane). The nuclear envelope is perforated by protein-lined channels called **nuclear pore complexes (NPCs)** (see figure 1.23). In a typical vertebrate animal cell, there can be as many as about 2000 NPCs that control the exchange of materials between the nucleus and the cytoplasm. Large molecules are transported into and out of the nucleus via the NPC nuclear pore complexes. Molecules that move from the nucleus to the cytoplasm include RNA and ribosomal proteins, while large molecules that move into the nucleus from the cytoplasm include proteins.

The most important molecule housed in the nucleus is the genetic material DNA. The DNA is usually dispersed within the nucleus. During the process of cell reproduction, however, the DNA becomes organised into a number of rod-shaped chromosomes. The nucleus also contains one or more large inclusions known as nucleoli (singular = nucleolus) (see figure 1.24).

nuclear envelope membrane surrounding the nucleus of a eukaryotic cell

nuclear pore complex (NPC) protein-lined channel that perforates the nuclear envelope

All cells have the same complement of DNA (except for reproductive cells), but the set of genes expressed in one type of cell differentiates these cells from cells of another type. For example, smooth muscle cells express the genes for the contractile proteins actin and myosin. These genes are silent in salivary gland cells that express the gene for the digestive enzyme amylase. Both cells, however, express the genes concerned with essential life processes, such as those involved with cellular respiration.

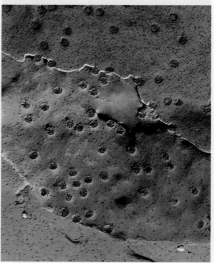

FIGURE 1.23 False-coloured TEM image of part of the nuclear envelope of a liver cell. The inner membrane (top blue) and the outer membrane (brown) are both visible. The rounded pores on the membrane allow large molecules to exit the nucleus and move into the cytosol.

The nucleus houses the genetic material and the functions of the nucleus include:
- control of DNA replication during cell division
- repair of the genetic material
- initiation of gene expression
- control of the metabolic activities of a cell by regulating which genes are expressed.

Nucleolus

The **nucleolus** (plural = nucleoli) is composed of **ribonucleic acid (RNA)**. The nucleolus is not enclosed within a membrane. The function of the nucleolus is to produce the ribosomal RNA (rRNA) that forms part of the ribosomes. It is often darker in appearance when stained and viewed under a light microscope.

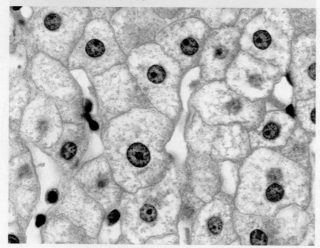

FIGURE 1.24 Liver cells (hepatocytes) seen with a light microscope. Their nuclei show a very large nucleolus stained in red.

nucleolus a small, dense, spherical structure in a nucleus that is composed of RNA and produces rRNA

ribonucleic acid (RNA) type of nucleic acid consisting of a single chain of nucleotide sub-units that contain the sugar ribose and the bases A, U, C and G

EXTENSION: Cells with multiple nuclei

Textbook diagrams typically show a cell as having a single nucleus. This is the usual situation, but it is not always the case. Some liver cells have two nuclei. Your bloodstream contains very large numbers of mature red blood cells, each with no nucleus. However, at an earlier stage — as immature cells located in your bone marrow — these cells did have a nucleus. Given the function of red blood cells, can you think of any advantage that the absence of a nucleus might give to them? Skeletal muscle is the voluntary muscle in your body that enables you to stand up, pick up a pen or kick a soccer ball. Skeletal muscle consists of long fibres that are formed by the fusion of many cells. Such structures contain many nuclei and are said to be multinucleate.

> **SAMPLE PROBLEM 4** Compare the nucleus and the nucleolus
>
> Compare the differences between the nucleus and the nucleolus. (2 marks)
>
> **THINK**
>
> **TIP:** In *compare* questions, it is useful to use a table to show the differences.
> Remember, the nucleolus is part of the nucleus and is responsible for making rRNA, which makes up part of the ribosomes.
> Nucleus contains DNA in the form of chromosomes.
> The nucleus controls the conversion of these genes into proteins.
>
> **WRITE**
>
> | Nucleus | Nucleolus |
> |---|---|
> | Enclosed in a membrane | Not enclosed in a membrane |
> | Contains DNA | Contains ribosomal RNA (rRNA) |
> | Controls gene expression | Produces rRNA |
>
> (1 mark for nucleus, 1 mark for nucleolus)

Mitochondria: the powerhouse of the cell

Living cells are using energy all the time. In eukaryotic cells, energy production occurs mainly in the cell organelles known as **mitochondria** (singular = mitochondrion, from the Greek, *mitos* = thread; *chondrion* = small grain; figure 1.25). The readily useable form of energy for cells is the chemical energy present in the compound known as **adenosine triphosphate (ATP)** (figure 1.26). The supplies of ATP in living cells are continually being used and so must continually be replaced.

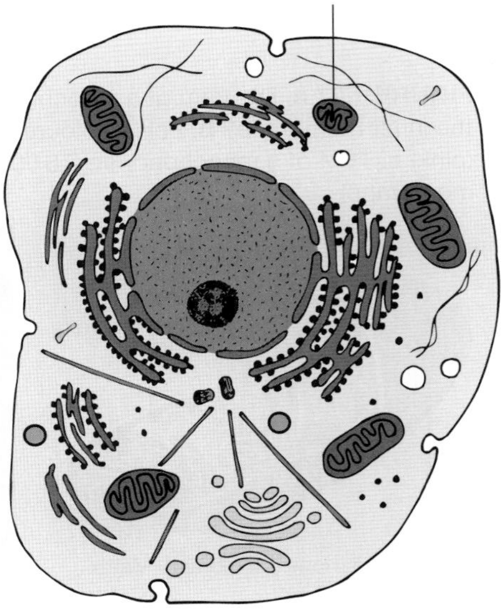

Plant and animal cells produce ATP in the process of **cellular respiration**.

Features of cellular respiration:
- A series of biochemical reactions that, in the presence of oxygen, transfer the chemical energy of sugars to the energy in chemical bonds in ATP. ATP is the readily useable form of energy for cells (see Extension box).
- Part of the process occurs in the cytosol but this series of reactions (glycolysis) produces only a small amount of ATP.
- The stages that occur in the mitochondria and make use of oxygen produce the greatest amounts of ATP — more than 95 per cent of the total yield — and this is why mitochondria are often referred to as the 'powerhouses' of the cell.

FIGURE 1.25 Diagram of a eukaryotic animal cell; the mitochondria are highlighted in brown

> The function of the mitochondrion is to convert chemical energy (glucose) into a usable form (ATP).

Mitochondria cannot be seen using a light microscope, but can easily be seen with an electron microscope. Each mitochondrion has a smooth outer membrane and a highly folded inner membrane (figure 1.27). Note that this structure creates two compartments within a mitochondrion. Carriers embedded in the folds or cristae (singular = crista) of the inner membrane are important in cellular respiration (see Extension box).

mitochondria in eukaryotic cells, organelles that are the major site of ATP production; singular = mitochondrion

adenosine triphosphate (ATP) the common source of chemical energy for cells

cellular respiration process of converting the chemical energy of food into a form usable by cells, typically ATP

FIGURE 1.26 Chemical structure of adenosine triphosphate (ATP). Note the three phosphate groups in this molecule, hence tri- (= 3) phosphate.

The number of mitochondria in different cell types varies greatly. In general, the more active the cell, the greater the number of mitochondria in that cell. For example, liver cells have one to two thousand mitochondria per cell. In contrast, mature red blood cells have no mitochondria. The difference between these two cell types is that the liver is a vital organ and its cells carry out various functions related to digestion, immunity, and the storage and release of nutrients, all of which require an input of energy. Red blood cells, however, are effectively just bags of haemoglobin that are carried passively around the bloodstream and therefore have low energy needs.

FIGURE 1.27 a. Diagram of a mitochondrion showing its two membranes. Which is more highly folded: the outer or the inner membrane? b. False-coloured scanning electron micrograph (SEM) of a section through a mitochondrion (pink) c. TEM of a mitochondrion (78 000× magnification); m = mitochondrion, cm = cell membranes (plasma membrane)

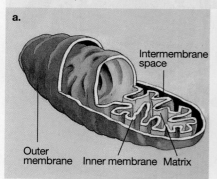
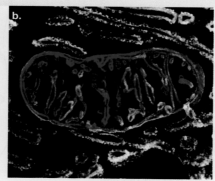
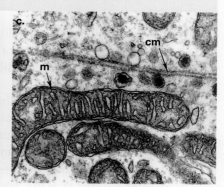

EXTENSION: Cellular respiration

Cellular respiration involves the conversion of chemical energy (glucose) to a usable form (ATP), summarised in the provided diagram.

Cellular respiration is an extension concept for the Units 1 & 2 course but is examinable in the Units 3 & 4 course.

To access more information on this extension concept please download the digital document.

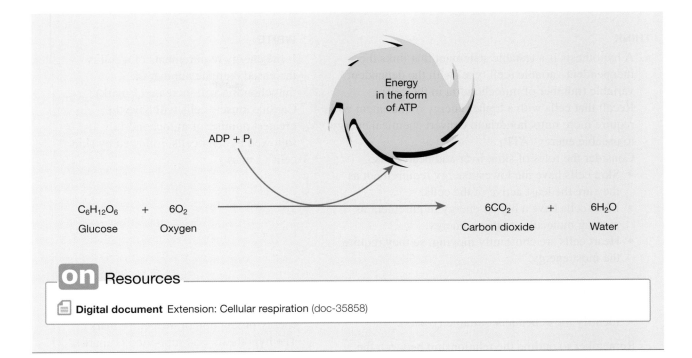

Resources

Digital document Extension: Cellular respiration (doc-35858)

SAMPLE PROBLEM 5 Writing a hypothesis and a conclusion

tlvd-1741

Different cells have different energy requirements. A cardiac muscle cell requires energy to contract and expand to pump blood around the body. Liver cells have three main roles: detoxification of toxins in the body; synthesis of proteins involved in blood clotting, and bile for digestion; and storage of glycogen (carbohydrates) and fat-soluble vitamins A, D, E and K. Skin cells provide a barrier between the internal and external environment of the organism.

a. Write a hypothesis predicting the relative number of mitochondria in different cell types. Include a comparative statement for each of the three cell types. (2 marks)

b. An experiment was conducted to investigate the relationship between the number of mitochondria and the cell's energy output. Below is the data collected for the three different cell types.
Write a suitable conclusion based on your hypothesis. (2 marks)

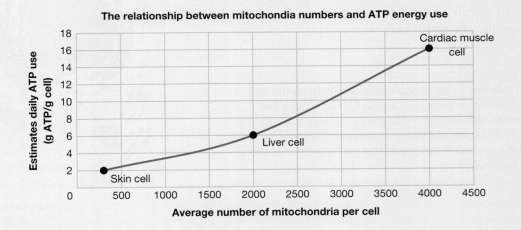

TOPIC 1 Cellular structure and function

THINK

a. A hypothesis is a testable statement that links the independent variable (cell type) with the dependent variable (number of mitochondria in the cell). Recall that cells with a higher energy requirement will require more mitochondria to convert chemical energy to useable energy (ATP).

Consider the roles of skin, liver and heart cells:
- Skin cells have the lowest energy requirement as they are the least active of the cells.
- Liver cells have a higher energy requirement as making molecules requires energy.
- Heart cells are constantly moving, so they require the most energy.

b. Analyse the graph carefully.
- Identify the independent variable (*x*-axis).
- Identify the dependent variable (*y*-axis).

Remember to outline the relationship between the independent variable and the dependent variable and state whether the hypothesis is supported or not supported.

WRITE

If the energy requirements of a cell is increased then the number of mitochondria will increase (1 mark). Cardiac muscle cells will have the greatest number of mitochondria, followed by liver cells, followed by skin cells (1 mark).

It was found that as the number of mitochondria increases, so does the energy production of the cell (1 mark). The hypothesis was supported (1 mark).

Endoplasmic reticulum

The endoplasmic reticulum (ER) is an interconnected system of membrane-enclosed flattened channels (figures 1.28 and 1.29). Where the ER has ribosomes attached to the outer surface of its channels, it is known as **rough endoplasmic reticulum**. When there are no attached ribosomes, it is referred to as the **smooth endoplasmic reticulum**. The rough ER and the smooth ER are separate networks of channels and they are not physically connected.

Both the rough ER and the smooth ER are involved in transporting different materials within cells, but they are not passive channels like pipes. As well as their roles in internal transport within a cell, the rough ER and the smooth ER have other important functions.

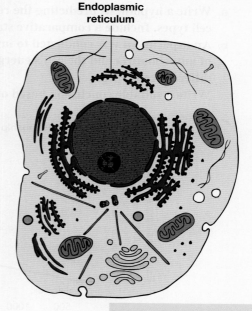

FIGURE 1.28 Diagram of a eukaryotic animal cell; the endoplasmic reticulum is highlighted in purple

rough endoplasmic reticulum endoplasmic reticulum with ribosomes attached

smooth endoplasmic reticulum endoplasmic reticulum without ribosomes

TIP: When referring to the endoplasmic reticulum for the first time in your writing, ensure you use the full name, with the acronym in parentheses. After that point, you may then use the acronym ER. If you are answering exam questions, you should write the full name out in each question (as different assessors may mark each question).

Rough ER

Through its network of channels, the rough ER is involved in transporting some of the proteins to various sites within a cell. Proteins delivered from the ribosome into the channels of the rough ER are also processed before they are transported. The processing of proteins within the rough ER includes:
- attaching sugar groups to some proteins to form glycoproteins
- folding proteins into their correct functional shape or conformation
- assembling complex proteins by linking together several polypeptide chains, such as the four polypeptide chains that comprise the haemoglobin protein.

> The functions of the rough endoplasmic reticulum are internal transport and protein modification.

Smooth ER

The smooth ER of different cells is involved in the manufacture of substances, detoxifying harmful products, and the storage and release of substances.

For example:
- the outer membrane surface of the smooth ER is a site of synthesis of lipids; these lipids are then enclosed in a small section of the smooth ER membrane that breaks off and transports the lipids to sites within the cell, where they are either used or exported from the cell
- in cells of the adrenal gland cortex and in hormone-producing glands, such as ovaries and testes, the smooth ER is involved in the synthesis of steroid hormones; for example, testosterone
- in liver cells, the smooth ER detoxifies harmful hydrophobic products of metabolism and barbiturate drugs by converting them to water-soluble forms that can be excreted via the kidneys. It stores glycogen as granules on its outer surface (see figure 1.30) and breaks it down into glucose for export from the liver.

The importance of both rough and smooth ER in their various cellular functions is highlighted by the fact that, in an animal cell, about half of the total membrane surface is part of the endoplasmic reticulum.

> The functions of the smooth endoplasmic reticulum are lipid synthesis and storage. It also has a role in detoxification of harmful substances.

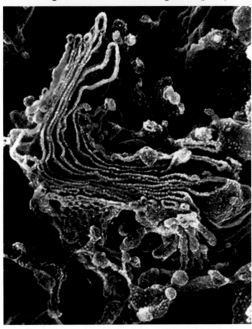

FIGURE 1.29 False-coloured SEM image showing the channels of rough ER (pink)

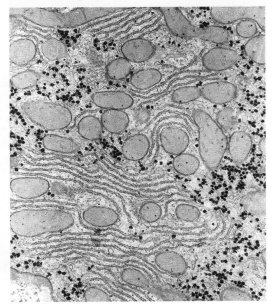

FIGURE 1.30 TEM image of an area of a liver cell with an abundant supply of smooth ER (parallel membrane-lined channels). Dark clusters of glycogen particles (black dots) are visible around the smooth ER. Also visible are sections of mitochondria (large circular shapes).

Golgi apparatus

Some cells produce proteins that are intended for use outside the cells where they are formed. Examples include the following proteins that are produced by one kind of cell and then exported (secreted) by those cells for use elsewhere in the body:
- the digestive enzyme pepsin, produced by cells lining the stomach and secreted into the stomach cavity
- the protein hormone insulin, produced by cells in the pancreas and secreted into the bloodstream
- protein antibodies, produced in special lymphocytes and secreted at an area of infection.

How do these substances get exported from cells? The cell organelle responsible for the export of substances out of cells is the **Golgi apparatus**, also known as the Golgi complex or Golgi body. The Golgi apparatus has a multilayered structure composed of stacks of membrane-lined channels (see figure 1.32).

Proteins from the rough ER that are intended for export must be transferred to the Golgi apparatus. The pathway for export is summarised below and in figure 1.33.
- There is no direct connection between the membranes of the ER and the Golgi apparatus.
- The proteins are shuttled to the Golgi apparatus in membrane-bound transition vesicles.
- These vesicles are taken into the Golgi apparatus, where the proteins are concentrated and packaged into secretory vesicles.
- These secretory vesicles break free from the Golgi apparatus and move to the plasma membrane of the cell where they merge with it, discharging their protein contents.

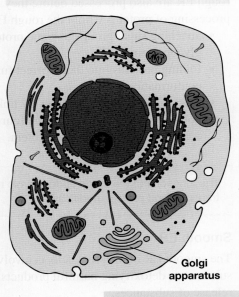

FIGURE 1.31 Diagram of a eukaryotic animal cell; the Golgi apparatus is highlighted in yellow

Golgi apparatus organelle that packages material into vesicles for export from a cell (also known as Golgi complex or Golgi body)

FIGURE 1.32 a. False-coloured TEM image of the Golgi apparatus (orange). Note the stacks of flattened membrane-lined channels with their wider ends that can break free as separate vesicles. **b.** 3D representation of the Golgi apparatus. Note the vesicles breaking off from the ends.

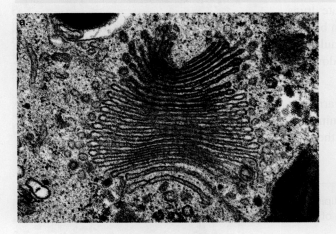
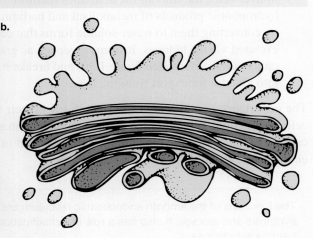

The function of the Golgi apparatus is to package and export substances.

FIGURE 1.33 The secretory export pathway for proteins from the rough ER in transition vesicles to the Golgi apparatus, and then in secretory vesicles to the plasma membrane of the cell, where they merge with the membrane and discharge their contents.

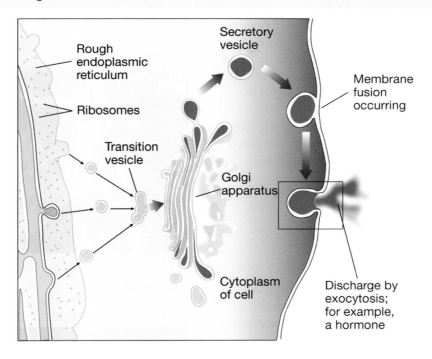

Vesicles

Vesicles refer to any membrane-bound sac in which substances are transported around, into or out of a cell.

Some vesicles have specific functions, such as lysosomes and peroxisomes. These are discussed below.

Lysosomes

The cytoplasm of animal cells contains fluid-filled sacs enclosed within a single membrane, known as **lysosomes** (from the Greek, *lysis* = destruction; *soma* = body) (see figure 1.34). They are typically spherical in shape with diameters in the range of 0.2 to 0.8 μm. The fluid in these cell organelles contains a large number (about 50) of digestive enzymes that can break down carbohydrates, proteins, lipids, **polysaccharides** and nucleic acids. These lysosomal enzymes are only active in an acidic environment (pH of about 4.8). Lysosomes were first identified as cell organelles in 1955 by Christian de Duve, a Belgian cytologist.

A similar function is carried out in plant cells by vacuoles containing a similar range of enzymes to those in the lysosomes of animal cells. Some experts call these lysosome-like vacuoles, while others simply call them plant lysosomes.

A lysosome is a specialised vesicle filled with digestive enzymes.

FIGURE 1.34 Diagram of a eukaryotic animal cell; the lysosomes are highlighted in white

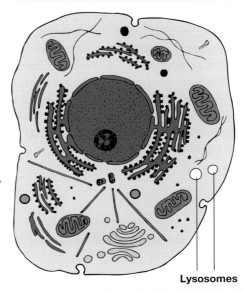

Lysosomes

lysosome vesicle filled with digestive enzymes

polysaccharides long chains of sugars joined together to form large molecules

Functions of lysosomes include:
- digestion of some of the excess macromolecules within a cell. The excess material is delivered to the lysosome where it is broken down into small sub-units by specific enzymes, such as proteins to their amino acid sub-units and polysaccharides to their sugar sub-units. The sub-units are released from the lysosome into the cytoplasm where they can be reused — an example of recycling at the cellular level.
- **autophagy** — the breakdown of non-functioning cell organelles that are old and/or damaged and in need of turnover (from the Greek, *auto* = self; *phagy* = eating)
- the breakdown of substances, such as bacteria, brought into the cell by phagocytosis.

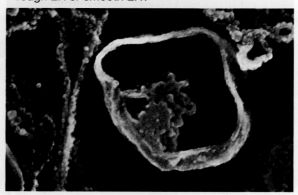

FIGURE 1.35 False-coloured SEM image of a lysosome (green) from a pancreatic cell. The material inside the membrane wall of the lysosome is undergoing digestion. Note the surfaces of the nearby membranes of the ER (pink) that are covered with ribosomes. Is this rough ER or smooth ER?

Peroxisomes

Another small cell organelle found in the cytoplasm of both plant and animal cells is the **peroxisome**. Peroxisomes were discovered in 1954 from electron microscopy studies of cell structure. They range in diameter from about 0.1 to 1.0 μm.

Peroxisomes have a single membrane boundary and contain a large number of enzymes; about 50 peroxisomal enzymes have been identified to date, including catalase and peroxidase.

Peroxisomes carry out several functions in cellular metabolism, including the oxidation of fatty acids, an important energy-releasing reaction in some cells. Peroxisomes also break down substances that are either toxic or surplus to requirements. For example, hydrogen peroxide (H_2O_2) is produced during normal cell metabolism. If it is not rapidly broken down, it poisons the cell. The peroxisome enzyme catalase breaks down hydrogen peroxide to water and oxygen, as shown in the following equation, preventing its accumulation and toxic effects.

FIGURE 1.36 Peroxisomes visible with a fluorescent protein (green)

$$2H_2O_2 \xrightarrow{\text{catalase}} 2H_2O + O_2$$

A peroxisome is a specialised vesicle filled with enzymes that break down substances toxic to the cell.

autophagy breakdown by lysosomes of non-functioning cell organelles that are old and/or damaged and in need of turnover

peroxisome small membrane-bound organelle rich in the enzymes that detoxify various toxic materials that enter the bloodstream

EXTENSION: Diseases associated with lysosomes and peroxisomes

Defects of lysosome enzymes and peroxisomes cause disruption of normal cell function. In lysosomes these are known as lysosome storage diseases, and include Pompe disease, Tay-Sachs disease and Hurler syndrome. In peroxisomes, a defect in a transport protein causes adrenoleukodystrophy (ALD).

To access more information on this extension concept please download the digital document.

FIGURE 1.37 Muscle tissue affected by Pompe disease showing glycogen within cell vacuoles

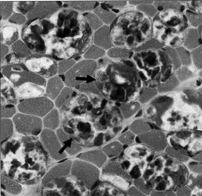

on Resources

Digital document Extension: Diseases associated with lysosomes and peroxisomes (doc-36011)

Weblink Pompe disease

SAMPLE PROBLEM 6 Designing a decomposition reaction

tlvd-1742

Liver cells have a high number of peroxisomes and therefore when liver is placed in a solution of hydrogen peroxide, a reaction takes place.

a. What is being produced by this decomposition reaction? **(1 mark)**

b. As the concentration of the hydrogen peroxide increases, the amount of products should increase. Design an experiment that would determine the effect of concentration of hydrogen peroxide on rate of production of oxygen. **(4 marks)**

THINK

a. Decomposition means 'breaking down'.

b. When designing a method, remember to consider the following:
- Identify the independent variable and how it will be changed. (The values should be incremental.)
- Identify the dependent variable (what is going to be measured).
- Identify what variables/conditions are going to be controlled. Control variables ensure the test is fair and valid.

WRITE

Hydrogen peroxide (H_2O_2) decomposes into water (H_2O) and oxygen (O_2) (1 mark).

1. Independent variable: concentration of hydrogen peroxide
   Dependent variable: amount of oxygen produced
   Controlled variables: amounts of liver and H_2O_2 solution, time period measured, increments of time measurement
2. Use four concentrations of hydrogen peroxide (H_2O_2): 0.5%, 1%, 1.5% and 2%.
3. Add 5 mL of each concentration of H_2O_2 to four test tubes.

- How will the test be repeated to make it reliable?
- What result would support the hypothesis?

4. Record the volume of oxygen produced every 20 seconds for 2 minutes.
5. If there was an increase in the rate of oxygen production with an increase in H_2O_2 concentration, the hypothesis is supported (1 mark for identification of IV and DV, 1 mark for controlled variables, 1 mark for reproducible and reliable method, 1 mark for comment related to hypothesis).

Cytoskeleton

Cell organelles are not floating freely like peas in soup. They are supported by the **cytoskeleton** of the cell. The cytoskeleton forms the three-dimensional structural framework of eukaryotic cells. The cytoskeleton and its components are important to cell functioning because they:
- supply support and strength for the cell
- determine the cell shape
- enable some cell mobility
- facilitate movement of cell organelles within a cell
- move chromosomes during cell division.

The cytoskeleton of the cell is a dynamic three-dimensional scaffold in the cell cytoplasm. Figure 1.38 shows two fluorescent microscope images that reveal the extent of the cytoskeleton within a cell. Different coloured fluorescent probes denote different components of the cytoskeleton.

cytoskeleton network of filaments within a cell

FIGURE 1.38 a. A prize-winning image of eukaryotic cells. The actin filaments (purple) and the microtubules (yellow) are part of the cytoskeleton of these cells. The nucleus is stained green. Note that the cell on the left appears to be in the process of dividing. (Image courtesy of Torsten Wittmann). **b.** Image demonstrating the 3D nature of the cytoskeleton

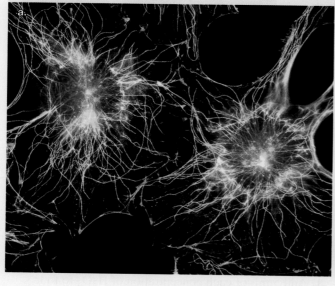
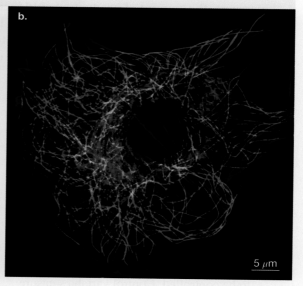

The function of the cytoskeleton is to provide structure and support for the cell.

The cytoskeleton has three components, as shown in figure 1.39:

a. **Microtubules**: Hollow tubes, about 25 nm in diameter with an average length of 25 μm (but may be much longer); the walls of the tube are composed of chains of a spherical protein, tubulin (see figure 1.39a). Microtubules move material within cells and give cilia and flagella their structure and motion.
b. **Intermediate filaments**: Tough threads, about 10 nm in diameter, woven in a rope-like arrangement; composed of one or more kinds of protein, depending on cell type. For example, in skin cells, intermediate filaments are made of the protein keratin that is also found in skin and nails (see figure 1.39b). Intermediate filaments give mechanical support to cells.
c. **Microfilaments**: Thinnest of threads ranging from 3 nm to 6 nm in diameter; composed mainly of the spherical protein actin (see figure 1.39c). Actin filaments are responsible for the movement of cells.

microtubules part of the supporting structure or cytoskeleton of a cell, made of sub-units of the protein tubulin

intermediate filaments one of the components of the cytoskeleton of a cell, composed of protein; they form a rope-like arrangement and give mechanical support to cells

microfilaments one of the components of the cytoskeleton of a cell; very thin threads composed of the protein actin; they allow cells to move and change in shape

FIGURE 1.39 Components of the cytoskeleton in eukaryotic cells **a.** Microtubule (side view above, top view below). Microtubules are built of the spherical protein tubulin. **b.** Side view of intermediate filament. Note the rope-like arrangement. **c.** Side view of microfilament. The building blocks of microfilaments are the protein actin.

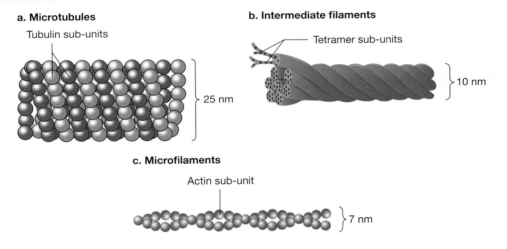

Cilia and flagella

For some unicellular eukaryotes, their ability to move depends on the presence of special cell structures: cilia and flagella, which share the same basic structure. These were introduced in section 1.2.4. Each cilium and flagellum are enclosed in a thin extension of the plasma membrane. Inside this membrane are microtubules arranged in a particular '9 + 2' pattern (see figure 1.40). Each microtubule is composed of 13 protein filaments forming a circular hollow tube.

- Flagella (singular = flagellum, from the Latin meaning 'whip') are generally singular, like a tail, and have a function in locomotion of some bacteria, some protists and the gametes of eukaryotes (see figure 1.12a).
- Cilia (singular = cilium, from the Latin meaning 'eyelash') are shorter than flagella and are more numerous. They have a variety of roles in different eukaryotes. They beat in a wavelike motion to move the cell through a watery environment, or to create water currents to move food and oxygen to specialised organs for uptake in sessile (fixed to one spot) organisms such as clams and sponges (see figure 1.12b).

FIGURE 1.40 Cross-section through a eukaryotic cilium. Both cilia and flagella have the same arrangement of microtubules in their structure, with 9 paired microtubules in an outer ring and 2 central microtubules. A microtubule consists of 13 protein filaments that form a hollow tube.

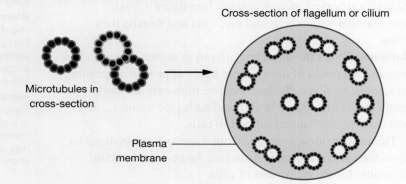

In the human body, the cells lining the trachea, or air passage, have cilia that project into the cavity of the trachea. Mucus in the trachea traps dust and other particles and even potentially harmful bacteria. The synchronised movement of the cilia moves the mucus up the trachea to an opening at the back of the throat. Cells lining the fallopian tubes also have large numbers of cilia. The beating of these cilia moves an egg from the ovary towards the uterus.

1.4.4 Differences between types of eukaryotic cells

When exploring eukaryotes it is important to note that there are some key differences between the different groups: **Protista**, **Plantae**, **Fungi** and **Animalia**.

Protista are a group of organisms that do not fit into any other group — the misfits if you like. They are varied in size and shape and are generally unicellular, although they can form **colonies**. They live in watery environments as they need to gain all their requirements from their direct surroundings.

Protista that form colonies act more like a multicellular organism, all working to the benefit of the group. Algae are a good example of protists that form colonies (figure 1.43).

> **Protista (protists)** a group of organisms similar in the fact that they do not fall into any other kingdom; generally unicellular; can have aspects of both plant and animal cells
>
> **Plantae** a group of organisms that include land plants and algae. Cells have a cell wall of cellulose, a large permanent vacuole and some contain chloroplasts.
>
> **Fungi** any of a wide variety of organisms that have a cell wall made of chitin and reproduce by spores, including the mushrooms, moulds, yeasts, and mildews
>
> **Animalia** a group of eukaryotic multicellular organisms whose cells lack a cell wall
>
> **colony** several individuals living together in close association

FIGURE 1.41 Eukaryotes can be single-celled and multicellular.

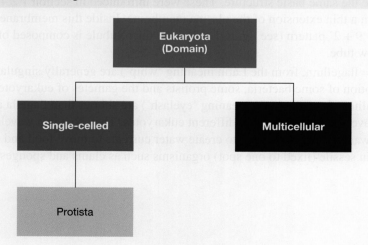

FIGURE 1.42 a. *Euglena*, **b.** *Amoeba* and **c.** *Paramecium* are all examples of protists that are unicellular.

FIGURE 1.43 Both **a.** *Volvox* and **b.** diatoms are algae that are plant-like protists that form colonies. *Volvox* form colonies of tens of thousands of individuals. Diatoms have ornate silica shells and are used as the abrasive component of toothpaste.

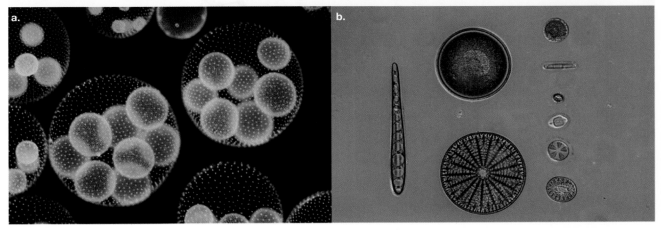

FIGURE 1.44 Multicellular eukaryotes can be subdivided into those without a cell wall (animal cells) and with a cell wall (plant and fungi cells).

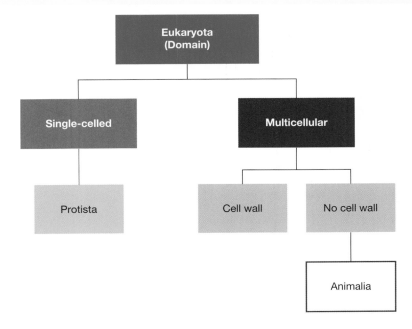

TOPIC 1 Cellular structure and function

1.4.5 Organelles for animal cells

Animals are multicellular organisms. What differentiates animal cells from plant and fungi cells is that they do not have a cell wall. Without a cell wall, animal cells have specialised organelles called **centrioles**, which are involved with cell replication (see topic 2).

centrioles a pair of small cylindrical organelles, used in spindle development in animal cells during cell division

Centrioles

Centrioles are organelles found near the nucleus in animal cells. They are part of the cytoskeleton and consist mainly of the protein tubulin (figure 1.45). Centrioles exist in pairs, which replicate and then migrate to the poles of the cell when it is time for cell division (see figure 1.46).

FIGURE 1.45 Centrioles are made of microtubules arranged in a specific star-like structure.

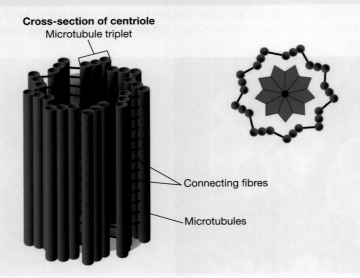

FIGURE 1.46 The centrioles replicate and migrate during the early stages of cell replication.

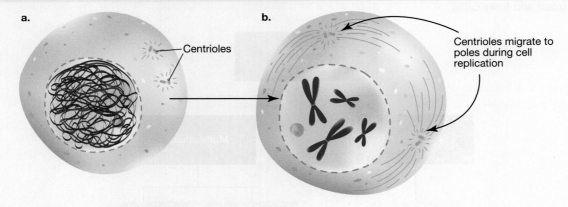

1.4.6 Organelles for plants

One of the defining difference between plants and fungi is the compositions of their cell wall. Plant cell walls are made of cellulose, while fungi cell walls are made of **chitin** (figure 1.47); this is discussed in more detail later in this section. However, there are another two organelles that are present in plant cells but are not usually found in other cell types. These are **chloroplasts** and a large, permanent **vacuole**.

FIGURE 1.47 Plants and fungi are differentiated by the composition of their cell wall.

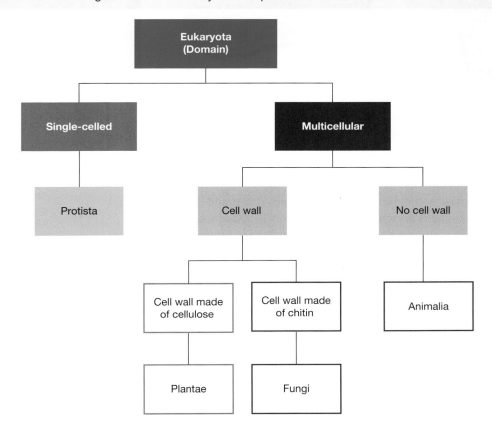

Chloroplasts

Solar cells are a relatively new technology. However, thousands of millions of years ago, some bacteria developed the ability to capture the radiant energy of sunlight and to transform it to chemical energy present in organic molecules such as sugars. This ability exists in algae and plants. The remarkable organelles present in the cells of plants and algae that can capture the energy of sunlight are the chloroplasts (figure 1.48a). The complex process of converting sunlight energy to chemical energy is known as **photosynthesis**.

chitin a fibrous substance, mainly composed of polysaccharides, used in the cells walls of fungi

chloroplast chlorophyll-containing organelle that occurs in the cytosol of cells of specific plant tissues

vacuole structure within plant cells that is filled with fluid-containing materials in solution, including plant pigments

photosynthesis process by which plants use the radiant energy of sunlight trapped by chlorophyll to build carbohydrates from carbon dioxide and water

FIGURE 1.48 a. Internal structure of a chloroplast showing the many layers of its internal membrane b. 3D representation of a chloroplast c. SEM (78 000× magnification) image of a fractured chloroplast from the cell of a red alga (scale bar = 1 μm)

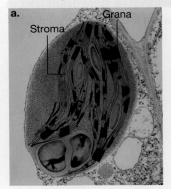
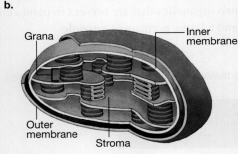
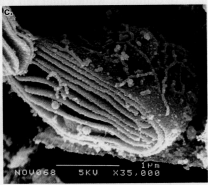

Features of chloroplasts:
- They are relatively large cell organelles that, when present in a plant cell, can be easily seen using a light microscope under certain stains.
- They have a green colour due to the presence of light-trapping pigments known as **chlorophylls**.
- Each chloroplast is enclosed in two membranes (outer and inner membrane).
- A third membrane is present internally and this is folded to create an intricate internal structure consisting of many flattened membrane layers called **grana**. The surfaces of the grana provide a large area where chlorophyll pigments are located.
- The region of fluid-filled spaces between the grana is known as the **stroma** (see figure 1.48).

chlorophyll green pigment required for photosynthesis that traps the radiant energy of sunlight

grana stacks of membranes on which chlorophyll is located in chloroplasts; singular = granum

stroma in chloroplasts, the semi-fluid substance between the grana, which contains enzymes for some of the reactions of photosynthesis

Chloroplasts are not present in *all* plant cells; they are found only in the parts of a plant that are exposed to sunlight, such as the cells in some parts of leaves (see figure 1.49) and in stems. Chloroplasts are not present in the cells of the upper and lower surface of a leaf.

FIGURE 1.49 False-coloured SEM image of a section of a leaf. Sandwiched between the upper and lower leaf surfaces are the cells that contain chloroplasts.

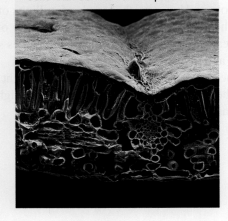

EXTENSION: Photosynthesis

Plants and algae transform the energy of sunlight into the chemical energy of organic molecules, such as glucose, through the process of photosynthesis. In a complex series of reactions, this allows them to use simple, inorganic molecules to create complex, energy-rich molecules required for survival.

Photosynthesis is an extension concept for the Units 1 & 2 course but is examinable in the Units 3 & 4 course.

To access more information on this extension concept please download the digital document.

on Resources

Digital document Extension: Photosynthesis (doc-35859)

Vacuole

In a mature plant cell a large central vacuole bound by a membrane is visible.

A plant vacuole can take up to 80 to 90 per cent of the cell volume and it pushes the rest of the cell contents up against the plasma membrane (see figure 1.50a). Plant vacuoles are fluid-filled and are separated from the rest of the cytosol by the vacuole membrane, or **tonoplast**, that controls the entry and exit of dissolved substances into the vacuole. As a result, the composition of the vacuole fluid is different from that of the cytosol.

Plant vacuoles serve a number of functions, including storage of nutrients and mineral salts, and are involved in waste disposal (see Lysosomes in section 1.4.3). Vacuoles may contain plant pigments, such as the anthocyanins that produce the purple and red colours of some flowers and fruits (see figure 1.50b).

FIGURE 1.50 a. TEM image of a mature plant cell showing its large vacuole **b.** The vacuole is the site of the anthocyanin pigments that give the red and purple colours to some fruits and flowers.

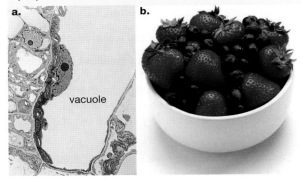

The cell wall

Let's move outside the plasma membrane to a structure that is present in most prokaryotic cells and in all the cells of plants, algae and fungi, but *not* in animal cells. This is the cell wall.

tonoplast in plant cells, the membrane of the plant vacuole; separates it from the rest of the cytosol

The cell wall is located outside and surrounds the plasma membrane. It provides strength and support to cells and acts to prevent the overexpansion of the cell contents if there is a net movement of water into cells. Cell walls are present in most prokaryotic cells, in some protists and in all plant, algal and fungal cells. The presence of a cell wall is *not* a diagnostic feature that allows you to decide if a cell is prokaryotic or eukaryotic. This decision can only be made based on knowledge of the major component(s) in the cell wall.

Table 1.4 summarises this situation in prokaryotes and eukaryotes. Note that animal cells do not have cell walls and this absence serves to distinguish animal cells from all other eukaryotic cells.

TABLE 1.4 Composition of cell walls in various cell types

Type of organism	Major compound(s) in cell walls
Prokaryote: most bacteria	Peptidoglycan, a polymer composed of long strands of polysaccharides, cross-linked by short chains of amino acids
Prokaryote: most archaea	Various compounds, including proteins or glycoproteins or polysaccharides; peptidoglycan *not* present
Eukaryote: all fungi	Chitin, a complex polysaccharide
Eukaryote: all plants	Cellulose in the primary cell wall; lignin in the secondary cell wall

Note: Bacteria of the genus *Mycoplasma* do not have cell walls. These parasitic bacteria are the smallest living organisms (just 0.1 µm, or 100 nm). Most are surface parasites growing on cells but some species, such as *M. penetrans*, are intracellular pathogens growing inside the cells of their hosts.

The function of the cell wall is to provide protection, shape and support to the cell.

EXTENSION: Why do some plants have wood?

Some plants have a second cell wall!

All plant cells have primary cell walls, which have the following features:

- They are made of fibrils of **cellulose** combined with other substances (see figure 1.51).
- They provide some mechanical strength for plant cells and allow them to resist the pressure of water taken up by osmosis.
- Cells with just a primary cell wall are able to divide and can also expand as the cells grow.

cellulose complex carbohydrate composed of chains of glucose molecules; the main component of plant cell walls

lignin a complex, insoluble cross-linked polymer

Secondary cell walls develop in woody plants, such as shrubs and trees, and in some perennial grasses.
- Like primary cell walls, they are composed of cellulose.
- Secondary cell walls are further strengthened and made rigid and hard by the presence of **lignin** — a complex, insoluble, cross-linked polymer. Secondary cell walls are laid down inside the primary cell walls (in the various cells of the xylem tissue).
- The middle lamella is a layer that forms between adjacent plant cells and this also contains lignin.
- Cytoplasmic connections exist between adjacent plant cells through pits in the cell wall. These connections are known as plasmodesmata.
- As their secondary cell walls continue to thicken, the cells die.
- This dead xylem tissue creates the 'woodiness' of shrubs and trees that gives increased mechanical strength to these plants, enabling them to grow high (figure 1.52).

FIGURE 1.51 The primary cell wall of plant cells is made mainly of cellulose fibrils. Secondary cell walls develop in woody plants, such as shrubs and trees, and in some perennial grasses.

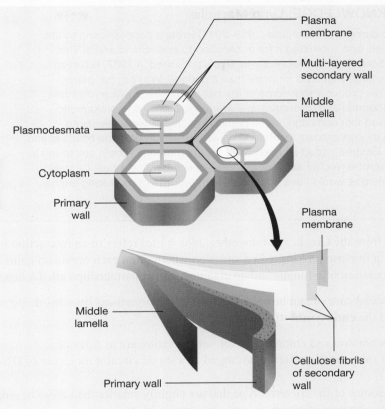

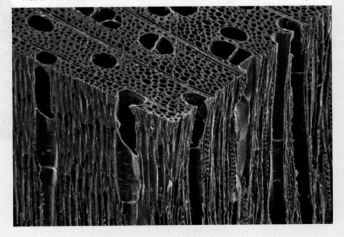

FIGURE 1.52 SEM image of a sample of hardwood. The wood is dead xylem tissue that consists mainly of the secondary cell walls, including tracheids — small elongated cells that make up most of the wood; and vessels — larger cells but much fewer in number. Soft woods from conifer trees can be distinguished from hard woods because their wood lacks vessels.

TOPIC 1 Cellular structure and function 47

1.4.7 From prokaryotic beginnings

The endosymbiotic theory

> **BACKGROUND KNOWLEDGE: Lynn Margulis**
>
> In her early academic career Lynn Margulis (1938–2011) wrote a paper relating to the origin of eukaryotic cells and submitted it for publication to scientific journals. This paper was rejected about 15 times before it was finally published in 1967, but even then its contents were largely ignored for more than a decade. Why? In her paper, Margulis put forward the proposal that some of the cell organelles in eukaryotic cells — in particular, mitochondria and chloroplasts — were once free-living prokaryotic microbes. She proposed that eons ago, some primitive microbes were taken into another cell, perhaps by endocytosis, conferring an advantage on both the host cell and its ingested microbe. Fanciful? Keep an open mind for the moment. A host cell taking in a sunlight-trapping microbe would have a new source of energy; and a host cell taking in an oxygen-using microbe would have a more efficient process of producing energy.

FIGURE 1.53 A pencil drawing of Lynn Margulis

The term 'symbiosis' (from the Greek, *syn* = together; *bios* = life) refers to an interaction between two different kinds of organism living in close proximity in a situation where often each organism gains a benefit — as in, for example, the close association of a fungus and an alga that form a partnership called a lichen (see figure 1.54).

Endosymbiosis is a special case of symbiosis where one of the organisms lives inside the other. Margulis' proposal is now termed the **endosymbiotic theory**.

A few facts about mitochondria and chloroplasts for consideration are as follows:
- Both contain their own genetic material, arranged as a single circular molecule of DNA, as occurs in bacteria.
- Both contain ribosomes of the bacterial type that are slightly smaller than those of eukaryotes.
- Both reproduce by a process of binary fission as occurs in bacteria.
- Both have sizes that fall within the size range of bacterial cells.

In addition, genomic comparisons indicate that chloroplasts are most closely related to modern cyanobacteria; these are photosynthetic microbes that possess chlorophyll, enabling them to capture sunlight energy and use that energy to make sugars from simple inorganic material. Genomic comparisons indicate that mitochondria are most closely related to modern *Rickettsia* bacteria.

FIGURE 1.54 Lichens are a symbiotic association between an alga and a fungus.

This evidence, and ongoing biochemical and genetic comparisons, support the view that mitochondria and chloroplasts share an evolutionary past with prokaryotes. The theory of endosymbiosis is now generally accepted as explaining, in part, the origin of eukaryotic cells, in particular the origin of their chloroplasts and mitochondria.

Endosymbiosis is not just a theoretical concept; we can see examples of endosymbiosis, including:
- nitrogen-fixing bacteria that live in the cells of nodules on the roots of legumes, such as clovers
- single-celled algae that live inside the cells of corals.

As more and more evidence was identified in support of the endosymbiotic theory, Margulis (see Background knowledge box) became recognised as an important contributor to the biological sciences and received several awards.

endosymbiosis a special case of symbiosis where one of the organisms lives inside the other

endosymbiotic theory see endosymbiosis; a theory proposed by Lynn Margulis

INVESTIGATION 1.3

online only

Viewing and staining cells

Aim

To view the organelles under a light microscope by using certain stains on cells

 Resources

- **eWorkbook** Worksheet 1.3 Cells and their organelles (ewbk-7190)
- **Video eLesson** Cells and organelles (eles-0054)

KEY IDEAS

- Ribosomes are cell organelles where proteins are manufactured.
- Eukaryotic cells are partitioned into several compartments, each bound by a membrane, with each compartment having specific functions.
- In eukaryotic cells, the nucleic acid DNA is enclosed within the nucleus, a double-membrane-bound organelle.
- The nucleolus is the site of production of RNA.
- The endoplasmic reticulum (ER) is made of a series of membrane-bound channels.
- Rough ER is so named because of the presence of ribosomes on the external surface of its membranes. It is involved in the processing of proteins and in their transport.
- Smooth ER lacks ribosomes and has several functions, including the synthesis of lipids and detoxifying harmful substances.
- The Golgi apparatus packages substances into vesicles for export from a cell.
- Mitochondria are the site of cellular respiration, converting sugars to ATP.
- Vesicles are membrane-bound organelles used for transport within and out of the cell.
- Lysosomes and peroxisomes are specialised vesicles used for the breakdown of substances and toxins.
- The cytoskeleton is the three-dimensional structural framework of eukaryotic cells that provides support for the cell organelles.
- Flagella and cilia have a similar structure and are concerned with movement, either of an organism or of fluids and other substances.
- Protists are unicellular organisms and some form colonies.
- Centrioles are found in animal cells and have a role in cell replication.
- Plants, fungi and protists have cell walls made of different compounds, allowing this to be a defining feature between these eukaryotes.
- Cell walls provide structure and support for the cell.
- Chloroplasts are cell organelles, bound by a double membrane and containing chlorophyll on layers of an internal folded membrane.
- The chloroplasts found in plants and protists are the site of photosynthesis, converting light energy to sugars.
- The vacuoles in plants store nutrients and mineral salts, are involved in waste disposal and may contain plant pigments.
- The function of the cell wall is to provide protection, shape and support to the cell.
- The endosymbiotic theory proposes that some cell organelles in eukaryotic cells, in particular mitochondria and chloroplasts, were once free-living microbes.
- Many lines of evidence exist in support of the endosymbiotic theory.
- Endosymbiosis is a special case of symbiosis where one organism lives inside the host organism.

1.4 Activities

To answer questions online and to receive **immediate feedback** and **sample responses** for every question, go to your learnON title at www.jacplus.com.au. A **downloadable solutions** file is also available in the resources tab.

| 1.4 Quick quiz | 1.4 Exercise | 1.4 Exam questions |

1.4 Exercise

1. What is the difference between the cytosol and the cytoplasm of a cell?
2. A scientist wished to examine ribosomes in a liver cell.
    a. Where should the scientist look: in the nucleus or the cytoplasm?
    b. What kind of microscope is likely to be used by the scientist: a light microscope or a transmission electron microscope? Explain your choice.
    c. Outline the function of the ribosome.
    d. Explain why the presence of ribosomes cannot be used to identify whether a cell is prokaryotic or eukaryotic.
3. Identify the organelle and outline the function of each of the following.
    a.
    b.
    c.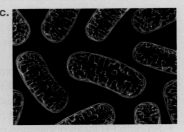
4. Where is the functional shape of a protein finally formed?
5. Compare the similarities and differences between the rough and smooth endoplasmic reticulum.
6. How do substances move from the endoplasmic reticulum to the Golgi apparatus?
7. The following diagram shows a cross-section of a leaf.

Cell structure of a leaf

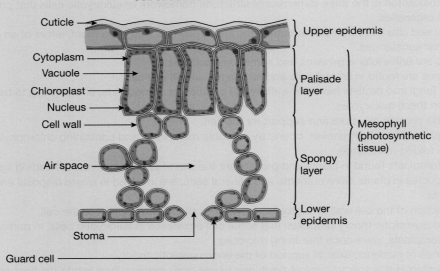

There are differing numbers of chloroplasts in different cells and not all cells are photosynthetic.
    a. Write a hypothesis relating the presence or absence of chloroplasts to the function of the cell.
    b. Design an experiment to test your hypothesis.

8. White blood cells engulf invaders and destroy them as shown in the diagram.

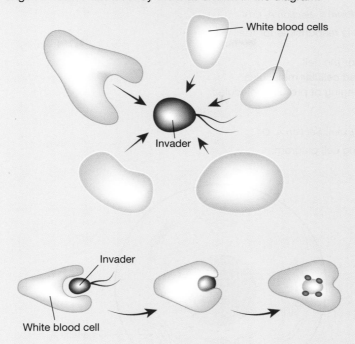

What organelle would you expect to find in abundance in these cells? Explain your response.
9. What features of chloroplasts and mitochondria indicate that they may have prokaryotic beginnings?
10. A cell has a cell wall. Can a conclusion be drawn about the kingdom the organism belongs to? Explain.
11. The liver is an important organ in the detoxification of harmful substances. What organelle in the liver is active in this process?

1.4 Exam questions

Question 1 (1 mark)
Source: *VCAA 2015 Biology Exam, Section A, Q12*

MC A student was investigating four cell types from different organisms. She recorded the results of her microscopic examination of the cells in the table below.

	Cell W	Cell X	Cell Y	Cell Z
Mitochondria	few	many	absent	few
Chloroplasts	present	absent	absent	present
Nucleus	present	present	absent	present

Which one of the following is the correct conclusion that can be drawn from this data?
A. Cell W could be a muscle cell from an insect.
B. Cell Y could be a living leaf cell from a corn plant.
C. Cell X could be a heart-muscle cell from a mammal.
D. Cell Z could be an underground root cell from a pea plant.

Question 2 (1 mark)
Source: VCAA 2011 Biology Exam 1, Section A, Q14

MC The Golgi apparatus is responsible for the
A. manufacture of lipids.
B. production of energy for the cell.
C. destruction of unwanted cellular molecules.
D. modification and packaging of protein molecules.

Question 3 (1 mark)
Source: VCAA 2014 Biology Exam, Section A, Q4

MC Consider the following cell diagram.

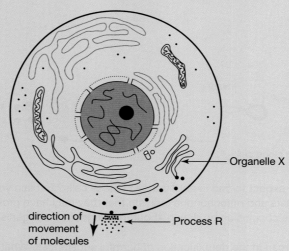

Source: www.cronodon.com

Organelle X
A. is the site of cellular respiration.
B. packages protein molecules for export from the cell.
C. absorbs sunlight and produces carbohydrates.
D. produces ribosomal RNA.

Question 4 (2 marks)
Source: VCAA 2020 Biology Exam, Section B, Q3c

In plants and algea, photosynthesis is carried out in chloroplasts. It is thought that chloroplasts originated from bacteria.

Describe two features of chloroplasts that support the theory that chloroplasts originated from bacteria.

Question 5 (6 marks)

Euglena is a genus of unicellular organisms found in aquatic habitats. Many *Euglena* species have both plant and animal cell characteristics.

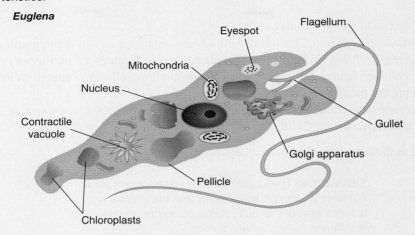

If light is available, they can manufacture their own food. In dark conditions they can ingest prey.

a. State evidence from the diagram that the species of *Euglena* shown has both plant and animal cell characteristics. **2 marks**

b. Complete the table showing either the name of an organelle or the function of an organelle in *Euglena*. **4 marks**

Organelle	Function in *Euglena*
	Site of ATP production (through aerobic respiration)
Nucleus	
	Expels excess water (taken in from hypotonic surroundings)
Eyespot	

More exam questions are available in your learnON title.

1.5 The plasma membrane

KEY KNOWLEDGE

- The structure and function of the plasma membrane in the passage of water, hydrophilic and hydrophobic substances via osmosis, facilitated diffusion and active transport

Source: VCE Biology Study Design (2022–2026) extracts © VCAA; reproduced by permission.

1.5.1 The structure of the plasma membrane: the fluid mosaic model

In the previous sections we have looked at cells and organelles bound by a plasma membrane. So what is this plasma membrane and why is it so important? The function of the plasma membrane (also sometimes referred to as the cell membrane) is to control the entry and exit of substances from one area to another; for example, from the external environment to the internal environment of the cell.

So how do all of these components fit together? The truth is we do not really know. The plasma membrane is too small to be resolve by our strongest microscopes; however, there is a model that fits with how the membrane functions. This is the **fluid mosaic model**. This model was proposed by Singer and Nicholson in 1972 and is widely accepted.

The plasma membrane has two major components:
1. *Phospholipid bilayer (two layers)*. The phospholipids are organised so that the hydrophobic tails face each other, creating a region that is hydrophobic. This arrangement separates the two watery environments inside and outside the cell (see figures 1.55 and 1.56). The phospholipid bilayer is able to stretch, contract, break and reform, giving the membrane its *fluid* nature.
2. *Proteins*. The proteins are embedded within this bilayer, giving the membrane a *mosaic* appearance. These proteins are able to move within the bilayer (see figure 1.56).

Cholesterol is found in the bilayer maintaining its fluidity.

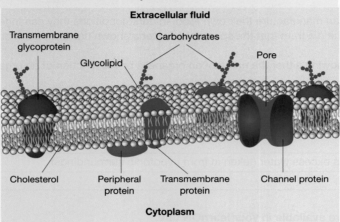

FIGURE 1.55 The structure of the plasma membrane. A bilayer of phospholipids separates the internal and external environments. Proteins embedded in the membrane have a variety of roles.

All carbohydrates occur on the external surface of the membrane, which allows for identification of the cytoplasm (intracellular environment) from the extracellular environment.

1.5.2 The components of the plasma membrane

The fluid mosaic model describes the nature of the plasma membrane, with a phospholipid bilayer with embedded proteins. In this section we will examine how these components control the entry and exit of substances from one area to another. As it only allows some substances across, the plasma membrane is said to be **selectively permeable (semipermeable)**.

> The plasma membrane controls the entry and exit of substances. The plasma membrane can be selectively permeable because of the components it is made of. These are: phospholipids, proteins, cholesterol and carbohydrates.

fluid mosaic model a model proposing that the plasma membrane and other intracellular membranes should be considered as two-dimensional fluids in which proteins are embedded

selectively permeable (semipermeable) allows some substances to cross but precludes the passage of others

Phospholipids

The plasma membrane consists of a double layer (bilayer) of **phospholipids**. Each phospholipid molecule consists of two fatty acid chains joined to a phosphate-containing group. The phosphate-containing group of a phospholipid molecule constitutes its water-loving (**hydrophilic**) head. The fatty acid chains constitute the water-fearing (**hydrophobic**) tail of each phospholipid molecule.

Examine figure 1.56. Notice that for the plasma membrane around a cell, the two layers of phospholipids are arranged so that the hydrophilic heads are exposed at both the external environment of the cell and at the cytosol (the internal environment of the cell). In contrast, the two layers of hydrophobic tails face each other in the central region of the plasma membrane. Water and lipids do not mix.

> **phospholipid** major type of lipid found in plasma membranes
> **hydrophilic** substances that dissolve easily in water; also called polar
> **hydrophobic** substances that tend to be insoluble in water; also called non-polar

FIGURE 1.56 a. Chemical structure of a phospholipid and stylised representations showing the hydrophilic head and the two fatty acid chains that make up its hydrophobic tail **b.** Part of the bilayer of phospholipid molecules in the plasma membrane. Notice that the tails face each other and are enclosed in the central region of the membrane, while the heads face outwards to the cell's external environment and inwards to its cytoplasm.

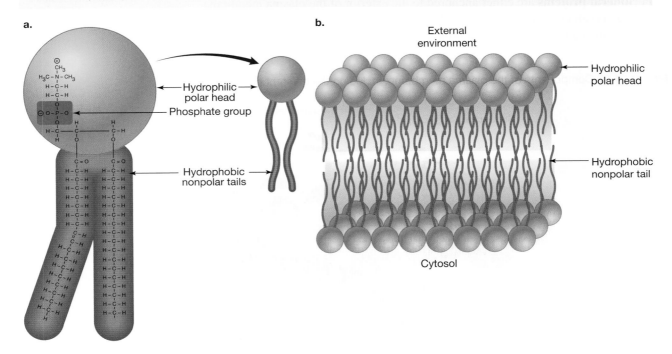

At human body temperature, the fatty acid chains in the inner portion of the plasma membrane are not solid. Instead, they are viscous fluids — think about thick oil or very soft butter — which makes the plasma membrane flexible, soft and able to move freely. This property of the plasma membrane is very important as it enables cells to change shape (provided they do not have a cell wall outside the plasma membrane). For example, red blood cells are about 8 μm in diameter. When circulating red blood cells reach a capillary bed, they must deform themselves by bending and stretching in order to squeeze through capillaries, some of which have diameters as narrow as 5 μm. Shape changes by animal cells are only possible because of the flexible nature of the lipids in the plasma membrane. Flexibility and shape changes are not possible for cells with cell walls.

The plasma membranes ability to change shape, break and reform is very important as it allows larger particles and substances to move into and out of cells (see section 1.5.4).

Proteins

Proteins form the second essential part of the structure of the plasma membrane. Many different kinds of protein comprise part of the plasma membrane. They can be broadly grouped into:
- integral proteins
- peripheral proteins.

Integral proteins, as their name implies, are fundamental components of the plasma membrane. These proteins are embedded in the phospholipid bilayer. Typically, they span the width of the plasma membrane, with part of the protein being exposed on both sides of the membrane (figure 1.57). Proteins like this are described as being **trans-membrane**. In some cases, carbohydrate groups such as sugars are attached to the exposed part of these proteins on the outer side of the membrane, creating a combination called a **glycoprotein**. Integral proteins can be separated from the plasma membrane only by harsh treatments that disrupt the phospholipid bilayer, such as treatment with strong detergents.

Peripheral proteins are either anchored to the exterior of the plasma membrane through bonding with lipids or are indirectly associated with the plasma membrane through interactions with integral proteins in the membrane.

> **integral proteins** fundamental components of the plasma membrane that are embedded in the phospholipid bilayer
>
> **trans-membrane** proteins spanning the plasma membrane but have parts exposed to the exterior and interior of the cells
>
> **glycoprotein** combination formed when a carbohydrate group becomes attached to the exposed part of a trans-membrane protein
>
> **peripheral** proteins either anchored to the exterior of the plasma membrane through bonding with lipids or indirectly associated through interactions with integral protein
>
> **biomacromolecules** large biological polymers such as nucleic acids, proteins and carbohydrates

FIGURE 1.57 Components of the plasma membrane, including the different proteins

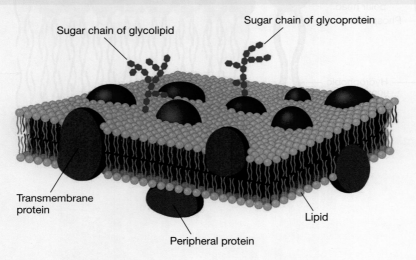

EXTENSION: Proteins are chains of amino acids

Proteins are important **biomacromolecules** that are each comprised of a chain of amino acids. The basic structure of an amino acid is a central carbon joined to a carboxyl group (COOH) on one end, an amino group (NH_2) on the other end and what is known as an 'R' group (figure 1.58a). An R group is a hydrocarbon chain (made of hydrogen and carbon atoms). These are different on each of the 20 amino acids that make up all proteins.

A chain of amino acids is also known as a polypeptide chain and once this chain is folded it becomes a functional protein (figure 1.58b).

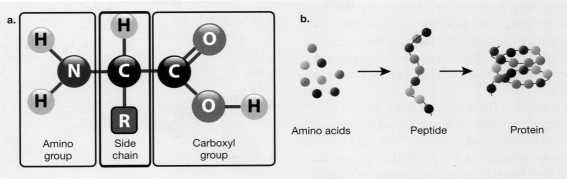

FIGURE 1.58 a. The chemical composition of an amino acid, consisting of a central C to which a carboxyl group (COOH), an amino group (NH$_2$) and an R group is attached **b.** Amino acids join together to form a peptide, which then folds to form a functional 3D protein.

The roles of proteins in the plasma membrane

There are six roles of proteins in the plasma membrane.

1. *Transport*. Trans-membrane proteins facilitate the movement of substances across the phospholipid bilayer. Molecules that are hydrophilic do not easily pass through the phospholipid bilayer and so require the help of transport proteins.
   - Channel proteins open to all molecules to diffuse into and out of the cell. These substance are usually small and charged like ions.
   - Carrier proteins are responsible for the movement of larger molecules or molecules that are moving against their concentration gradient (from low concentration to high). If movement is against the concentration gradient then ATP is required. These proteins and the transport processes involved will be explored further in section 1.5.4.
2. *Reception*. Trans-membrane proteins on the outer surface of the plasma membrane are the receptors for signalling molecules (peptide, or amino acid-based, molecules); each cell has many different kinds of receptor protein. When the signal binds to the receptor protein it alters the shape of the receptor protein and starts a specific response in the cell.

Cell signalling is important as it tells the cell when to make, or stop, certain cellular functions such as breaking down harmful products, uptake of glucose or even undergoing cell death. If cell processes aren't tightly controlled through cell signalling, diseases like cancer can result.

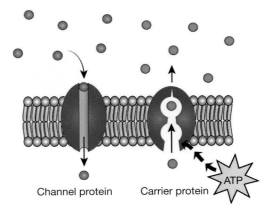

FIGURE 1.59 Transport proteins allow the movement of substances.

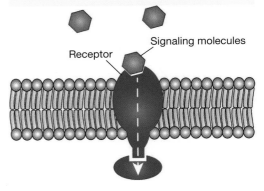

FIGURE 1.60 Proteins bind to receptor sites.

3. *Anchorage*. Proteins connect the intracellular cytoskeleton to the extracellular matrix, holding cells in place. Cells need to be anchored to nearby cells to prevent them from floating away, particularly in animals. To do this, proteins on the cell surface are linked to collagen fibres and peptidoglycan filaments that have been excreted by the cells. These extracellular components are then linked to the extracellular matrices of nearby cells.
4. *Cell identity*. These proteins mark the cell as belonging to 'self'. If the cell is not recognised, a defence is mounted against it. They are known as glycoproteins, antigens or cell identity tags. Each cell type has a different combination of surface markers. In mammals, for example, these markers enable the immune system to identify the cells as 'self' and distinguish them from foreign cells. Glycolipids on the plasma membrane play a role in tissue recognition.

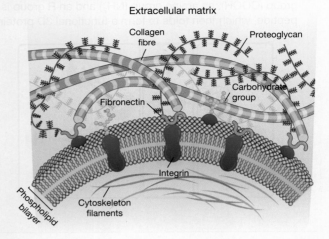

FIGURE 1.61 Proteins anchor the cells in place.

FIGURE 1.62 Proteins identify foreign cells.

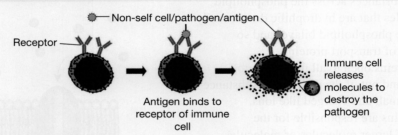

5. *Intercellular joinings*. These proteins are involved with plasmodesmata and tight junctions. They join cells together and facilitate communication between cells, which is necessary for the efficient functioning of tissues. The function of the intracellular joining is determined by type. For example, tight junctions between endothelial cells (figure 1.63) prevent the movement of substances between cells. Plasmodesmata, found in plant cells, link the cell cytosols of plant cells together, facilitating molecular movement.

FIGURE 1.63 Proteins join cells and allow communication between them.

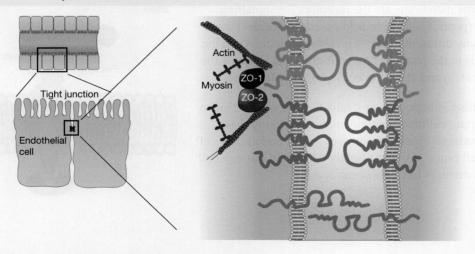

6. *Enzymatic activity.* For reactions to occur, there must be enough energy to initially break some chemical bonds (activation energy). Enzymes are proteins that catalyse (speed up) reactions by lowering this activation energy. Without enzymes, most chemical reactions would occur too slowly to sustain life. Enzymes lower the activation energy by binding to the substrate/s and weakening the bonds to initiate the reaction, thereby speeding up the rate of reaction. Enzymes are specific as they only bind to one or two substrates and they are not altered in the process, so they can be used over and over again. Enzymes in the membrane facilitate a number of important biochemical pathways, such as the production of ATP.

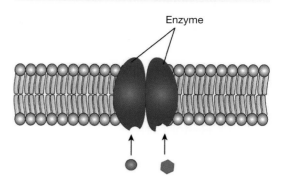

FIGURE 1.64 Proteins act as enzymes in biochemical pathway reactions.

The roles of proteins in the plasma membrane

The roles of proteins in the plasma membrane can be remembered by using the acronym TRACIE:
- **T** — Transport
- **R** — Reception
- **A** — Anchorage
- **C** — Cell identity
- **I** — Intracellular joinings
- **E** — Enzymatic activity

Cholesterol

Cholesterol is a lipid (fat) that is present in the plasma membrane. Its role is to ensure the flexibility of the membrane by reducing membrane fluidity in high temperatures and increasing it in cold temperatures.

Cholesterol maintains space between the fatty acid tails, preventing them from binding strongly together under cold conditions, and conversely when hot, it prevents the phospholipids from separating too far apart as molecular movement increases.

Recall that as the temperature increases, the kinetic energy of molecules increases (according to the kinetic theory of matter), which increases the distance between the molecules. This is demonstrated in figure 1.65.

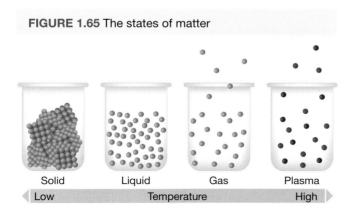

FIGURE 1.65 The states of matter

cholesterol sterol compound important in the composition of cell membranes

Carbohydrates

Some phospholipids and proteins have small chains of carbohydrates (sugars) attached to them. These are then known as **glycolipids** and glycoproteins respectively.

Glycolipids have a role in cell recognition, so that our bodies can tell if the cell belongs to us ('self' cells) or are foreign cells and therefore may cause harm.

Glycoproteins are also utilised in the recognition of 'self' cells (cell identity) as well as intracellular joinings (connections between cells) and anchorage.

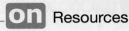

> **eWorkbook** Worksheet 1.4 Structure of the plasma membrane (ewbk-7192)

1.5.3 Functions of the plasma membrane

The plasma membrane is the site of many interactions. It is an **active boundary**. It is a place where some, but not all, substances are transported between the internal and external environments. The plasma membrane acts as a selective barrier between the cell membrane and the outside of the cell, and the cytosol and the internal compartments of organelles. This allows for many different reactions that require different conditions to occur in the same cell.

As described in the previous section, the plasma membrane is also a place where *messages* (signalling molecules) that originate outside the cell are received. These signals lead to changes in cell function, allowing an organism to change as the environment changes. *Cell identity* is worn like a badge on the surface of the cell. If a cell is not displaying the correct molecules, an immune response is mounted and the cell dies (see Extension box).

Arguably the most important role of the membrane is its ability to *transport* materials between the intracellular and extracellular environments, and this will be explored in detail in section 1.5.4.

Membranes are constantly changing. For example, to facilitate the uptake of a particular molecule, more transport proteins are added. This is the case for glucose uptake. When cell receptors bind to insulin, glucose transports stored in vesicles in the cell move to the cell membrane to increase glucose uptake (figure 1.66).

Functions of the plasma membrane

The plasma membrane carries out several important functions for a cell.

The plasma membrane:
1. is an active and selective boundary
2. denotes cell identity
3. receives external signals
4. transports materials.

---

glycolipids small chains of carbohydrates (sugars) attached to the phospholipids and proteins of the plasma membrane; aid in cell recognition

active boundary a barrier that is constantly changing and responsive to the environment

FIGURE 1.66 An external message (insulin) is received by the cell; as a result, glucose transporters (GLUT-4) are moved to the cell membrane to increase glucose uptake.

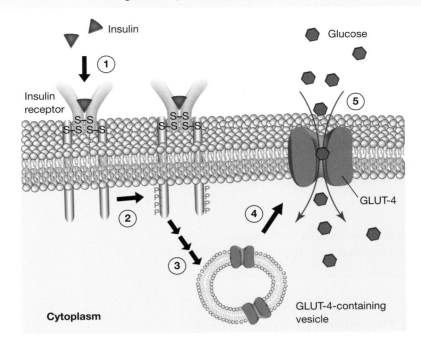

EXTENSION: Major histocompatibility complex

The major histocompatibility complex (MHC) are very important surface proteins due to their role in cell identity and immune response.

The immune response is an extension concept for the Units 1 & 2 course but is examinable in the Units 3 & 4 course.

To access more information on this extension concept please download the digital document.

FIGURE 1.67 Structure of the MHC (class I and II)

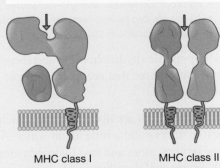

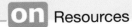 Resources

Digital document Extension: Major histocompatibility complex (doc-35860)

SAMPLE PROBLEM 7 Identifying features of the plasma membrane

tlvd-1743

The plasma membrane is described as an active and selective boundary. Explain what this means? **(2 marks)**

THINK

Describe what each term means in relation to the plasma membrane.

Active means:
- not passive
- constantly changing, including the movement of substances, glucose transporters
- responding to its environment, such as changes in uptake.

Selective means:
- only some substances move through the boundary to maintain conditions within the cell/organelle.

WRITE

Active boundary:
- It is constantly changing with no fixed components.
- It responds to its environment by receiving external signals and changing cell function (1 mark).

Selective boundary:
- Only some substances can pass through the membrane (1 mark).

1.5.4 Crossing the plasma membrane

Crossing the membrane can be achieved in a number of ways, depending on:
- the type of molecule (hydrophilic or hydrophobic)
- the size of the molecule
- whether there is a difference in concentration gradient.

Substances can be roughly divided into two groups:
- *Hydrophilic (water-loving)* — dissolve in water. These are often **lipophobic**.
- *Hydrophobic (water-fearing)* — dissolve in lipids (**lipophilic**).

Small lipophilic (hydrophobic) substances can pass directly through the phospholipid bilayer; substances that are hydrophilic or too large to fit through the bilayer must find another way.

Which substances are lipophilic and which are hydrophilic? In order to understand this, it is best to understand a little about water. As shown in figure 1.68, a water molecule consists of one oxygen atom and two hydrogen atoms. Due to the bonding in water (see Extension box), an unequal distribution of electrons means the oxygen atom is slightly more negative than the hydrogen atoms. This means the water molecule has a slightly positive end and a slightly negative end, so is a polar molecule. This slight difference in charge means that the negative ends of the water molecule will be attracted to any part of another molecule that is positive (as opposite charges attract), and the positive parts will be attracted to any part of a molecule that is negative. This results in attraction with ions (charged atoms) such as Cl^-, Na^+ or K^+, as well as any other polar molecules.

Molecules that dissolve in water (hydrophilic) must therefore also be polar molecules.

Molecules without a difference in charge, such as O_2, CO_2, urea and ethanol, are not attracted to water and are all lipophilic.

FIGURE 1.68 Water is a polar molecule with a negative end and positive ends.

H_2O

lipophobic 'lipid-fearing' molecules that do not dissolve in lipids (hydrophilic)

lipophilic 'lipid-loving' molecules that dissolve in lipids (hydrophobic)

Additionally, despite being polar, water can slowly pass through the plasma membrane due to its small size. Although the tails of the phospholipids are hydrophobic, movement of these tails allows enough space for the small water molecules to pass through without being repelled.

FIGURE 1.69 The semipermeable nature of a phospholipid bilayer membrane. The membrane is fully permeable to some substances, partially permeable to others and impermeable to yet other substances.

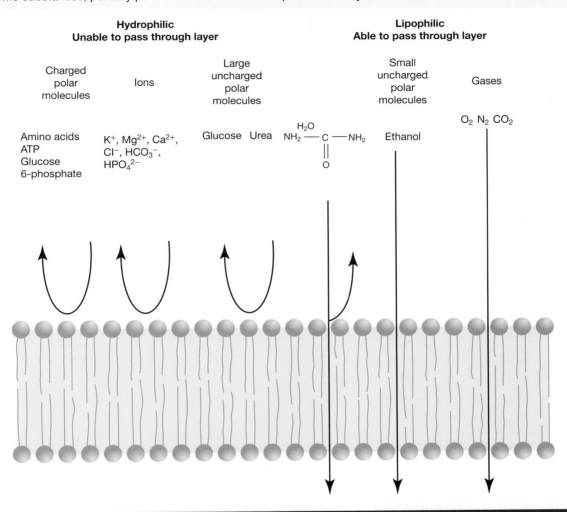

elog-0780

INVESTIGATION 1.4 — online only

Modelling the properties of the plasma membrane

Aim

To model the properties of plasma membranes and show how this allows for the membrane to function

on Resources

eWorkbook Worksheet 1.5 Function of the plasma membrane (ewbk-7194)

EXTENSION: Covalent bonding of water

Covalent bonds are formed between non-metal elements when they share electrons. Atoms share electrons to become more stable by filling their outer electron shells. For example, hydrogen has one electron in its outer shell (and its only shell).

In some molecules, one of the atoms has a greater affinity for the bonding electrons and so the electrons bond closer to that atom, creating an uneven distribution of charge (a dipole) in the molecule. When this occurs, it is called a polar molecule. The slight negative and positive regions cause the molecule to attract other polar molecules.

As water contains hydrogen–oxygen bonds, a particularly strong intermolecular (between molecules) attraction occurs known as a hydrogen bond. As hydrogen has only one electron, when it moves towards the oxygen atom (figure 1.70) no other electrons are left to shield the hydrogen nucleus from the negative region it is approaching. This allows water molecules to attract other molecules very closely, resulting in a stronger bond than would be the case if hydrogen were not involved. Hydrogen bonding is fundamental for life processes; these interactions hold the two strands together in the double helix that forms our DNA, for example.

FIGURE 1.70 The shared electrons are closer to the oxygen (O) nucleus than the hydrogen (H) nucleus, making the oxygen more negative than the hydrogen.

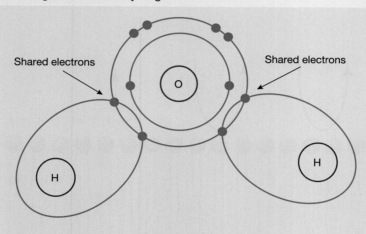

1.5.5 Methods of membrane transport

Passive versus active transport

Whether the transport can occur passively (does not require energy to occur) or actively (requires energy to occur) depends on whether a **concentration gradient** exists. A concentration gradient exists when there is a difference in concentration on each side of the plasma membrane.

If movement occurs from a region of high concentration to low concentration, the movement is *down or along a concentration gradient*. If the movement is from a region of low concentration to high concentration, movement is *against the concentration gradient*.

Very large molecules (like proteins) need to be transported via vesicles through the plasma membrane in processes that are termed **bulk transport**.

covalent bonds a type of bond between atoms where electrons are shared. The bond is very strong.

concentration gradient occurs when there is a difference in solute concentration from one area to another

bulk transport the movement of material into a cell (endocytosis) or out of a cell (exocytosis)

FIGURE 1.71 Flow chart summarising transport across the plasma membrane

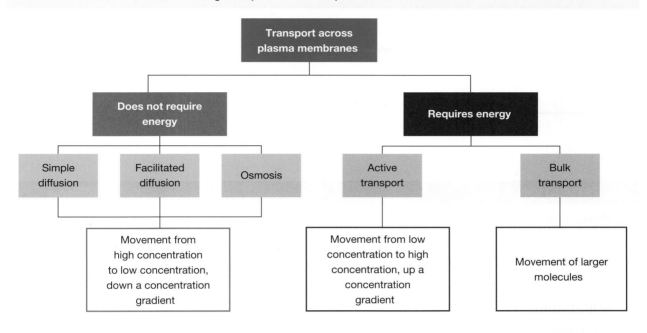

Mixture terminology

Solute = substance that is dissolved

Solvent = liquid in which a solute dissolves

Solution = liquid mixture of the solute in the solvent

Net movement is the overall direction that movement has occurred. It should be noted that movement will occur in both directions across the membrane.

Simple diffusion

Simple diffusion is the movement of **solutes** across the phospholipid bilayer from a region of higher concentration to one of lower concentration of that solute; that is, *down* its concentration gradient (see figure 1.72a).

Movement down a concentration gradient by simple diffusion does *not* require any input of energy. It is the gradient that drives the diffusion (like letting a ball roll down a slope).

Simple diffusion occurs in stages (figure 1.72b):
1. Initially, substance X starts to move into the cell because of random movement that results in some collisions with the membrane (figure 1.72bi).
2. Midway, molecules of substance X are moving both into and out of the cell, but the net movement is from outside to inside (figure 1.72bii).
3. The end point of simple diffusion is reached when equal concentrations of substance X are reached on both sides of the plasma membrane (figure 1.72biii). Movement in both directions will continue to occur across the membrane, but the concentration of the substance will remain constant.

solute substance that is dissolved

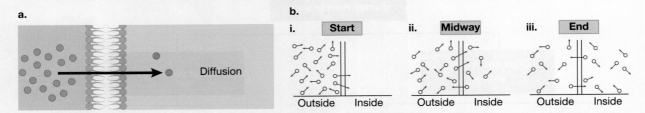

FIGURE 1.72 a. Simple diffusion involves the movement of substances through the phospholipid bilayer of the plasma membrane down the concentration gradient of the diffusing substance (from high to low concentration). **b.** The stages of simple diffusion

elog-0782

INVESTIGATION 1.5 — online only

Diffusion across the membrane

Aim

To observe the diffusion of substances across a semipermeable membrane

Facilitated diffusion

Facilitated diffusion is an example of protein-mediated transport. Facilitated diffusion is so named because the *diffusion* across the membrane is enabled or *facilitated* by special protein transporters in the plasma membrane.

Like simple diffusion, facilitated diffusion does not require an input of energy and it moves substances down their concentration gradients. However, facilitated diffusion of dissolved substances requires the action of protein transporters that are embedded in the cell membrane. Facilitated diffusion enables molecules that cannot diffuse across the phospholipid bilayer to move across the plasma membrane through the agency of transporter proteins. These transporters are either **channel proteins** or **carrier proteins**. Are transporters required in simple diffusion? (See figure 1.73.)

> **channel proteins** trans-membrane proteins involved in the transport of specific substances across a plasma membrane by facilitated diffusion
> **carrier protein** protein that binds to a specific substance and facilitates its movement through the membrane

Channel proteins

Channel proteins are transport proteins involved in facilitated diffusion. Different channel proteins are specific for the diffusion of charged particles and polar molecules.

Each channel protein consists of a central, narrow, water-filled pore in the plasma membrane through which substances can move down their concentration gradients (see figure 1.73b). By providing water-filled pores, channel proteins create a hydrophilic passage across the plasma membrane that bypasses the phospholipid bilayer and facilitates the diffusion of charged particles, such as sodium and potassium ions as well as small polar molecules.

Carrier proteins

Carrier proteins are another group of transport proteins involved in facilitated diffusion. Carrier proteins are specific, with each kind of carrier enabling the diffusion of one kind of molecule across the plasma membrane. After binding to its specific cargo molecule, the carrier protein undergoes a change in shape as it delivers its cargo to the other side of the plasma membrane (see figure 1.73c).

Carrier proteins are important in the facilitated diffusion of hydrophilic uncharged substances, such as glucose and amino acids. In the absence of carrier proteins, these hydrophilic molecules cannot cross the plasma membrane directly.

FIGURE 1.73 a. Simple diffusion. In contrast, facilitated diffusion requires either b. channel proteins or c. carrier proteins to enable certain dissolved substances to diffuse down their concentration gradients. Note the change in shape of the carrier protein as it operates.

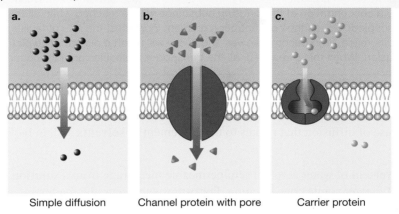

Simple diffusion   Channel protein with pore   Carrier protein

EXTENSION: Rates of diffusion

In the previous sections, two types of diffusion of dissolved substances across plasma membranes have been explored: simple diffusion and facilitated diffusion. Do these two types of diffusion differ in the rate at which substances can move down their concentration gradients in either direction across a plasma membrane?

In simple diffusion, the rate at which substances move across the plasma membrane by simple diffusion is determined by their concentration gradient; that is, the difference in the concentration of the substance inside and outside the cell. The higher its concentration gradient, the faster a substance will move by simple diffusion down this gradient and across a plasma membrane. (Think of this as like rolling a ball downhill — the steeper the incline, the faster the ball rolls.)

In facilitated diffusion, the rate of movement of a substance is also influenced by the steepness of its concentration gradient on either side of the plasma membrane. The steeper the concentration gradient, the faster the rate of facilitated diffusion, *but only up to a point* (see figure 1.74).

FIGURE 1.74 Graph showing rates of simple and facilitated diffusion with increasing concentration gradient of the diffusing substance. Note that the rate of simple diffusion continues to increase as the concentration gradient increases. In contrast, the rate of facilitated diffusion is initially linear, but begins to taper off and finally reaches a plateau. The maximum rate is reached when all the transporters are fully occupied.

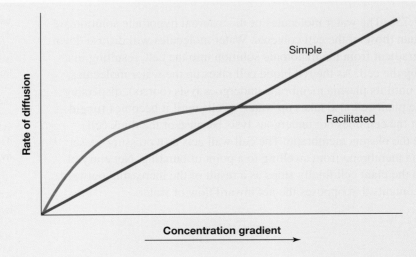

Why the difference between simple diffusion and facilitated diffusion? Facilitated diffusion requires the involvement of transporters, either channel proteins or carrier proteins, for the movement of substances. These channel and carrier proteins are present in limited numbers on the plasma membrane. Because their numbers are limited, eventually a concentration of the diffusing substance will be reached when all the transporters are saturated (fully occupied). So, as the concentration gradient increases, the rate of facilitated diffusion of a substance will at first increase, then become slower, and finally will reach a plateau. The plateau is the maximum rate of facilitated diffusion. When this is reached, all the transporter molecules are fully occupied.

Osmosis

Osmosis is a special case of diffusion that relates to the movement of **solvents** and, in biological systems, that solvent is water.

Osmosis is the net movement of water across a semipermeable membrane from a **solution** of lesser solute concentration to one of greater solute concentration. For osmosis to occur, the membrane must be permeable to water but not to the solute molecules. Solutions that have a high concentration of dissolved solute have a lower concentration of water, and vice versa. The net movement of water molecules in osmosis from a solution of high water concentration to one of lower concentration is known as **osmotic flow**.

When an external solution is compared with the dissolved contents of a cell, the external solution may be found to be either:
- **isotonic** — having an equal solute concentration to that of the cell contents
- **hypotonic** — having a lower solute concentration than the cell contents
- **hypertonic** — having a higher solute concentration than the cell contents

Osmosis can be seen in action when cells are immersed in watery solutions containing different concentrations of a solute that cannot cross the plasma membrane. Remember that as the concentration of solute molecules increases, the concentration of the water molecules decreases.

Isotonic

Look at figure 1.75a. The water molecules of the external isotonic solution are at the same concentrations as those in the cell contents. Since there is no concentration gradient, no net uptake of water molecules occurs in either cell; in a given period, the same number of water molecules will diffuse into the cell as will diffuse out.

Hypotonic

Now look at figure 1.75b. The water molecules of the external hypotonic solution are more concentrated than those of the cell contents. Water molecules will diffuse down their concentration gradient from the hypotonic solution into the cell, resulting in a net uptake of water by the cell. As the red blood cell takes up the water molecules, it continues to swell until its plasma membrane undergoes **lysis** (bursts), dispersing the cell contents. The plant cell also takes up water, swells until it becomes **turgid** (rigidly swollen), but the cell does not undergoes lysis because of the thick cell wall that lies outside the plasma membrane. The cell wall acts as a pressure vessel preventing the plasma membrane from swelling to a point of bursting. Net entry of water molecules into the plant cell finally stops as a result of the increasing outward pressure of the cell contents that opposes the net inward flow of water.

osmosis net movement of water across a partially permeable membrane without an input of energy and down a concentration gradient
solvent liquid in which a solute dissolves
solution liquid mixture of the solute in the solvent
osmotic flow the net movement of water molecules from a solution of high water concentration to lower water concentration (or alternatively, from a region of low to high solute concentration)
isotonic having the same concentration of dissolved substances as the solution to which it is compared
hypotonic having a lower concentration of dissolved substances than the solution to which it is compared
hypertonic having a higher concentration of dissolved substances than the solution to which it is compared
lysis bursting of a cell
turgid swollen and distended

Hypertonic

Finally, look at figure 1.75c. The water molecules of the external hypertonic solution are at a lower concentration than those in the cell contents. Water molecules will diffuse down their concentration gradient from the cells into the external solution, resulting in a net loss of water from the cells. The red blood cell shrinks, becoming **crenated**. The plant cell membrane shrinks away from the cell wall, leading to **plasmolysis**.

> **crenation** shrinking of cell due to water loss
> **plasmolysis** shrinking of the cytoplasm away from the wall due to water loss

FIGURE 1.75 Osmosis in action: behaviour of animal (top) and plant cells (bottom) in solutions of different concentrations of a dissolved substance (solute molecules = red dots). Note the presence of a cell wall outside the plasma membrane in the plant cells. The solute molecules are too large to cross the plasma membrane, but water molecules will move down their concentration gradient in a process called osmotic flow.

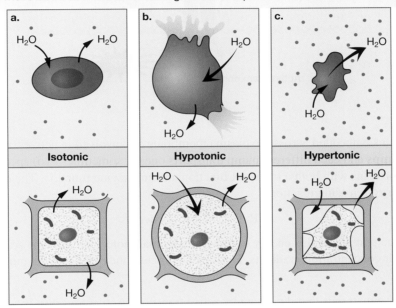

Water is a polar molecule and while it can and does move through the phospholipid bilayer, it also moves through special channel proteins called aquaporins. Therefore, osmosis is really a mix of simple diffusion and facilitated diffusion for water.

Passive transport

Passive transport across the plasma membrane does not require the input of energy. Movement occurs from high solute concentrations to lower solute concentrations by simple diffusion, facilitated diffusion and osmosis.

---

elog-0784

INVESTIGATION 1.6

online only

Osmosis in action

Aim
To observe osmosis in plant cells

Active transport

Active transport is the process of moving substances from a region of low concentration to a region of high concentration of those substances (that is, in the opposite direction to diffusion). This can only occur with an input of energy — the energy source is typically adenosine triphosphate (ATP).

FIGURE 1.76 A simplified representation of active transport. A pump is a trans-membrane protein that is both a carrier and an ATPase enzyme. The enzyme component of the pump catalyses the energy-releasing reaction that powers active transport.

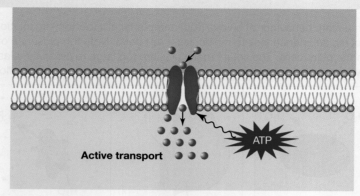

Special transport proteins embedded across the plasma membrane are involved in active transport. The proteins involved are called **pumps** and each different pump transports one (or sometimes two) specific substance(s). Important pumps are proteins with both a transport function and an enzyme function. The enzyme part of the pump catalyses an energy-releasing reaction:

$$ATP \rightarrow ADP + Pi + energy$$

The transport part of the pump uses this energy to move small polar molecules and ions across the plasma membrane *against* their concentration gradients. During this process, the protein of the pump undergoes a shape change (see figure 1.76).

Cells use pumps to move materials that they need by active transport. Active transport is essential for the key function of cells, including pH balance, regulation of cell volume and uptake of needed nutrients. Examples of active transport include:
- uptake of dissolved mineral ions from water in the soil by plant root hair cells against their concentration gradient
- production of acidic secretions (pH of nearly 1) by stomach cells that have a low internal concentration of hydrogen ions (H^+) but produce secretions (gastric juice) with an extremely high concentration of hydrogen ions
- uptake of glucose from the small intestine into the cells lining the intestine against its concentration gradient, using the glucose–sodium pump
- maintenance of the difference in the concentrations of sodium and potassium ions that exist inside and outside cells (see table 1.5) by action of the **sodium–potassium pump**, which actively transports these ions against their concentration gradients (figure 1.77).

The sodium–potassium pump

Some pumps actively transport a single dissolved substance against its concentration gradient. Other pumps transport two substances simultaneously; for example, the sodium–potassium pump. The sodium–potassium pump is particularly important in the human body as about 25 per cent of the body's ATP is expended in keeping this pump operating. For brain cells, the figure is even higher — about 70 per cent.

pumps special transport proteins embedded across the plasma membrane that carry out the process of active transport

sodium–potassium pump protein that transports sodium and potassium ions against their concentration gradients to maintain the differences in their concentrations inside and outside cells

The sodium–potassium pump plays a key role in excitable cells, such as nerve and muscle cells. During the transmission of a nerve impulse, sodium ion channels open and sodium ions rapidly flood into the nerve cell by facilitated diffusion. After the nerve impulse has passed, the sodium channels close and the sodium–potassium pump then restores the concentrations of sodium and potassium ions to their resting levels by actively pushing sodium ions across the membrane out of the cell, and dragging potassium ions into the cell (see table 1.5). Restoring these concentrations involves active transport against the concentration gradients of these ions.

TABLE 1.5 Approximate concentrations of sodium and potassium ions in the cytosol of cells and in the surrounding extracellular fluid

Ion	Inside cell	Outside cell
Sodium (Na^+)	10 mM	142 mM
Potassium (K^+)	150 mM	5 mM

FIGURE 1.77 Sodium–potassium pump. Sodium is pumped against its concentration gradient and potassium is pumped against its concentration gradient. This requires energy. ATP becomes ADP and Pi and energy is released.

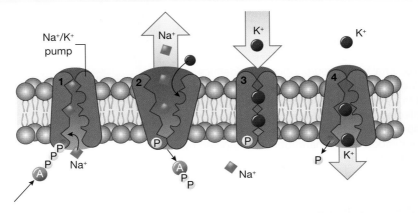

CASE STUDY: When transport goes wrong

The importance of transport proteins in moving substances becomes apparent if they do not operate as expected, as may be seen in cystic fibrosis and in cholera infections.

Cystic fibrosis

Cystic fibrosis affects the following:
- *Sweat glands in skin*. In the skin, the chloride ion channel is normally involved in reabsorbing salt (NaCl) from fluid within the sweat glands before it is released as sweat. When the chloride ion (Cl^-) transporter is blocked, reabsorption does not occur and very salty sweat is produced.
- *Pancreas*. Pancreatic enzymes are normally involved in the digestion of some foods. In cystic fibrosis, these enzymes are unable to enter the gut because abnormally thick mucus blocks the narrow duct that connects the pancreas to the small intestine. Without these enzymes, food cannot be fully digested. However, replacement enzymes in the form of tablets, powders or capsules can replace the enzymes normally released by the pancreas (see figure 1.78).

FIGURE 1.78 Capsules and tablets containing the missing pancreatic enzymes (lipase, protease and amylase) are taken by persons affected by cystic fibrosis. Note the enzyme granules in the capsules.

Resources

Digital document Extension: Ion concentrations in cystic fibrosis (doc-35856)

Cholera

The pores of some channel proteins are permanently open, while the pores in other channel proteins open only in response to a specific signal. The chloride ion channel on the plasma membranes of cells lining the intestine is *not* always open. Normally chloride ions are retained within the cells lining the intestine. Only when this channel is opened can chloride ions move from the cells into the cavity of the intestine.

Cholera results from a bacterial infection of *Vibrio cholerae*. A toxin produced by these bacteria causes the chloride ion channels in the cells lining the intestine to be locked in the 'open' position. This results in a flood of chloride ions into the intestinal space that is followed by a flow of sodium ions (down the resulting electrochemical gradient that is created). In turn, the increased concentration of salt in the gut creates a hyperosmotic environment that draws water into the gut by osmosis. The continuous secretion of water into the intestine causes the production of large volumes of watery diarrhoea. If left untreated, the water loss caused by this diarrhoea can be fatal within hours. Cholera epidemics have caused many deaths. Cholera can be spread by water that is contaminated by contact with untreated sewage or by the faeces of an infected person. Cholera can be spread by food that is washed in or mixed with water contaminated by cholera bacteria, or by food that is inappropriately handled by a person infected with cholera.

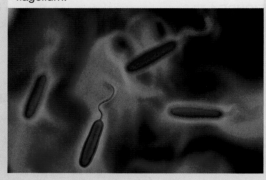

FIGURE 1.79 These rod-shaped *Vibrio cholerae* bacteria have a single polar flagellum.

Bulk transport of solids and liquids

FIGURE 1.80 Endocytosis — a summary

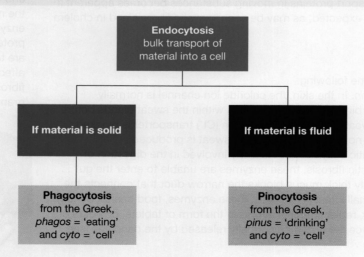

To this point, we have been concerned with movement of dissolved substances across the plasma membrane. In addition, small solid particles and liquids in bulk can be moved across the plasma membrane into or out of cells. Figure 1.80 gives a summary of how bulk material can enter cells.

Endocytosis: getting in

Endocytosis is the process of bulk transport of material into a cell. Part of the plasma membrane encloses the material to be transported and then pinches off to form a membranous vesicle that moves into the cytosol (figure 1.81a).

For example, one kind of white blood cell is able to engulf a disease-causing bacteria cell and enclose it within a lysosome sac, where it is destroyed. Unicellular protists, such as those in the genera *Amoeba* and *Paramecium*, obtain their energy for living in the form of relatively large 'food' particles, which they engulf and enclose within a sac, where the food is digested (see figure 1.81b). When the material being transported is a solid food particle, the type of endocytosis is called **phagocytosis**.

FIGURE 1.81 a. Endocytosis occurs when part of the plasma membrane forms around food particles to form a phagocytic vesicle (or phagosome). This vesicle then moves into the cytosol where it fuses with a lysosome, a bag of digestive enzymes (a phagolysosome). The same digestive process can also occur to microbes. **b.** Transport of a solid food particle across the membrane of *Amoeba proteus* (common amoeba).

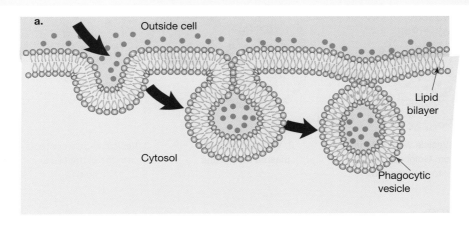

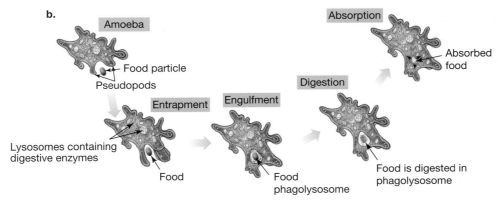

Although some cells are capable of phagocytosis, most cells are not. Most eukaryotic cells rely on **pinocytosis**, a form of endocytosis that involves material that is in solution being transported into cells. The process of endocytosis is an energy-requiring process and requires an input of ATP.

endocytosis bulk movement of solids or liquids into a cell by engulfment

phagocytosis bulk movement of solid material into cells

pinocytosis bulk movement of material that is in solution being transported into cells

Video eLesson Phagocytosis (eles-2444)

Exocytosis: getting out

Exocytosis is the bulk transport out of cells, such as the export of material from the Golgi apparatus (section 1.4.3). In exocytosis, vesicles formed within a cell fuse with the plasma membrane before the contents of the vesicles are released from the cell (see figure 1.82). If the released material is a product of the cell (for example, the contents of a Golgi vesicle), then 'secreted from the cell' is the phrase generally used. If the released material is a waste product after digestion of some matter taken into the cell, 'voided from the cell' is generally more appropriate. The process of exocytosis requires an input of energy in the form of ATP.

exocytosis movement of material out of cells via vesicles in the cytoplasm

FIGURE 1.82 Exocytosis (bulk transport out of cells) occurs when vesicles within the cytosol fuse with the plasma membrane and vesicle contents are released from the cell.

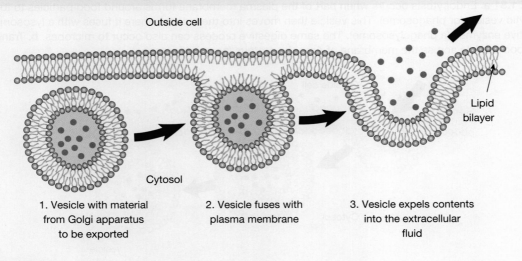

1. Vesicle with material from Golgi apparatus to be exported
2. Vesicle fuses with plasma membrane
3. Vesicle expels contents into the extracellular fluid

Resources

- **eWorkbook** Worksheet 1.6 Transport across the membrane (ewbk-7196)
- **Video eLesson** Mechanisms of membrane transport (eles-2463)
- **Interactivity** Movement across membranes (int-0109)

KEY IDEAS

- Functions involving the plasma membrane include:
  - creating a compartment that separates the cell from its external environment
  - receiving external signals as part of communication between cells
  - providing cell surface markers that identify the cell
  - transporting materials across the plasma membrane.
- Transport proteins in the plasma membrane enable movement of substances that cannot cross the phospholipid bilayer of the membrane.
- Simple diffusion moves dissolved substances across the plasma membrane down their concentration gradient and requires no input of energy.
- Osmosis is a special case of diffusion, being the movement of water across the plasma membrane down its concentration gradient.
- Facilitated diffusion moves dissolved substances across the plasma membrane down their concentration gradients, but this movement occurs through involvement of transport proteins — either channel or carrier proteins — and requires no input of energy.
- Active transport moves dissolved substances across the plasma membrane *against* the concentration gradient, a process that can occur only via the action of protein pumps.

- Active transport requires an input of energy that commonly comes from ATP, catalysed by the ATPase enzyme that is part of some protein pumps.
- Endocytosis is the bulk transport of material into cells; if solids are being moved, the process is termed phagocytosis, while if liquids are being moved, the process is termed pinocytosis.
- Exocytosis is the bulk movement of materials via secretory vesicles out of cells.

1.5 Activities

learn on

To answer questions online and to receive **immediate feedback** and **sample responses** for every question, go to your learnON title at **www.jacplus.com.au**. A **downloadable solutions** file is also available in the resources tab.

| 1.5 Quick quiz on | 1.5 Exercise | 1.5 Exam questions |

1.5 Exercise

1. What are the four components of the plasma membrane?
2. What is meant by the *fluid mosaic model*?
3. What are the six functions of proteins in the cell membrane?
4. Compare and contrast the differences between passive and active transport in terms of energy, direction and part of the membrane passed through.
5. Look at the following cells. Each of them were placed in solutions of differing salt concentrations.

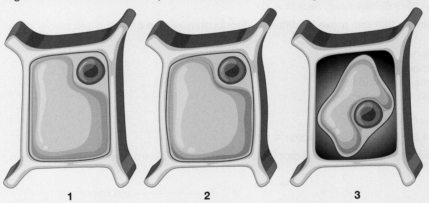

1   2   3

Identify which type of solutions each cell was placed in and why you came to this conclusion.
6. Draw a labelled diagram of the process of endocytosis.
7. By which process do cells of the stomach lining manage to move hydrogen ions out of the cells to produce a highly acidic gastric secretion?
8. Sucrose cannot cross the plasma membranes of red blood cells, but glucose can. Red blood cells are immersed in the following solutions:
   - a hypertonic sucrose solution
   - a hypertonic glucose solution
   - a hypotonic sucrose solution
   - a hypotonic glucose solution.
   a. Which solution would be expected to cause the greatest water loss and shrinkage of the red blood cells? Explain.
   b. Which solution, if any, might cause the red blood cells to burst? Explain.
9. Suggest a possible explanation for the following observations.
   a. Proteins can move laterally across the plasma membrane.
   b. A person with cystic fibrosis is at high risk of lung infections.
   c. Lipophilic substances cross the plasma membrane by simple diffusion, but not charged particles.
   d. A baby with cystic fibrosis produces abnormally salty sweat.
   e. Persons with a cholera infection suffer severe diarrhoea.

10. An artificial membrane, composed of a phospholipid bilayer only, was manufactured. Its behaviour was compared with that of a natural plasma membrane.
Predict if these two membranes might behave in a similar or a different manner when tested for their ability to allow the following dissolved substances to cross them:
   - small lipophilic substances
   - charged particles, such as sodium ions
   - glucose
   - proteins.

Briefly justify each of your decisions.

1.5 Exam questions

Question 1 (1 mark)
Source: VCAA 2019 Biology Exam, Section A, Q10

MC Large hydrophilic molecules cannot easily cross a plasma membrane due to the presence of which one of the following molecules in the membrane?
A. proteins
B. cholesterol
C. glycoproteins
D. phospholipids

Question 2 (1 mark)
Source: VCAA 2018 Biology Exam, Section A, Q1

MC Substances that can move by diffusion directly through the phospholipid bilayer of the plasma membrane include
A. sodium ions.
B. oxygen molecules.
C. polar protein molecules.
D. ribonucleic acid molecules.

Question 3 (1 mark)
Source: VCAA 2019 Biology Exam, Section A, Q1

MC Consider the movement of macromolecules across the plasma membrane, as shown in the diagram below.

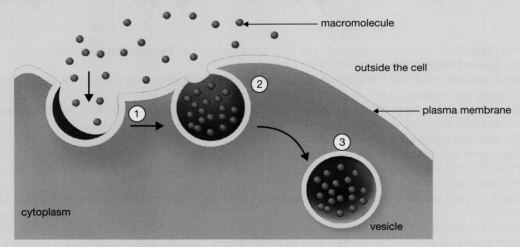

Source: Designua/Shutterstock.com

What type of transport is shown?
A. facilitated diffusion
B. simple diffusion
C. endocytosis
D. exocytosis

Question 4 (1 mark)
Source: VCAA 2017 Biology Exam, Section B, Q1bi

Consider the diagram of a plasma membrane below.

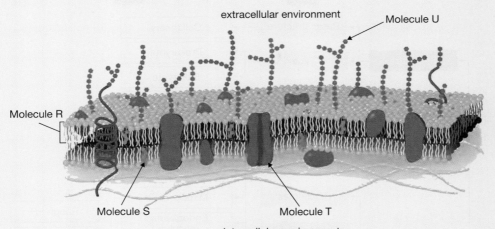

Source: Jamilia Marini/Shutterstock.com

One of the molecules – Molecule R, Molecule S, Molecule T or Molecule U – contains many amino acids. Circle the molecule below that contains many amino acids.

Molecule R   Molecule S   Molecule T   Molecule U

Question 5 (1 mark)
Source: Adapted from VCAA 2016 Biology Exam, Section A, Q6

MC Many eukaryotic cells have proteins as part of their plasma membranes. An experiment was performed on two different animal cells. The diagrams show the positions and shapes of two proteins on the plasma membranes of the two different cells.

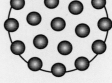

These cells were then fused. After one hour, the plasma membrane of the resulting living cell was observed. The changed positions of the proteins are shown.

The redistribution of proteins on the plasma membrane can be explained by
A. the fluid mosaic model.
B. movement due to osmosis.
C. the presence of cholesterol in the plasma membrane.
D. the active transport of proteins across the plasma membrane.

More exam questions are available in your learnON title.

1.6 Review

1.6.1 Topic summary

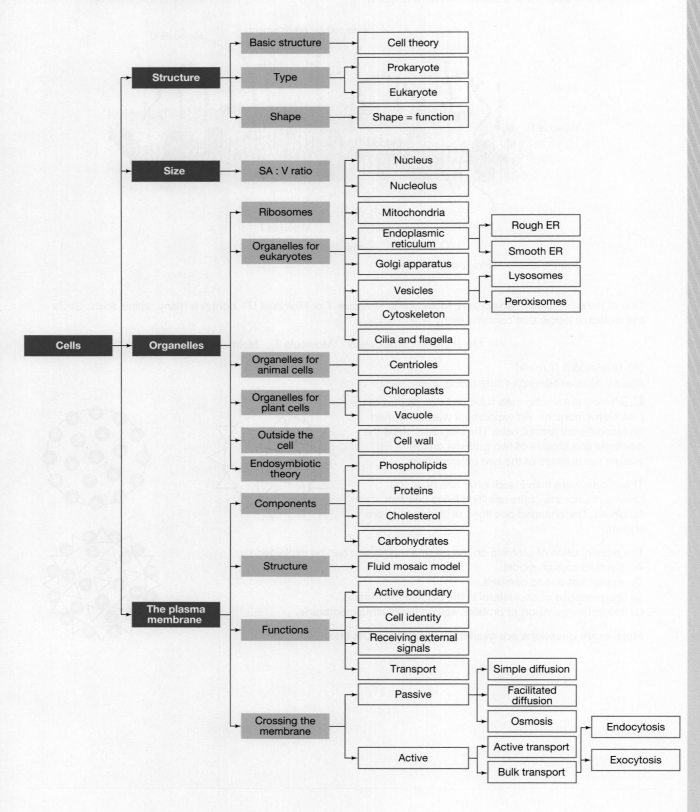

Resources

- **eWorkbook** — Worksheet 1.7 Reflection — Topic 1 (ewbk-7200)
- **Practical investigation eLogbook** — Practical investigation eLogbook — Topic 1 (elog-0156)
- **Digital documents** — Key terms glossary — Topic 1 (doc-34649)
  Key ideas summary — Topic 1 (doc-34660)

1.6 Exercises

To answer questions online and to receive **immediate feedback** and **sample responses** for every question, go to your learnON title at **www.jacplus.com.au**. A **downloadable solutions** file is also available in the resources tab.

1.6 Exercise 1: Review questions

1. Describe the difference between *prokaryotic* and *eukaryotic* cells.
2. Identify what components all cells share.
3. Outline the key differences between fungi, plant, protist and animal cells.
4. By identifying the key differences and similarities, compare cilia and flagella.
5. Deep in the ocean are superheated thermal vents that reach temperatures of over 120 °C. The pressures at these depths are extremely high, yet life can exist here.
   a. What kind of organism would you expect to find in these extreme environments?
   b. What conditions must exist for life to survive here and how are they met?
6. Suggest explanations for the following observations:
   a. Cardiac heart muscle cells have very high numbers of mitochondria.
   b. Cells of the skin oil glands that produce and export lipid-rich secretions have large amounts of smooth endoplasmic reticulum.
   c. Mature red blood cells cannot synthesise proteins.
   d. One cell type has a very prominent Golgi apparatus, while another cell type appears to lack this organelle.
7. The table shows data for two sets of cells of identical shape but of decreasing sizes.

TABLE Dimensions of various cells

Cell	Shape	Dimensions	Surface area	Volume	SA : V ratio
P	Flat sheet	10 × 10 × 0.1	204	10	20.4
Q	Flat sheet	5 × 5 × 0.05	51	1.25	40.8
R	Flat sheet	1 × 1 × 0.01	2.04	0.01	204
H	Sphere	Diameter: 10	314.2	523.6	0.6
K	Sphere	Diameter: 5	78.5	65.5	1.2
L	Sphere	Diameter: 1.0	3.14	0.52	6

Note: Relative to the first shape in each set, the dimensions of other members of the set are scaled down by a factor of 2 and by a factor of 10.

a. A student stated the same shape scaled down should retain the same SA : V ratio, their reason being 'the shapes stay the same'. Do you agree with this student? Explain your decision.
b. With regard to the information in the table, identify how scaling a shape (up or down) affects the SA : V ratio of a given shape by completing the following sentences:
   i. If the size of a given shape is doubled, its SA : V ratio is …
   ii. If the size of a given shape is halved, its SA : V ratio is …

c. A particular shape has an SA : V ratio of 10.
   i. What would happen to this ratio if this shape were scaled up by a factor of 5?
   ii. What would happen to this ratio if this shape were scaled down by a factor of 2?
d. Another cell, M, is sphere-shaped and has a diameter of 0.5 units. Refer to the table and predict its SA : V ratio.
e. Consider a different shape, such as a cube or a pyramid, that is changed in scale. Would its SA : V ratio be expected to follow a similar or a different pattern to that shown by the flat sheets and the spheres?

1.6 Exercise 2: Exam questions

  Resources

▶ **Teacher-led videos** Teacher-led videos for every exam question

Section A — Multiple choice questions
All correct answers are worth 1 mark each; an incorrect answer is worth 0.

▶ **Question 1**
A student is given samples that contain different cells. These are shown in the figure.

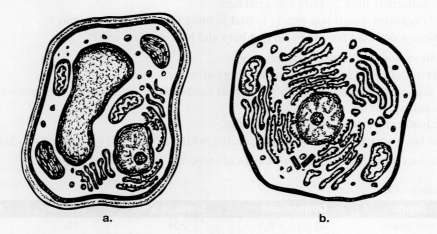

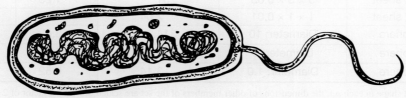

The kingdoms that cells a, b and c belong to, respectively, are

A. Plantae, Animalia and Prokaryotae.
B. Fungi, Protista and Prokaryotae.
C. Protista, Plantae and Animalia.
D. Fungi, Protista and Animalia.

Use the following information to answer Questions 2–4.

The following diagrams (1, 2, 3 and 4) illustrate different ways that substances move into or out of a cell. W, X, Y and Z represent structures in the plasma membrane. The concentration of a substance is shown by the number of particles inside and outside the cell. An arrow shows the direction of movement of the substance. The properties of each substance are listed in the table.

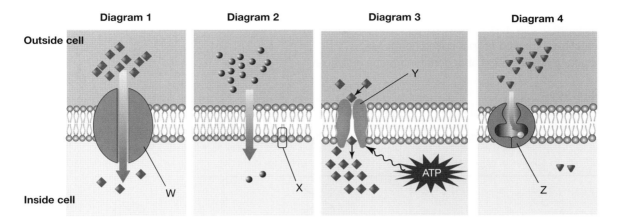

Symbol for substance	♦	•	△
Properties of each substance	hydrophilic small non-polar charged	hydrophilic small non-polar uncharged	hydrophilic small polar uncharged

▶ Question 2
Source: Adapted from VCAA 2020 Biology Exam, Section A, Q1

Structure W represents a

A. phospholipid bilayer.
B. channel protein.
C. carrier protein.
D. glycoprotein.

▶ Question 3
Source: Adapted from VCAA 2020 Biology Exam, Section A, Q2

It is reasonable to state that

A. diagram 1 could represent carbon dioxide exiting a leaf cell during photosynthesis.
B. diagram 2 could show facilitated diffusion of urea out of an active muscle cell.
C. diagram 3 could represent pumping potassium ions into a cell.
D. diagram 4 could represent oxygen diffusing into a blood cell.

▶ **Question 4**

Source: Adapted from VCAA 2020 Biology Exam, Section A, Q3

Consider the fluid mosaic model of the plasma membrane.

It is correct to state that

A. structure W changes shape to actively transport substances.
B. multiples of structure X move during endocytosis.
C. structure Y transports lipid-based hormones.
D. structure Z is the most abundant membrane molecule.

▶ **Question 5**

Source: VCAA 2016 Biology Exam, Section A, Q5

The diagrams below represent three of the major macromolecule groups in living things.

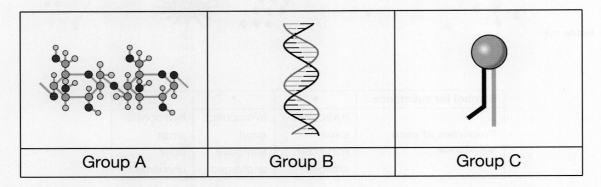

If a sample of a macromolecule from Group C were chemically analysed, you would expect to find that it contains

A. amino acids and phosphate.
B. fatty acids and phosphate.
C. amino acids only.
D. glucose.

▶ **Question 6**

Source: VCAA 2016 Biology Exam, Section A, Q7

In animal cells, tight junctions are multi-protein complexes that mediate cell-to-cell adhesion and regulate transport through the extracellular matrix. Proteins that form these complexes are made within the cell.

One pathway for the production of protein for these junctions is

A. nucleus – ribosome – Golgi apparatus – vesicle – endoplasmic reticulum.
B. nucleus – ribosome – endoplasmic reticulum – vesicle – Golgi apparatus.
C. nucleus – vesicle – endoplasmic reticulum – Golgi apparatus – ribosome.
D. nucleus – vesicle – Golgi apparatus – ribosome – endoplasmic reticulum.

Question 7

Source: VCAA 2010 Biology Exam 1, Section A, Q21

Solute Q moves across the plasma membrane via protein channels. Solute R moves across the plasma membrane by simple diffusion. The rate of movement of each solute into a cell is recorded and graphed. The results are shown in the following graph.

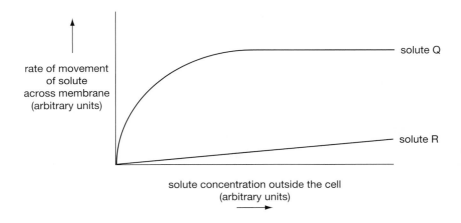

It would be reasonable to conclude that

A. solute Q is lipid soluble.
B. solute Q saturates the protein channels.
C. the gradient of the graph for solute R will increase as the temperature decreases.
D. as solute R is metabolised within a cell, its rate of movement across the membrane decreases.

Question 8

Nerve impulses involve several movements of sodium ions in different directions across the plasma membrane of a nerve cell, according to the following steps:

i. Before a nerve impulse occurs, sodium ions are more concentrated in the extracellular fluid outside the cell than inside the cell.
ii. During transmission of the nerve impulse, sodium ions flood into the nerve cell from the extracellular fluid.
iii. After the impulse has passed, the original concentration of sodium ions is restored to its high concentration outside the cell by a process that moves sodium ions out of the cell.

Which process is responsible for environments i, ii and iii, respectively?

A. Simple diffusion, facilitated diffusion, active transport
B. Facilitated diffusion, active transport, facilitated diffusion
C. Active transport, simple diffusion, active transport
D. Active transport, facilitated diffusion, active transport

Question 9

A scientist carried out an experiment to determine the time it took for a cell to manufacture proteins from amino acids. The scientist provided the cell with radioactively labelled amino acids and then tracked them through the cell to establish the time at which protein synthesis commenced. He monitored the cell at 5 minutes, 20 minutes and 40 minutes after production started, in order to track the proteins from the site of synthesis to a point in the cell from which they were discharged from the cell.

The scientist made an image of the cell at each of these times but forgot to mark each image with its correct time. The images are provided. The location of the radioactivity is shown by the green spots.

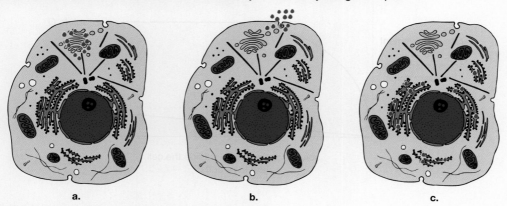

Which of the following gives the correct order of the images?

A. a, b, c
B. c, b, a
C. c, a, b
D. b, a, c

Question 10

Endosymbiosis is the theory that explains how eukaryotic cells originated from prokaryotic cells. This process is illustrated in the figure.

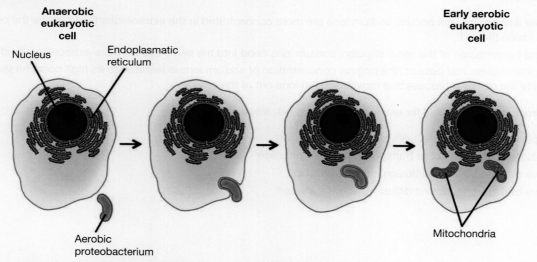

The evidence that would *not* support the theory of endosymbiosis would be

A. mitochondria reproduce through a process called binary fission, like prokaryotes.
B. mitochondria have DNA very similar to nuclear DNA.
C. mitochondria, when removed from a eukaryotic cell, are still able to reproduce independently to the cell.
D. mitochondria have a double membrane.

Section B — Short answer questions

Question 11 (5 marks)

a. Describe how the distribution of genetic material in prokaryotic cells differs from eukaryotic cells. **3 marks**

b. A student said that the cell they were viewing using a microscope was a photosynthetic plant cell. What two features of the cell would the student be able to see that enabled them to reach this conclusion? **2 marks**

Question 12 (3 marks)

An experiment was carried out on three groups of identical cells. All cells were immersed in isotonic solution at the start of the experiment. The solution that two of the groups of cells were immersed in was then changed.

The results of the experiment are shown.

	Group 1 cells	Group 2 cells	Group 3 cells
Before the experiment			
After the experiment			

Explain the appearance of the three groups of cells at the end of the experiment.

Question 13 (6 marks)

The image is of an organelle found in cells.

a. Name the organelle. **1 mark**

b. What is its function? **1 mark**

c. Give an example of a mammalian cell that would:
   i. possess this organelle in high numbers **1 mark**
   ii. not have any of these organelles. **1 mark**

d. Would this organelle be found in prokaryotic cells? Explain your response. **2 marks**

Question 14 (5 marks)

The following table shows data for three different shapes, each having the same volume. (Where necessary, figures have been rounded.)

TABLE Dimensions, surface area and volume of three different cells

Cell	Shape	Dimensions	Surface area	Volume	SA : V ratio
A	Flat sheet	10 × 10 × 0.1	204	10	20.4
B	Cube	2.15 × 2.15 × 2.15	28	10	2.8
C	Sphere	Diameter: 1.67	22	10	2.2

a. Which cell (A, B or C) would be most efficient in moving required materials into and removing wastes from the cell? Explain. **2 marks**

b. What biological consequence could a low SA : V ratio have on a cell? **2 marks**

c. Identify one way in which a cell might retain its overall shape, but greatly increase its surface area with a minimal increase in volume. **1 mark**

Question 15 (10 marks)

Sterile saline (NaCl) solutions, with or without glucose, may be used to treat a person in certain circumstances. The treatment may be delivered either orally or directly into a vein by intravenous infusion. The normal salinity level of body cells and the surrounding extracellular fluid is 0.9 per cent sodium chloride.

Fluids that might be administered include:
- normal-strength saline solution with 0.9 per cent saline, with solutes in balance with normal body fluids, making the solution isotonic
- half-strength saline solution with 0.45 per cent salt, with fewer electrolytes, making it hypotonic to body fluids; or
- double-strength saline, with greater than 0.9 per cent dissolved solutes, making it hypertonic to body fluids.

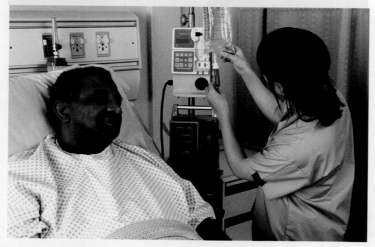

Infusion of an intravenous saline drip is performed by experienced medical professionals who decide which saline strength should be used, whether the solution should also include glucose, and calculate the correct rate of flow for the particular patient.

A patient (MM) is in urgent need of treatment following blood loss. To increase the circulating blood volume and raise the blood pressure, an emergency treatment — while waiting on blood typing results — might be the intravenous infusion of a saline solution.

a. Would you expect the saline solution selected for this purpose to be:
   A. isotonic
   B. hypotonic
   C. hypertonic?
   Explain your decision. 2 marks

b. The treatment was given intravenously. Would it be equally effective if given by mouth (orally)?
   Briefly explain. 2 marks

c. i. Another patient (NN) is suffering from severe dehydration and salt loss. A possible treatment in this case involves administration of a sterile saline solution.
   Would you predict the saline solution used in this case to be:
   A. isotonic
   B. hypotonic
   C. double strength?
   Explain your decision. 2 marks
   ii. Is glucose likely to be included in the saline? Explain your decision. 2 marks

d. Another patient (PP) is immobilised in bed recovering from an operation. PP is given an infusion of an intravenous saline drip in order to prevent edema; that is, an accumulation of excess extracellular fluid in his body tissues.
   Would you predict the saline solution to be:
   A. normal strength
   B. half strength
   C. double strength?
   Explain your decision. 2 marks

1.6 Exercise 3: Biochallenge online only

Resources

 eWorkbook Biochallenge — Topic 1 (ewbk-8071)

 Solutions Solutions — Topic 1 (sol-0646)

teachon

Test maker
Create unique tests and exams from our extensive range of questions, including past VCAA questions.
Access the assignments section in learnON to begin creating and assigning assessments to students.

Online Resources

Below is a full list of **rich resources** available online for this topic. These resources are designed to bring ideas to life, to promote deep and lasting learning and to support the different learning needs of each individual.

eWorkbook

- 1.1 eWorkbook — Topic 1 (ewbk-2290)
- 1.2 Worksheet 1.1 Exploring cells (ewbk-7186)
- 1.3 Worksheet 1.2 Surface area to volume ratio (ewbk-7188)
- 1.4 Worksheet 1.3 Cells and their organelles (ewbk-7190)
- 1.5 Worksheet 1.4 Structure of the plasma membrane (ewbk-7192)
  - Worksheet 1.5 Function of the plasma membrane (ewbk-7194)
  - Worksheet 1.6 Transport across the membrane (ewbk-7196)
- 1.6 Worksheet 1.7 Reflection — Topic 1 (ewbk-7200)
  - Biochallenge — Topic 1 (ewbk-8071)

Solutions

- 1.6 Solutions — Topic 1 (sol-0646)

Practical investigation eLogbook

- 1.1 Practical investigation eLogbook — Topic 1 (elog-0156)
- 1.2 Investigation 1.1 Cells under the microscope (elog-0774)
- 1.3 Investigation 1.2 What limits the size of cells? (elog-0776)
- 1.4 Investigation 1.3 Viewing and staining cells (elog-0778)
- 1.5 Investigation 1.4 Modelling the properties of the plasma membrane (elog-0780)
  - Investigation 1.5 Diffusion across the membrane (elog-0782)
  - Investigation 1.6 Osmosis in action (elog-0784)

Digital documents

- 1.1 Key science skills — VCE Biology Units 1–4 (doc-34648)
  - Key terms glossary — Topic 1 (doc-34649)
  - Key ideas summary — Topic 1 (doc-34660)
- 1.4 Extension: Protein production (doc-35857)
  - Extension: Cellular respiration (doc-35858)
  - Extension: Diseases associated with lysosomes and peroxisomes (doc-36011)
  - Extension: Photosynthesis (doc-35859)
- 1.5 Extension: Major histocompatibility complex (doc-35860)
  - Extension: Ion concentrations in cystic fibrosis (doc-35856)

Teacher-led videos

- Exam questions — Topic 1
- 1.2 Sample problem 1 Identifying bacterium cells (tlvd-1737)
  - Sample problem 2 Identifying types of cells (tlvd-1738)
- 1.3 Sample problem 3 Surface area to volume ratios of cells (tlvd-1739)
- 1.4 Sample problem 4 Compare the nucleus and the nucleolus (tlvd-1740)
  - Sample problem 5 Writing a hypothesis and a conclusion (tlvd-1741)
  - Sample problem 6 Designing a decomposition reaction (tlvd-1742)
- 1.5 Sample problem 7 Identifying features of the plasma membrane (tlvd-1743)

Video eLessons

- 1.2 Living organisms are made of cells (eles-4165)
- 1.4 Cells and organelles (eles-0054)
- 1.5 Phagocytosis (eles-2444)
  - Mechanisms of membrane transport (eles-2463)

Interactivities

- 1.5 Movement across membranes (int-0109)

Weblink

- 1.4 Pompe disease

Teacher resources

There are many resources available exclusively for teachers online

To access these online resources, log on to **www.jacplus.com.au**

AREA OF STUDY 1 HOW DO CELLS FUNCTION?

2 The cell cycle and cell growth, death and differentiation

KEY KNOWLEDGE

In this topic you will investigate:

The cell cycle and cell growth, death and differentiation
- binary fission in prokaryotic cells
- the eukaryotic cell cycle, including the characteristics of each of the sub-phases of mitosis and cytokinesis in plant and animal cells
- apoptosis as a regulated process of programmed cell death
- disruption to the regulation of the cell cycle and malfunctions in apoptosis that may result in deviant cell behaviour: cancer and the characteristics of cancer cells
- properties of stem cells that allow for differentiation, specialisation and renewal of cells and tissues, including the concepts of pluripotency and totipotency.

Source: VCE Biology Study Design (2022–2026) extracts © VCAA; reproduced by permission.

PRACTICAL WORK AND INVESTIGATIONS

Practical work is a central component of learning and assessment. Experiments and investigations, supported by a **practical investigation eLogbook** and **teacher-led videos**, are included in this topic to provide opportunities to undertake investigations and communicate findings.

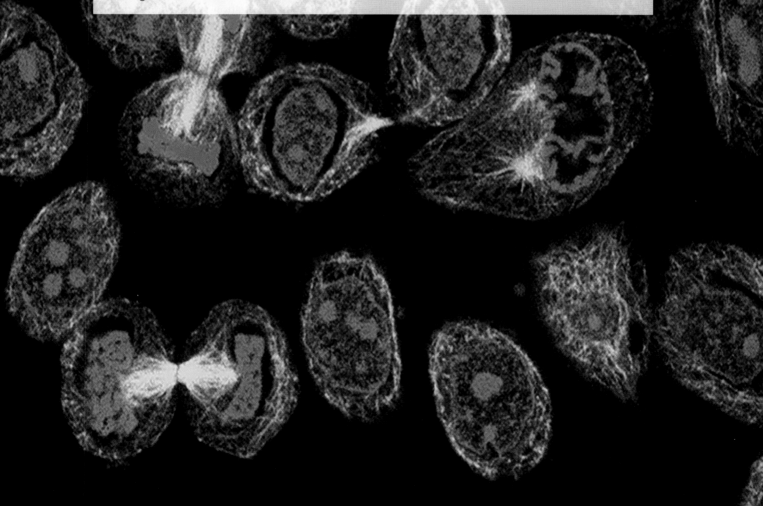

2.1 Overview

Numerous **videos** and **interactivities** are available just where you need them, at the point of learning, in your digital formats, learnON and eBookPLUS at **www.jacplus.com.au**.

2.1.1 Introduction

The replication of cells in the body is a particularly important process for all organisms. In multicellular organisms it allows an organism to grow and repair damage as it occurs. In unicellular organisms, cell replication enables the survival of the species, producing progeny that are genetically identical to the parent cell. Cells need to be carefully monitored and checked for flaws that will affect the functioning of the organism. When they no longer function correctly, they need to be destroyed and replaced through the process of programmed cell death.

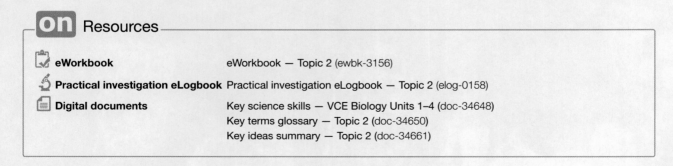

FIGURE 2.1 Cell embryo undergoing cell division

The processes of cell replication in both multicellular and unicellular organisms are explored in this topic, along with the mechanism that is responsible for the destruction of cells when they become damaged. We also delve into what happens when these processes are not controlled, which leads to diseases.

A fascinating area of cell growth that will be explored in this topic is the differentiation and specialisation of cells. This allows the one cell present at conception to form all the different cells required to make an organism, and research into this area may allow us to treat many genetic diseases.

LEARNING SEQUENCE

2.1 Overview 90
2.2 Cell replication — an asexual process 91
2.3 Regulation of the cell cycle 103
2.4 Cell differentiation 118
2.5 Review 130

on Resources

eWorkbook eWorkbook — Topic 2 (ewbk-3156)

Practical investigation eLogbook Practical investigation eLogbook — Topic 2 (elog-0158)

Digital documents Key science skills — VCE Biology Units 1–4 (doc-34648)
Key terms glossary — Topic 2 (doc-34650)
Key ideas summary — Topic 2 (doc-34661)

2.2 Cell replication — an asexual process

KEY KNOWLEDGE

- Binary fission in prokaryotic cells
- The eukaryotic cell cycle, including the characteristics of each of the sub-phases of mitosis and cytokinesis in plant and animal cells

Source: VCE Biology Study Design (2022–2026) extracts © VCAA; reproduced by permission.

Cell **replication** is an **asexual** process. The contents of one parent cell are copied to make two genetically identical daughter cells. Both resulting daughter cells have the same functionality as the cell that produced them. There is no combining of DNA material so they are, in effect, clones of the parent cell.

Cells undergo cell replication for a number or reasons:
- **Growth**. In order to grow, an organism must have more cells. Cells can simply not just get bigger as they are limited in size (see subtopic 1.3).
- **Repair**. Old and damaged cells must be replaced so they do not cause disease.
- **Procreation**. In the case of bacteria, archaea, protozoa and even some multicellular eukaryotes, the survival of the species depends on cell replication.

2.2.1 Cell replication in prokaryotes

Prokaryotes (archaea and bacteria) reproduce through a process of **binary fission**. This process is simpler and faster than asexual reproduction in eukaryotic organisms.

The steps of binary fission

1. Replication of the circular DNA **chromosome** and **cell elongation** occurs.
2. The two circular chromosomes **migrate** to either ends of the cell.
3. The cell membrane pinches in two and a **septum** (a new cell wall) forms along the middle of the cell, which extends and finally breaks in half to form two cells.

replication copying or reproducing something

asexual reproduction that only requires one parent, leading to the production of a clone

growth the process of increasing in size

repair to restore something damaged or faulty to a good condition

procreation the production of offspring

binary fission process of cell multiplication in bacteria and other unicellular organisms in which there is no formation of spindle fibres and no chromosomal condensation

chromosome a thread-like structure composed of DNA and protein

cell elongation any permanent increase in size of a cell

migrate to move from one part of something to another

septum a wall, dividing a cavity or structure into smaller ones

FIGURE 2.2 Binary fission of a prokaryote, showing the replication and migration of the circular chromosomes, the formation of the septum and the final division to two individual cells

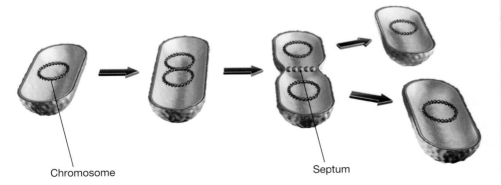

Binary fission in bacterial cells can be completed in about 20 minutes at room temperature. This means that — if resources are available — one bacterial cell, through successive binary fissions over an 8-hour period, could produce 16 million descendants! This is an example of **exponential growth** and it reminds us why a bacterial infection, if not treated, can have serious outcomes. Figure 2.3 shows a cell of the bacterial species *Listeria* dividing by binary fission.

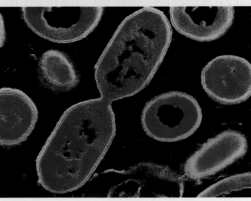

FIGURE 2.3 Cells of the bacterial species *Listeria*, one of which is dividing by binary fission. The circular outlines are cross-sections through bacterial cells. The DNA of the bacterial circular chromosome appears as darkly stained material.

exponential growth population growth that follows a J-shaped curve but cannot continue indefinitely

tlvd-1744

SAMPLE PROBLEM 1 Binary fission

a. Outline the three key steps in binary fission in a bacterial cell. **(3 marks)**
b. Under certain conditions, *E. coli* can replicate by binary fission once every 20 minutes. If there is one initial bacteria, how many bacteria would you expect after two hours? **(1 mark)**

THINK

a. When you answer an *outline* question you should include the names of the key steps and a brief statement of what occurs during these.

The question also mentions bacterial cells, so you need to ensure you reference this in your response. For example, you should reference to the circular chromosome and cell wall.

1. Cell elongation and replication — cell grows, DNA is copied; this is semi-conservative
2. Migration — migration of DNA to opposite ends of the cell
3. Separation — cell membrane pinches and septum forms to create a new cell wall

TIP: This type of question may be answered in either dot points or a short paragraph.

b. 1. Consider how many binary fission events would have occurred in the two-hour period.

WRITE

The first stage is cell elongation and replication — the cell grows and DNA is replicated in a semi-conservative nature, to create two identical circular chromosomes (1 mark).

The second stage is migration, where the two circular chromosomes move to either end of the cell (1 mark).

The third stage is separation, where a septum and a new cell wall forms to create two new cells (1 mark).

2 hours = 120 minutes

$$\frac{120}{20} = 6$$

6 binary fission events

2. Calculate how many bacteria there would be. Two ways you may do this are by:
   - calculating each binary fission event through doubling (occurring every 20 minutes)
   - using an exponential growth calculation: 2^x, where x is the number of divisions.

   After 1 event: 2 bacteria
   After 2 events: 4 bacteria
   After 3 events: 8 bacteria
   After 4 events: 16 bacteria
   After 5 events: 32 bacteria
   After 6 events: 64 bacteria
   $2^6 = 64$ bacteria

3. Clearly answer the question, ensuring that you link back to *E. coli*

   There would be 64 *E. coli* bacteria after two hours (1 mark).

Resources

eWorkbook Worksheet 2.1 Binary fission in prokaryotes (ewbk-7211)
Video eLesson Binary fission (eles-2464)

EXTENSION: DNA replication is semi-conservative

DNA replication is an important concept as it occurs in both mitosis and meiosis.

The process of DNA replication is an extension concept for the Units 1 & 2 course. However, understanding of this concept is required in the manipulation of DNA, which is examinable in the Units 3 & 4 course.

To access more information on this extension concept please download the digital document.

FIGURE 2.4 DNA replication is semi-conservative.

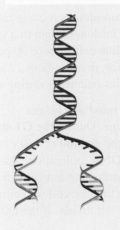
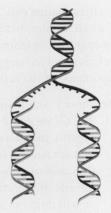

Resources

Digital document Extension: DNA replication is semi-conservative (doc-35876)

INVESTIGATION 2.1

Observing binary fission

Aim

To observe the process of binary fission as a method of reproduction

2.2.2 Cell replication in eukaryotes

The **cell cycle** in eukaryotes is slightly more complicated than that of prokaryotes, due in part to the amount of DNA in each cell. In eukaryotes, DNA is organised into chromosomes. There are multiple chromosomes, which must all be carefully copied. In comparison, prokaryotes have one circular chromosome.

cell cycle the series of events of cell growth and reproduction that results in two daughter cells

Cell cycle

The key events that occur during a cell cycle are summarised in simple terms in table 2.1. These events occur in three distinct phases of the cell cycle: **interphase**, **mitosis** and **cytokinesis**.

TABLE 2.1 A simplified summary of key events during the cell cycle

Cell cycle	What happens	Phase of cell cycle
Step 1	Replication of DNA of parent cell	Interphase
Step 2	Organisation of chromosomes, followed by their separation into two identical groups at different poles of the parent cell, leading to two identical nuclei	Mitosis
Step 3	Division of parent cell into two cells through the splitting of the membrane	Cytokinesis

These stages are shown in figure 2.5. Let's explore each of these steps in some detail.

Step 1: Interphase — period of DNA replication

An essential process in the cell cycle is the replication of DNA, the genetic material. DNA replication occurs during a stage of the cell cycle known as interphase. (This stage was once called the 'resting phase', but the cells are far from resting during interphase.) If you looked through a light microscope at cells during interphase you would see the cell nucleus, but you would not see any discrete chromosomes. In interphase, the chromosomes are decondensed and distributed through the nucleus. However, if you could watch the uptake of the nucleic acids that are the building blocks of DNA, you would see that the cells were busily copying their DNA and performing many other biochemical activities. Other organelles, such as mitochondria and ribosomes, also replicate and divide during interphase to prepare for cell division.

In a mammalian cell, a complete cell cycle takes about 24 hours. The time spent by a cell in interphase is far longer than that spent in any other stage of the cell cycle. For example, in mammalian cells, about 90 per cent of the time of a complete cell cycle is spent in interphase; that is, about 22 hours. This highlights the importance of the activities occurring during interphase.

Interphase is subdivided into three stages:
1. *The G1 or gap 1 stage.* During the **G1 stage of interphase** a cell undergoes growth, increasing the amount of cell cytosol. The cell also synthesises proteins that are needed for DNA replication. The mitochondria of the cell divide and, in the cases of photosynthetic plant cells, their chloroplasts also divide. It is near the end of this stage that the cell will either commit to continuing the cell cycle or will drop out and not divide. If the latter occurs, the cell enters a non-dividing quiescent G0 stage.
2. *The S or synthesis stage.* During the **S stage of interphase** the parent cell synthesises or replicates its DNA, the genetic material of the cell. At the end of the S stage, the parent cell contains two identical copies of its original DNA.
3. *The G2 or gap 2 stage.* During the **G2 stage of interphase** further growth of the cell occurs in preparation for **cell division**. In addition, the synthesis of proteins occurs, including those that form the microtubules of the **spindle**. By the end of interphase, the cell has doubled its size.

For a typical human cell that requires 24 hours to complete one cell cycle, the time spent in the various stages might be: G1 stage about 11 hours, S stage about 7 hours, G2 stage about 4 hours and the remainder (mitosis and cytokinesis) about 2 hours. This is in contrast to the rapid process of binary fission in prokaryotes that can produce two daughter cells within a period of 20 to 40 minutes.

interphase a stage in the cell cycle that is a period of cell growth and DNA synthesis

mitosis process involved in the production of new cells genetically identical with the original cell; an essential process in asexual reproduction

cytokinesis division of the cytoplasm occurring after mitosis

G1 stage of interphase the first stage of interphase in the cell cycle where the cell grows, increasing the amount of cell cytosol

S stage of interphase the stage where the parent cell replicates its DNA; at the end of the S stage the parent cell contains two identical copies of its original DNA

G2 stage of interphase the third stage of interphase where proteins are synthesised and the cell continues to grow in preparation for division

cell division division of a cell into two genetically identical daughter cells

spindle fine protein fibres that form between the poles of a cell during mitosis and to which chromosomes become attached

FIGURE 2.5 Stages of the cell cycle. Most of the cell cycle is taken up by the three stages of interphase (G1, S and G2). The M stage is the division stage that includes the division of the nucleus (mitosis) and the division of the remainder of the cell (cytokinesis).

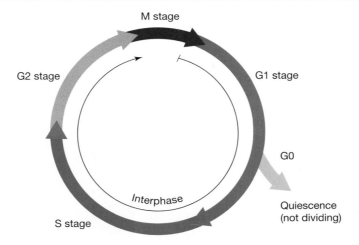

Step 2: Mitosis — organising and separating chromosomes

The appearance of chromosomes, initially thin and long, and the disappearance of the nuclear membrane mark the start of the part of the cell cycle known as mitosis — the M stage.

The stages of mitosis

1. **Prophase**: Chromosomes gradually condense, becoming shorter and thicker, and become visible as double-stranded structures (figure 2.7a). The spindle forms and the nuclear membrane breaks down.
2. **Metaphase**: The double-stranded chromosomes, also called dyads, line up around the equator of the cell.
3. **Anaphase**: The sister **chromatids** separate and are pulled to opposite ends of the spindle by the contraction of spindle fibres (figure 2.7b).
4. **Telophase**: A nuclear membrane forms around each separate group of single-stranded chromosomes and the chromosomes gradually decondense.

Mitosis completes the division of the nucleus. Figure 2.6 provides details of the different stages of mitosis.

FIGURE 2.6 The four main stages of mitosis

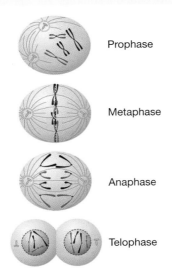

Prophase

Metaphase

Anaphase

Telophase

prophase stage of mitosis in which the chromosomes contract and become visible, the nuclear membrane begins to disintegrate and the spindle forms

metaphase stage of mitosis during which chromosomes align around the equator of a cell

anaphase stage of mitosis during which sister chromatids separate and move to opposite poles of the spindle fibre within a cell

chromatid one of two identical threads in a replicated DNA molecule

telophase stage of mitosis in which new nuclear membranes form around the separated groups of chromosomes

TOPIC 2 The cell cycle and cell growth, death and differentiation

Remember that mitosis is a continuous process. The stages of mitosis identify key changes in the appearance and the position of chromosomes. Remember also that chromosomes are not routinely visible when viewing cells through a light microscope. Only cells that are capable of division will ever show chromosomes and this will be for only a short period during the cell cycle. The disappearance of discrete chromosomes does not mean that the genetic material has disappeared; rather, the DNA is present as **chromatin** granules dispersed throughout the nucleus.

FIGURE 2.7 a. False-coloured SEM image of a human chromosome that has condensed. The chromosome is double-stranded and can be called a dyad. **b.** A dividing cell of a newt (*Notophthalmus* sp.) at anaphase of mitosis. The chromosomes (stained blue) are attached to the microtubules that form the spindle fibres (stained green). Keratin fibres (stained red) surround the spindle.

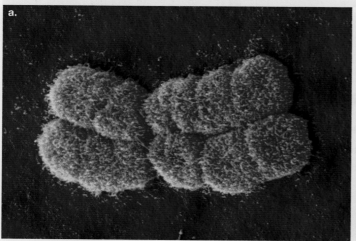
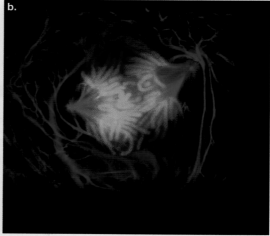

Let us consider the changes to chromosomes during mitosis in detail:
- Individual chromosomes first become visible as double, thread-like structures held together in a constricted region. Each of these threads is called a chromatid and the position where they are held together is called a **centromere**. The fact that the chromosomes are double-stranded and therefore contain two molecules of DNA indicates that the genetic material in the parent cell has already been replicated.
- The chromosomes continue to shorten and thicken, and the nuclear membrane disintegrates. At the same time, the very fine protein fibres, or microtubules, in the cytosol move towards the nucleus. The function of the fibres is to guide the movement of the chromosomes in the cell. The fibres become arranged in the cell, rather like the lines of longitude on a globe, to form a structure called a spindle. The chromosomes become attached by their centromeres around the 'equator' of the spindle.
- Two things then happen: the centromeres split, so that there are pairs of chromosomes, and the **spindle fibres** contract.
  - The contraction of the spindle fibres is responsible for the movement of the chromosomes towards the poles of the spindle. The movement of the new chromosomes is very ordered. One of the new chromosomes from each pair moves to one end of the spindle; its identical pair moves towards the opposite pole.
  - The end result is a set of chromosomes at each end of the spindle. Because the new chromosomes behave in an orderly way, the set of chromosomes at one end of the spindle is identical to the set of chromosomes at the other end.
- The chromosomes at each end of the spindle begin to lengthen and become less visible as distinct structures. At the same time, the protein fibres disperse back into the cytosol and a nuclear membrane develops around each group.

chromatin a mass of genetic material composed of DNA and proteins that condense to form chromosomes during eukaryotic cell division

centromere the position where the chromatids are held together in a chromosome

spindle fibres clusters of microtubules, composed of the contractile protein actin, that grow out from the centrioles at opposite ends of a spindle

> **on Resources**
> ▶ **Video eLesson** Mitosis (eles-4215)
> ✥ **Interactivity** The stages of mitosis (int-3028)

Step 3: Cytokinesis

At the end of mitosis, the division of the nucleus into two new identical nuclei is complete. However, the cell cycle is completed only after the cytosol, and the organelles in the cytosol, distribute around the new nuclei and become enclosed within an entire plasma membrane. This final process of the cell cycle is called cytokinesis.

As the two new nuclei form at the end of mitosis, the cytosol and organelles, such as mitochondria and chloroplasts, surround each nucleus and cytokinesis occurs. Minor differences occur during cytokinesis in different organisms. Generally, in animal cells, the bridge of cytoplasm between the two new nuclei narrows as the plasma membrane pinches in to separate the nuclei and cytoplasm into two new cells (figure 2.8a). In plant cells, a cell plate forms between the two groups of chromosomes and develops into a new cell wall for each of the newly produced cells (figure 2.8b).

Mitosis is essentially the same in plant and animal cells. The small differences that do exist are not related to the genetic material, nor do they have an impact on the biological significance of the process.

FIGURE 2.8 Minor differences are visible in plant and animal cells during mitosis and cytokinesis. **a.** An animal cell has a pair of centrioles at each pole of the spindle and a ring of contracting filaments that separates the cytosol and organelles during cytokinesis. **b.** In a newly replicating plant cell, a cell plate forms between the two groups of chromosomes and gives rise to a new cell wall for each new cell.

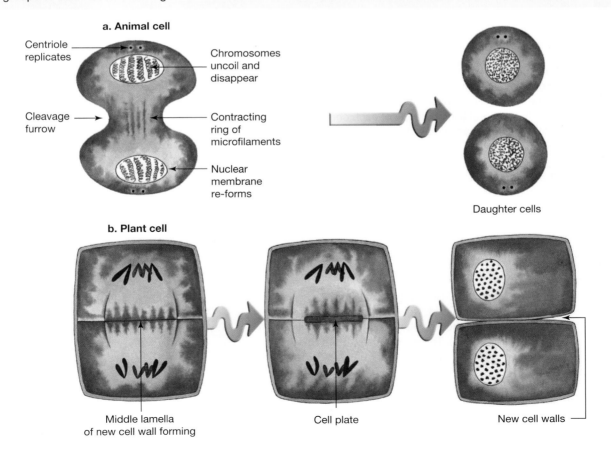

FIGURE 2.9 Summary of interphase, mitosis and cytokinesis. The light micrographs show mitosis in the seed of an African blood lily, *Scadoxus katherinae*. Chromosomes are stained purple and microtubules are stained pink.

STARTING POINT: One cell containing four single-stranded chromosomes

	Description	Drawing	Light micrograph
INTERPHASE	i. Chromosomes are not visible. In the G1 stage, the cell undergoes growth, the amount of cytosol increases and the mitochondria and chloroplasts divide. In the S stage, DNA replication occurs. In the G2 stage, further growth of the cell occurs and microtubules that form the spindle are formed (in animal cells, these form near the centrioles).	Interphase	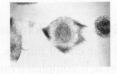 Interphase
MITOSIS	ii. Chromosomes become visible early in mitosis. At first they appear thin and long but gradually become thicker and shorter. Later, the chromosomes can be seen to be double-stranded, held together at the centromere. The replicated centrioles move apart; microtubules of the mitotic spindle continue to extend from the centrioles.	 Prophase	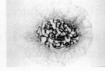 Prophase
	iii. The mitotic spindle is fully formed between the pairs of centrioles at the two poles of the spindle. The double-stranded chromosomes (each strand is called a chromatid) line up around the equator of the cell. From the side, they form a line across the middle of the cell. The nuclear membrane has disappeared.	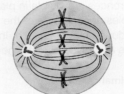 Metaphase	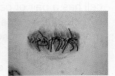 Metaphase
	iv. Each centromere divides, so that the single-stranded copies of each chromosome move to opposite ends of the cell as the tubules shorten. This migration is orderly and results in one copy of each chromosome moving toward each end of the spindle.	 Anaphase	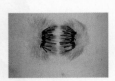 Anaphase
	v. The chromosomes become thinner and less obvious. A new nuclear membrane begins to form around each group of chromosomes. This completes the process of mitosis.	 Telophase	 Telophase
CYTOKINESIS	vi. Division of the cytoplasm by a process called cytokinesis is completed. New membranes form, enclosing each of the two new cells (and cell walls in the case of plants). These cells then return to interphase, and may divide again or enter G0.	Cytokinesis	 Cytokinesis

END POINT: Two cells, each containing four single-stranded chromosomes

Remembering the order

	Letter	Name of stage	Short description	What occurs
	I	Interphase	Intermission/ Invisible	This is the **intermission** between cell divisions. DNA replication occurs. Chromosomes have not condensed and are **invisible** under a light microscope.
Mitosis	P	**P**rophase	**P**lump	Chromosomes condense, becoming **plump** and visible. Spindle fibres form.
Mitosis	M	**M**etaphase	**M**iddle	Chromosomes line up along the **middle** of the cell.
Mitosis	A	**A**naphase	**A**part	Chromosomes are pulled **apart**.
Mitosis	T	**T**elophase	**T**wo nuclei	Nuclear membranes form around the two sets of chromosomes, forming **two nuclei**.
	C	**C**ytokinesis	**C**ut/**C**leave/ **C**omplete	The cell membrane is **cleaved** to form two cells. In plant cells, a cell plate extends from the middle to the sides to **complete** the separation.

tlvd-1745

SAMPLE PROBLEM 2 Identifying the stages of mitosis

The following image shows mitosis.

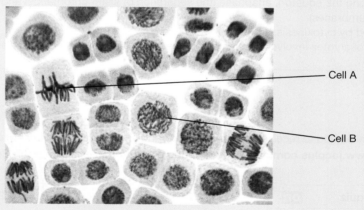

Identify the stage of mitosis of:
a. cell A
b. cell B.
Explain your reasoning.

(4 marks)

THINK

a. Recall the stages of mitosis:
- Metaphase — chromosomes line up along the middle
- Anaphase — chromosomes are pulled apart
- Telophase — two nuclei form
- Prophase — chromosomes appear.

Observe cell A — the chromosomes are aligned along the middle (equator) of the cell.

b. Observe cell B — the chromosomes are visible but are not ordered.

WRITE

Cell A is in metaphase (1 mark) as the chromosomes are lined up along the equator of the cell (1 mark).

Cell B is in prophase (1 mark) as the chromosomes have condensed but have not organised (1 mark).

INVESTIGATION 2.2

Observing the cell cycle under the microscope

Aim

To observe the different stages of the cell cycle under the microscope

 Resources

 eWorkbook Worksheet 2.2 The cell cycle (ewbk-7213)

 Interactivity Labelling the stages of mitosis (int-8128)

KEY IDEAS

- Cell replication can be used for growth, repair and procreation.
- Eukaryotic cells divide during the cell cycle, giving rise to two genetically identical daughter cells.
- The three main parts of the cell cycle are interphase, mitosis and cytokinesis.
- An essential early process in the cell cycle is the replication of DNA, which occurs during interphase.
- Mitosis is the carefully governed separation of sister chromatids and is another essential step in cell division.
- The four stages of mitosis are prophase (the condensing of chromosomes), metaphase (the lining up of chromosomes along the equator), anaphase (the separation of sister chromatids) and telophase (the formation of two nuclear membranes).
- Mitosis is followed by cytokinesis.
- Cell division in prokaryotes involves a relatively simple and rapid process of binary fission.

2.2 Activities

To answer questions online and to receive **immediate feedback** and **sample responses** for every question, go to your learnON title at **www.jacplus.com.au**. A **downloadable solutions** file is also available in the resources tab.

| 2.2 Quick quiz | 2.2 Exercise | 2.2 Exam questions |

2.2 Exercise

1. What are the stages of interphase?
2. What is the key event of the S stage of interphase?
3. What is the average time for a complete cell cycle:
   a. in a mammal
   b. by binary fission in a microbe?
4. Why is binary fission considered simpler than cell division in eukaryotes?

5. Consider the image provided with the labels A to F.

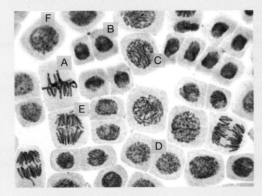

Place the stages A to F in order, giving the stage name and the key features you observed to support your answer.

2.2 Exam questions

▶ Question 1 (1 mark)
Source: VCAA 2015 Biology Exam, Section A, Q20

MC Which one of the following is true of prokaryotic cell division?
A. There is an equal division of the cytoplasm.
B. Daughter cells of varied genetic composition are formed.
C. Replicated chromosomes consist of two sister chromatids joined by a centromere.
D. Membrane-bound organelles are randomly distributed between the daughter cells.

▶ Question 2 (1 mark)
Source: VCAA 2010 Biology Exam 2, Section A, Q1

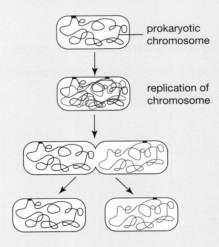

MC The diagram above is a representation of
A. mitosis.
B. apoptosis.
C. binary fission.
D. gamete formation.

Question 3 (1 mark)
Source: VCAA 2014 Biology Exam, Section A, Q20

MC Which one of the following is a correct statement about mitosis?
A. The spindle forms during prophase.
B. Chromatids separate to opposite poles of the spindle during metaphase.
C. Homologous chromosomes separate to opposite poles of the spindle during anaphase.
D. Homologous chromosomes line up at the equator of the cell during telophase.

Question 4 (1 mark)
Source: Adapted from VCAA 2008 Biology Exam 2, Section B, Q1a

When a cell replicates it goes through a series of events that can be summarised by the following diagram. The cycle moves in a clockwise direction and includes mitosis. Note four points, labelled A, B, C and D.

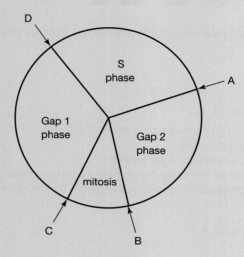

Given that two cells are formed as a result of replication, a cell must replicate its DNA during the cycle.

Using crosses, mark on the graph below the relative amount of DNA present at each of the points B and D in the cycle.

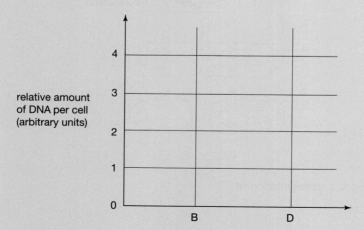

Question 5 (1 mark)
Source: VCAA 2012 Biology Exam 2, Section A, Q12

The following images show plant cells from a tissue that is undergoing mitosis.

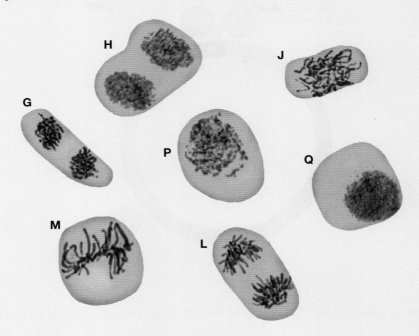

MC The order of the cells in a single mitotic phase would be
A. Q P J M L G H.
B. P Q M G J L H.
C. P G L M J Q H.
D. Q H J P L G M.

More exam questions are available in your learnON title.

2.3 Regulation of the cell cycle

KEY KNOWLEDGE

- Apoptosis as a regulated process of programmed cell death
- Disruption to the regulation of the cell cycle and malfunctions in apoptosis that may result in deviant cell behaviour: cancer and the characteristics of cancer cells

Source: VCE Biology Study Design (2022–2026) extracts © VCAA; reproduced by permission.

2.3.1 Regulation of the cell cycle

As examined in subtopic 2.2, the cell cycle is a complex series of events that, when error-free, produces two identical daughter cells.

During the cell cycle, there are several checkpoints that function to ensure that a complete and damage-free copy of the genome is transmitted to the two daughter cells. Each checkpoint can detect a particular kind of error. If an error is detected, then depending on the type of error, the cell cycle is either aborted or it is delayed, allowing time for the error to be corrected.

FIGURE 2.10 The G1, G2 and M checkpoints check for any problems in the DNA sequence that may have occurred prior to replication, during the initial replication or during mitosis. When errors such as unreplicated or damaged DNA are found, the DNA sequence may be repaired or the cell may be targeted for destruction.

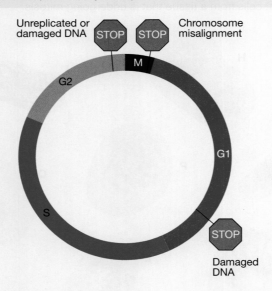

Figure 2.10 shows the location of three checkpoints in the cell cycle. These are known as the G1, G2 and M checkpoints.

- The **G1 checkpoint** occurs at the G1 (gap 1) stage of interphase, when the cell is ready to undergo division. The DNA of the cell is checked and if it is found to be damaged or incomplete, the cell is stopped from continuing through the cell cycle.

  Instead, the cell may enter a non-dividing quiescent stage called G0, or it may be targeted for destruction. (The 'security guard' at the G1 checkpoint is a protein known as **p53**, a tumour-suppressor protein. What do you think might happen if a mutation occurred in the p53 gene that codes for this protein, so that it could not carry out its normal function?)

  If the cell passes the G1 checkpoint, it proceeds into the cell cycle and enters the S stage of interphase. During the S stage, the cell replicates its DNA so that, by the end of the S stage, the cell should have double the amount of DNA. This DNA should be two complete and accurate copies of its genome.

- The cell now moves to the G2 stage of interphase, where it must pass the **G2 checkpoint**. Here, the replicated DNA of the cell is checked for completeness and lack of damage. If the cell passes this checkpoint, it can then advance to the mitosis stage of the cell cycle.
- The **M checkpoint** (or spindle assembly checkpoint) occurs at the metaphase stage of mitosis. A check is carried out to ensure that the sister chromatids (that is, the two strands of each double-stranded chromosome) are attached to the correct microtubules of the spindle and are pulled in opposite directions to different poles of the spindle. If an error is detected, the cell cycle is delayed until the error is fixed.

G1 checkpoint a check that occurs during G1 of interphase that makes sure the DNA is not damaged and is ready to undergo replication

p53 a protein that is coded for by a gene of the same name and regulates the cell cycle, hence functioning as a tumour suppressor

G2 checkpoint a check that occurs during G2 of interphase where the replicated DNA of the cell is checked for completeness and lack of damage; if the cell passes this checkpoint, it can then advance to mitosis

M checkpoint a check that occurs during mitosis where the connection between chromatid and spindle fibres is checked and corrected

The mitotic spindle

The focus in mitosis is typically on chromosomes. However, the positioning and the movement of the chromosomes depend on the presence of a microtubule framework, the spindle.

In animal cells, once mitosis starts, the paired **centrioles** move to opposite ends of the cell where they form the poles of the spindle. Clusters of microtubules grow out from the centrioles towards the middle of the cell. These microtubule clusters are called spindle fibres. At metaphase, these fibres anchor the double-stranded chromosomes around the equator of the cell. Each chromatid has a special attachment site (a protein complex) called a **kinetochore**, which bind each chromosome to the microtubules of the spindle (figure 2.11).

Spindle fibres from one pole attach to one sister chromatid and fibres from the opposite pole attach to its partner chromatid. (What would happen if the two sister chromatids of one chromosome became linked to fibres from the same pole of the spindle?) As mentioned earlier, this attachment is checked during the M checkpoint.

Spindle fibres are composed of actin, a contractile protein. At anaphase, the orderly migration of each pair of sister chromatids is achieved by contractions of the fibres that pull these now single-stranded chromosomes to the opposite poles of the spindle.

> **centrioles** a pair of small cylindrical organelles, used in spindle development in animal cells during cell division
>
> **kinetochore** a special attachment site of a chromatid by which it links to a spindle fibre
>
> **cancer** a disease in which cells divide in an uncontrolled manner, forming an abnormal mass of cells called a tumour

FIGURE 2.11 a. The attachment of the spindle fibres to the kinectochore **b.** Microscope image of a HeLa cell treated with various stains undergoing mitosis. The pericentrin stain shows the centrioles (orange). The ACA stain shows the kinetochores (purple). The α-tubulin stain shows the microtubules of the spindle (green). (Image courtesy of A Loynton-Ferrand, IMCF, University of Basel)

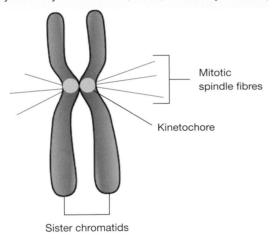

CASE STUDY: HeLa cells and the evolution of bioethics

Henrietta Lacks was a 31-year-old mother of five with terminal cervical **cancer**. She died in 1951. Prior to her death she attended a cancer clinic, where her surgeon took a tissue biopsy of her cancer cells without her consent. Her cells have an astonishing ability to continuously replicate and have been fundamental to cell human cell research. Her cells are known as HeLa cells.

For more information on HeLa cells and the subsequent evolution of bioethics, please see the digital document.

FIGURE 2.12 Lacks' story led to significant changes in the practice of medical research.

Resources

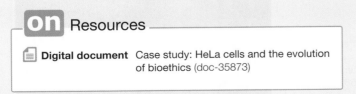

Digital document Case study: HeLa cells and the evolution of bioethics (doc-35873)

The role of proto-oncogenes and tumor suppressor genes in the cell cycle

Like all processes in a cell, cell replication is under the control of signaling molecules.

Proto-oncogenes are genes contained in the DNA of a cell that lead to the production of proteins that initiate the cell cycle. These signals, of course, can be switched off, but this is not the only control in switching off the cell cycle so that excessive production of cells does not occur. Mutations to these genes can turn these proto-oncogenes into **oncogenes**, which lead to excessive production of cells (a tumour), or cancer (see section 2.3.4).

Tumour-suppressor genes also exist. The proteins made as a result of these genes signal the cell to reduce cell division, repair DNA mistakes or initiate programmed cell death (see section 2.3.3). Again, mutations to these genes can also lead to tumours and cancer (see section 2.3.4).

Mitochondria and chloroplasts also replicate

We have seen that mitosis is followed by cytokinesis. This is essential so that the two new nuclei formed can each be combined with cytosol to give two new cells. The organelles such as mitochondria and chloroplasts within the cytosol must also be replicated during the cell cycle (during interphase), otherwise cells would contain an ever-decreasing number of these structures. This is also a carefully regulated process.

 Resources

 eWorkbook Worksheet 2.3 Regulating the cell cycle (ewbk-7215)

2.3.2 The cell cycle in action

Cell cycle in humans

In mammals, such as a human adult, actively dividing cells are found in several tissues, such as the **epidermis** of the skin, the epithelial lining of the gut and the bone marrow. Cell division normally occurs at a tightly regulated rate, so that the production of new cells matches or balances the rate of cell loss. Tissues with a population of actively dividing stem cells are tissues that have a high and continual level of cell loss or cell death. See subtopic 2.4 for more information about the mechanisms of cellular division in humans.

proto-oncogene a gene that leads to the production of proteins which initiate the cell cycle

oncogene a gene that signals cells to continue dividing

tumour-suppressor gene a type of gene that produces a protein that signals for cells to stop dividing

epidermis the outer layer of cells; in human skin it consists of three layers (outer region of dead cells, layers of living keratinocytes, and a basal layer of melanocytes and constantly dividing stem cells)

CASE STUDY: Saving burns victims — Professor Fiona Wood

Professor Fiona Wood is an Australian scientist and Australian National Living treasure. Her work on burns victims, particularly following the 2003 Bali bombings, has been focused on 'spray-on skin cells'. This treatment allows regeneration of skin cells using the patients' own cells. This ability to grow new skin cells is due to the replication by the cell cycle.

For detailed information on this process, please see the digital document.

FIGURE 2.13 Professor Fiona Wood and her colleagues developed an improved method of skin-cell regeneration for burns victims.

 Resources

 Digital document Case study: Saving burns victims — Professor Fiona Wood (doc-35874)

Cell cycle in other animals

Planaria, phylum Platyhelminthes, are flatworms that live in water. They are one of the few animals that can reproduce asexually by regeneration. The parent breaks into two or more pieces and each piece grows into a new planarian. The new parts are produced by mitosis of cells and each new planarian is an exact copy of the parent.

If a starfish loses some of its 'arms', new ones are regenerated by mitosis (figure 2.14).

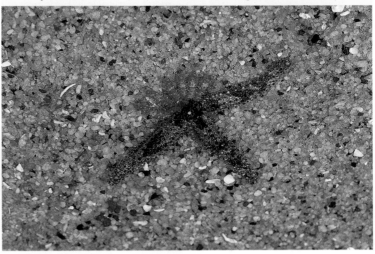

FIGURE 2.14 If a starfish loses some of its 'arms', they regrow. Here you can see six new 'arms' on a damaged starfish.

Cell cycle in plants

In vascular plants, only the cells in **meristematic tissue** can complete cell cycles and divide to produce identical daughter cells. The cells in permanent plant tissues cannot divide. Meristematic tissue is present in several locations, including root tips and stems. Figure 2.15 shows a section through the meristematic tissue of a root tip. As a region of active cell division, many rows of cells can be seen, including some cells in the mitosis stage of the cell cycle.

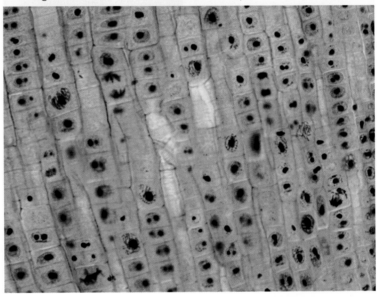

FIGURE 2.15 Light microscope image of a longitudinal section through the meristematic tissue of a root tip

Epicormic shoots after a bushfire

Bushfires are common in many areas of Australia, but the 2019–2020 summer season saw bushfires ravage many parts of Australia (see Background knowledge box). Although trees may appear to be burnt to a point that one might think they are dead, a picture such as the one in figure 2.16 (taken in February 2020 near Mallacoota, Victoria, six weeks after the area was devastated by bushfire) shows this is not the case. It is clear from the photograph that the fire has completely destroyed the undergrowth of grasses, shrubs and herbs. Fire-blackened trees with their scorched dead canopy of leaves are in the background, while in the foreground the burnt trunk of a rough-barked eucalypt tree is visible. The tree is already showing signs of regrowth; its thick outer layer of protective bark has insulated the underlying living tissues from the effects of the fire.

meristematic tissue plant tissue found in tips of roots and shoots that is made of unspecialised cells that can reproduce by mitosis

FIGURE 2.16 The new shoots from the trunk of a burnt eucalypt tree develop as a result of mitosis in buds present beneath the bark.

The trunk of a eucalypt tree does not usually show growing shoots. However, if the normal leaf canopy is destroyed, as happened in this fire, buds that are present beneath the bark grow and reproduce new green leafy shoots, known as **epicormic shoots**. The growth of epicormic shoots involves the production of new cells. The buds below the bark contain tissue called meristem or meristematic tissue, which is made up of cells that are able to reproduce to give rise to new cells. These new cells are identical to each other and to the parent cell.

epicormic shoot growth occurring from dormant buds under the bark after crown foliage is destroyed

BACKGROUND KNOWLEDGE: The Australian bushfire season of 2019–2020

Australia experienced its worst bushfire season on record in the 2019–2020 summer. It has been estimated that:
- 18.6 million hectares of land was burnt
- over 5900 buildings were destroyed
- at least 34 people were killed
- 1 billion animals were killed.

This was accompanied by hazardous air pollution over large areas of the country. Multiple states of emergency were declared in New South Wales, Victoria and the Australian Capital Territory.

Historically, much of Australia has been prone to bushfires due to a dry climate, high temperatures and low humidity. However, the intensity and extent of this season's fires has been subject to considerable debate surrounding the effects of fire management practices and the role of climate change. At the start of the fire season, much of eastern Australia was experiencing severe drought, and when combined with record high temperatures, the fires were both intense and widespread. Although land management practices — including prescribed burning and fuel loads — have been debated, experts have suggested that these have been difficult to achieve due to longer summer conditions, and have also been skeptical of their effectiveness, given that many fires burnt through agricultural land.

CASE STUDY: New liverworts from cells in a cup

Liverworts, genus *Hepatica*, are small plants that have a flat, fleshy, leaf-like structure from which **rhizoids** extend into the soil. The name liverwort is derived from the shape of the organism — rather like that of a liver — and the Anglo-Saxon word for herb — wort. In addition to reproducing sexually, liverworts reproduce asexually by means of fragmentation of parts of the plant. Also, liverworts produce gemmae — small, multicellular bodies produced in special cup-like structures called gemma cups (figure 2.17). When rain falls, the gemmae are splashed out of the cup. Gemmae are produced from cells of the parent plant by mitosis. When they grow into new plants they do so by mitosis. The new liverwort plants produced by growth of the gemmae are genetically identical to the parent plant from which they were derived.

FIGURE 2.17 A new plant develops from each of the small bodies that splash out of the gemma cups on a liverwort plant. The new plants are genetically identical to the parent plant.

Cell cycle in fungi

The cell cycle plays an important role in the reproduction of fungi.

The fungus or mould you see on bread or fruit grows by mitosis. A single cell, a **fungal spore**, lands on food and grows into a mass of threads called **hyphae**. Specialised stalks — each with a spore case at its tip — grow up from the mass of hyphae (see figure 2.18). Mitosis occurs within the spore case and thousands of black spores are formed. On maturing, the spore case splits open and the tiny, light spores are scattered. When conditions are favourable, each spore germinates and grows into a new hyphal mass.

rhizoids fine, root-like structures present in some plants, such as mosses

fungal spore microscopic biological particle that allows fungi to reproduce asexually

hyphae long, branching, filamentous structures of a fungus that make up mycelium; singular = hypha

FIGURE 2.18 The fungus on a rotting tomato **a.** comprises a mass of white threads or hyphae. Asexual reproduction occurs at the tips of some hyphae and **b.** large numbers of black spores are formed, each genetically identical to the parent.

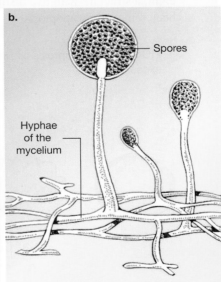

2.3.3 Programmed cell death

At any time, the cells in many organs and tissues of our bodies are in a state of turnover — old cells are dying and new cells are being formed through the cell cycle. This turnover is happening, for example, in cells that form the lining of the gut, in cells of the epidermis of the skin and in cells of the blood. In adults, the rates of the two processes are closely related so that, normally, a balance exists between the rate of new cell production and the rate of cell loss; that is: rate of cell renewal = rate of cell death.

Problems in cells result in abnormal cell behaviour. Damage to the cell's DNA is detected by the organism and leads to a process called **apoptosis** or programmed cell death (figure 2.19). Apoptosis is a genetically controlled and highly regulated process of cell self-destruction. Apoptosis plays an essential role in healthy body functioning.

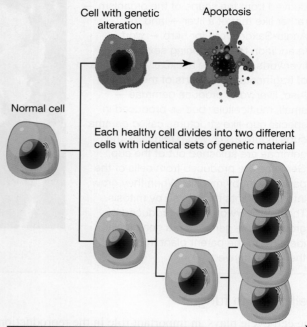

FIGURE 2.19 A healthy cell will continue replicating, whereas one that can no longer replicate undergoes apoptosis.

The signal pathway of apoptosis brings about the planned death of various cells. These include the following:
- *Cells at the end of their natural life*, such as gut lining cells and skin cells
- *Dysfunctional, damaged or diseased cells*, including:
  - cells infected with a virus — the programmed death of these cells prevents the virus from replicating and spreading to other cells
  - cells with irreparable issues detected during the checkpoints in the cell cycle.
- *Excessive cells:*
  - Every day, the bone marrow produces millions of new immune cells (white blood cells) and red blood cells. Over the same period, this production of new cells must be balanced by the loss of a similar number of cells. This is achieved in an orderly manner through apoptosis.
  - During embryonic development, the final shaping of organs depends on the programmed cell death of excess cells. Figure 2.20 shows the role of apoptosis in digit formation in a human embryo. By week eight, a structure with five distinct digits is clearly visible, with the 'webbing' between the digits having been removed by apoptosis.

FIGURE 2.20 Apoptosis allows for the formation of five digits by removing excessive cells.

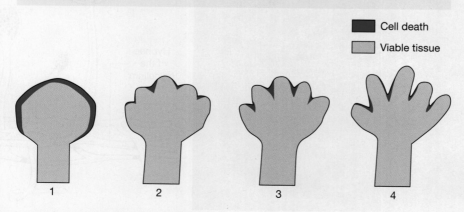

apoptosis the programmed death of cells that occurs as a normal and controlled part of an organism's growth or development

Pathways leading to apoptosis

Mechanisms of apoptosis

The two mechanisms by which apoptosis can be achieved are:
- an **intrinsic** (internal) signal within the cell.
  This is also known as the mitochondrial pathway. The intrinsic pathway is used when cells come under stress, such as through infection or damage. When cells are damaged during the cell cycle and the damage cannot be repaired, they undergo apoptosis via the intrinsic pathway.
- an **extrinsic** (external) signal, where the signal is from a source external to the cell.
  This is also known as the death receptor pathway. One way this may be generated is by a signal from the immune system.

Both mechanisms set off a biochemical pathway (shown in figure 2.21) that results in:
- cell shrinkage
- the formation of **blebs** (protrusions of the cell membrane)
- the eventual formation of **apoptotic bodies**
- the clean-up of apoptotic bodies through phagocytosis (after apoptosis).

FIGURE 2.21 A cell receives the signal to undergo apoptosis and undergoes shrinkage and blebbing to form apoptotic bodies. Phagocytosis occurs to clean up and recycle the contents of the apoptotic bodies, but is not a part of apoptosis.

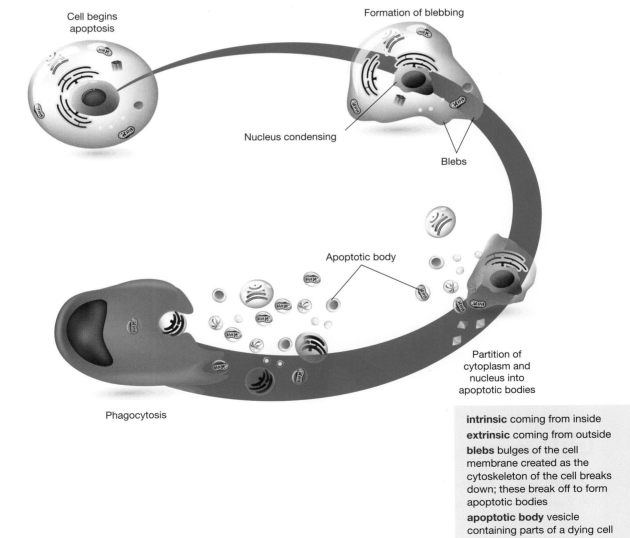

intrinsic coming from inside
extrinsic coming from outside
blebs bulges of the cell membrane created as the cytoskeleton of the cell breaks down; these break off to form apoptotic bodies
apoptotic body vesicle containing parts of a dying cell

EXTENSION: Mechanisms of apoptosis

Delving deeper into the control of apoptosis, we must divide the mechanisms based on where it originates within the cell (intrinsic pathway) or outside the cell (extrinsic pathway).

The protein p53, as mentioned in section 2.3.1, is a tumour-suppressor protein. When problems in the DNA arise, p53 leads to leakage of a protein called **cytochrome c** from the mitochondria. This results in the formation of the **apoptosome**, a large protein that in turn activates the **caspase cascade** (figure 2.22). **Caspases** are a set of proteins that are sequentially activated (from a pro-caspase) and break down proteins within the cell. This cell breakdown results in the formation of apoptotic bodies.

Sometimes the intrinsic pathway is not activated but the cell can still be marked for destruction. A **ligand** binds to a **death receptor** on the surface of the cell. This then activates the caspase cascade, leading to apoptosis. The extrinsic pathway, once activated, can also lead to the activation of the intrinsic pathway. This pathway is activated by signals such as tumour necrosis factors (TNFs). An example of a TNF is the ligand Fas. This intiates FADD, a specialised protein that activates a pro-caspase.

cytochrome c a protein that has a role in the formation of ATP in mitochondria; its leakage from the mitochondria leads to apoptosis

apoptosome a large protein formed during apoptosis; its formation triggers a series of events that leads to apoptosis

caspase cascade a group of proteins that are sequentially activated to bring about apoptosis

caspases protease enzymes that break down proteins during apoptosis

ligand a substance that forms a complex with a biomolecule to serve a biological purpose, such as the production of a signal upon binding to a signal

death receptor receptors on the surface of the cell that, when activated, lead to apoptosis of the cell

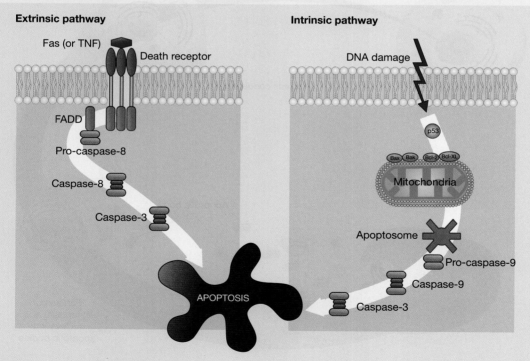

FIGURE 2.22 The intrinsic and extrinsic pathways leading to apoptosis. Both these pathways involve the activation of caspases.

Resources

eWorkbook — Worksheet 2.4 Apoptosis (ewbk-7217)

Video eLesson — Apoptosis (eles-3694)

2.3.4 When things go wrong: deviant cell behaviour

As discussed earlier in this topic, the cell cycle in various tissues is normally regulated so that, in a mature organism, the rate of production of new cells balances the rate of loss of cells.

Psoriasis

If the rate of cell production exceeds that of cell loss, a build-up of cells results. This can be seen in the skin condition **psoriasis** (figure 2.23). Psoriasis is a chronic autoimmune condition in which skin cells are overproduced, resulting in raised patches of red inflamed skin, often covered in a crust of small silvery scales.

Cancer: control of the cell cycle gone awry

More serious consequences of errors in the regulation of the cell cycle in a tissue are cancers.

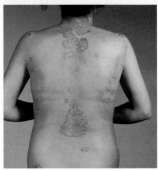

FIGURE 2.23 Psoriasis on the skin of a person's back. This condition is a result of the overproduction of skin cells.

Cancers may result from a breakdown of the normal regulation of the cell cycle, when the cycle becomes uncontrolled. In cancerous tissue, cells reproduce at a rate far in excess of the normal regulated rate of the cell cycle and produce masses of cells called tumours. Some tumours are malignant, such as **melanomas**, which are cancers derived from the pigment-producing cells, or **melanocytes**, of the skin epidermis. In malignant tumours, individual cells can break free from the primary tumour and migrate throughout the body, establishing sites of secondary cancers.

A clue to what goes wrong in cancer comes from studying cells growing in culture in a Petri dish in a laboratory. In culture, normal (non-cancerous) cell numbers increase through regulated cell divisions and form a single, orderly layer attached to the base of plastic dishes. These cells do not crowd; they are said to show contact inhibition. In addition, normal, non-cancerous cells typically undergo a limited number of cell cycles.

In contrast, cancerous cells in culture continue to divide in an unregulated manner. These cells show no contact inhibition, become crowded and form masses of cells in disorganised multiple layers. In addition, the number of cell cycles that cancerous cells can undergo is unlimited.

psoriasis chronic autoimmune condition in which skin cells are overproduced, resulting in raised patches of red, inflamed skin, often covered in a crust of small silvery scales

melanoma cancer derived from the pigment-producing cells (melanocytes)

melanocytes the pigment-producing cells in the basal layer of the epidermis

What causes the breakdown in the control of the cell cycle in cancerous cells?

TABLE 2.2 Comparing the cell cycle in normal and cancerous cells

Normal cells	Cancerous cells
The rate of cell division is regulated so that, in a mature organism, cell production matches cell loss.	Mutations in genes that control the cell cycle occur, causing the cell cycle to continue in an unregulated manner.
Checkpoints exist in normal cells to ensure that the DNA that is to be transmitted to daughter cells is complete and error free (see section 2.3.1).	Checkpoints are overridden or fail.
Chemical signals convey information to cells about when to divide faster and when to slow down or stop dividing. Two kinds of genes are involved in this signalling: proto-oncogenes that signal cells to continue dividing, and tumour-suppressor genes that signal cells to stop dividing.	Mutations in the proto-oncogenes and tumour-suppressor genes disrupt the control of the cell cycle. Mutated proto-oncogenes that lead to cancer are known as oncogenes.
Contact inhibition occurs, which stops cell division if overcrowding occurs.	Contact inhibition does not occur — the cells continue to grow and masses of cells form.

In cancer cells, these various controls of the cell division cycle are lost and no error detection or error correction takes place. The cells continue to divide even in the presence of significant DNA damage and do not respond to signals to stop dividing. When this happens, abnormal cells with errors in their DNA continue through the cell cycle, passing these errors on to their daughter cells, and these cells in turn will pass the errors on to their daughter cells. This results in the formation of a tumour. While some tumours are benign, others are malignant because cells from these tumours can enter the bloodstream or lymphatics and spread to other regions of the body — a process known as **metastasis** (figure 2.24)

metastasis a process where malignant tumours spread throughout the body

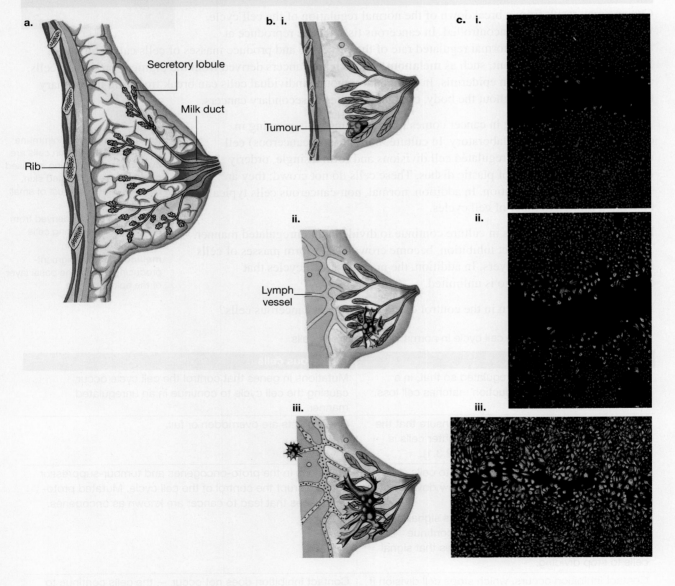

FIGURE 2.24 a. Longitudinal section of a breast showing normal tissues **b. i.** Development of a discrete tumour or 'lump in the breast' **ii.** Development of tumour into cancer **iii.** Spread of cancer within the breast, and migration of cancer cells from the breast through lymph vessels and blood vessels to new sites where secondary cancers develop — the process of metastasis **c.** Spread of breast cancer cells *in vitro*; that is, in a cell culture in the laboratory. Migrating cancer cells show the presence of a specific protein, vimentin, that stains green and is not present in normal cells. Note how the cancer cells (green) migrate and multiply, filling the space and crowding out the normal cells (red). (Image courtesy of Professor Leigh Ackland)

To illustrate the overriding of checkpoints, as introduced in section 2.3.1, the 'security guard' that operates the G2 checkpoint is a protein called p53. The normal p53 protein binds to DNA, and this sets up a sequence of events that stops cells from continuing through the cell cycle and enables checks to be carried out. However, when a mutation of the controlling gene occurs, the abnormal p53 protein cannot bind to DNA, so the cell cycle cannot be stopped. As a result, cells can divide in an uncontrolled manner and form tumours.

tlvd-1746

SAMPLE PROBLEM 3 Effect of drugs on mitosis and cell replication

Some drugs used in the treatment of cancers act on microtubules. They act by interfering with the normal contraction and extension capabilities of microtubules. Explain the effect you would expect such drugs to have on mitosis and cell replication. (4 marks)

THINK

Recall that microtubules are a component of centrioles and spindle fibres are made of microtubules that extend from the centrioles.

Consider the distinct points required to gain four marks. If any of the microtubules are affected by the drug:
- the spindle will not form correctly
- the spindle will not adhere to the kinetochore; or
- the spindle will not function (contract) to pull the chromosomes apart at anaphase. This will be picked up at the M checkpoint.
- then any of the above mean that mitosis cannot be completed, so there will be no cellular reproduction.

TIP: Use key terms: mitosis, cell replication, contraction, expansion, microtubules.

WRITE

Interference with the ability of microtubules to expand and contract will affect the formation and contraction of the mitotic spindle (1 mark), therefore affecting its ability to separate chromosomes during anaphase (1 mark).

Without a functioning spindle, mitosis cannot occur (1 mark); therefore there will be no cell replication (1 mark).

Many cancers have a genetic component. Some cancers are inherited; for example, retinoblastoma, a cancer of the eye. In other cases, the presence of a particular gene can increase the risk of occurrence of a cancer. The *BRCA1* and *BRCA2* genes are rare, but between 45 and 90 per cent of women with one of these genes develop breast cancer. These genes also increase the risk of ovarian cancer. The *BRCA2* gene in men increases their risk of developing breast or prostate cancer. However, it should be noted that these genes are involved in only a small percentage of cancers of a particular organ; for example, less than three per cent of breast cancers are caused by a faulty gene.

eWorkbook Worksheet 2.5 Disruption of the cell cycle (ewbk-7219)

Weblink Apoptosis in cancer

KEY IDEAS

- Checkpoints occur at various points in the cell cycle. Some checkpoints identify damaged or missing DNA and delay or stop the cell cycle.
- During the G1 checkpoint, the DNA of the cell is checked for damage before being replicated.
- During the G2 checkpoint, DNA replicated in the S stage of interphase is checked.
- The spindle is essential for chromosome arrangement and precise movement during mitosis.
- Sister chromatids must become linked to spindle fibres from opposite poles of the spindle to pass the M checkpoint.
- Actively dividing human tissues include the epidermis of the skin, the epithelium of the gut and the bone marrow.
- The meristematic tissue of plants contains cells that can complete the cell cycle and produce identical daughter cells.
- In vascular plants, meristematic tissue is present in root tips, shoots and stems.
- Cell division in epicormic shoots is important in the recovery of trees damaged by bushfire.
- Some cells produced by the cell cycle have a reproductive function, but offspring from this process are genetically identical.
- Apoptosis is programmed cell death for cells that are infected, no longer needed or no longer able to reproduce correctly.
- Apoptosis requires either an internal or external signal to proceed.
- Apoptosis results in cell shrinkage and blebbing to form apoptotic bodies.
- The cell cycle is normally regulated so that, in a mature organism, the rate of production of new cells balances the rate of loss of cells.
- Cancers may result from the breakdown of the normal control of the cell cycle.
- Cancerous cells are characterised by unregulated rates of cell division.

2.3 Activities

To answer questions online and to receive **immediate feedback** and **sample responses** for every question, go to your learnON title at www.jacplus.com.au. A **downloadable solutions** file is also available in the resources tab.

| 2.3 Quick quiz | 2.3 Exercise | 2.3 Exam questions |

2.3 Exercise

1. What is the role of the M checkpoint?
2. In which plant tissues would you expect to find dividing cells?
3. Consider the gemmae of liverworts. Would the next generation of plants that are derived from the gemmae of one liverwort be genetically identical or genetically dissimilar?
4. How do epicormic shoots contribute to the survival of fire-damaged trees in the Australian bush?
5. After a cell with 10 chromosomes completed the cell cycle, its daughter cells were examined. One daughter cell was found to contain 11 chromosomes and the other daughter cell had only 9 chromosomes. Suggest a possible explanation in biological terms for this observation.
6. What are the roles of proto-oncogenes and tumour-suppressor genes in controlling the cell cycle?
7. An error occurs in the DNA that leads to the formation of p53, a tumour-suppressor gene. Outline what will occur as a result of this damage.

2.3 Exam questions

Question 1 (1 mark)
Source: VCAA 2014 Biology Exam, Section A, Q9

MC In apoptosis
A. a cell rapidly divides and releases antibodies.
B. an inflammatory response is initiated by cell fragments.
C. nuclear material and organelles are broken down.
D. DNA is replicated.

Question 2 (1 mark)
Source: VCAA 2013 Biology Section A, Q12

MC In multicellular organisms, cells have receptors for death-signalling molecules. These death-signalling molecules play a role in apoptosis.

The death-signalling molecules
A. allow for tumour formation.
B. act by decreasing the rate of mitosis.
C. initiate a response causing a cell to swell and burst.
D. result in destruction of cells that are no longer required.

Question 3 (2 marks)
The p53 gene is a tumour-suppressor gene that checks for damage to DNA during interphase of the cell cycle.

A scientist investigated the action of the p53 gene by irradiating cells during the G2 phase of the cell cycle. The radiation treatment caused damage to the cells' DNA. He then observed the cells' behaviour through the rest of the cell cycle.

The scientist had two groups of cells in his experiment:
- Group A — cells that possessed the normal p53 gene
- Group B — cells that lacked the p53 gene.

In group A cells, the G2 phase stopped until the DNA damage was repaired. Following the repair, mitosis commenced and the cell cycle continued.

State two observations that the scientist would make for the irradiated group B cells.

Question 4 (2 marks)
Source: VCAA 2009 Biology Exam 2, Section B, Q1b, c

Diagram X outlines a mitotic cell cycle. Image D shows the appearance of a chromosome during one of these cycles.

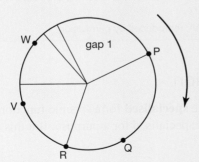

Diagram **X**: Outline of the mitotic cell cycle

Image **D**

a. Explain at which labelled point (P, Q, R, V, W) in the cycle image **D** would be found. **1 mark**
b. Explain why apoptosis sometimes occurs during the cell cycle represented in the above diagram. **1 mark**

▶ **Question 5 (2 marks)**
The diagram shows cells in a cancerous tumour.

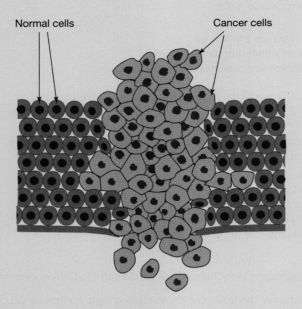

a. How does a tumour form? **1 mark**
b. If a tumour invades healthy tissue it can harm a person's health. Explain why invasive tumours are harmful. **1 mark**

More exam questions are available in your learnON title.

2.4 Cell differentiation

KEY KNOWLEDGE

- Properties of stem cells that allow for differentiation, specialisation and renewal of cells and tissues, including concepts of pluripotency and totipotency

Source: VCE Biology Study Design (2022–2026) extracts © VCAA; reproduced by permission.

2.4.1 Differentiation and specialisation

Cell **differentiation** is the process by which cells become **specialised** for a specific function. In this process, the gene expression (proteins produced by the cell) becomes specialised for a purpose. But this comes at a cost — these cells lose the ability to reproduce themselves.

As a result, some cells must remain undifferentiated to replenish cells when needed. These cells are called stem cells.

Consider the development of a human from one solitary cell to the 37.2 trillion cells and numerous different cell types of a human adult.

differentiation the process by which cells, tissues and organs acquire specialised features

specialisation the adaptation of something for a specific function

In the transition from a single-celled zygote to a newborn baby, remarkable changes will take place:
- Many mitotic cell divisions occur that, by the time of birth, will increase the total number of cells to many billions. Estimates of the number of cells in a newborn vary; however, a reliable indication that this figure must be in the billions comes from one study that identified, at birth, the number of cells in *just the forebrain* is 38 billion.
- A process of cell differentiation occurs, which will produce over 200 different cell types.
- A process of organisation of these differentiated cells of various types into tissue organs and systems occurs.

2.4.2 Stem cells

Stem cells are undifferentiated or unspecialised cells that have the ability to differentiate into organ or tissue-specific cells with specialised functions, such as nerve cells, blood cells, bone cells, heart cells, skin cells and so on. These terminal cells with specialised functions, such as a liver cell or a muscle cell, are differentiated and, once differentiated, cannot normally revert to an undifferentiated state.

A second feature of stem cells is that they are capable of dividing and renewing themselves over long periods. Figure 2.27 shows mouse stem cells that have been stained to show the presence of one of the proteins (Oct4) that is essential to keep the stem cells in an undifferentiated state.

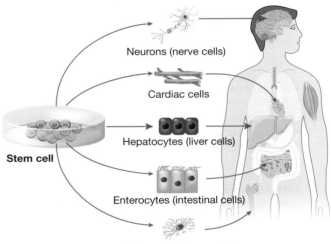

FIGURE 2.25 Stem cells are unspecialised and can differentiate into specialised cells

FIGURE 2.26 The ear and hand of a 20-week-old male foetus. Twenty weeks prior it was a single fertilised egg.

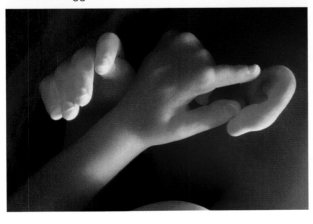

FIGURE 2.27 Mouse stem cells. The yellow staining indicates protein Oct 4 that prevents the cells from differentiating.

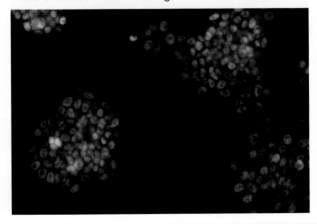

Some stem cells in your body are constantly dividing to replace tissues. Examples of these are the stem cells in the basal layer of your skin (refer to section 2.4.4, figure 2.32) and the stem cells in the **crypts** of your intestine (refer to section 2.4.4, figure 2.33). Each of these stem cells divides to produce a specialised differentiated cell and a replacement stem cell (see figure 2.28). This is how stem cells self-renew.

crypts tube-like depressions of the mucosa located in the intestine and the site of glandular cells

FIGURE 2.28 The division of a stem cell by mitosis gives rise to two daughter cells, one of which differentiates to become a specific cell type and the other that replaces or renews the original stem cell. Why is this self-renewal important?

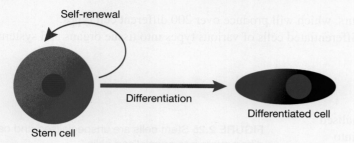

Different kinds of stem cells occur and they can be distinguished in terms of their potency to produce different cell types.

Different stem cell potencies

- **Totipotent** (from the Latin, *totus* = entire). These cells have the potential to give rise to all cell types. Totipotent cells include the fertilised egg and embryonic cells of a two, four or eight-cell **embryo**.
- **Pluripotent** (from the Latin, *plures* = several). These are the cells of the primary germ layers: **ectoderm**, **mesoderm** and **endoderm**. These cells can differentiate into many cell types. Examples include embryonic stem cells from the inner cell mass of the embryonic blastocyst.
- **Multipotent**. These cells have the ability to differentiate into a closely related family of cells; for example, a multipotent blood stem cell can develop into a red blood cell, a white blood cell or platelets (all specialised cells).
- **Oligopotent**. These cells have the ability to differentiate into a few cell types; for example, adult (somatic) lymphoid or myeloid stem cells.
- **Unipotent**. These cells have the ability to produce only cells of their own type, but because they can self-renew they are termed stem cells. Examples include adult (somatic) muscle stem cells.

totipotent a cell that is able to give rise to all different cell types

embryo early stage of a developing organism; in humans this includes the first eight weeks of development

ectoderm the most external primary germ or cell layer that differentiates into epithelial tissue, which covers the outer surfaces of the body

mesoderm the middle primary germ layer that differentiates into various tissues and organs, including the heart

endoderm the innermost primary germ layer that differentiates into digestive lining and organs like the lungs

multipotent a cell that can differentiate into a number of closely related cell types

oligopotent a cell that has the ability to differentiate into a few different cell types

unipotent a cell that has the ability to produce only cells of their own type

2.4.3 Sources of stem cells

Stem cells can be obtained from the sources shown in figure 2.29.

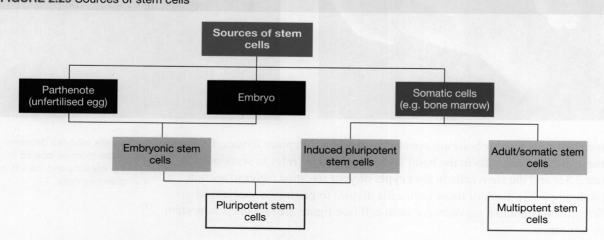

FIGURE 2.29 Sources of stem cells

These types of stem cells can be examined in more detail.

- **Embryonic stem cells (ESCs)** may be obtained from the inner cell mass of an early embryo at the blastocyst stage (see figure 2.30); that is, the clump of cells adhered to the inside surface of a blastocyst (see figure 2.31a). A single cell is isolated from this inner cell mass and is grown in culture, dividing by mitosis to produce a culture of stem cells. These ESCs are obtained from extra embryos created as part of IVF procedures that are in excess of requirements. Taking these cells from the inner mass of a blastocyst destroys an embryo and therefore this procedure has raised ethical issues.
- **Parthenotes** are another potential source of ESCs. These are derived from unfertilised human eggs that are artificially stimulated to begin development. Such an egg, of course, may start development, but it is not capable of developing into a human being.
- **Adult stem cells** (more accurately called **somatic stem cells**) can be obtained from various sources throughout the body, such as bone marrow, skin, the liver, the brain, adipose tissue and blood. In addition, another source of stem cells is cord blood, which can be harvested from the umbilical cord of a baby after birth (see figure 2.31b). Samples of some of these tissues are more accessible than others, such as blood. Bone marrow and adipose tissue are harder to obtain, and can be harvested by drilling into bones (usually the iliac crest or the femur) or through liposuction respectively.

  Somatic stem cells are multipotent. This means that they can give rise to particular cell types such as different kinds of blood cells or skin cells. Cord blood, for example, contains mainly stem cells that give rise to various blood cells.
- **Induced pluripotent stem cells (iPSCs).** Research by Shinya Yamanaka in Japan in 2006 led to the discovery that some specialised adult somatic (skin) cells could be genetically reprogrammed to return to an undifferentiated embryonic state. This reprogramming was achieved by the addition to these cells of four specific embryonic genes, which encode proteins that are known to keep stem cells in an undifferentiated state. One of these genes is the *OCT4* gene that encodes the Oct4 protein (see figure 2.27). The creation of iPSCs does not involve the ethical issues related to the embryo deaths that necessarily accompany ESCs derived from blastocysts.

embryonic stem cell (ESC) an undifferentiated cell obtained from early embryonic tissue that is capable of differentiating into many cell types

parthenote potential source of embryonic stem cells, derived from unfertilised human eggs that are artificially stimulated to begin development

adult stem cells undifferentiated cells obtained from various sources and capable of differentiating into related cell types; also known as somatic stem cells

somatic stem cells undifferentiated cells obtained from various sources and capable of differentiating into related cell types; also known as adult stem cells

induced pluripotent stem cell (IPSC) a stem cell that has been genetically reprogrammed to return to an undifferentiated embryonic state

cell-based therapies the use of stem cells in the treatment of human disorders or conditions to repair the mechanisms of disease initiation or progression

The ability to produce iPSCs is supporting new lines of research into disease and drug development. For example, iPSCs can be made from skin samples of patients with Parkinson's disease, and these cells show signs of that disease. This means that aspects of the disease can be studied in detail in cell cultures in the laboratory, allowing the effectiveness of new drugs to be explored using these iPSCs.

Cell-based therapies using iPSCs are not practical at present. The current procedure for reprogramming of somatic cells involves genetic modification, which can sometimes cause cells to produce tumours.

FIGURE 2.30 The development of a fertilised egg to a blastocyst stage

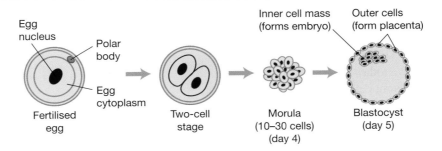

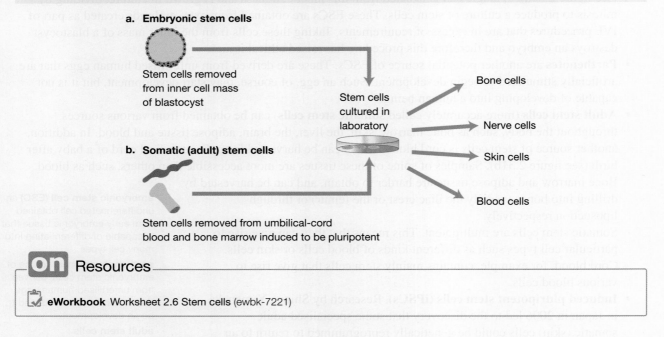

FIGURE 2.31 Stem cell lines can be created from various sources. a. One source of stem cells is embryonic stem cells from the inner cell mass of a blastocyst. b. Somatic stem cells can be extracted from bone marrow and from umbilical cord blood. *Somatic stem cells* is the preferred term for adult stem cells. In order to differentiate into multiple cell types like embryonic stem cells, they need to be induced to be pluripotent (iPSCs).

Resources

eWorkbook Worksheet 2.6 Stem cells (ewbk-7221)

2.4.4 Somatic stem cells

There are a number of cell types that somatic stem cells can be derived from. The following are some examples.

Basal stem cells of the epidermis

In human skin, surface cells are constantly being shed and are being replaced by daughter cells produced by division of basal stem cells. Each basal stem cell that undergoes cell division produces two daughter cells. Of these two daughter cells, one becomes a keratinocyte and the other remains in the basal layer as a basal stem cell, replacing the original parent cell. The other daughter cell progressively moves upwards through the epidermis, differentiates into a keratinocyte and is shed from the skin surface (figure 2.32). Within a period of about 48 days, the entire epidermis is replaced by new cells. This means that the skin that you have today is made of completely different cells from the skin that you had two months ago.

FIGURE 2.32 Cell division in the epidermis of the skin. The basal stem cells divide to produce two cells, one of which replaces the parent stem cell, while the other will differentiate and progressively move to the skin surface and be lost. The cells at the surface become filled with keratin and die.

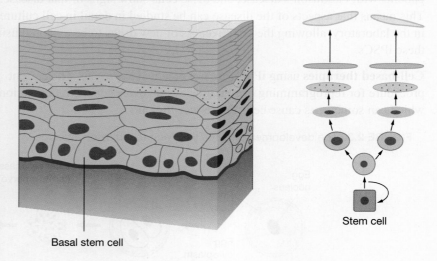

For a newly produced cell to move from the base of the epidermis where it was formed to the base of the dead layer of cells takes about two weeks. To move through the layer of dead cells and be shed takes a further four weeks. An estimate of the rate of loss of dead skin cells from an adult person is 30 to 40 thousand per hour. This makes the cell cycle activity of basal stem cells of the epidermis very important.

The ability of the skin to heal after considerable damage, as exemplified by the recovery of burns patients, is due to the presence of stem cells in the basal layer of the epidermis and the stem cells in the **dermis**.

Intestinal stem cells of the gut

The epithelial lining of the small intestine is regenerated every four to five days. This means that a person aged 18 years will have experienced more than 1000 replacement cycles of the lining of the small intestine.

dermis underlying part of the skin

FIGURE 2.33 a. Longitudinal section through the small intestine showing the upward-projecting villi with the downward-projecting crypts. Intestinal stem cells that are responsible for the regeneration of the intestinal lining are located in these crypts. **b.** Diagram showing the progression of the cells produced by the intestinal stem cells. Note that of the two cells produced by a stem cell, one will differentiate into a cell on the villus and the other replaces the stem cell.

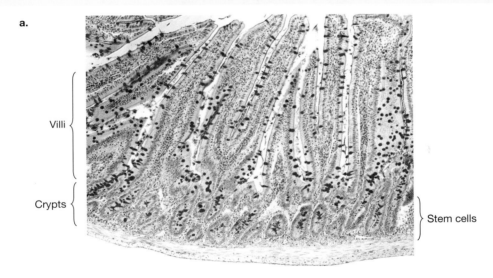

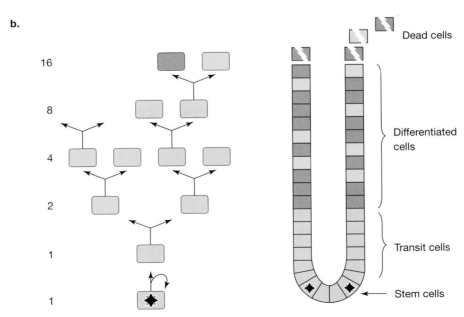

As intestinal cells die, they are replaced by new cells produced by intestinal stem cells. These stem cells are located at the base of infoldings, known as crypts, that are located between intestinal villi (singular = villus) (see figure 2.33). The replacement cells formed by division of the stem cells take two to seven days to move from the crypts to the tip of the villi from where they are lost.

Haematopoietic stem cells

Haematopoietic stem cells, located in the bone marrow, divide to give rise to cells that subsequently differentiate into the various types of blood cells, including red blood cells, white blood cells of various kinds and platelets (figure 2.34). Bone marrow is a spongy tissue found in the core of most bones, including the ribs, hips and spine. Most blood cells are short-lived and must be constantly replaced.

FIGURE 2.34 All blood cells develop from stem cells in bone marrow. Stem cells continually reproduce by mitosis and then differentiate.

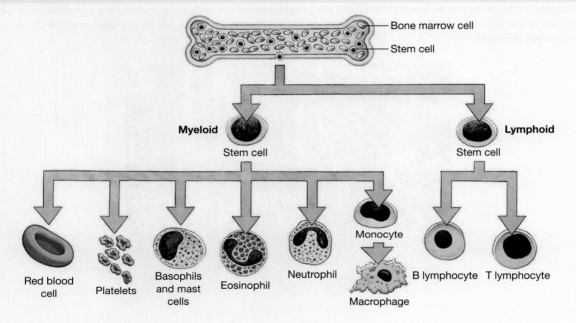

2.4.5 Stems cells in medicine

Stem cells in regenerative medicine

As people age, a number of degenerative disorders appear more commonly, such as Parkinson's disease. This particular disorder results from the death of certain brain cells that normally produce a chemical (dopamine) that controls muscle movements. People with Parkinson's disease show impairment of their motor movements, balance and speech. Early treatment for Parkinson's disease involved administering dopamine to affected persons. This treatment gave only short-term improvement.

Is there a way in which the lost dopamine-producing cells can be replaced? Experimental work is now proceeding on the potential use of stem cells to replace the lost cells in the brain.

Because stem cells have the unique ability to regenerate damaged tissue, research is being carried out on the potential of their use in the treatment of a large number of human disorders or conditions. Potential uses of stem cells for these purposes are called cell-based therapies, and the field of research is termed **regenerative medicine**.

regenerative medicine an experimental field of research involving stem cells in medicine that raises promise for the treatment of degenerative conditions and severe trauma injuries

In Australia, the only proven cell-based therapies are corneal (eye), skin grafting and blood stem cell transplants for the treatment of some cancers and autoimmune diseases, as well as some blood, inherited immune and metabolic disorders. These therapies are traditional cell therapies in which cells or tissues are taken from a donor and transplanted into or onto a recipient. As detailed in the Case study, significant developments are being made in the culturing of stem cells and their differentiation into specific cell types under laboratory conditions. This will significantly increase our understanding of genetic diseases and ultimately allow for more targeted (patient-specific) treatments to be developed.

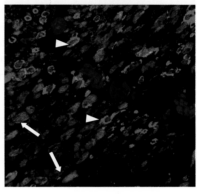

FIGURE 2.35 Injured spinal cord of mouse following injection of human stem cells. These stem cells developed into myelin-producing cells that form a wrapping (green) around nerve cells (red) (see the areas marked by arrowheads). Other nerve cells remained without a myelin wrapping (see the areas indicated with arrows).

Regenerative medicine using stem cells has extraordinary potential to reverse or alleviate conditions once thought to be permanent. Scientists at the University of California reported that, following the injection of human stem cells from nerve tissue into the spinal cords of paralysed mice, the test group of mice displayed better mobility than the non-injected controls after just nine days and, after four months, the test group of mice could walk. The stem cells migrated up the spinal cord and developed into different kinds of cells, including those cells that form insulating layers of myelin around nerve cells. Figure 2.35 shows the growth of myelin around nerve cells in the damaged region of a mouse spinal cord following injection of stem cells. Unfortunately, use of stem cells to treat spinal cord injuries in humans is problematic, largely due to a significant lack of trials in humans, which precludes strong evidence that the treatment is effective and safe over the long-term. However, in 2019, Japan approved stem cell treatment for spinal-cord injuries in a therapy known as Stemirac. In this case, the therapy uses stem cells from the patient's bone marrow, which are cultured externally and then returned to the patient, making double-blind studies impossible.

CASE STUDY: Stem cell research in Melbourne — Murdoch Children's Research Institute

The Murdoch Children's Research Institute (MCRI) in Melbourne is home to the Stem Cell Medicine program, which is a world-leading research program in induced pluripotent stem cells (iPSCs). The program is currently conducting research into a number of diseases, including kidney and heart diseases, blood, immune, brain and muscle disorders.

For detailed information on the stem cell research of the institute, please see the digital document.

on Resources

Digital document — Case study: Murdoch Children's Research Institute (doc-35875)

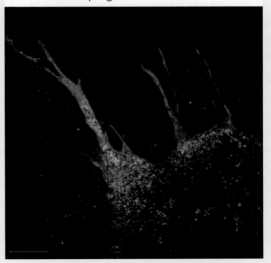

FIGURE 2.36 Small round blood cells (blue) in close association with linear, branching blood vessels (endothelium) (red). All cells were differentiated *in vitro* from human pluripotent stem cells as part of the Stem Cell Medicine program at MCRI.

Resources

Weblinks Human fetal forebrain
Murdoch Children's Research Institute
Stem Cells Australia
Stem cells

Therapeutic cloning

The purpose of **therapeutic cloning** is to produce stem cells for use in treatment.

Therapeutic cloning involves the creation of an embryo, through the technique of somatic nuclear transfer, for the purpose of obtaining stem cells from that embryo. These stem cells are intended for use in treating a patient who has a degenerative disease. The cell that provides the nucleus in therapeutic cloning is a healthy cell from the patient who is to receive treatment. As a result, the embryo that is created is a genetic match to the patient and these cells do not cause an immune response. Figure 2.37 shows the process of therapeutic cloning.

FIGURE 2.37 Therapeutic cloning involves the creation of an embryo that is genetically identical to a patient. The patient's cell is fused with an enucleated egg cell and develops into an early embryo. Stem cells are then taken from the inner cell mass of the early embryo (blastocyst) and grown in culture as pluripotent stem cells.

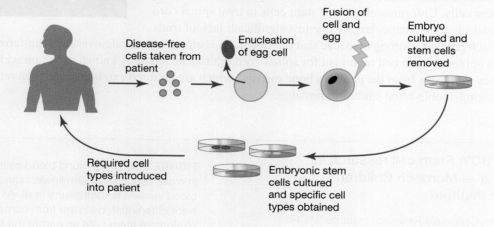

The use of early embryos as a source of stem cells raises **ethical issues** because establishing an embryonic stem cell line destroys an embryo. Likewise, ethical issues arise for therapeutic cloning because this procedure involves the artificial creation of an embryo solely for the purpose of obtaining stem cells, a process that then destroys the embryo.

In December 2002, the *Research Involving Human Embryos Act 2002* was passed in the Australian Parliament. This Act established a framework that regulated the use of 'excess' embryos. Provisions of this Act included the statement that 'embryos cannot be created solely for research purposes'. Under the provisions of this Act, therapeutic cloning was not permitted in Australia. However, in December 2006, the legislation was amended, with Parliament lifting the ban of the cloning of human embryos for stem cell research and allowing therapeutic cloning to be undertaken.

Reproductive cloning technologies are discussed in topic 9.

therapeutic cloning cloning carried out to create an embryo from which stem cells can be harvested

ethical issue a problem or situation that requires a person or organisation to choose between alternatives that must be evaluated as right (ethical) or wrong (unethical)

SAMPLE PROBLEM 4 Ethical considerations in stem cell research

A team of medical researchers has successfully developed a new drug that has shown preliminary signs that it might be effective in reducing the severity of symptoms in Parkinson's disease. In order to test its effectiveness, the pharmaceutical company funding the research has opted to do further testing in the laboratory using embryonic stem cells before proceeding to clinical testing using human volunteers. Suggest an ethical consideration that may influence the decision in using stem cells for this purpose. (1 mark)

THINK

Ethical considerations are those that look at the benefit versus the harm of the research. Consider the following:
- If stem cells are used to test the drug, then no animal or human is harmed by the drug.
- Should embryos be used to procure stem cells or is this a violation of life?
- What would happen to the embryos if they were not used for stem cells?
- Further testing could delay treatment for individuals with Parkinson's disease.

WRITE

Any one of the following (1 mark):
- Is it safe to test the new drugs on stem cells rather than animals or humans?
- Is it ethical to use embryos to procure stem cells?
- Should the treatment be tested on those with Parkinson's disease rather than stem cells so treatment is not delayed?

INVESTIGATION 2.3

Debating issues on stem cells

Aim

To explore the different ethical issues related to stem cells and use these to formulate an argument

 Resources

 eWorkbook Worksheet 2.7 Case studies in stem cells (ewbk-7223)

KEY IDEAS

- Stem cells are undifferentiated or unspecialised cells that have the ability to differentiate into organ or tissue-specific cells with specialised functions and to self-renew.
- Stem cells include embryonic stem cells and somatic (adult) stem cells.
- Stem cells from different sources differ in their potency or ability to produce differentiated cells of various types.
- Totipotent stem cells can give rise to all cell types. Pluripotent stem cells can give rise to most cell types, whereas multipotent stem cells give rise to closely related cells.
- Pluripotent stem cells can be derived from the inner cell mass of an embryo at the blastocyst stage. They may also be obtained from parthenotes or induced pluripotent stem cells.
- Somatic (adult) stem cells are multipotent, and can be gained from various locations such as the bone marrow or cord blood.
- Stem cells carry out the cell divisions that are responsible for tissue regeneration.

2.4 Activities

To answer questions online and to receive **immediate feedback** and **sample responses** for every question, go to your learnON title at **www.jacplus.com.au**. A **downloadable solutions** file is also available in the resources tab.

| 2.4 Quick quiz | 2.4 Exercise | 2.4 Exam questions |

2.4 Exercise

1. What is the difference between the members of the following pairs?
   a. Totipotent and pluripotent cells
   b. Undifferentiated and differentiated cells
   c. Embryonic and somatic (adult) stem cells
2. Identify one source of embryonic stem cells.
3. List two sources that could be used to obtain somatic (adult) stem cells.
4. What is a parthenote?
5. Identify where you would find the following:
   a. Skin stem cells
   b. A red blood cell precursor
   c. Keratinocytes
   d. Haematopoietic stem cells.

2.4 Exam questions

Question 1 (1 mark)
Source: Adapted from VCAA 2011 Biology Exam 2, Section A, Q5

MC Stem cells
A. are used in human reproductive cloning and can differentiate into a limited number of cell types.
B. are also called adult (somatic) cell types and are used in human reproductive cloning.
C. can differentiate into a limited number of cell types and can be obtained from a 2 or 4-cell embryo.
D. are also called adult (somatic) cell types and can differentiate into a limited number of cell types.

Question 2 (1 mark)
MC A special property of stem cells is that they are able to
A. develop into many different cell types.
B. divide by meiosis only.
C. divide once only.
D. live forever.

Question 3 (1 mark)
MC Consider the following cells:
I bone marrow stem cell
II fertilised egg cell
III embryo cell (8-cell stage)
IV blastocyst inner mass cell

Which of the cells listed is a totipotent stem cell?
A. II only
B. II and III only
C. III and IV only
D. IV only

Question 4 (4 marks)
a. What is the difference between totipotent, pluripotent and multipotent stem cells? **3 marks**
b. Which stem cells have the greater potency (power of differentiation) — embryonic stem cells or adult (somatic) stem cells? **1 mark**

Question 5 (2 marks)

The diagram summarises the steps involved in therapeutic cloning. An embryo that is a genetic clone of the patient is created by somatic nuclear transfer technique. Stem cells from this embryo are then cultured to produce the cell types needed to replace the diseased cells in the patient.

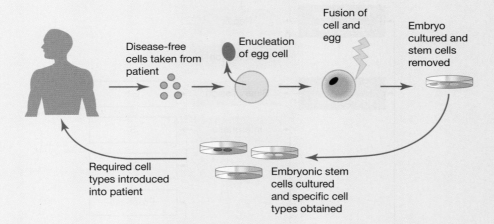

a. From which part of the embryo would the stem cells be obtained?   **1 mark**
b. Why are the patient's own cells used to create the embryo?   **1 mark**

More exam questions are available in your learnON title.

2.5 Review
2.5.1 Topic summary

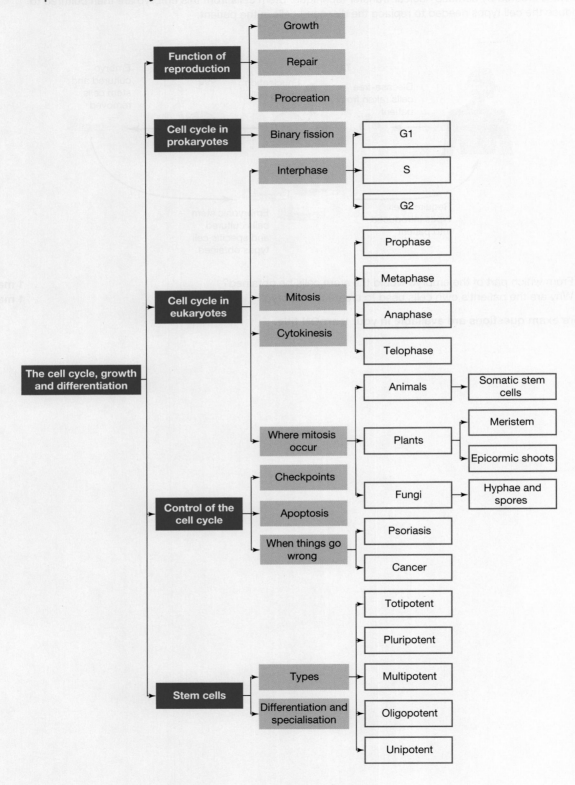

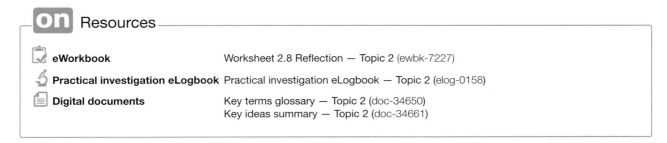

2.5 Exercises

To answer questions online and to receive **immediate feedback** and **sample responses** for every question, go to your learnON title at **www.jacplus.com.au**. A **downloadable solutions** file is also available in the resources tab.

2.5 Exercise 1: Review questions

1. Do you agree or disagree with each of the following claims about mitosis? Justify your response.
   a. The nuclear envelope is visible throughout the process.
   b. Mitosis would occur in the developing limb of a larval frog.
   c. Mitosis in plants is significantly different from mitosis in animals.
   d. Mitosis is preceded by replication of cell organelles, such as mitochondria and ribosomes.

2. The figure provided shows a series of drawings, all of the same cell at some stage during mitosis.
   a. Starting with cell A, place the drawings in the sequence that the stages would occur during mitosis.
   b. Draw what you would expect to see next in the sequence.
   c. Name each stage.

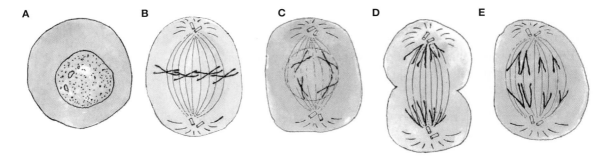

3. The photograph shows the winner of the 2012 Healthcare Cell Imaging Competition Microscopy category. Various fluorescent stains have been used to highlight different components of a cell that is progressing through the cell cycle.
   a. Suggest a possible identity for each of the following:
      i. The blue-stained structures
      ii. The green-stained structures
      iii. The red-stained structures.
   b. At what stage of the cell cycle was this image taken?

4. A cell containing 24 chromosomes reproduced by mitosis. A genetic accident occurred and one of the resulting cells had only 23 chromosomes.
   a. How many chromosomes would you expect to be in the other cell produced? Explain why.
   b. At what stage of cell reproduction do you think the genetic accident occurred?
5. Stem cell therapy is a treatment that uses stem cells, or cells that come from stem cells, to replace or repair damage to a patient's cells or tissues. The stem cells might be put into the blood or transplanted into the damaged tissue directly, or even recruited from the patient's own tissues for self-repair.
   a. What differentiated cells might come from pluripotent stem cells?
   b. What differentiated cells might come from unipotent stem cells?
   c. Outline one procedure by which a patient's own cells might be recruited for self-repair.
   d. Immunological reactivity is when the immune system recognises cells as foreign, which may lead to their destruction. Which, if any of these procedures, would not entail the problem of immunological reactivity?
6. A particular gene mutation affects a protein that is a key part of the special attachment site, the kinetochore, that allows a chromatid to be linked to spindle fibres. This mutation is present in cell B and it disables the function of the kinetochore.
   a. Would this mutation be expected to affect the progress of cell B through the cell cycle?
   b. If so, what effect would you predict? If not, give a reason for your decision.
7. The cell cycle in eukaryotes is highly regulated so that cell production in a tissue occurs at a rate that balances cell loss.
   a. What is a possible outcome if a breakdown in the regulation of the cell cycle occurs?
   b. A disorder known as polycythemia vera is a result of the overactivity of the bone marrow, resulting in the production of too many red blood cells. This condition results in a thickening of the blood and the common treatment is the regular removal of a fixed amount of blood. The cause of polycythemia vera is a mutation in the *JAK2* gene. What is the probable function of the normal *JAK2* gene?
8. The figure shows the number of scientific publications in the domain of stem cell research in the period from 1996 to 2012.
   (*Note:* ESC = embryonic stem cells; hESCs = human embryonic stem cells; iPSCs = induced pluripotent stem cells)

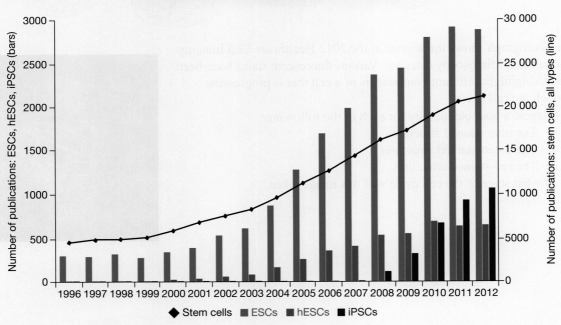

a. Explain why global research publications on iPSCs appear much later than research publications on hESCs.
b. Identify which area of stem cell research dominates the publications.
c. By what approximate factor has the number of research publication on iPSCs increased in the period from 2008 to 2012?
d. In what year did the total number of publications on stem cells of all types first exceed 10 000?

9. Suggest one ethical consideration involved with the use of stem cells in therapeutic cloning.

10. Cytokinesis is not part of the cell cycle but it is important in the production of new cells.
    a. What does the term *cytokinesis* refer to?
    b. How does cytokinesis differ in plants compared to animals?
    c. What feature of the cell membrane allows this process?

11. Binary fission is a simple form of cell replication where the primary purpose is reproduction. It does not use the formation of the mitotic spindle, instead relying on the replicated DNA to attach to the cell membrane (or nuclear membrane in eukaryotes) before separation.
    a. What consequences might occur if there was no mitotic spindle used in the cell cycle?
    b. Why would organisms use binary fission rather than mitosis?

2.5 Exercise 2: Exam questions

on Resources

▶ **Teacher-led videos** Teacher-led videos for every exam question

Section A — Multiple choice questions

All correct answers are worth 1 mark each; an incorrect answer is worth 0.

Use the following information to answer Questions 1–3.

The diagram below illustrates two of the early stages of the apopstosis in a human cell.

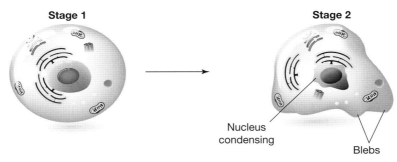

Question 1

Source: VCAA 2020 Biology Exam, Section A, Q15

The human cell will begin apoptosis when

A. a toxin is released from the cell.
B. an increase in the number of cells is required.
C. the cell has been irreversibly damaged by very low temperatures.
D. a signalling molecule attaches to a death receptor on the plasma membrane of the cell.

Question 2

Source: VCAA 2020 Biology Exam, Section A, Q16

Consider the change in appearance of the human cell that occurs from Stage 1 to Stage 2.

The change in appearance can be explained by

A. inflammation in the area surrounding the cell.
B. increased activity of caspases within the cell.
C. the release of histamine from nearby mast cells.
D. decreased activity of DNA polymerase within the nucleus.

Question 3

Source: VCAA 2020 Biology Exam, Section A, Q17

After the condensation of the nucleus and the production of many blebs, the human cell will

A. swell and burst open, and its contents will be released into the surrounding environment.
B. produce apoptotic bodies that are engulfed by phagocytes.
C. become larger and be absorbed by adjacent cells.
D. be completely broken down by lysosomes.

Use the image to answer Questions 4 and 5.

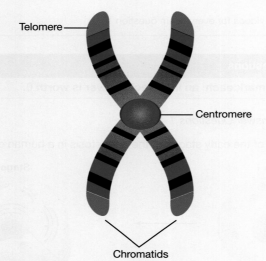

Question 4

In which phase(s) is the structure shown visible during the cell cycle?

A. Interphase and prophase
B. Prophase and metaphase
C. Metaphase and anaphase
D. Telophase and cytokinesis

▶ **Question 5**

Which part of the figure shown joins to the spindle fibres during the cell cycle?

A. Centromere
B. Telomere
C. Chromatid
D. None of the above

▶ **Question 6**

Which one of the following is a characteristic of embryonic stem cells?

A. They are difficult to grow in culture.
B. They are more differentiated than tissue stem cells.
C. They have the capacity to differentiate into all different cell types.
D. They will develop into cell types that are closely associated with the tissue they are found in.

▶ **Question 7**

The diagram shows how three receptors (A, B and C) on the surface of a stem cell, if activated, can bring about different responses. The type of response triggered is dependent on the number of and which receptors are activated.

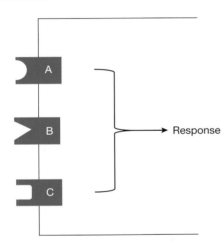

Receptor(s) activated	Type of response
A	Increased growth of cell
B	Cell replication
C	Cytokinesis
A and B	Differentiation into skin cells
A and C	Differentiation into muscle cells
B and C	Apoptosis of cell
A, B and C	Differentiation into nerve cells

For a stem cell to differentiate into muscle cells if must receive which of the following signals?

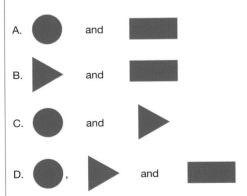

Use the following information to answer Questions 8–10.

Bynoe's gecko (*Heteronotia binoei*) is a lizard found around Australia that has caught the attention of scientists, as populations of the gecko can be made up entirely of females. The geckos lay eggs from which baby geckos hatch, and are said to use parthenogensis, or 'virgin birth', as a reproductive strategy.

Question 8
Parthenogenesis is a type of

A. binary fission.
B. asexual reproduction.
C. sexual reproduction.
D. cloning.

Question 9
All offspring would be genetically

A. identical, as they are all clones of the mother lizard.
B. different, as two lizards must be involved to make offspring, thereby combining their DNA.
C. different, as offspring will only have half the DNA of the mother lizard.
D. identical, because it is part of the mother that grows into a new individual.

Question 10
A researcher captured ten Bynoe's geckos from one location and genetically tested them all before re-releasing them.

Which of the following hypotheses was the researcher testing?

A. That all Bynoe's geckos of this population eat the same food
B. That all Bynoe's geckos are genetically identical
C. That all Bynoe's geckos of this population are genetically identical
D. That all Bynoe's gecko's give birth

Section B — Short answer questions

Question 11 (11 marks)

Study the diagram showing cells growing in a cancerous tumour, after staining and viewed under a light microscope.

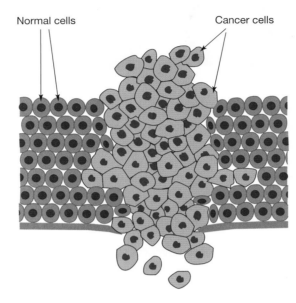

a. Construct a table listing three differences between normal cells and cancer cells that can be observed in the diagram. **3 marks**

b. Cancer occurs after a series of events that would not normally happen:
   1. Abnormal cells survive.
   2. Abnormal cells multiply rapidly to form a tumour.
   3. Abnormal cells invade other tissues.
   i. What is meant by the term *abnormal cell* in a genetic sense? **1 mark**
   ii. State two main reasons for a genetic abnormality being present in a cell. **2 marks**

c. The body usually gets rid of abnormal cells or stops them from dividing. There are two classes of genes involved in this control: proto-oncogenes and tumour-suppressor genes. Describe the role of each type of gene in controlling the rate of cell division. **4 marks**

d. What disease occurs if the proto-oncogenes or tumour-suppressor genes themselves are damaged or mutated? **1 mark**

Question 12 (8 marks)

Figures A to D show some important stages of the mitotic phase of the cell cycle.

a. Identify the stages shown, giving the evidence from the diagrams that supports your conclusion.   4 marks
b. Arrange the letters A to D to show the correct sequence of stages.   1 mark
c. Give two pieces of evidence that support that the cell shown is an animal cell and not a plant cell.   2 marks
d. Describe how cytokinesis would occur in this cell.   1 mark

Question 13 (9 marks)

Cell A has four pairs of chromosomes with a total DNA content of 12 units. Cell A undergoes one cell cycle.

a. List, in order, the stages that cell A would proceed through, starting from the earliest.   3 marks
b. How many cells would be present at the end of this cell cycle: one, two or three?   1 mark
c. How many copies of DNA would be present in cell A at the following points in the cell cycle?
    i. G2 stage of interphase   1 mark
    ii. Anaphase of mitosis   1 mark
    iii. G1 stage of interphase   1 mark
d. How many units of DNA would be present in one daughter cell of cell A?   1 mark
e. How many chromosomes would be present in this daughter cell?   1 mark

Question 14 (9 marks)

Cells that undergo a mutation to the gene that codes for the p53 protein are unable to regulate their cell cycle. As a result, their replication continues in an uncontrolled manner.

a. What is the function of the p53 protein?   1 mark
b. In what stage of the cell cycle is this protein important?   1 mark
c. Would the mutation prevent the cell from undergoing apoptosis? Explain.   3 marks
d. Chemotherapy is used to treat people with cancer. It involves the administration of anti-cancer drugs. There are many different classes of drugs. One type is an antimetabolite. These drugs interfere with the metabolism (all chemical reactions involved in maintaining the living state of cells) of the cell. Methotrexate is one of these drugs. It interferes with the cell's ability to make nucleic acids such as DNA and RNA.
    i. What effect would this have on the cell cycle?   2 marks
    ii. Given the drug is administered through the blood, would it have an effect on non-cancerous cells? Explain.   2 marks

Question 15 (12 marks)

The photograph shows cells at different stages of the cell cycle. Cells A, B and C are undergoing different phases of mitosis. Cell N is in interphase.

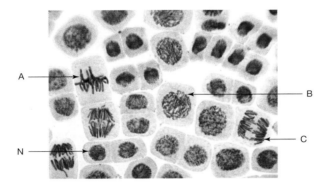

a. Complete the table for cells A, B and C to show which phase of mitosis is occurring and list two events that happen during each stage. **6 marks**

Cell	Phase of mitosis	Two events occurring during this phase
A		1 Replicated chromosomes attach by centromeres to spindle fibres. 2 Replicated chromosomes line up across the equator of the cell.
B	Prophase	1 2
C		1 2

b. i. List two observations from the photograph that support the statement that cell N is at interphase. **2 marks**
   ii. State three events that would occur in cell N during interphase. **3 marks**
c. For some time, biologists included interphase as a substage in mitosis. This is no longer the case; interphase is now considered to be a separate stage of the cell cycle, significantly different from mitosis. Suggest one significant difference between interphase and mitosis. **1 mark**

2.5 Exercise 3: Biochallenge online only

on Resources

eWorkbook Biochallenge — Topic 2 (ewbk-8072)

Solutions Solutions — Topic 2 (sol-0647)

teach on

Test maker
Create unique tests and exams from our extensive range of questions, including practice exam questions.
Access the assignments section in learnON to begin creating and assigning assessments to students.

Online Resources

Below is a full list of **rich resources** available online for this topic. These resources are designed to bring ideas to life, to promote deep and lasting learning and to support the different learning needs of each individual.

eWorkbook

- 2.1 eWorkbook — Topic 2 (ewbk-3156)
- 2.2 Worksheet 2.1 Binary fission in prokaryotes (ewbk-7211)
  Worksheet 2.2 The cell cycle (ewbk-7213)
- 2.3 Worksheet 2.3 Regulating the cell cycle (ewbk-7215)
  Worksheet 2.4 Apoptosis (ewbk-7217)
  Worksheet 2.5 Disruption of the cell cycle (ewbk-7219)
- 2.4 Worksheet 2.6 Stem cells (ewbk-7221)
  Worksheet 2.7 Case studies in stem cells (ewbk-7223)
- 2.5 Worksheet 2.8 Reflection — Topic 2 (ewbk-7227)
  Biochallenge — Topic 2 (ewbk-8072)

Solutions

- 2.5 Solutions — Topic 2 (sol-0647)

Practical investigation eLogbook

- 2.1 Practical investigation eLogbook — Topic 2 (elog-0158)
- 2.2 Investigation 2.1 Observing binary fission (elog-0786)
  Investigation 2.2 Observing the cell cycle under the microscope (elog-0788)
- 2.4 Investigation 2.3 Debating issues on stem cells (elog-0790)

Digital documents

- 2.1 Key science skills — VCE Biology Units 1–4 (doc-34648)
  Key terms glossary — Topic 2 (doc-34650)
  Key ideas summary — Topic 2 (doc-34661)
- 2.2 Extension: DNA replication is semi-conservative (doc-35876)
- 2.3 Case study: HeLa cells and the evolution of bioethics (doc-35873)
  Case study: Saving burns victims — Professor Fiona Wood (doc-35874)
- 2.4 Case study: Murdoch Children's Research Institute (doc-35875)

Teacher-led videos

- Exam questions — Topic 2
- 2.2 Sample problem 1 Binary fission (tlvd-1744)
  Sample problem 2 Identifying the stages of mitosis (tlvd-1745)
- 2.3 Sample problem 3 Effects of drugs on mitosis and cell replication (tlvd-1746)
- 2.4 Sample problem 4 Ethical considerations in stem cell research (tlvd-1848)

Video eLessons

- 2.2 Binary fission (eles-2464)
  Mitosis (eles-4215)
- 2.3 Apoptosis (eles-3694)

Interactivities

- 2.2 The stages of mitosis (int-3028)
- 2.2 Labelling the stages of mitosis (int-8128)

Weblinks

- 2.3 Apoptosis in cancer: from pathogenesis to treatment
- 2.4 Human fetal forebrain
  Murdoch Children's Research Institute
  Stem Cells Australia
  Stem cells

Teacher resources

There are many resources available exclusively for teachers online

To access these online resources, log on to **www.jacplus.com.au**

UNIT 1 | AREA OF STUDY 1 REVIEW

AREA OF STUDY 1 How do cells function?

OUTCOME 1
Explain and compare cellular structure and function and analyse the cell cycle and cell growth, death and differentiation.

PRACTICE EXAMINATION

STRUCTURE OF PRACTICE EXAMINATION		
Section	Number of questions	Number of marks
A	20	20
B	5	30
	Total	50

Duration: 50 minutes
Information:
- This practice examination consists of two parts. You must answer all question sections.
- Pens, pencils, highlighters, erasers, rulers and a scientific calculator are permitted.

SECTION A — Multiple choice questions

All correct answers are worth 1 mark each; an incorrect answer is worth 0.

1. Consider the diagram of the cell shown.

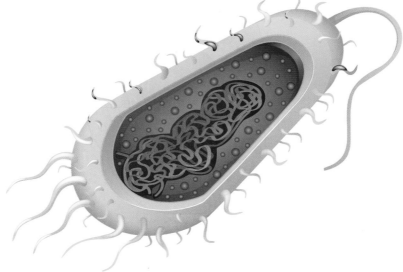

   The cell can be identified as a prokaryotic cell due to the
   A. presence of ribosomes.
   B. presence of a cell wall.
   C. absence of genetic material.
   D. absence of membrane-bound organelles.

2. What is the functional unit of life?
   A. A cell
   B. An organelle
   C. The nucleus
   D. The genetic material
3. Consider the diagram of the cell shown.

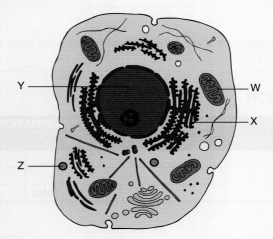

   Folding proteins into their correct functional shape would occur in the organelle labelled
   A. W.
   B. X.
   C. Y.
   D. Z.
4. Consider the diagram of the cell shown.

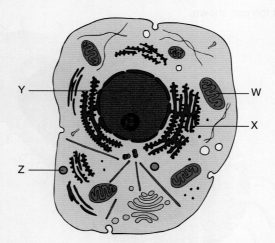

   The energy required for the synthesis of molecules within a cell would be produced in the organelle labelled
   A. W.
   B. X.
   C. Y.
   D. Z.

5. What is the organelle that is bound by a double membrane and can trap light energy?
   A. Mitochondrion
   B. Chloroplast
   C. Lysosome
   D. Ribosome

6. Which one of the following methods of transport across a plasma membrane requires the cell to use energy?
   A. Facilitated diffusion
   B. Simple diffusion
   C. Active transport
   D. Osmosis

7. Facilitated diffusion differs from simple diffusion as it
   A. uses energy.
   B. occurs more rapidly.
   C. requires a transporter protein.
   D. moves molecules against a concentration gradient.

8. A cell is placed in an isotonic solution. The movement of water across the plasma membrane would
   A. not occur.
   B. be faster into the cell than out of the cell.
   C. be faster out of the cell than into the cell.
   D. occur at the same rate into and out of the cell.

9. What is the name given to the bulk transport of a molecule dissolved in a liquid across the plasma membrane and into the cell?
   A. Phagocytosis
   B. Pinocytosis
   C. Exocytosis
   D. Facilitated diffusion

10. Consider the information in the following table.

Cell	SA : V ratio
A	20 : 1
B	8 : 1
C	30 : 1
D	9 : 1

Which cell is least likely to have a role in absorbing digested food from the human small intestine?
   A. Cell A
   B. Cell B
   C. Cell C
   D. Cell D

11. In which stage of mitosis are sister chromatids separated?
   A. Prophase
   B. Metaphase
   C. Anaphase
   D. Telophase

12. Consider the cell cycle as shown.

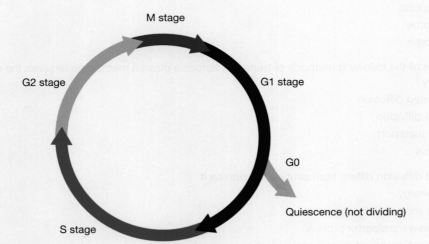

Cytokinesis occurs during which stage?
   A. M
   B. G1
   C. S
   D. G2

13. The number of molecules of DNA in a cell in the G1 stage of the cell cycle is 20. The expected number of molecules in the same cell in the G2 stage of the cell cycle would be
   A. 10.
   B. 20.
   C. 30.
   D. 40.

14. Consider a plant cell undergoing cytokinesis. What difference would be visible in the plant cell compared to an animal cell?
   A. The presence of centrioles
   B. The formation of a cell plate
   C. Nuclear membranes reappearing
   D. Chromosomes uncoiling and disappearing

15. Programmed cell death results in the death of many cells. What major group of molecules help regulate this process?
   A. Polymerases
   B. Caspases
   C. Synthases
   D. Ligases

16. The intrinsic pathway of programmed cell death
   A. results in an inflammatory response.
   B. causes the rupture of the plasma membrane.
   C. involves the release of cytochrome C from a mitochondrion.
   D. is initiated when a signalling molecule attaches to a receptor on the plasma membrane.

17. A mutation in a gene can increase the chance of a person having cancer. A gene in which a mutation results in cancer would most likely be associated with the
    A. control of the cell cycle.
    B. rate of cellular respiration.
    C. synthesis of glycoproteins.
    D. production of phospholipids.

18. What is the role of a tumour suppressor gene?
    A. To cause cells to become cancerous
    B. To initiate the start of cell division
    C. To stimulate the growth of a cell
    D. To slow cell division

19. Stem cells from umbilical cord blood are considered to be
    A. unipotent.
    B. totipotent.
    C. multipotent.
    D. pluripotent.

20. What is the name of the process in which dividing cells change their type and become specialised?
    A. Proliferation
    B. Differentiation
    C. Determination
    D. Transformation

SECTION B — Short answer questions

Question 21 (5 marks)

The diagram shows the structure of the plasma membrane.

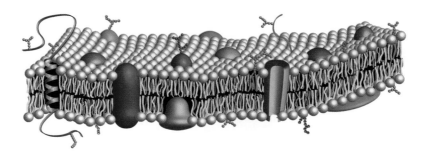

a. On the diagram:
   i. label one protein molecule — 1 mark
   ii. shade the section of the membrane that is hydrophobic — 1 mark
   iii. draw an arrow to indicate the pathway a potassium ion may take as it crosses the membrane. — 1 mark

b. Compare and contrast channel proteins and carrier proteins. — 2 marks

Question 22 (3 marks)

A carrot was peeled and placed in a hypotonic solution for one hour.

a. State whether the cells of the carrot after one hour would be flaccid, turgid or plasmolysed. **1 mark**

b. Describe what has happened to the carrot cells in the hour that they were placed in the hypotonic solution. **2 marks**

Question 23 (3 marks)

Consider the diagram of the cell shown.

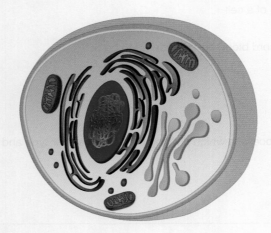

a. The cell is divided into many membrane-enclosed compartments. What is the advantage to the cell of having many different membrane-enclosed compartments? **1 mark**

b. A structure within the cell is marked X.

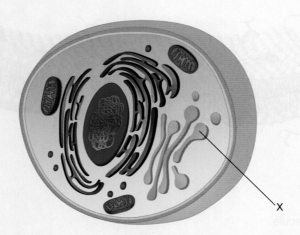

    i. Identify the structure labelled X. **1 mark**
    ii. Describe the role played by X in the functioning of the cell. **1 mark**

Question 24 (4 marks)

Consider the cells that make up a multicellular organism.

a. Name two conditions required by cells to sustain life. **2 marks**

b. Explain the importance of both the surface area and volume of a cell in terms of meeting the cell's requirements for sustained life. **2 marks**

Question 25 (5 marks)

Prokaryotes reproduce through binary fission. Draw a series of diagrams to show how the process of binary fission occurs. Label at least three important steps in the process.

Question 26 (4 marks)

Consider the diagrams of the three cells shown. The two cells on the right of the functional living cell show different ways a cell may die.

Functional living cell | Cell becoming smaller and dying | Cell becoming bigger and dying

a. State the name given to each type of cell death shown. **2 marks**

b. Give an example of when a cell may undergo each of the different types of cell death. **2 marks**

Question 27 (3 marks)

Describe three differences between functioning liver cells and cancer cells found within a liver. **3 marks**

Question 28 (3 marks)

A person who had both leukaemia and HIV was recently treated with a bone marrow transplant. The bone marrow contained stem cells from a healthy donor who is resistant to HIV.

a. What is a stem cell? **1 mark**

b. The bone marrow transplant was successful in treating the leukaemia and HIV. The patient no longer needs to take the prescribed antiretroviral drugs that they took before the bone marrow transplant. Describe what has happened to the blood cells that allows the patient to stop taking the antiretroviral drugs. **2 marks**

END OF EXAMINATION

UNIT 1 | AREA OF STUDY 1

PRACTICE SCHOOL-ASSESSED COURSEWORK

ASSESSMENT TASK — Modelling the eukaryotic cell cycle

This task is a modelling or simulation activity on the eukaryotic cell cycle, including the characteristics of each of the subphases of mitosis and cytokinesis in plant and animal cells.
- This SAC comprises a modelling activity and 11 questions; you must complete ALL question sections.
- Pens, pencils, highlighters, erasers, rulers and a scientific calculator are permitted.
- Mobile phones and/or any other unauthorised electronic devices including wrist devices are NOT permitted.

Total time: 50 minutes

Total marks: 42 marks

PART A: MODELLING INTERPHASE (12 marks)

Mitosis is the division of the nucleus of somatic cells with the intent of making two exact copies of the parent cell. During this activity we will be using pipe cleaners to simulate the chromosomes during mitosis.

Materials

3 × whole pipe cleaner (colour 1)
3 × whole pipe cleaner (colour 2)
A4 paper
Marker

Procedure and results

- Cut one pipe cleaner of each colour in half.
- Take 2 × whole pipe-cleaner chromosome (1 of each colour) and 2 × $\frac{1}{2}$ pipe-cleaner chromosome (1 of each colour).

1. How will you know that your chromosomes are homologous? **1 mark**
2. The A4 paper will represent your cell. Demonstrate the G_1 phase using the A4 paper and your pipe-cleaner chromosomes.
   a. Draw your representation. **2 marks**
   b. Describe what is occurring within the cell during the G_1 phase. **2 marks**
3. Demonstrate the **S** phase using the A4 paper and your pipe-cleaner chromosomes.
   a. Draw your representation. **2 marks**
   b. Describe what is occurring within the cell during the **S** phase. **1 mark**
4. Demonstrate the G_2 phase using A4 paper and your pipe-cleaner chromosomes.
   a. Draw your representation. **2 marks**
   b. Describe what is occurring within the cell during the G_2 phase. **2 marks**

PART B: MODELLING MITOSIS (12 marks)

5. Now you are going to use your cell model to demonstrate mitosis. Starting with prophase, demonstrate each stage of mitosis and illustrate relevant organelles by drawing them on the A4 paper. Take a photo and insert or draw your representation. Annotate each photo or drawing to highlight key parts of each stage.

PART C: CYTOKINESIS (3 marks)

6. Answer the following questions about cytokinesis.
   a. What is the outcome of cytokinesis? **1 mark**
   b. Describe how cytokinesis differs in plant and animal cells. **2 marks**

PART D: CONTROL OF THE CELL CYCLE (15 marks)

The cell cycle is tightly controlled. When errors are detected, particularly in the replicated DNA of a cell that cannot be repaired, the cell undergoes programmed cell death.

7. What is the name of the cell death that is activated when unrepairable errors in the cell cycle are detected? **1 mark**
8. Outline the key points in this process of cell death. **3 marks**
9. Sometimes an error occurs in the DNA that codes for a specific protein that activates the programmed cell death biochemical pathway. If this occurs, the cell cannot activate this intrinsic pathway.
   a. What could be the outcome of this error? **4 marks**
   b. Identify two roles of the immune system in apoptosis. **2 marks**
10. DNA damage detected during the cell cycle is one way that a cell will undergo programmed cell death. Identify three other reasons why a cell may need to initiate programmed cell death. **3 marks**
11. Some drugs known as mitotic inhibitors prevent the spindle fibres from forming.
    a. Which part of the cell cycle would be interrupted by a mitotic inhibitor? **1 mark**
    b. What would this prevent the cell from doing? **1 mark**

Resources

Digital document Unit 1 Area of Study 1 School-assessed coursework (doc-35077)

AREA OF STUDY 2 HOW DO PLANT AND ANIMAL SYSTEMS FUNCTION?

3 Functioning systems

KEY KNOWLEDGE

In this topic you will investigate:

Functioning systems
- specialisation and organisation of plant cells into tissues for specific functions in vascular plants, including intake, movement and loss of water
- specialisation and organisation of animals cells into tissues, organs and systems with specific functions: digestive, endocrine and excretory.

Source: VCE Biology Study Design (2022–2026) extracts © VCAA; reproduced by permission.

PRACTICAL WORK AND INVESTIGATIONS

Practical work is a central component of learning and assessment. Experiments and investigations, supported by a **practical investigation eLogbook** and **teacher-led videos**, are included in this topic to provide opportunities to undertake investigations and communicate findings.

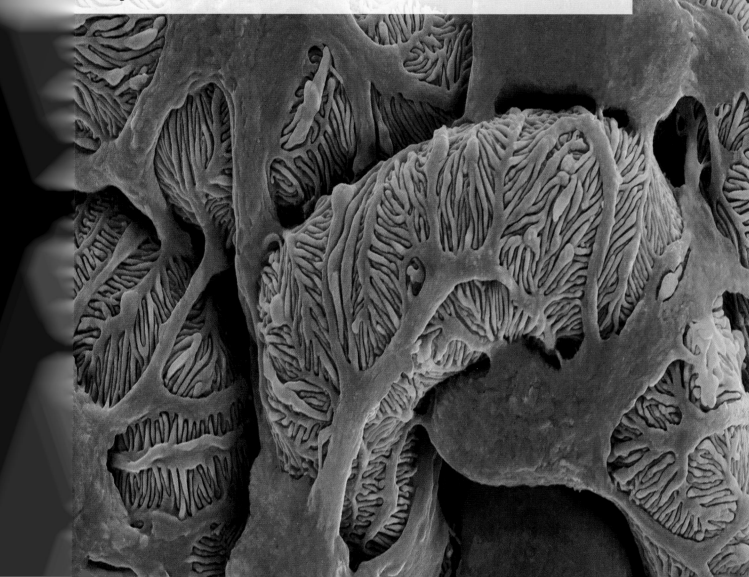

3.1 Overview

Numerous **videos** and **interactivities** are available just where you need them, at the point of learning, in your digital formats, learnON and eBookPLUS at **www.jacplus.com.au**.

3.1.1 Introduction

The first known organisms that appeared on Earth were neither plant nor animal, but were unicellular microbes. Microscopic microbial fossils, found in 3465 million year old rocks in the Pilbara region of Western Australia appear to be the earliest direct evidence of life on Earth. Primitive life existed for another 2800 million years until the appearance of multicellular life on Earth, around 600 million years ago (Mya). Striking evidence of the first complex multicellular organisms came from the diverse fossils unearthed in the Ediacaran Hills of the Flinders Ranges in South Australia (figure 3.1). The nature of these strange Ediacaran creatures was the subject of debate for decades. In 2018, research into preserved steroid lipids recovered from *Dicksonia* fossils provided clear evidence that these Ediacaran organisms were animals.

FIGURE 3.1 Impressions of *Dicksonia*, an Ediacaran fossil animal, preserved in sedimentary rocks from the Ediacaran Hills in South Australia

The evolution of multicellular organisms meant huge increases in the diversity and level of complexity of organisms in addition to extensions in lifespans, as organisms continue to live, even when individual cells die. In a unicellular organism, the single cell must perform all the functions needed to stay alive and their size is limited by the ability of diffusion to provide nutrients and remove wastes. In multicellular organisms, each cell type performs a distinct and specialist role, which gives rise to different levels of organisation: at the cellular level; at the tissue level where cells of one type are organised to specialise in one function; at the organ level where tissues of different types can be aggregated to form organs to carry out a particular function; and the system level where several different organs can be aggregated to form a system that performs a major life-supporting function. The topic opening image shows one such tissue, the glomerulus of the kidneys, showing blood vessels (green) enveloped in podocytes (brown).

LEARNING SEQUENCE

3.1 Overview	152
3.2 Specialisation and organisation of plant cells	153
3.3 Specialisation and organisation of animal cells	164
3.4 Digestive system in animals	174
3.5 Endocrine system in animals	205
3.6 Excretory system in animals	227
3.7 Review	243

Resources

- **eWorkbook** — eWorkbook — Topic 3 (ewbk-3158)
- **Practical investigation eLogbook** — Practical investigation eLogbook — Topic 3 (elog-0160)
- **Digital documents** — Key science skills — VCE Biology Units 1–4 (doc-34648)
  - Key terms glossary — Topic 3 (doc-34651)
  - Key ideas summary — Topic 3 (doc-34662)

3.2 Specialisation and organisation of plant cells

> **KEY KNOWLEDGE**
>
> - Specialisation and organisation of plant cells into tissues for specific functions in vascular plants, including intake, movement and loss of water
>
> **Source:** VCE Biology Study Design (2022–2026) extracts © VCAA; reproduced by permission.

3.2.1 Plant systems, organs and tissues

The multicellular land plants with which people are familiar are the **vascular plants**. Their cells are organised at the system, organ and tissue levels.

Plants have two systems:
- an above-ground **shoot system**
- a below-ground **root system**.

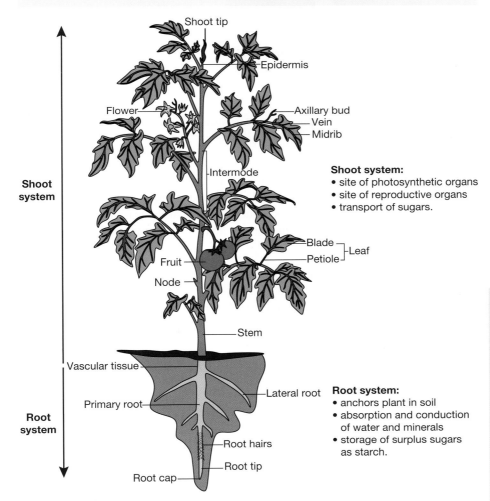

FIGURE 3.2 A typical flowering plant showing its two systems, the shoot and root systems and some of the functions performed by these systems

vascular plants plants with xylem and phloem tissue — the majority being flowering plants and conifers

shoot system the above-ground system of plants, the site of photosynthesis, transport of sugars and the site of reproductive organs

root system the below-ground system of plants which anchors the plant in the soil, is responsible for the absorption and conduction of water and minerals, and the storage of excess sugars (starch)

These root and shoot systems consist of different organs:
- shoot system organs — the stem and leaves, plus the reproductive organs, the flowers and fruit
- root system organs — the root, lateral roots and the root hairs.

Plant organs consist of different tissues. A simple classification is shown in figure 3.3.

FIGURE 3.3 A classification scheme of the major types of tissue in vascular plants

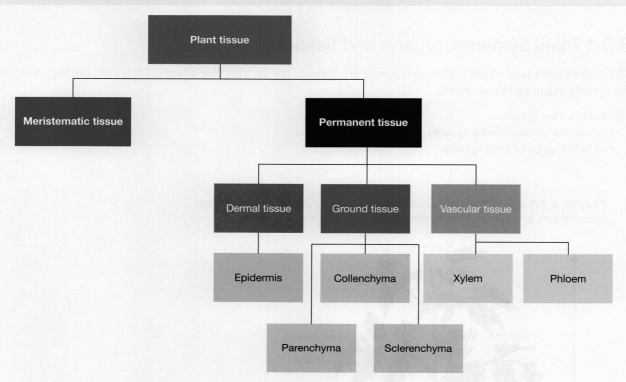

Meristematic tissues are made of cells that can undergo cell division and can continue to divide for the life of the plant. Meristematic tissue is usually found in the tips of roots and shoots and is responsible for an increase in the length of the plant stems and roots. (In grass, however, this tissue is located at the leaf base where the leaf blade joins the roots. Meristematic tissue is also present along the length of stems and roots and is responsible for the increase in girth of a plant.)

Permanent tissues are made of cells that can no longer divide. Permanent tissues include several tissue types that differ in their function:
- **Dermal tissue** protects plants and minimises water loss. For example, epidermal tissue made of flattened cells forms the outer layer of stems and leaves. A non-cellular layer of wax on top of the epidermal tissue adds further waterproofing.

meristematic tissues plant tissue found in tips of roots and shoots that is made of unspecialised cells that can reproduce by mitosis
permanent tissues plant cells that can no longer divide and includes dermal tissue, ground tissue (parenchyma, collenchyma and sclerenchyma) and vascular tissue (xylem and phloem)
dermal tissue tissue that protects the soft tissues of plants and controls water balance

- **Ground tissue** includes the following simple tissues:
  - **parenchyma** tissue — the most common plant tissue and is composed of living thin-walled cells. In leaves, the parenchyma tissue is the site of photosynthesis. In roots, tubers and seeds, the parenchyma tissue is the site of storage of starch (figure 3.4) or oils.
  - **collenchyma** tissue — the main supporting tissue of elongating stems; this tissue is composed of elongated living cells with thick but flexible primary cell walls.
  - **sclerenchyma** tissue — confers rigidity and strength to many plant organs and is composed of cells with thickened secondary cell walls; at maturity, sclerenchyma cells are dead.
- **Vascular tissue**, includes the complex tissues:
  - **xylem** tissue — consists mainly of hollow dead cells with thick cell walls hardened by lignin. This tissue transports water and dissolved minerals throughout a plant.
  - **phloem** tissue — consists of living cells that transport sugars, in the form of dissolved sucrose (table sugar), and other organic compounds, including hormones, throughout a plant.

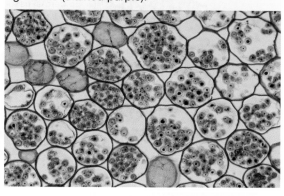

FIGURE 3.4 Cross-section through the root of a buttercup (*Ranunculus* sp.) showing the parenchyma tissue. Note the stored starch granules (stained purple).

Each plant organ (root, stem and leaf) contains all three tissue types. Figure 3.5 shows the typical arrangements of these various tissue types in vascular plants, with the different types of tissue denoted by different colours. Note the dermal tissue (shown in orange) forms a protective outer boundary of the plant's organs.

FIGURE 3.5 Arrangement of various tissue types in the organs of a vascular plant

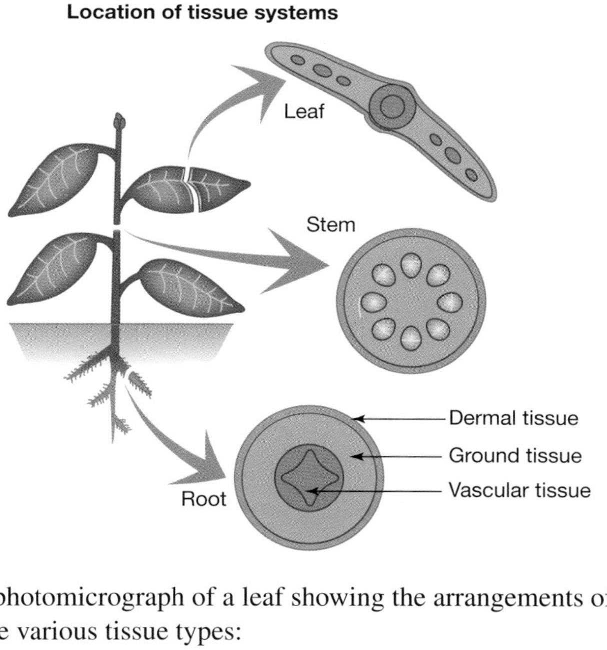

Figure 3.6 is a photomicrograph of a leaf showing the arrangements of different tissues. Note the various tissue types:
- Parenchyma tissue: an example of ground tissue, forms the bulk of the internal structure of the root and the leaf
- Vascular tissue: in the xylem and phloem tissue in both root and leaf
- Epidermal tissue: an example of dermal tissue, forms the upper and lower boundaries of the leaf blade.

ground tissue all plant tissues that are not dermal or vascular

parenchyma thin-walled living plant cells; in leaves the site of photosynthesis and in roots, tubers and seeds the site of stored starch and oils

collenchyma thick, flexible walled cells in plants; the main supporting tissue of stems

sclerenchyma dead cells with thickened walls for strength and rigidity

vascular tissue plant tissue composed of xylem and phloem

xylem the part of vascular tissue that transports water and minerals throughout a plant and provides a plant with support

phloem complex plant tissue that transports sugars and other organic compounds

FIGURE 3.6 Photomicrograph of a cross-section of part of the leaf of the bread wheat (*Triticum aestivum*). Note the parenchyma tissue that forms the bulk of the leaf.

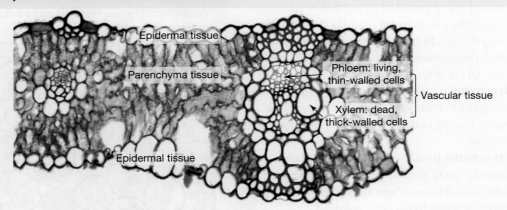

INVESTIGATION 3.1

elog-0254

A closer look at vascular tissue

online only

Aim

To observe vascular tissue under the microscope

3.2.2 Tissues involved in intake of water

The source of water for terrestrial plants is water in the soil. Soil consists mainly of solid mineral particles and, in dry conditions, the spaces between these particles are largely filled with air. After rain, this space is mainly filled with water.

The absorption and uptake of liquid water by plants takes place through **root hairs**. Root hairs are extensions of cells of the epidermal tissue that forms the outer cellular covering of the root (figure 3.7). Water enters the root hairs from the soil solution by osmosis. From the root hair cells, water moves across the cells of the cortex to the xylem in the vascular bundle from where it will be transported as a fluid to all the living cells of the plant.

root hairs extensions of cells of the epidermal tissue that forms the outer cellular covering of the root; responsible for absorption and uptake of liquid water

FIGURE 3.7 Diagram showing the root hairs that are extensions of the epidermal cells. Water (blue lines) from around the soil particles passes into the root hair cells by osmosis.

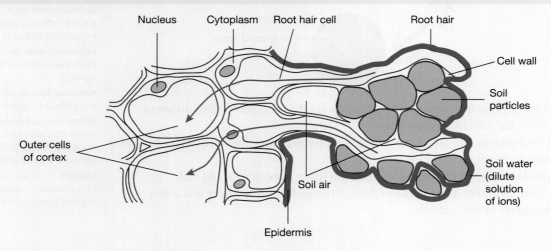

Plant tissues involved in the intake of water

- Uptake of water by plants occurs through osmosis by the root hairs.
- Root hairs are extensions of cells of the epidermal tissue that forms the outer cellular covering of the root.

elog-0256

INVESTIGATION 3.2 — online only

Root hairs

Aim

To observe the growth of roots and investigate the structure and function of root hairs

3.2.3 Tissues involved in movement of water

The xylem tissue is responsible for the transport of water and dissolved minerals taken up by root hair cells to the rest of the plant. Unlike most other plant tissues, xylem tissue contains more than one type of cell and, for this reason, xylem is said to be a complex tissue. Present in xylem tissue are the major water-conducting cells called **tracheids** and **vessels**. Both cell types have thickened cell walls and this adds structural support to plants. Other cells types found in this tissue are fibres and parenchyma cells, with the latter being the only living cells in xylem tissue.

tracheids major water conducting cells in the xylem of all vascular plants

vessels major water conducting cells in the xylem of angiosperms

FIGURE 3.8 a. A simplified drawing of xylem vessels (joined end-to-end) and a cluster of tracheids
b. Longitudinal section of a stem of cottonwood (*Populus* sp.) showing the xylem tissue. The xylem vessels are dead cells with thick walls and lignin thickening in the form of spirals or rings.

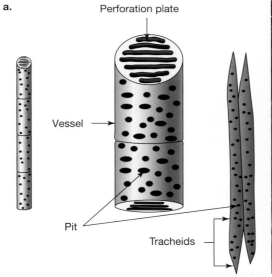

The major water-conduction cells in vascular plants are tracheids and vessels.
- Tracheids:
  - are present in all vascular plants
  - are long thin tubular cells with tapered ends
  - have no cell contents and so are no longer living
  - have both a primary and a secondary cell wall
  - have pits in their secondary cell wall enabling lateral movement of water to nearby tissues.
- Vessels:
  - are found only in flowering plants
  - lose their cell contents and are dead
  - form a continuous wide pipe-like structure by the joining of vessels end-to-end
  - have perforation plates that are typically present at the junction between individual vessels in the pipe
  - have both a primary and a secondary cell wall.
  - The secondary cell walls are made even stronger by thick deposits of lignin that forms spirals or rings (figure 3.8b).
  - The walls are perforated by pits — these are spots where the secondary wall of the vessel is thin or absent so that water can move laterally and exit the xylem.

Xylem tissue

- Xylem tissue is responsible for the movement of water and nutrients in a plant.
- It is composed of tracheids, fibres and parenchyma cells, and, in flowering plants only, vessels.

tlvd-1850

SAMPLE PROBLEM 1 Comparing and contrasting tracheids and vessels

Tracheids and vessels are both cells involved in the movement of water. Compare and contrast these cell types. **(2 marks)**

THINK	WRITE
1. In a compare and contrast question, you need to ensure that you address similarities (compare) and differences (contrast).	
2. Write a statement identifying similarities between the tracheids and vessels.	Both tracheids and vessels are water-conducting cells that are present in xylem tissue, adding structural support (1 mark).
3. Write a statement identifying differences between tracheids and vessels. When identifying differences, ensure you refer to both of the cells.	Tracheids are present in all vascular plants and are long and tubular, whereas vessels are only present in flowering plants and have a wide pipe-like structure (1 mark).

EXTENSION: Growing plants without soil

Hydroponics is a method of growing plants without planting them in soil. Instead, plants are grown with their roots exposed to mineral nutrients in a solution of water.

To access more information on this extension concept please download the digital document.

FIGURE 3.9 Lettuce is a common hydroponic crop.

on Resources

📄 **Digital document** Extension: Growing plants without soil (doc-35878)

3.2.4 Tissues involved in the loss of water

Plants use less than 5 per cent of the water absorbed by roots for cell functions. The remainder simply passes out of plants directly into the atmosphere in a process called **transpiration**. The loss of water from plant leaves by transpiration can be extensive; for example, one estimate is that a large oak tree can transpire more than 150 000 litres in a year. However, the exact amount of water lost by a plant through transpiration is influenced by temperature, relative humidity of the air, wind movements and the availability of water in the soil. Can you suggest what temperature (hot or cold), wind movements (blowing or still) and relative humidity of the air (high or low) would be expected to result in higher transpiration rates?

Water loss by transpiration occurs in the leaves of plants, and the tissues that are involved include:
- the air spaces in the spongy mesophyll tissue of leaves are usually saturated with water vapour
- the **stomata** (singular = stoma) located mainly in the lower epidermis of leaves are the exit points from where water vapour may diffuse out from the leaves into the air.

A loss of water vapour will occur, provided:
- a concentration gradient exists between the water content in the leaf spaces (high) and in the air outside the leaf (low)
- the leaf stomata are open.

> **transpiration** loss of water from the surfaces of a plant
> **stomata** pores, each surrounded by two guard cells that regulate the opening and closing of the pores

FIGURE 3.10 Longitudinal section thorough a leaf showing water (blue arrows) being pulled from xylem vessels (orange) into the mesophyll tissue of the leaf. This input of water replaces the water lost from the leaf as vapour through the open stomata.

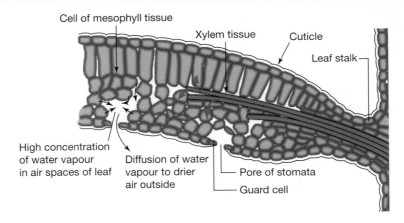

Action of the stomata

Given the extensive water loss that occurs in plants through transpiration, the question might be asked: Why does such an apparently wasteful process of water loss occur? To answer this, you must recall that leaves are the site of the life-sustaining process of photosynthesis. To make sugars during photosynthesis, carbon dioxide is an essential requirement, and this is captured from the atmosphere. So, the entry point of carbon dioxide to a leaf is via the open stomata (figure 3.11). It is not a case of one water molecule out as one carbon dioxide molecule enters a leaf — one estimate is that for each molecule of carbon dioxide taken into a leaf, an average of about 400 water molecules are lost from the leaf. Stopping transpiration means stopping photosynthesis, but when stomata are open, a plant faces the risk of dehydration. However, controls exist over the opening and the closing of plant stomata.

The stomata in the epidermal tissue are pores and each is surrounded by two guard cells. Figure 3.11b shows the inner wall of a guard cell is significantly thicker than its outer wall. In sunlight, when photosynthesis is occurring, guard cells take up water and swell, resulting in the opening of the stomata. This water uptake is triggered by an inflow of potassium ions (K^+) into the guard cells. In the dark, when carbon dioxide is not needed, the reverse reaction occurs: the guard cells lose water and shrink, the stomata close, and water loss is prevented.

FIGURE 3.11 a. Photomicrograph of the epidermal tissue of the lower surface of a leaf where five stomata are visible **b.** Line diagram showing a stoma in the open condition (at left) and the closed condition (at right)

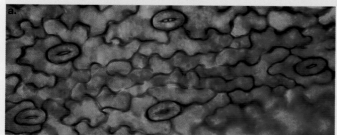
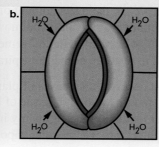
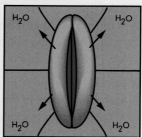

Stomata

- The stomata are the sites of carbon dioxide uptake and loss of water in a plant.
- Stomata are mainly located in the lower epidermis of leaves.
- Water is lost as vapour when a concentration gradient exists between the water content in the leaf spaces (high) and in the air outside the leaf (low) and the leaf stomata are open.

elog-0258

INVESTIGATION 3.3 — online only

Leaf epidermal layer

Aim

To observe guard cells, stomata and epidermal cells in plants under the microscope

Getting to the top

Plants move water through their xylem tissue from delicate root hairs to the highest leaves in a plant; in the case of trees, this can be quite a height (figure 3.12). How is this achieved? Osmosis certainly can't do it.

The energy provided by the Sun plays a role in this process. When the Sun shines, stomata open and water vapour moves through the stomata into the air surrounding leaves. This water is replaced by water that moves out of the xylem into a leaf, and the water column remaining in the xylem moves up. In effect, water vapour moving out through the stomata sets up a chain reaction in which water in the xylem is moved up through the xylem vessels by the pulling or sucking movement of water ahead of it. This pulling action means that *water in xylem vessels is under tension that affects the entire water column*. The journey of water molecules from the root hairs via the xylem to the leaves depends on intact continuous columns of water in the xylem tracheids and vessels. This journey is shown in figure 3.13, starting with the uptake of water by root hair cells, continuing with this water being transported via the xylem to the leaves and then diffusing as water vapour via the stomata to the air.

FIGURE 3.12 Water moves through xylem tissue from roots to the highest leaves of plants.

FIGURE 3.13 Water is absorbed from the soil by roots and passes into xylem tissue, which carries it to all parts of the plant. Some of the water is used by cells but much passes through stomatal pores in leaves and is lost by transpiration.

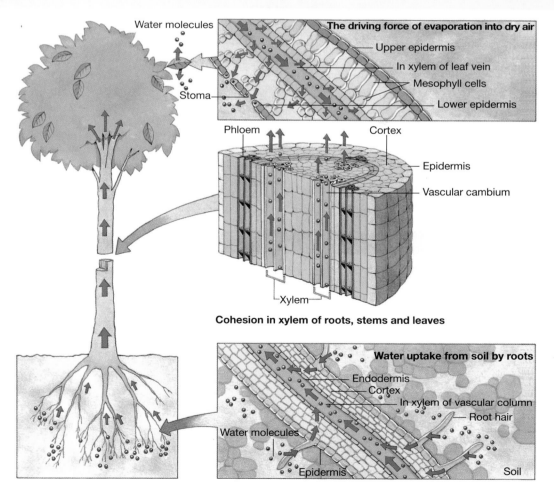

TOPIC 3 Functioning systems

Why does the water column in the xylem not break? Two properties of water molecules help keep the water column unbroken. Water molecules tend to stick together — a property called **cohesion**. Water molecules also tend to stick to the walls of a container — a property called **adhesion**. These combined forces prevent the water column from breaking apart and from pulling away from the walls of the xylem vessels as water is sucked up the many thin tubes in the xylem tissue (figure 3.14). If the water column in one set of xylem tubes breaks, the water is replaced by air and that set of tubes becomes useless for any further water transport.

FIGURE 3.14 Movement of an unbroken column of water through the xylem vessels. This movement is produced by tension in the water column that is kept unbroken by the cohesion and adhesion of water molecules. Blue arrows denote the direction of movement of the water column towards the leaves, and the yellow symbols denote the perforation plates that separate individual xylem vessels.

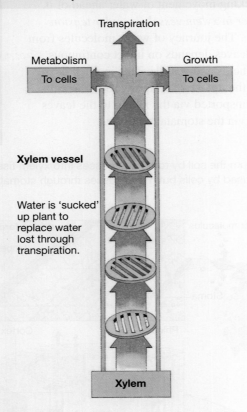

cohesion tendency of water molecules to stick together through the formation of hydrogen bonds between one another

adhesion attraction of water molecules to other kinds of charged molecules, such as the walls of xylem tubes

INVESTIGATION 3.4

online only

Water movement in the plant

Aim
To observe the movement of water through the xylem in celery and carrots

Resources

eWorkbook   Worksheet 3.1 Systems, organs and tissues in plants (ewbk-3486)

KEY IDEAS

- Plants have two systems: a root system and a shoot system.
- Root system organs (root, lateral roots and root hairs) are responsible for anchoring the plant in the soil, absorbing water and minerals and storing starch.
- Shoot system organs (stem, leaves, flowers and fruit) are responsible for photosynthesis, transport of sugars and water, and reproduction.
- Meristematic tissue can undergo cell division and permanent tissue are cells that can no longer divide.
- Permanent tissue includes dermal tissue, ground tissue and vascular tissue.
- Dermal tissue, such as the plant epidermis, has functions in protection and waterproofing.
- Ground tissues are parenchyma, collenchyma and sclerenchyma and their various functions include photosynthesis, storage, support and protection.
- Dermal tissue and ground tissue are simple tissues, each composed of one cell type only.
- Vascular tissue includes the complex tissues of xylem and phloem, each composed of several different cell types.
- Plant organs are structured with an outer layer of dermal tissue, and inner vascular tissue that is embedded in ground tissue.
- Intake of water and minerals occurs through root hairs by osmosis.
- Root hairs are thin extensions of the epidermal cells that form the outer covering of the root.
- Water and dissolved minerals are transported throughout a plant through the xylem tissue.
- Water loss occurs via transpiration of water vapour through the open stomata of leaves.
- Water loss occurs when the concentration of water vapour within the leaf spaces is higher than that in the air outside the leaves.
- The opening and closing of stomata is controlled by the guard cells that enclose the stomatal pores.
- Transpiration of water from leaves produces the suction or tension in water columns in xylem that can move water up the xylem to the top of trees.
- The cohesive and adhesive properties of water prevent the water columns in xylem breaking.

3.2 Activities

learn

To answer questions online and to receive **immediate feedback** and **sample responses** for every question, go to your learnON title at **www.jacplus.com.au**. A **downloadable solutions** file is also available in the resources tab.

| 3.2 Quick quiz | 3.2 Exercise | 3.2 Exam questions |

3.2 Exercise

1. Explain why root hairs are important to the water transport system of a vascular plant.
2. Describe the mechanism by which water reaches the tops of tall trees.
3. Identify where in a vascular plant you would expect to find:
   a. tracheids
   b. stomata
   c. meristematic tissue
   d. starch granules.
4. Explain the following observation: The epidermis of a root lacks the waterproof cuticle found on the epidermis of most leaves.
5. Identify:
   a. one biological similarity and one difference between collenchyma and sclerenchyma tissue
   b. the mechanism by which water enters root hair cells
   c. the structure that enables water to move laterally out of the water column in xylem tissue.
6. Briefly describe the arrangement of the major types of permanent tissue in a vascular plant.
7. Identify the expected effect — increase or decrease — on the rate of transpiration in a plant of the following:
   a. an increase in the humidity in the air
   b. an increase in wind speed.
   c. the closing of the stomata.

3.2 Exam questions

Question 1 (1 mark)

MC Functional dead cells observed in the vascular tissue of a living plant may be
A. sieve tube elements.
B. parenchyma cells.
C. xylem vessels.
D. companion cells.

Question 2 (2 mark)

Describe functional differences between xylem tissue and phloem tissue.

Question 3 (1 mark)

MC The movement of water through a plant
A. requires the plant cells to use the energy in ATP.
B. relies on the forces of attraction between water molecules.
C. slows down when the environment temperature increases.
D. takes place in the phloem tissue.

Question 4 (4 marks)

Consider the following photograph of a cross-section of the stem of a vascular plant.

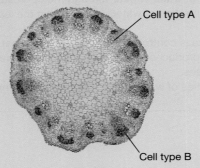

Name cell type A and describe how the cell is different to cell type B.

Question 5 (4 marks)

In 2010 and 2013 vandals ringbarked a eucalyptus tree referred to as the 'The Separation tree' growing in the Melbourne Botanical Gardens. Ringbarking removes the bark and phloem of the tree.
a. What you would predict to be the fate of the tree in the following years? **1 mark**
b. Explain your prediction by referring to the functioning of the plant. **3 marks**

More exam questions are available in your learnON title.

3.3 Specialisation and organisation of animal cells

KEY KNOWLEDGE

- Specialisation and organisation of animals cells into tissues, organs and systems

Source: Adapted from VCE Biology Study Design (2022–2026) extracts © VCAA; reproduced by permission.

3.3.1 Cellular level of organisation

Like plants, multicellular animals are arranged at the levels of cells, tissues and organs. Animals with the most simple level of organisation are the sponges, members of phylum *Porifera* (figure 3.15a). Sponges have a structure that is organised at the cellular level — they possess many cell types, but sponges have no definite tissues, organs or systems. Sponges consist of two layers of cells that enclose a space, the spongocoel, that has a single main opening. A non-living gel-like material (mesoglea) separates the two cell layers (figure 3.15b) and a few mobile cells may be present there. Among the specialised cell types in sponges are collar cells that beat their flagella and create incoming water currents that enter the sponge via cells called porocytes. In one study, it was shown that a 10-centimetre-tall sponge filtered about 100 litres of water each day. Tiny plankton and floating organic particles in the water drawn into the sponge are its food and this is digested by enzymes produced by cells of the inner cell layer.

The simplicity of the structure of sponges is illustrated by the fact that if the sponge cells are disaggregated and kept alive, within weeks, the different cell types will re-aggregate into the original structure of the sponge.

FIGURE 3.15 a. A living sponge. Note the single large opening that leads into an internal central space or enteron. Water and wastes both exit via this large opening. **b.** A longitudinal section of the wall of a sponge showing its two cell layers separated by non-living mesoglea. Note that the water enters the enteron via special cells called porocytes.

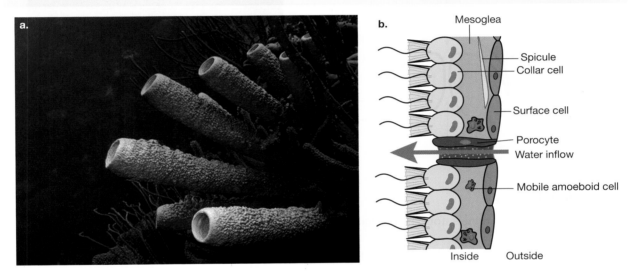

3.3.2 Tissues in animals

Tissues are formed by groups of cells of similar type — or even a single type — that act in a coordinated manner to perform a common function.

When did the first tissues appear in animals? The first animals to show the tissue level of organisation were the ancestors of jellyfish, corals, sea anemones and hydras that are members of phylum *Cnidaria*. Figure 3.16a shows the basic body plan of a jellyfish that consists of two cell layers, an outer ectoderm (shown in bright pink) and an inner endoderm (shown in blue). These two cell layers are separated by a non-cellular, gel-like mesoglea (yellow). Jellyfish have a single opening that serves as both the entry point for food to the gut and as the exit point for wastes.

Jellyfish have a number of tissues:
- **epithelial tissue**, which forms the ectoderm that lines the outside of the body
- **gastrodermal tissue**, another epithelial tissue, that lines the inner cavity of the body, called the enteron; cells lining this cavity produce digestive enzymes
- nervous tissue in the form of a network of nerve cells in the outer **ectoderm** that can respond to sensory and motor inputs — some researchers liken this to a nervous system

epithelial tissue sheets of cells that cover the external surface of the body and also line internal surfaces that connect to the external environment

gastrodermal tissue epithelial tissue that lines the inner cavity of the body

ectoderm epithelial tissue that covers the outer surface of the body

- muscle tissues that enable jellyfish to contract their tentacles and pulsate their bodies; they do this by rhythmically opening and closing their bell-shaped bodies.

Figure 3.16b shows one species of box jellyfish (*Chironex fleckeri*), commonly called the sea wasp, which is considered to be the most venomous animal on this planet. Along the length of its tentacles, the box jellyfish has very large numbers of stinging cells, each of which can discharge a microscopic harpoon-like structure that injects venom into its prey. Contact with two metres or more of the tentacles of this box jellyfish may cause death within a few minutes if the victim is not treated with anti-venom. *C. fleckeri* is distributed in coastal waters in northern Australian, ranging from Agnes Water in Queensland to Exmouth in Western Australia.

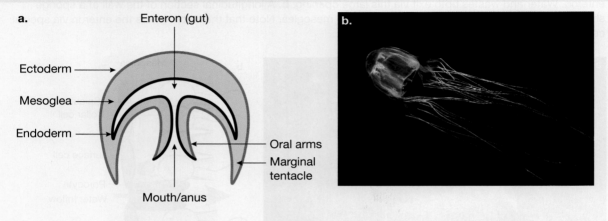

FIGURE 3.16 a. Body plan of a typical jellyfish, note the tentacles are not shown **b.** Photo of a box jellyfish (*Chironex fleckeri*), which has tentacles up to three metres long that are covered in millions of stinging cells

Mammalian tissues

In mammals, as in other animals, four major kinds of tissues are recognised: epithelial, muscle, connective and nervous tissues.

1. *Epithelial tissues*. These are sheets of cells that cover external and internal surfaces — for example, the outermost layer of your skin and the internal lining of tubes and cavities within the body that connect to the external environment, such as the lining of the bladder, lungs and alimentary canal. Epithelial tissues also form the secretory part of glands and their ducts. The functions of these tissues include protection (e.g. skin), absorption (as in the small intestine) and secretion (e.g. glands).
2. *Muscle tissues*. These tissues contract and enable movement — for example, the cardiac muscle tissue of your heart (figure 3.17a), the skeletal (also termed striated) muscles of your arms and legs, and the smooth muscles within the wall of your gut and the walls of your arteries.
3. *Connective tissues*. These provide structural support — for example, the loose **connective tissue** that holds the outer layers of the skin to the underlying muscle layers and the fibrous connective tissue of your bones and cartilage. These tissues also act as energy stores — for example, adipose tissue stores fat (figure 3.17b); and connective tissues transport substances within the body, such as the fluid tissue of your blood.
4. *Nervous tissues*. These tissues are made of different kinds of nerve cells (neurons), such as motor neurons and sensory neurons. Nervous tissue is found in the brain and the spinal cord of the central nervous system and in structures of the peripheral nervous system. An example of nervous tissue is the retina of your eye.

connective tissue diverse solid tissues that connect and support other tissues and organs, or store fat deposits, and fluid tissues that transport materials such as nutrients, wastes and hormones

FIGURE 3.17 a. Cardiac muscle tissue responsible for the pumping action of the heart. The purple ovoid shapes are nuclei. **b.** Fat or adipose tissue — an example of connective tissue. The white areas that are fat globules that occupy most of the volume of the fat cells, which are called adipocytes.

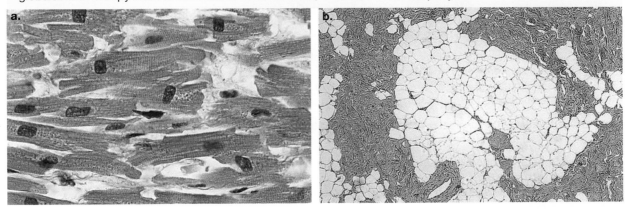

The further classification of these four major tissue groups is shown in figure 3.18.

FIGURE 3.18 A classification scheme of the four major tissue types in mammals. What type of tissue is cartilage?

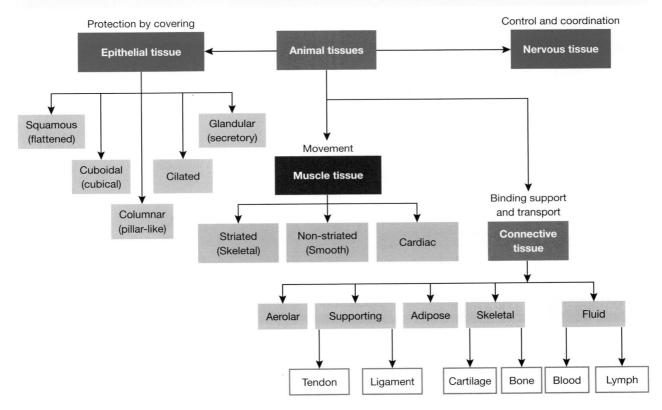

CASE STUDY: Structure and function of connective tissue

Connective tissue is the most abundant tissue type and it is widely distributed throughout the body. The various kinds of connective tissue include areolar tissue, bone, cartilage, ligament, fat tissue and blood.

Connective tissue has three main components: cells, fibres and ground substance.

The ground substance is non-cellular and is composed of proteins, polysaccharides and extra-cellular fluid. It may be solid as in bone and cartilage, or viscous as in the gut, or liquid as in blood. The non-cellular fibres include collagen fibres and elastic fibres. Cells present in the matrix include fibroblasts, adipocytes (fat cells) and stem cells; the fibroplasts produces the ground substance.

Connective tissues perform various functions: they bind and support, they protect and insulate, they store reserve energy as fat, and they transport materials.

- Loose (**areolar**) connective tissue provides structural support and binds the cells of tissues and organs together (figure 3.19).
- Dense connective tissues of bone and cartilage provide structural support for the body.
- Fluid connective tissues, blood and lymph, are involved in transport of substances — water, nutrients, wastes and hormones.
- Adipose connective tissue stores fat, an energy reserve.

areolar loose, irregularly arranged connective tissue

FIGURE 3.19 Microscopic views of **a.** areolar connective tissue. Note the nuclei of fibroblasts (red dots), thin elastic fibres and thicker collagen fibres (both pink). **b.** Human cartilage showing chondrocytes (cartilage cells) in a dense ground substance (light purple) that is rich in chondroitin sulfate.

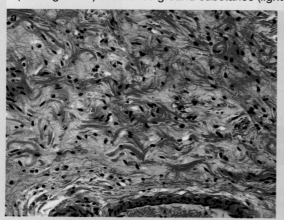
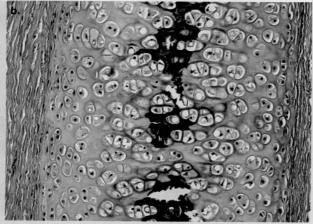

SAMPLE PROBLEM 2 Tissue types in mammals

tlvd-1851

Some of the earliest mammals were shrew-like creatures from the *Morganucodon* genus. Like other mammals, they possessed four main types of tissues.

a. Identify each of these tissue types. (1 mark)
b. Select two of these tissue types. Outline the function and structure of each of these tissues. (4 marks)

THINK

a. This question is asking you to identify. When you need to identify, you just need to state the relevant term.
As this is only worth one mark, you need to identify all four tissues types for the single mark.

WRITE

- Epithelial tissue
- Muscle tissue
- Connective tissue
- Nervous tissue (1 mark)

b. This question is asking you to outline. This requires a brief summary statement. In this case, you need to outline both the function and structure of two tissues. As this is worth 4 marks, it is one mark for each point. Outline the structure and function, clearly showing each part of your answer.	Epithelial tissue: The *structure* of the epithelial tissue involves sheets of cells covering surfaces (1 mark). The *function* of these cells is for protection, absorption and secretion (1 mark). Muscle tissue: The *structure* of muscle tissue involves tissue and cells that are able to contract (1 mark). The *function* of muscle tissue is to enable movement and contraction, such as in skeletal, cardiac and smooth muscle (1 mark). *Note:* you may have selected connective or nervous tissue instead in your response.

3.3.3 Organs in animals

An organ is a group of different kinds of tissue grouped together to form a discrete structure that works cooperatively to perform a specific function.

The first animals to show organisation at the organ level are flatworms of the phylum *Platyhelminthes*. These animals include free-living marine flatworms (figure 3.20) and parasitic tapeworms and flukes. In contrast to jellyfish with their two-layer body plan, flatworms were the first to have a three-layer body plan: outer ectoderm, inner endoderm and, between them, a cellular mesoderm. Different tissues in flatworms are arranged into organs, as, for example, the brain and the excretory organs (nephridia).

FIGURE 3.20 Marine flatworm (*Pseudobiceros gloriosus*). Its common name is the glorious flatworm. Ancestors of the flatworms were the first animals to show a three-layer body plan.

Organs in mammals

Heart, lungs, stomach, kidneys, liver and eyes — these are just some of the organs in not only the human body, but also across a range of organisms. An organ is composed of groups of different tissues that work cooperatively to perform a specific function — the heart pumps blood, the stomach churns and digests food, the kidneys excrete nitrogenous wastes and the eye receives visual signals that are passed to the brain for processing.

The photomicrographs of some organs in figure 3.21 show the different types of tissue that are present in each organ:
- Trachea (windpipe): The cross-section of this organ shows several tissues including ciliated epithelial tissue and cartilage tissue (figure 3.21a).
- Eye: The cross-section of the eye shows the multiple layers of nervous tissue that forms the retina and, below the retina, the connective tissue of the choroid of the eye, including dilated blood vessels, can be seen (figure 3.21b).
- Skin: The section of the skin reveals the epithelial tissue, both dead and living, that forms the outer lining of the skin and the connective tissue of the dermis that lies below it (figure 3.21c).

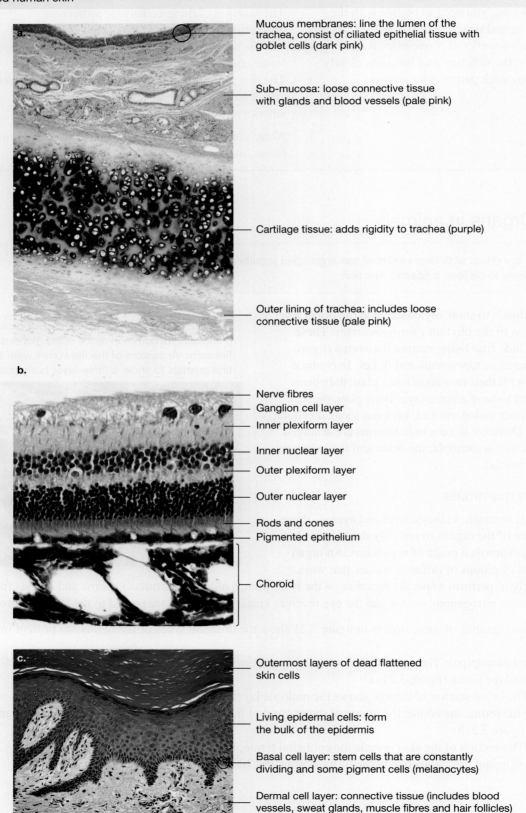

FIGURE 3.21 Photomicrographs showing different tissues that form the structure of various mammalian organs. **a.** Cross-section of the trachea (windpipe) **b.** Cross-section of the back wall of the eye **c.** Longitudinal section of stained human skin

3.3.4 Systems in animals

> A system is composed of a group of organs that cooperate to carry out a single life-sustaining function, such as excretion or digestion.

The flatworms of the phylum *Platyhelminthes* not only showed the first organ level organisation, but also the first system level organisation. Flatworms have a nervous system that includes a brain, eyespots and two ventral nerve cords (figure 3.22a). Free-living flatworms have a simple digestive system composed of several different organs, including a pharynx and a highly branched digestive tract (figure 3.22b). (Parasitic flatworms, such as tapeworms and flukes, have no digestive system, so how do they survive?) Flatworms also have organs that form an excretory system and a reproductive system, but they have no respiratory, skeletal or circulatory systems.

FIGURE 3.22 Diagram of a flatworm showing **a.** the nervous system **b.** the digestive system **c.** cross-section showing the three cell layers that form all its tissues, organs and systems. Within the mesoderm are various structures, including muscle fibres and gonads.

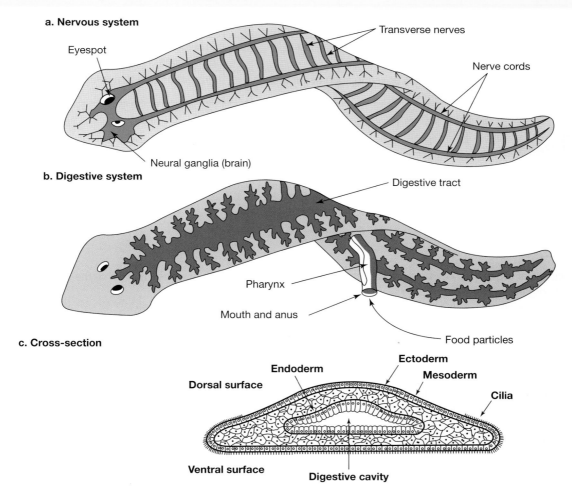

The mammalian systems and a brief summary of the major functions that they enable are shown in table 3.1.

In the following sections of this topic, we will look in more detail at the digestive, endocrine and excretory systems. It is important to recognise that systems do not operate in isolation. Instead, to carry out its major function, a system depends on the operation of other systems; for example, the kidneys of the excretory system cannot remove N-containing wastes from the body unless the circulatory system transports them to the kidneys; the lungs cannot excrete carbon dioxide unless it is transported from body cells to the lungs by the circulatory system.

TABLE 3.1 Summary of mammalian systems, some organs and tissues that they contain and their functions

System	Organs and tissues	Major function
Digestive	Oesophagus, stomach, pancreas, duodenum, liver, intestines	Breakdown of food to small products for absorption
Circulatory	Heart, arteries, veins, capillaries and blood	Transport of nutrients to cells and wastes from cells
Endocrine	Hypothalamus and glands (e.g. pituitary, thyroid, adrenal)	Secretion of hormones that regulate cells and organs
Respiratory	Larynx, trachea, bronchi and lungs	Input of oxygen to body and removal of carbon dioxide
Excretory (urinary)	Kidneys, ureters, bladder and urethra	Removal of nitrogenous wastes and compounds in excess from the body
Nervous	Brain, spinal cord, sensory organs (e.g. eye, ear, tongue)	Input of information from senses and coordination of activity
Musculo-skeletal	Muscles, bones, ligaments and tendons	Supports body and enables movement
Reproductive	Ovaries, uterus, vagina testes, penis	Production of gametes, fertilisation and, in females, gestation
Immune	Bone marrow, spleen, thymus	Protection against foreign bodies and pathogens
Integumentary	Skin, and covering such as fur and hair, and nails	Protection and temperature regulation

 Resources

 eWorkbook Worksheet 3.2 Systems, organs and tissues in animals (ewbk-3488)

KEY IDEAS

- Multicellular organisms have specialised cells used to perform different functions that serve the needs of the whole organism.
- Multicellular animals may show organisations at the levels of cell, tissue and organ.
- Sponges show the simplest level of organisation — the cellular level.
- Tissues are formed by groups of cells of similar or single type that act in a coordinated manner to perform a common function.
- Jellyfish, corals, sea anemones, hydra (phylum *Cnidaria*) have a two-layer body plan and show tissue level organisation.
- Major tissue types are epithelial, muscle, connective and nervous tissues.
- An organ is a group of different kinds of tissue grouped together to form a discrete structure that works cooperatively to perform a specific function.
- Flatworms have a three-layer body plan and show the first organ level organisation.
- A system is composed of a group of organs that cooperate to carry out a single life-sustaining function, such as excretion or digestion.
- Systems do not act in isolation but are dependent on each other.

3.3 Activities

To answer questions online and to receive **immediate feedback** and **sample responses** for every question, go to your learnON title at **www.jacplus.com.au**. A **downloadable solutions** file is also available in the resources tab.

3.3 Quiz quiz	3.3 Exercise	3.3 Exam questions

3.3 Exercise

1. Consider the following organisms: sponge, jellyfish and flatworm.
   a. Identify, giving a brief reason for your choice, which has the most complex organisation.
   b. Identify, which, if any, has a digestive system and identify whether it is a one-way or two-way system. Explain your answer.
   c. What is the body plan structure, in terms of cellular layers, of each of these animals?
2. Animal M has tissues, but no organs. What kind of animal might it be? Explain your answer.
3. Give an example of a tissue in animals:
   a. that forms a lining
   b. that is an energy store.
4. Identify an organ in which you would expect to find the following:
   a. cartilage tissue
   b. epithelial tissue with outer layers of dead flattened cells
   c. a layer of photoreceptor cells.
5. By which two features can epithelial tissues be classified?

3.3 Exam questions

Question 1 (1 mark)

MC A group of similar cells that perform the same function is referred to as
A. a tissue.
B. an organ.
C. a body system.
D. an organism.

Question 2 (1 mark)

MC An example of an organ in an animal is
A. a muscle cell.
B. a group of epithelial cells.
C. bone tissue.
D. a stomach.

Question 3 (1 mark)

MC Mammalian cells capable of carrying oxygen include
A. platelets.
B. white blood cells.
C. red blood cells.
D. macrophages.

Question 4 (4 marks)

Name four human tissue types and describe the function of each type.

> **Question 5 (2 marks)**
> Consider the following diagram that shows several different cell types.
>
>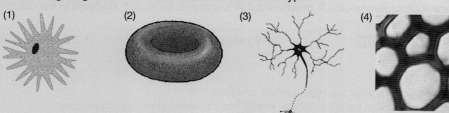
>
> Choose which cell type would be most suited to transmitting messages in an organism. Explain your choice.
>
> **More exam questions are available in your learnON title.**

3.4 Digestive system in animals

KEY KNOWLEDGE

- Specialisation and organisation of animals cells into tissues, organs and systems with specific functions: digestive

Source: Adapted from VCE Biology Study Design (2022–2026) extracts © VCAA; reproduced by permission.

3.4.1 What is the digestive system?

All animals require a source of chemical energy for living and a supply of organic molecules that are essential for both their structure and their function. The source of chemical energy and organic molecules for animals is their food. Animals exploit a varied range of food sources – paper for silverfish, wool for the larvae of clothes moths, grass and hay for cattle, blood for fleas, skin cells shed by people for dust mites, and eucalyptus leaves for koalas (see figure 3.23).

FIGURE 3.23 Koalas (*Phascolarctus cinereus*) are mainly limited to eating the leaves from 40 to 50 of the 900 species of eucalyptus trees.

> To be food for an animal, a substance must contain organic matter that can be broken down and absorbed by the animal and be used to supply it with chemical energy for living, and the organic matter to build and repair its own structures.

Why is a digestive system important for animals? With few exceptions, food that animals ingest is mainly in the form of macromolecules, such as proteins, carbohydrates and lipids (fats). These large molecules must be broken down to small sub-units or monomers before they can be absorbed into the internal environment of an animal's body for use in energy production, and as building blocks for growth, maintenance and repair. That's where the breakdown crew, the digestive system, comes in! The digestive system breaks down large food molecules, chemically and physically, into sub-units small enough to be absorbed into the body. It should be noted that small water-soluble vitamins and minerals that are part of an animal's diet are already small enough so that they do not need to be digested.

The digestive system breaks down macromolecules: proteins, carbohydrates, lipids (fats) into smaller sub-units that can be absorbed and used for energy production, growth, maintenance and repair.

In terms of structure, the digestive system of almost all animals — except sponges, jellyfish and flatworms — is an open hollow tube that takes in food at one end, the mouth, and releases wastes at the other end, the anus. The least complex animals with this one-way digestive system are nematode worms, also called roundworms, of the phylum *Nematoda* that includes both free-living and parasitic species. The parasitic nematodes include heartworms that infect dogs, and pinworms and hookworms that can infect small children. The digestive system of nematodes is a tube that consists of mouth, pharynx, intestine, rectum and anus.

Now let's look at the vertebrates. In terms of structure, the digestive system is made up of two components:

1. The **alimentary canal**, also called the **gastrointestinal tract** (GI tract) or the gut, is an open tube extending from mouth to anus that varies in diameter along its length, and consists of a series of hollow organs including the oesophagus, the stomach and the small intestine.
2. Accessory organs: these are solid organs that include the salivary glands, the liver and gall bladder, and the pancreas. These solid organs release secretions, including enzymes, that are delivered via ducts into the lumen of the alimentary canal.

alimentary canal the whole passage from mouth to anus; see gastrointestinal tract

gastrointestinal tract the whole passage from mouth to anus; see alimentary canal

Accessory organs are essential partners with the alimentary canal in achieving the functions of the digestive system. Refer to figure 3.24, which shows a stylised alimentary canal with its accessory organs (black text) that together form the digestive system (white text).

FIGURE 3.24 A stylised representation of the vertebrate digestive system as a one-way passage for the processing of food

int-3030

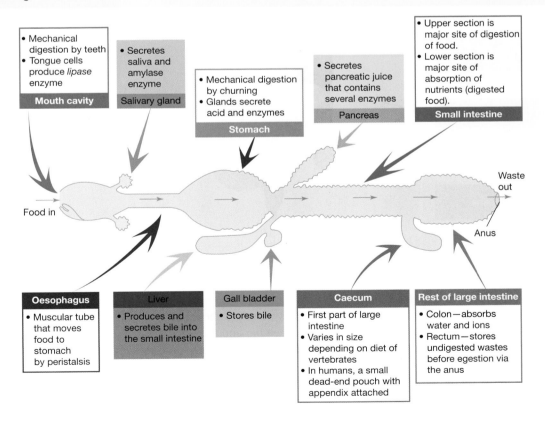

However, in vertebrates, the digestive system is not arranged in a straight line within the body. Because of their length, parts of the alimentary canal, such as the small and the large intestines, are coiled or bent in order to be accommodated with the body. This is not surprising, since the total length of the alimentary canal in an adult person is about 9 metres, and most of this involves the small intestine with a length ranging from 3 to 5 metres. The small intestine is not small in length, but it is small in diameter, almost 3 centimetres. The large intestine is about 1.5 metres in length, with a diameter of nearly 5 centimetres. Figure 3.25 shows the major parts and the relative positions of the hollow and solid organs of the human digestive system. Note that the liver has been folded back to show the stomach, pancreas and gall bladder. In reality, the liver is the largest organ within the body and it covers these organs. We will explore the organs of the digestive system in more detail later in this subtopic (section 3.4.4).

FIGURE 3.25 Line diagram showing the hollow organs of the alimentary canal (GI tract), starting from the mouth and ending at the anus, and its accessory solid organs, the salivary glands, liver and gall bladder, and pancreas

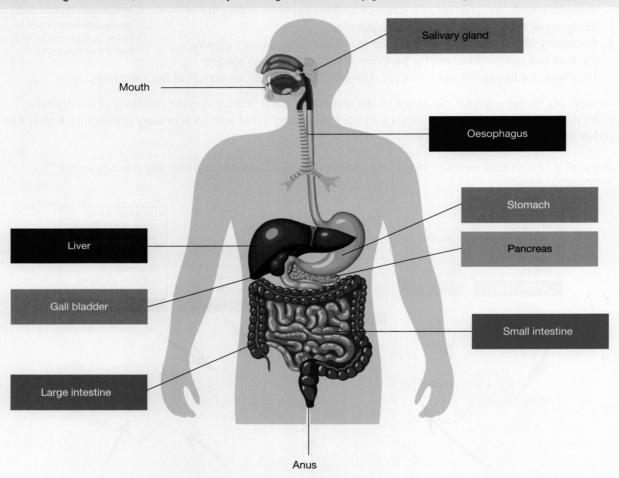

In terms of function, the key activities of the digestive system are shown in table 3.2. **Mechanical digestion** is achieved through chewing and through the churning action of the stomach. Breaking down food into smaller pieces produces an increase in its surface area. As a result, chemical digestion through enzyme action occurs more rapidly on these small food pieces than would occur if the pieces of food were larger. **Chemical digestion** is achieved through the action of enzymes.

> **mechanical digestion** digestion that uses physical factors such as chewing with the teeth
>
> **chemical digestion** the chemical reactions changing food into simpler substances that are absorbed into the bloodstream for use in other parts of the body

TABLE 3.2 The processes of the human digestive system that make the energy and nutrients in food available for use

Process	Where it occurs	Result
Ingestion	Mouth	Food is taken into the mouth.
Mechanical digestion of food	Begins in mouth; continues in stomach	Food is broken down into small pieces, increasing its surface area.
Chemical digestion of food	Begins in mouth; continues in stomach and small intestine	Macromolecules of food are broken down to smaller and smaller sub-units through the action of enzymes secreted by various glands.
Absorption of digested food	Small intestine	End products of digestion cross the tissue layer that lines the gut and enter the internal environment of the body.
Elimination of undigested wastes	Rectum and anus	Storage and elimination as faeces

Let us now look at the tissues that are part of the digestive system and become aware of how these tissues enable the digestive system to perform its essential functions.

3.4.2 Tissues in the digestive system

The width of the tubular canal of the digestive system differs along its length — it is widest at the stomach and narrowest at the oesophagus. However, the wall of the tubular canal has a similar tissue organisation along its entire length, with the wall of the gut having the same four concentric tissue layers.

Four tissue layers form the gut wall, starting from the inside of the gut cavity or **lumen**. The same four tissue layers form the wall of all the hollow organs of the digestive system as follows:

- Epithelial tissue forms the innermost lining of the digestive system, and it is part of the first tissue layer that is called the **mucosa**. The remainder of this layer is tissue that underlies and supports the epithelial tissue.
- Connective tissue forms the second layer, and it includes blood vessels, lymphatic vessels and nerves; this tissue layer is termed the **sub-mucosa**.
- Muscle tissue forms the third layer; this layer is called the **muscularis**.
- Connective tissue forms the fourth and outside layer that encloses the gut; this tissue layer is called the **serosa**.

lumen the inside space of a tubular structure

mucosa the innermost lining of the digestive system

sub-mucosa connective tissue forming the second layer of the gut lining

muscularis muscle tissue of the gut

serosa outer connective tissue which encloses the gut

Figure 3.26 shows a stylised version of the basic pattern of the four tissue layers of the gut wall.

FIGURE 3.26 The stomach, like all the hollow organs of the digestive system, has the arrangement of tissue layers as shown in inset.

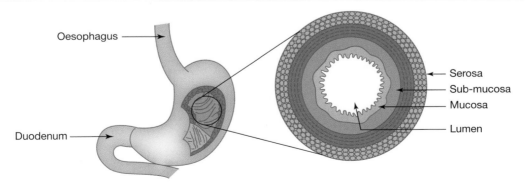

Don't worry about the official anatomical names of these tissue layers. They are included here because they are routinely used as labels on diagrams and other images, and they are a convenient shorthand. It is easier to write 'mucosa' than 'the innermost layer of the gut wall' and 'muscularis' rather than 'a muscle layer consisting of two rings of smooth muscle, the inner ring in a circular arrangement and the ring arranged longitudinally'.

Tissue layers of the digestive tract

The four layers of tissue that form the digestive system are the muscosa (epithelial tissue), sub-muscosa (connective tissue), muscularis (muscle tissue) and serosa (connective tissue).

Variation in the basic pattern

Different regions of the digestive system are involved in different functions: moving food, churning the gut contents, secretion, digestion and absorption. So, although the basic arrangement of tissue layers in the gut wall is the same throughout the gut, variations exist between different regions of the gut, such as thickness of tissue layers, presence (or absence) of features such as villi (outfoldings) or crypts (infoldings), as well as the specific cell types present in the tissue. In particular, the epithelial tissue of the mucosa that is in direct contact with the gut contents shows some striking regional differences.

Let's now look at a photomicrograph that shows these tissue layers in one segment of the gut. Figure 3.27 is a photomicrograph of a cross-section through part of a mammal's small intestine, the jejunum. The four tissue layers are clearly visible. In the small intestine, the epithelial tissue is a single layer of cells that forms the outer surface of the mucosa that is in direct contact with the contents of the gut. Note the extensive folding of the surface of the mucosa to form narrow deep infoldings (called crypts) and finger-like projections (called villi; singular = villus).

FIGURE 3.27 Photomicrograph of a section through the wall of the small intestine in a mammal. Note the four tissue layers that form the wall. In the small intestine, the epithelial tissue of the mucosa forms the gut lining; it is just one cell thick, but it is folded, forming villi and crypts.

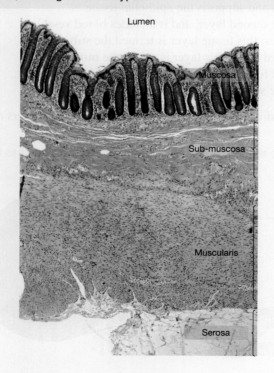

Epithelial tissues: Various types

Some facts about the epithelial tissue in the gastrointestinal tract:
- Epithelial tissue is undergoing constant renewal. Cells are shed in large numbers from epithelial surfaces into the gut lumen, but are replaced just as fast by mitosis of epithelial stem cells. It has been estimated that the epithelial lining of the stomach and that of the small intestine are regenerated every four to five days.
- Epithelial tissue can be classified in terms of the general shape of its cells and also the number of cell layers present.
  - In terms of shape, common cell shapes are squamous, cuboidal and columnar (figure 3.28a).
  - In terms of number of layers, an epithelium can be simple or stratified.
- Simple epithelium is composed of a single layer of cells.
- Stratified epithelium is composed of two or more layers of cells (figure 3.28b).
- Epithelial tissues can be named in terms of these features so, for example, we can talk about 'simple columnar epithelial tissue' or 'stratified squamous epithelium'.
- Different types of epithelial tissues are found in the wall of the alimentary canal, for example:
  - in the oesophagus, the epithelium is stratified and squamous
  - in the stomach, the epithelium is simple columnar and smooth with many infoldings (crypts)
  - in the small intestine, the epithelium is simple columnar with many outfoldings (villi) and infoldings (crypts).
- Epithelial tissues can also be classified in terms of whether they form a lining (lining epithelium) or whether they form a gland — that is, a solid structure below the surface that secretes substances (glandular epithelium).
  - the innermost lining of the GI tract is an example of a lining epithelium
  - the **gastric glands** of the stomach and the intestinal glands (crypts) of the small intestine are examples of glandular epithelium.
- Epithelial tissue itself contains no blood vessels. The epithelial cells obtain their essential requirements by diffusion from blood vessels in the underlying connective tissue of the sub-mucosa.

gastric glands glands of the stomach that contain various epithelial secretory cells, producing neutral mucus, stomach acid and enzymes

FIGURE 3.28 a. Types of simple epithelial tissue that are composed of a single layer of cells **b.** Two types of stratified epithelial tissue that is formed by two or more layers of cells

a.

Simple squamous Simple cuboidal Simple columnar

b.

Stratified squamous Stratified cuboidal

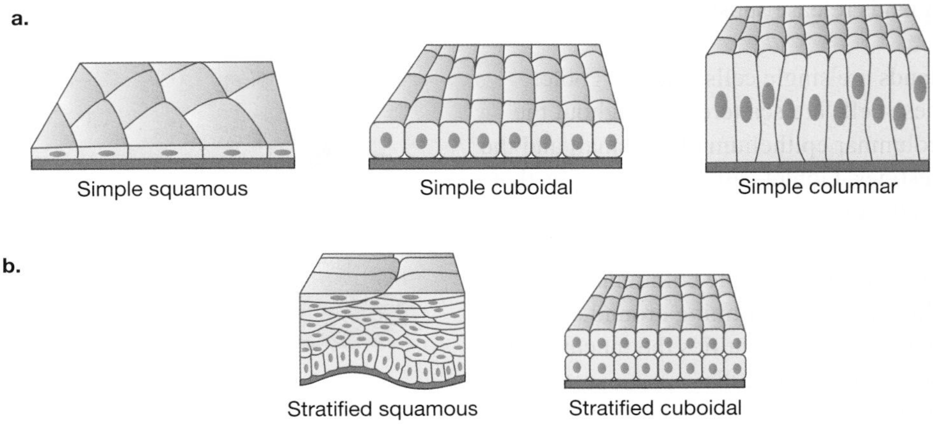

Epithelial tissue: Functions in the digestive system

The functions of the epithelial tissue of the alimentary canal include protection, absorption and secretion.

Protection

The protective role of epithelial tissue is to prevent physical or chemical damage to tissues. For example, the stratified squamous epithelium is present as the inner lining of several regions of the gut — the mouth, the oesophagus and the anal canal — and it protects the underlying tissues from damage. Figure 3.29 shows part of the wall of the oesophagus with its distinctive stratified squamous epithelium, showing many layers of cells. At the base of the tissue are living cells, including stem cells that are constantly producing new cells that push existing cells upwards.

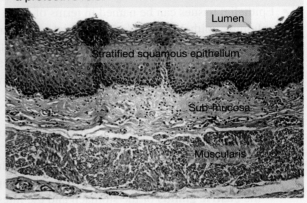

FIGURE 3.29 Photomicrograph of a cross-section of the wall of the human oesophagus (30 x magnification). The squamous stratified epithelium forms the inner lining of the oesophagus and plays a protective role.

Absorption

The end products of digestion are taken up from the gut lumen and transferred across the epithelial tissue of the mucosa to blood and lymph vessels for distribution. Most absorption takes place in the small intestine, where the muscosa is folded into villi, which greatly increases the surface area. The absorptive function is carried out by a layer of simple columnar epithelial cells called enterocytes which cover each villus (figure 3.30).

Secretion

Secretion is a function of glandular epithelial tissue and occurs when glands release substances into the lumen of the gut.

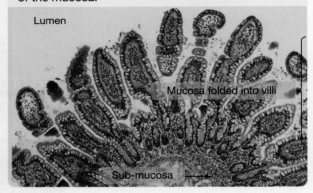

FIGURE 3.30 Cross-section of part of the wall of the small intestine showing the mucosa folded into villi. and part of the sub-mucosa. Note the finger-like villi projecting into the gut lumen (at top). Also visible are the narrow channels (crypts) at the base of the mucosa.

- Some glands are single cells, such as goblet cells. These cells are scattered throughout the simple columnar epithelium of the small and the large intestines. Goblet cells secrete mucus that lubricates the inner surface of the gut. Figure 3.31 shows a cross-section through seven villi, each covered by a layer of simple columnar epithelial tissue which forms outer lining around each villus. The bright pink stain identifies the mucus in the mucus-secreting goblet cells.
- Other glands are multicellular and they lie below the epithelial surface to which they are connected by ducts. Various types of gland exist, including the simple tubular glands in the small intestine (figure 3.32), which forms the crypts of the intestinal mucosa. Note the glandular epithelial tissue (shown in dark purple) at the base of the gland that produces the secretion. Secretions from the gland pass to the surface of the gut lining via the duct.

FIGURE 3.31 High power micrograph of a cross-section through seven villi, each covered by simple columnar epithelial tissue that forms the outer lining around each villus. The bright pink stain highlights the mucus secretion present in goblet cells.

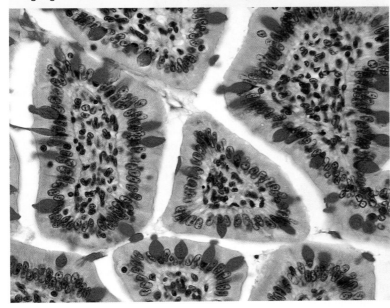

FIGURE 3.32 A tubular gland. The glandular epithelial tissue (purple) produces the secretions that pass to the surface of the gut lining via the duct.

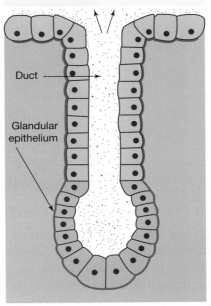

Connective tissue: Functions in the digestive system

In the alimentary canal, connective tissue forms the second layer of the gut wall, the sub-mucosa, where it occurs as areolar connective tissue with blood vessels, lymphatics and nerves; it also forms the outermost layer, the serosa. Connective tissue is also present in the solid organs of the digestive system.

Connective tissue in the hollow organs of the alimentary canal provides structural support to other tissues. For example, the connective tissue of the sub-mucosa of the gut wall supports the overlying mucosa and connects it to the underlying muscularis (figure 3.33).

Blood and lymph are also connective tissues. In the sub-mucosa and mucosa, they provide metabolic support by providing nutrients and oxygen, and remove cell wastes. They also transport the end products of digestion after they have been absorbed across the epithelial lining of the gut.

Connective tissue in the solid organs of the digestive system forms a thin surrounding capsule around and within solid organs. For example, a thin layer of connective tissue surrounds the liver and also encloses each functional unit in this organ (see figure 3.34).

FIGURE 3.33 Microscopic view of loose connective tissue in the sub-mucosa of the stomach showing wavy bundles of collagen fibres. Nuclei of fibroblasts appear black.

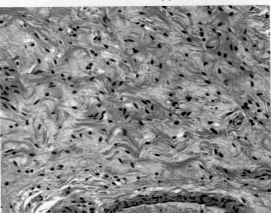

FIGURE 3.34 Photomicrograph of cross-section through the liver showing each of its functional units (lobules) surrounded by connective tissue (white). Note that the bile duct and various blood vessels are also enclosed by connective tissue.

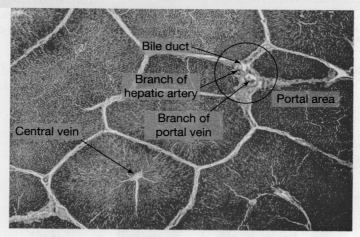

Muscle tissue: Functions in the digestive system

Muscle tissue is important in the structure and function of the digestive system. In almost all parts of the alimentary canal, the muscle tissue is composed of smooth involuntary muscle cells. The exceptions are the muscle tissues of the throat and upper region of the oesophagus and that at the end of the anal canal. In these locations, the muscle tissue is striated (skeletal) muscle that is under voluntary control. As you will be aware, this means that you have conscious control of swallowing food at the start of the gut, and of eliminating faeces at the other end of the gut, but no voluntary control in between.

Smooth muscle tissue forms the third layer (muscularis) of the gut wall in all segments of the gut. Except for the stomach, the muscularis consists of two layers of muscle: an inner smooth muscle that is arranged in a circular pattern and an outer smooth muscle with a longitudinal arrangement (figure 3.35). In the stomach, a third layer of smooth muscle is present with an oblique arrangement (figure 3.36). These three muscle layers act in a coordinated manner to produce the churning movement of food in the stomach.

FIGURE 3.35 Micrograph showing the muscularis layer of the small intestine with of its two layers of smooth muscle

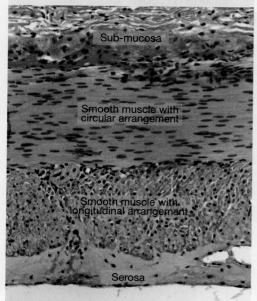

FIGURE 3.36 Illustration of the stomach showing the three layers of muscle tissue that form the musculosa of the stomach wall (Note the pyloric sphincter; you will shortly meet it.)

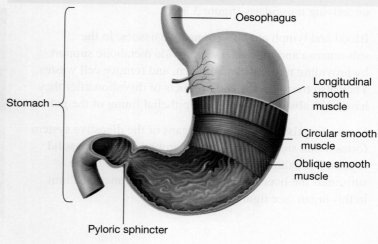

Muscles have the ability to contract (and relax) and they are responsible for various movements in the alimentary canal including:
- coordinated contraction of the musculosa to create the peristaltic movement of the gut content starting at the oesophagus and ending at the anal canal. These movements propel the digested food through the gut and cause mixing.
- regulation of the movement of the gut contents from one segment of the gut to the next. Regulation of the flow of the gut content from segment to segment is achieved through **sphincters**, which are thickened rings of muscle located at the junction of different segments of the alimentary canal. When relaxed and open, the sphincter allows food to pass and, when contract and closed, prevents the forward movement of food.

Sphincters

- A key sphincter is the **pyloric sphincter**, which is present at the junction of the stomach with the duodenum of the small intestine. The stomach contents are large in volume and highly acidic. In contrast, the duodenum is a small segment of the gut and the enzymes that are active in the duodenum require alkaline conditions. An unregulated free flow of **chyme** from the stomach into the duodenum would have negative results. Figure 3.37 shows an endoscopic photo of a normal pyloric sphincter in the closed position.
- Another sphincter in the alimentary canal is present at the junction of the lower end of the oesophagus and the upper region of the stomach — this is the lower oesophageal sphincter that keeps stomach acid from moving up into the oesophagus.
- Another sphincter is located at the anus. This sphincter is in fact, two sphincters: an inner sphincter of smooth muscle that is involuntary and an outer sphincter of striated muscle that is under voluntary control.

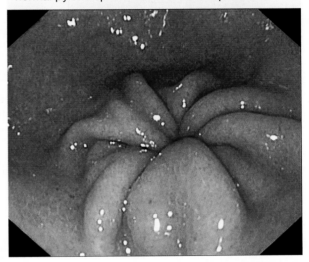

FIGURE 3.37 View through an endoscope of a normal pyloric sphincter in the closed position

3.4.3 Organs involved in digestion

Food goes through a series of processes in the period after it enters the mouth and leaves as indigestible wastes via the anus. Each mouthful of food passes from mouth via the oesophagus to the stomach, then through the 3- to 5-metre length of small intestinal tubing into the 1.5 metre of the large intestine before the undigested waste leaves via the anus. The time for the passage from mouth to anus shows variability among healthy persons and is affected by the composition of the meal eaten. However, on average, to totally empty the stomach takes 4 to 5 hours, and the entire trip through the gut may take from 2 to 4 days.

On its journey from mouth to anus, the digestion of food relies on many enzymes to progressively break down large complex nutrients in the food to small monomers or components that can be absorbed from the gut and used by body cells. If this does not happen, the item will be eliminated from the body as undigested waste. Table 3.3 is a list of the enzymes that are key players in the chemical break down of food. The majority of these enzymes operate under alkaline conditions, but the enzymes in the stomach are active under acidic conditions.

sphincters thickened rings of muscle which control the opening and closing of a tube

pyloric sphincter a sphincter at the join of the stomach and the duodenum of the small intestine; it controls the flow of acidic chyme into the alkaline duodenum.

chyme slurry of partially digested food produced in the stomach which passes into small intestine

TABLE 3.3 Enzymes of the digestive system, their sites of production and sites of action, and the result of enzyme activity. End products of digestion that can be absorbed by the enterocytes that line the small intestine are shown in red. The enzymes shown in *italics* are the so-called 'brush border' enzymes.

Macromolecule	Digestive enzyme	Site of enzyme production	Site of enzyme action	Enzyme action
Carbohydrates: broken down to simple sugars (monosaccharides)	Salivary amylase	Salivary glands	Mouth	Starch → sugars + dextrins
	Pancreatic amylase	Pancreas	Duodenum	
	α-dektrinase	Enterocytes of small intestine	Brush border	Oligosaccharides → glucose
	Maltase			Maltose → glucose
	Sucrase			Sucrose → glucose + fructose
	Lactase			Lactose → glucose + galactose
Proteins: broken down to amino acids	Pepsin	Gastric glands	Stomach	Protein → polypeptides & peptides
	Trypsin Chymotrypsin	Pancreas	Duodenum	
	Endopopetidases Aminopeptidases Carboxypeptidases	Enterocytes of small intestine	Brush border	Peptides → amino acids
Fats (lipids): broken down to fatty acids and glycerol or monoglycerides	Lingual lipase	Tongue	Mouth and stomach	Triglyceride → diglyceride + fatty acid
	Gastric lipase	Gastric glands	Stomach	Triglyceride → diglyceride + fatty acid
	Pancreatic lipase	Pancreas	Duodenum	Triglyceride → a monoglyceride + fatty acids
Nucleic acids: broken down to 5-C sugars, phosphate and bases	Deoxyribonuclease	Pancreas	Duodenum	DNA → nucleotides
	Ribonuclease	Pancreas	Duodenum	RNA → nucleotides
	Nucleosidase Phosphatase	Enterocytes of small intestine	Brush border	Nucleotides → sugars + phosphates + bases

The mouth — intake of food and start of mechanical and enzymic digestion

Teeth begin the mechanical breakdown of ingested food into smaller particles — this fragmentation increases the surface area of the food that will be exposed to enzymes and other secretions of the digestive system. Saliva contains the enzyme salivary amylase that begins the digestion of carbohydrates such as starch. A fat digesting enzyme, lingual lipase, produced by glandular cells of the tongue, begins the digestion of lipids.

Food leaves the mouth by the conscious act of swallowing — after this, all activity along the digestive tract is involuntary until the anus, where egestion of faeces involves conscious action. As food passes from the mouth into the throat (pharynx), a small flap of tissue called the epiglottis closes over the entry to the trachea (windpipe). Why is this important?

Glands of the mouth include:
- salivary glands that secrete a watery saliva that contains the carbohydrate-digesting enzyme, amylase
- lingual glands on the surface of the tongue that secrete the fat-digesting enzyme, lipase.

The oesophagus — transport of food into stomach

The oesophagus is a narrow tube about 25 centimetres long. Bands of circular muscle tissue are present in the oesophagus wall — near the mouth this muscle is striated muscle but further along the oesophagus, the muscle changes to smooth muscle. Waves of contraction of this muscle progressively move food to the stomach — a process called **peristalsis** (figure 3.38) Peristalsis involves a zone of muscle contraction immediately above the food and a zone of muscle relaxation immediately below the food. Because this movement is due to muscular action and not gravity, a person can swallow food while upside down.

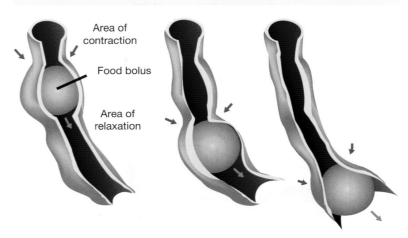

FIGURE 3.38 Diagram showing the movement of food down the oesophagus to the stomach through peristalsis

Glands of the oesophagus include mucus-secreting glands, with the mucus acting as a lubricant and as a neutraliser of acidic **gastric juices** that may flow back up from the stomach.

The stomach — more enzymic and mechanical digestion

The stomach (figure 3.39) is an expandable region of the gut that serves as a temporary holding chamber for ingested food – when empty, in humans, its volume is about 50 millilitres, but it can expand up to several litres when food is present. This ability of the stomach to expand is due to the presence of folds (called rugae) in the stomach wall.

The stomach contains three layers of smooth muscle tissues — longitudinal, circular and oblique — these enable it to actively churn its contents; and contribute to mechanical digestion, which produces a slurry of partly digested food called chyme. Food remains in the stomach for up to four hours, and alcohol and aspirin are two of a very small number of substances that can be absorbed in the stomach.

Enzymic digestion occurs with the release from secretory epithelial cells in gastric glands of gastric juice that is a mixture of mucus, hydrochloric acid, pepsinogen (an inactive **proenzyme**) and the enzyme, gastric lipase. The hydrochloric acid lowers the pH of the lower stomach contents to about a pH of 2 and, in these acidic conditions, inactive pepsinogen is converted to the active enzyme pepsin, that starts the digestion of proteins. Gastric lipase continues the digestion of lipids, and the lingual lipase secreted in the mouth also continues its enzymic breakdown of fats. The passage of chyme from the stomach into the small intestine (duodenum) is controlled by the pyloric sphincter located in the junction.

Glands of the stomach are the gastric glands that contain various secretory cells including:
- **parietal cells** that secrete hydrochloric acid, which creates the acid conditions of the stomach (pH 2–3)
- **chief cells** that secrete the enzyme gastric lipase and the inactive proenzyme, pepsinogen. In the acidic conditions of the stomach, pepsinogen is converted to the active enzyme pepsin where it begins the digestion of proteins.

peristalsis involuntary constriction and relaxation of muscles in the alimentary canal to push food to the stomach

gastric juice acid fluid secreted by the stomach glands for digestion in the stomach

proenzyme a precursor of an enzyme that must be activated to form the functional enzyme

chief cells in the parathyroid gland, secretory cells that produce parathyroid hormone (PTH) (Note: different cells with same name are present in the gastric pits of stomach.)

FIGURE 3.39 a. Diagram of stomach showing the creases or rugae (singular = ruga) that enable the stomach to expand, and the pyloric sphincter that controls the exit of chyme **b.** A section of the stomach wall showing its four tissue layers. Note in particular the gastric glands located in the mucosa that secrete gastric juice.

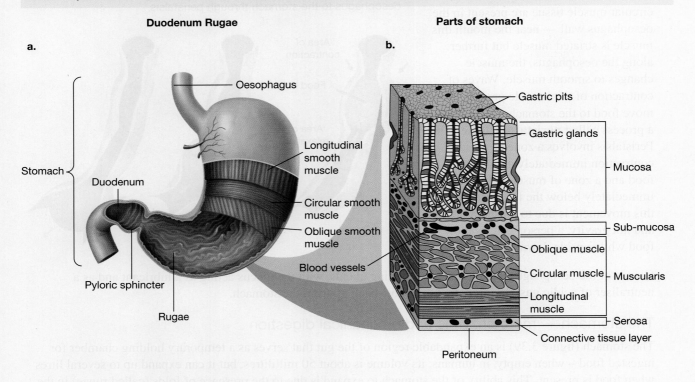

The liver — an accessory organ and producer of bile

Bile is secreted by liver cells (hepatocytes) and is transferred via small ducts to the **gall bladder** where bile is either stored or is released via the major bile duct into the duodenum, the first part of the small intestine. The components of bile that are important in digestion are the bile salts. Bile salts are emulsifying agents, not enzymes that disperse fat into smaller particles.

FIGURE 3.40 Illustration of the liver (upper left) and a liver lobule (lower right). Blood vessels are shown as veins (blue) and arteries (red), and the bile-carrying ducts and the expanded gall bladder are shown in green. The liver is organised into many lobules, each built of chains of liver cells (hepatocytes) radiating from a central vein.

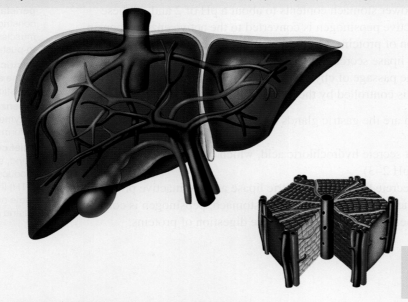

gall bladder sac shaped organ which stores the bile after it has been secreted by the liver

The pancreas — an accessory organ and producer of enzymes

The pancreas secretes and releases pancreatic fluid. Pancreatic fluid leaves the pancreas via the main pancreatic duct. This duct joins the common bile duct from the liver that releases bile and pancreatic fluid simultaneously into the duodenum, the first section of the small intestine (figure 3.41).

Pancreatic fluid contains several enzymes — lipase, amylase and inactive proteases (protein-digesting enzymes) — and it also releases bicarbonate ions into the duodenum that help neutralise the acidic chyme from the stomach.

Glands of the pancreas contain **acinar cells** that secrete:
- trypsinogen, the inactive precursor (proenzyme) of the protein-digestingenzyme, trypsin
- chymotrypsin, the inactive precursor (proenzyme) of the protein-digesting enzyme, chymotrypsin
- pancreatic lipase, a fat-digesting enzyme
- pancreatic amylase, a carbohydrate-digesting enzyme.

FIGURE 3.41 The pancreas showing the junction of the major pancreatic duct with the common bile duct that opens to the duodenum

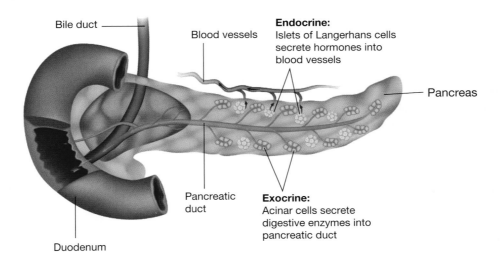

The small intestine — final stage of digestion and absorption of nutrients

The small intestine has three parts: the duodenum (about 25 centimetres long), followed by the jejunum (up to 2.5 metres long), and then the ileum (about 3 metres long).

The mucosa of the small intestine can be distinguished from that of all other regions of gut by the presence of villi, crypts and microvilli.
- **Villi** are finger-like structures that project into the lumen of the small intestine; villi are covered by absorptive cells (enterocytes) and mucus-secreting goblet cells.
- **Crypts** are narrow depressions that descend into the mucosa and they are intestinal glands (see figure 3.42).
- **Microvilli** are minute outfoldings of the plasma membranes of enterocytes. When viewed using a light microscope, microvilli cannot be individually resolved, and the microvilli-covered surface of these cells is referred to as a 'brush border' (figure 3.43).

acinar cells cells of the pancreas that produce and transport enzymes that are passed into the duodenum where they assist in the digestion of food

villi outfoldings or projections of the mucosa of the small intestine

crypts tube-like depressions of the mucosa located in the intestine and the site of glandular cells

microvilli sub-microscopic outfoldings of the plasma membrane of enterocytes of the small intestine that form the so-called 'brush border' of these cells

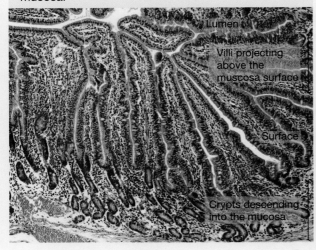

FIGURE 3.42 Photomicrograph of a cross-section of the small intestine wall showing its distinctive mucosa.

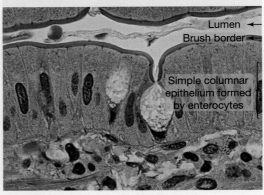

FIGURE 3.43 High power photomicrograph of simple columnar epithelium made of enterocytes that line the villi. The thick upper border of these cells is a brush border (bright purple) of tightly packed outfoldings of microvilli. Note the central mucus-secreting goblet cell.

Compared with a smooth cylinder of similar dimensions, the epithelial surface of the small intestine has a much greater surface area. This increased surface area is due to the presence of regular **circular folds** in the wall of the small intestine, the large numbers of villi on these folds, and the microvilli on the outer surface of enterocytes (figure 3.43). These combined features provide a large surface area over which absorption of digested nutrients can occur. Furthermore, the circular folds means that the intestinal contents flow in a slower spiral manner rather than flowing straight through more rapidly.

Areas of the small intestine that secrete mucus:
- Brunner's gland in the crypts (intestinal glands) of the small intestine, which produces an alkaline mucus that coats the epithelial lining of the duodenum. This alkaline mucus protects the small intestine from the strongly acidic chyme coming from the stomach. This mucus also creates the alkaline conditions required by the enzymes that operate in the duodenum.
- Goblet cells that are part of the surface epithelium also secrete mucus as a protective lubricant.

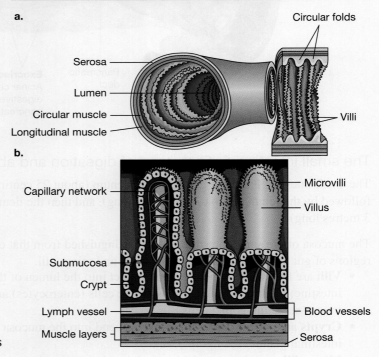

FIGURE 3.44 a. Diagram showing the regular circular folds present in the inner lining of the small intestine. **b.** Diagram at higher magnification showing the large numbers of finger-like villi that are located on either side of these folds

circular folds permanent macroscopic folds (ridges) in the mucosa of the small intestine

Functions of the small intestine: digestion

The duodenum, the first segment of the small intestine, is the site of most of the digestive action that occurs in the alimentary canal. The enzymes involved are:
- pancreatic enzymes that are released by the pancreas into the lumen of the duodenum
- brush border enzymes that are produced by the enterocytes of the duodenum epithelium.

Let's look at a summary of the digestion of fats, carbohydrates and proteins.

Digestion of fats

The duodenum receives bile containing bile salts from the liver via the gall bladder. In the duodenum, bile salts emulsify water-insoluble fats by forming coatings around large droplets of fat and dispersing them into smaller and smaller particles — this process of emulsification increases the surface area of the fat droplets that is available for enzyme action.

The duodenum also receives pancreatic fluid that contains the enzyme, pancreatic lipase. This enzyme acts on the emulsified fats, breaking them down to monoglycerides and fatty acids that are the end products of fat digestion.

Digestion of carbohydrates

The duodenum receives pancreatic fluid that contains the enzyme pancreatic amylase. This enzyme continues the digestion of carbohydrates that began in the mouth and it breaks carbohydrates down to dextrins (branched oligosaccharides) and various disaccharide sugars — these are all too large to be absorbed.

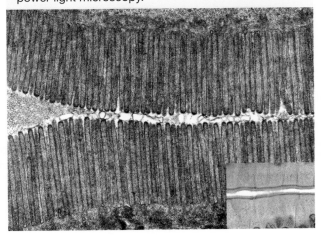

FIGURE 3.45 TEM showing microvilli on the plasma membrane of two epithelial cells forming the brush border that has enzymes embedded within it. The coloured inset shows the brush border using high power light microscopy.

Various 'brush border' enzymes, including dextrinase, sucrase, maltase and lactase complete the digestion of carbohydrates. These enzymes breakdown dextrins and disaccharide sugars to monosaccharides that are small enough to be absorbed. Brush border enzymes are not free in solution in the intestinal lumen as is the case for other digestive enzymes. They are embedded in the plasma membranes of microvilli of the enterocytes that line the intestinal wall and act on food while in place on the microvilli (figure 3.45).

Digestion of proteins

The duodenum receives pancreatic fluid that contains two inactive proenzymes, trypsinogen and chymotrypsinogen. These proenzymes are converted to the active protein-digesting enzymes, trypsin and chymotrypsin. Proteins are broken down by these enzymes to polypeptides and peptides.

The finishing touches in protein digestion involve the actions of specific brush border enzymes that can breakdown polypeptides and peptides to amino acids that can be absorbed across the epithelial boundary. These brush border enzymes include endopeptidases and aminopeptidases. Endopeptidases cut peptide bonds between amino acids at sites within the chain, while exopeptidases, such as aminopeptidases and carboxypeptidases, cut amino acids off from one end of the chain, either the amino end or the carboxyl end.

Functions of small intestine: absorption

The small intestine is the primary site of absorption of the end products of digestion — that is, the digested nutrients, in addition to water, salts and vitamins.

The various secretions produced by digestive glands — saliva, gastric juice, bile, pancreatic fluid — have a high water content and, each day, they contribute on average about 7 litres of water to the alimentary canal. Our daily fluid and food intake adds about another 2 litres of water. Most, about 80 per cent, of this water is reabsorbed in the small intestine. The structure of the small intestine is well equipped to carry out this function because of the expanded surface area for absorption created by circular folds, villi and microvilli. However, different regions of the small intestine absorb different nutrients:
- the duodenum is the primary site of absorption for fatty acids and water-soluble vitamins
- the jejunum is the major region where glucose and amino acids are absorbed
- the ileum absorbs bile salts and vitamin B_{12}.

Absorption of the end products of digestion is a function of the enterocytes that line the small intestine and its villi. Table 3.4 and figure 3.46 shows the various end products of digestion that are able to cross into the enterocytes.

TABLE 3.4 End products of the digestion of carbohydrates, proteins and fats that are able to be absorbed across the epithelial layer of the wall of the small intestine

Input to mouth	Absorbable components
Carbohydrates	Glucose
	Fructose
	Galactose
Proteins	Amino acids
	Dipeptides
	Tripeptides
Fats	Monoglycerides
	Fatty acids

FIGURE 3.46 Examples of absorbable end products of digestion. Note the long hydrophobic fatty acid chain that is part of the monoglyceride. Which of these end products of digestion are fat soluble and able to diffuse into enterocytes?

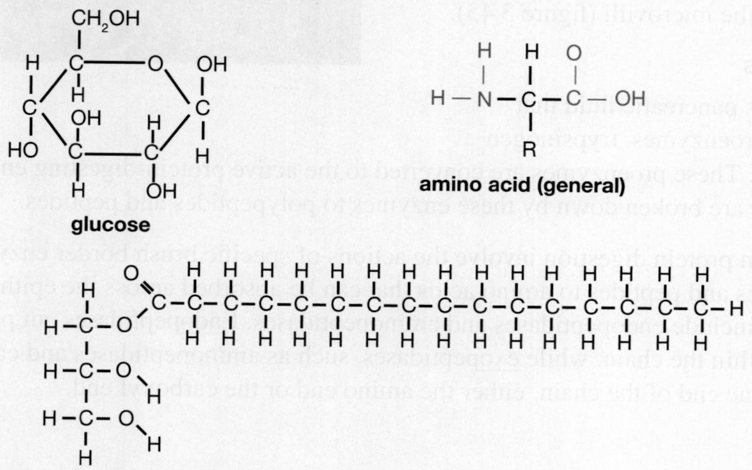

glucose

amino acid (general)

a monoglyceride

Both diffusion and active transport are needed to absorb digested nutrients into the enterocytes that line the villi. Fat-soluble end products of digestion, such asmonoglycerides and fatty acids, can diffuse across the plasma membranes into the enterocytes. However, the absorption of water-soluble digested nutrients, such as amino acids and monosaccharides, requires some form of energy-requiring active transport.

The amino acids and monosaccharides, such as glucose, fructose and galactose, move the enterocytes into the blood capillary network within the villi. From there, these nutrients are eventually distributed to the rest of the body via the circulatory system.

The absorbed monoglycerides and fatty acids are re-formed into fats and packaged into particles that are moved from the enterocytes into **lacteals**. Lacteals are blind-ending vessels of the lymphatic system (figure 3.47). The fat particles initially travel in the lymphatic system but finally enter the blood circulatory system.

Large intestine — the final stages

The large intestine is the last segment of the digestive tract, and it consists of caecum, colon, rectum and anus. In total, the large intestine is about 1.5 metres long and has a diameter of about 5 centimetres. The longest segment of the large intestine is the colon.

The wall of the large intestine has the same four tissue layers as other regions of the alimentary canal. In contrast to the small intestine with its epithelial layer folded into many villi, the large intestine lacks villi, and is characterised by an abundance of mucus-secreting goblet cells. Figure 3.48 shows the mucosa of the colon with its numerous crypts that lie below the epithelial surface of the mucosa. Goblet cells in the crypt walls can be identified by the clear circular areas — these clear areas occur because the staining technique used here removes the mucus.

Functions of the large intestine

Reabsorption of water and ions

The large intestine receives mainly water with undigested material from the small intestine. Almost all of the remaining water is removed by the colon along with various ions through osmosis. (We are reminded of the importance of this orderly reabsorption of water if the process goes wrong, as occurs in diarrhoea.)

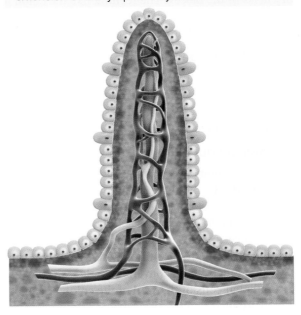

FIGURE 3.47 Longitudinal section through a villus in the small intestine showing its epithelial layer of absorptive enterocytes, a few goblet cells and, within the villus, thin capillaries (shown in red and blue) and a lacteal (shown in yellow) that is a blind extension of the lymphatic system.

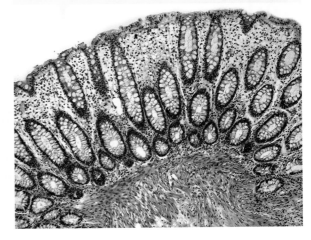

FIGURE 3.48 Micrograph of the mucosa of the colon of the large intestine showing the large number of crypts (intestinal glands) that are lined with enterocytes and mucus-secreting goblet cells. Can you suggest why the crypts might have the various shapes visible in this image?

lacteals the vessels of the lymphatic system which absorb digested fats

Formation and storage of faeces

Water is reabsorbed from the wastes as they are moved along the colon by peristalsis. As this happens, the contents of the colon become more concentrated and solid. By the time they reach the rectum, this solid material has formed faeces. The rectum serves as temporary storage place for the faeces that consist largely of indigestible material, such as cellulose, mucus and bacteria, the last mentioned making up more than half of the dry mass of the faeces. The rectum, like the colon has an epithelium of simple columnar cells with an abundance of goblet cells. Mucus secreted by goblet cells in both the colon and the rectum serves to lubricate the passage of waste material and protect the underlying tissue of the large intestine from damage.

Elimination of faeces

Faeces pass from the rectum and out of the body through the anus. In contrast to the simple columnar epithelium of the colon and rectum, the inner lining of the anus is a stratified squamous epithelium that serves for protection. The external anal sphincter is a band of striated muscle that enables voluntary control of the elimination of faeces.

Maintaining gut bacteria

The colon is home to enormous numbers of bacteria of hundreds of different species – these bacteria form our **gut microbiota** and they are at home in the oxygen-free environment of the colon. People, as hosts, and bacteria, as gut residents, gain benefits from each other. For the bacteria, the benefit is a warm environment and a supply of undigested food that the bacteria can ferment to meet their energy and nutrient needs. For the human hosts, the benefit is access to some of the metabolites that the bacteria produce through fermentation, such as amino acids, short-chain fatty acids that can be converted to glucose, and K and B vitamins that can be used by epithelial cells of the wall of the colon or can be absorbed into the blood and distributed to other tissues.

gut microbiota the population of organisms which live in the gut and play a crucial role in maintaining immune and metabolic homeostasis and protecting against pathogens

INVESTIGATION 3.5

online only

Digestive systems

Aim

To investigate the role of the digestive system in breaking down food to obtain nutrients

 Resources

 eWorkbook Worksheet 3.3 Structure and function of the digestive system (ewbk-3490)
Worksheet 3.4 The ins and outs of digestion (ewbk-3492)

3.4.4 Digestive system in different animals

Earlier in this topic, the digestive cells, tissues or systems of invertebrate animals were introduced, starting with sponges, jellyfish and flatworms. From there, we briefly met the nematodes, the first animals to show the mouth-to-anus one-way digestive tract that is present in all other animals. We have also explored in detail the human digestive system in section 3.4.3. In this section, we will look at the digestive system in some other mammals and birds.

Who eats what?

In terms of the nature of their primary and preferred food source, mammals can be classified as:
- *herbivores.* Primary food source is plant material. Herbivores may be further subdivided into groups, such as fruit eaters, seed eaters, nectar feeders or leaf eaters. Among native Australian fauna, the honey possum (*Tarsipes rostratus*) is a nectar feeder (figure 3.49a) while the eastern grey kangaroo (*Macropus giganteus*) is a leaf eater.

- *carnivores*. Primary food source is the flesh of other animals either hunted as prey or scavenged.
  - the Tasmanian devil (*Sarcophilus harrisii*) that lives on small prey and the carcasses of dead animals (carrion) is Australia's largest carnivorous marsupial. The numbat (*Myrmecobius fasciatus*) is an insectivorous carnivore whose food intake is almost exclusively termites (see figure 3.49b). Like other insect-eating mammals, the numbat has a long tongue that is coated with a sticky saliva secreted by its salivary glands.
- *omnivores*. These animals eat and survive on a mixed diet sourced from both plants and animals.

FIGURE 3.49 a. A honey possum, a herbivore that feeds on nectar, feeding on a *Banksia* flower head **b.** A numbat, an insectivorous carnivore, whose preferred food is termites

To survive, a mammal must be able to access and ingest its preferred food and it must have the means — mechanical and/or enzymic — to break down its food to supply the energy and nutrients it needs for living. The main driver that has influenced the evolution of different features in the digestive system of mammals is the nature of the food on which they rely for their energy and nutrients. Dietary preferences of mammals are indicated in the organisation of their digestive system, such as the length of the digestive tract, the size of the caecum and the characteristics of the teeth, such as types present, sizes, and shapes. Table 3.6 shows some of the variation that exists in the digestive systems of different mammals.

TABLE 3.6 Variation the digestive systems of mammals. Combinations of these features enable mammals to exploit gaining and using particular food sources.

	Variations in mammals	
Mouth	Canine teeth present Continuously growing incisors or molars Gap between incisors and premolars Molar(s) with grinding surfaces High crowned cheek teeth	Canine teeth absent Fixed-size incisors or molars No gap No grinding surfaces on molars Low-crowned cheek teeth
Digestive tract	Long	Short
Stomach	Single-chambered stomach	Multi-chambered stomach
Large intestine	Long caecum	Short or absent caecum
Gut microbiota	Microbiota in foregut	Microbiota in hindgut

Teeth in different mammals

Teeth play an important role in the digestive system. They are a major means by which the mechanical digestion of food is achieved — chewing creates food particles with an increased surface area for enzymic action. Mammals have four kinds of teeth: incisors, canines, premolars and molars. Mammalian teeth can indicate whether a mammal is *most likely* to be a carnivore, a herbivore or an omnivore. Figure 3.50 shows the teeth of four mammals.

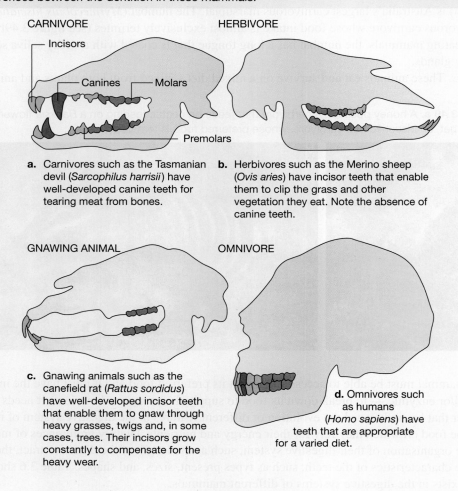

FIGURE 3.50 Diagram showing the teeth of **a.** a carnivore **b.** a herbivore **c.** a gnawing animal **d.** an omnivore. Note the differences between the dentition in these mammals.

a. Carnivores such as the Tasmanian devil (*Sarcophilus harrisii*) have well-developed canine teeth for tearing meat from bones.

b. Herbivores such as the Merino sheep (*Ovis aries*) have incisor teeth that enable them to clip the grass and other vegetation they eat. Note the absence of canine teeth.

c. Gnawing animals such as the canefield rat (*Rattus sordidus*) have well-developed incisor teeth that enable them to gnaw through heavy grasses, twigs and, in some cases, trees. Their incisors grow constantly to compensate for the heavy wear.

d. Omnivores such as humans (*Homo sapiens*) have teeth that are appropriate for a varied diet.

Features of carnivores:
- The presence of large canine teeth and the absence of large grinding molars indicate a flesh-eating carnivore. One such carnivore, now extinct, with very impressive canine teeth was the sabre-toothed cat (*Smilodon fatalis*) (see figure 3.51a).
- Many carnivores, including members of the cat and dog families, have sharp-edged premolars and molars that do not meet on top of each other when the jaw closes. Instead, these teeth, called **carnassial teeth**, produce a shearing or slicing action as their sharp edges move past each other like the blades of scissors. (If you watch a cat eating a strip of fresh meat, it will move the meat to the side of its mouth as it uses its carnassial teeth to slice it.) Figure 3.51b shows a lioness using her carnassial teeth to slice flesh — notice how she has positioned the flesh at the side of her mouth.
- Carnivores cannot move their jaws from side to side when they are chewing.

Features of herbivores:
- The presence of large molars with grinding surfaces indicates that a mammal is herbivorous.
- Herbivores have sharp front teeth (incisors) that are equipped to tear off or nip vegetation. Some species only have lower incisors.
- Canine teeth are often absent from the dentition of herbivores — rabbits and macropods (kangaroos and wallabies) have none.
- They can move their jaws from side to side when chewing.

carnassial teeth paired upper and lower premolars and molars which do not meet, and hence allow a shearing action to tear food

FIGURE 3.51 a. Skull of sabre-toothed cat, an extinct carnivore, showing its massive canine teeth b. A lioness eating flesh which is being sheared by her sharp-edged carnassial teeth (the last upper premolar and the lower molar) as they move past each other

Omnivores, such as pigs, dogs and people, typically have sharp incisors, canines and grinding surfaced molars reflecting a diet that is both animal and plant-based.

Teeth size: fixed size or continuously growing?

Most herbivores have a diet of tough, fibrous plant material, often with abrasive soil particles attached. Consuming this material over extended periods damages and erodes the structure of their teeth. Various herbivores show adaptations that enable them to deal with this situation.

FIGURE 3.52 The wombat (*Vombatus ursinus*) has a diet of mainly highly fibrous and tough native grasses, sedges and roots.

- *Constantly growing teeth:* Rabbits, hares, rats (but not mice) and wombats have teeth that grow continuously. In rabbits, hares and wombats, it is all their teeth (incisors, premolars and molars), while in rats, it is just the incisors. The roots of these teeth have stem cells that produce dentine and enamel tissues for the growth of new teeth. Continuously growing teeth mean that these herbivores can for their entire lifetimes replace worn and eroded teeth. As teeth wear away at their tips, new teeth material is added at their roots.
- *Fixed size teeth:* Other herbivores also have constant wear and tear on their teeth from their fibrous plant diet, but they do not have constantly growing teeth. Instead, their teeth are fixed in size, but some or all of their teeth have very high crowns. Figure 3.53a shows a comparison of a low-crowned molar tooth, such as occurs in people, cats and dogs, with a high-crowned molar tooth, such as may be seen in horses and in **ruminants**, such as cattle. In the case of the horse, all of its permanent teeth have high crowns, with its premolars and molars being up to 10 centimetres in length. Sounds unlikely? The bulk of these teeth is hidden within the jaw bones and only a small segment is visible (see figure 3.53b). Over time, the grinding surfaces of molars and premolars wear away at the rate of about 2 to 3 millimetres per year. As exposed tooth tissue is eroded, teeth concealed in the jaw bones move equally slowly and replace the lost tooth tissue.

ruminants animals that absorb nutrients by fermenting food in a specialised stomach prior to digestion

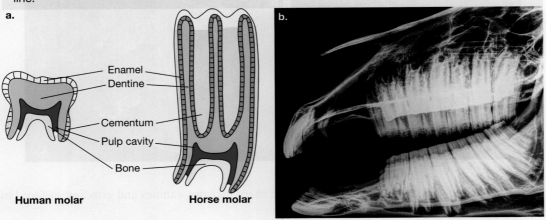

FIGURE 3.53 a. A low-crowned molar (as in a person) and a high-crowned molar (as in a horse). Only the pale section at the top of high-crowned molar is visible above the gum margin. b. X-ray of the side view of the skull of a horse showing the upper and lower jaws with the prominent cheek teeth (premolars and molars). Only a few centimetres of these teeth appear above the jaw line.

Hollow organs of the digestive tract in different mammals

Figure 3.54 shows drawings of the alimentary canal in several different mammals. In each case, the digestive tract has features that cater for the particular dietary input of the mammal and that ensure that the mammal can obtain the energy and nutrients it needs. Examine this figure and note the differences.

FIGURE 3.54 The alimentary canals of a cow, a koala, a Tasmanian devil and a honey possum. Notice the differences in the comparative sizes of some segments of the gut, such as the colon and the caecum.

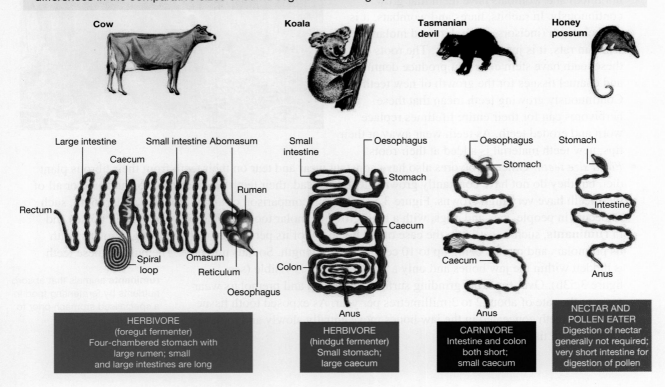

Length of the alimentary canal

- The shortest gut is that of the honey possum with its diet of nectar and pollen. Nectar is essentially a solution of sugars, (glucose, fructose and sucrose), and only the sucrose disaccharide requires digestion.
- The longest alimentary canals are those of the herbivorous cow and koala. The digestion of plant material such as the cellulose in fibrous grass and eucalyptus leaves is a more complex process than the digestion of protein. *Mammals do not themselves produce enzymes that can digest cellulose.* Instead, they depend of the action of their gut microbiota to do this for them in a mutually beneficial arrangement.
- The alimentary canal of the carnivore, in comparison with that of the herbivores is very short. The protein of animal flesh is more easily digested than the cellulose in plants since mammals are equipped with a family of protein-digesting enzymes.
- The length of the alimentary canal indicates the difficulty or otherwise of digesting the preferred food of a mammal — the shorter the canal, the simpler the digestive processes; the longer the canal, the more elaborate the digestive processes.

Number of chambers in the stomach

Many mammals have a stomach with a single chamber that is like an expandable sac. However, the group of mammals termed ruminants (from Latin = 'to chew again') have a stomach with a more complex structure consisting four chambers, with the first of these chambers being the **rumen**. The more than 200 species of ruminant mammals include cattle, sheep, deer, antelopes, giraffes and gazelles.

Microbial fermentation

The grass-eating cow and the leaf-eating koala each have a chamber in their alimentary canals where their gut microbiota carry out fermentation. We all have microbes in our gut, but in the case of herbivores, there is typically a very large concentration of bacteria and other microbes in one particular segment of their alimentary canals.

For example:
- Cows: fermentation site is the very large first chamber of its stomach, the rumen. The rumen has a capacity of up to 150 litres and is said to be 'one of the most dense microbial habitats in the world'. In comparison, the human stomach when distended has a capacity of 2 to 4 litres.
- Koalas: fermentation site is the caecum. Note the long caecum in the koala that is about 2 metres in length and houses the microbiota that are essential for leaf digestion. In comparison, the adult human caecum is about 6 centimetres long and can be removed surgically with no effect on digestion.

In the fermentation chambers of herbivores, gut microbes ferment ingested cellulose to produce volatile fatty acids, microbial proteins and the gases hydrogen and methane.

Mammals that accommodate their microbiota in the stomach are said to be *foregut fermenters*, with the biggest group being the ruminant mammals. Mammals with their microbiota housed in the caecum are said to be *hindgut fermenters*. As well as the koala, this group of mammals includes horses, rhinos, elephants, rodents and rabbits.

> **rumen** first part of the stomach of a ruminant where digestion occurs with the aid of gut microbiota

---

CASE STUDY: Digestion in the cow

Like all other ruminants, cows depend on their gut microbiota to convert the grasses and hay that they eat into compounds that they can used to meet their energy and nutrient needs, such as proteins.

Cows have a stomach with four separate compartments (figure 3.55), each having its own role in the digestive process.

The four compartments are:

- the rumen
  - houses the enormous numbers of microbes (mainly bacteria and protozoa) and serves as a large fermentation vat
  - can hold 50 to 150 litres of food and fluid, depending on the size of a cow
  - is the site where the breakdown (digestion) of grass and hay occurs through fermentation by rumen microbes
- the **reticulum** (from French *reticule* 'small purse')
  - is a pouch-like chamber that is the smallest of the four chambers. As the cow eats, any dense feed and any heavy foreign objects, such as screw or nails, that the cow might ingest in its hay, drop into this chamber.
- the **omasum** (from Latin *omasum* 'tripe') is the third chamber with a volume of about 8 litres. This compartment is where much of the water is absorbed from the stomach contents, and it also acts as a filter that allows only fine particles of food to pass to the final compartment.
- the **abomasum** (*ab* = away from *omasum*) is the final chamber of the stomach, with a volume of about 27 litres. The abomasum is the only chamber of the stomach where the cow's own enzymes are secreted and used in digestion. The inner lining of the abomasum is a glandular epithelium that secretes hydrochloric acid and enzymes, including the precursor (proenzyme) of the protein-digesting pepsin.

reticulum second part of the stomach of a ruminant that is the smallest of the four chambers; collects any heavy objects

omasum third part of the stomach of a ruminant which acts as a filter and where water is absorbed

abomasum final part of the stomach of a ruminant where enzymes are secreted and used in digestion; food then passes to the small intestine.

FIGURE 3.55 The four-chambered stomach of a cow. This illustrates the mass of the rumen, which is the fermentation chamber, and shows the size of the rumen relative to the size of the cow.

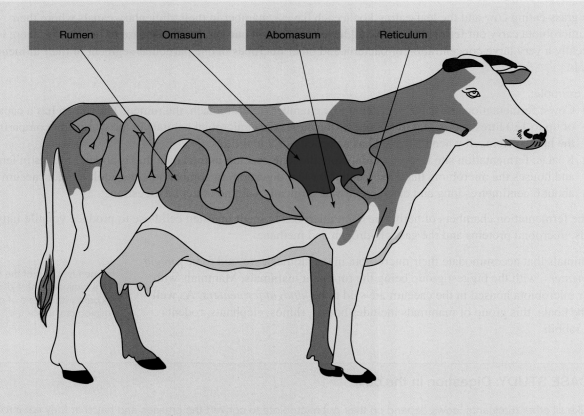

Let's follow the path of a mouthful of hay that a cow has just ingested (figure 3.56).

An outline of the steps in the pathway of a mouthful of ingested hay is as follows:

1. Mechanical breakdown of the hay begins in the cow's mouth where the hay mixes with saliva and is fragmented by the cow's side-to-side chewing action.
2. The hay moves down the oesophagus to the rumen. Mechanical digestion continues through the churning action of the rumen walls.
3. Microbes in the rumen start the breakdown (digestion) of the hay through a process of fermentation. All digestion in the rumen is carried out by microbes.
4. Partly digested hay, called cud, is returned to the mouth for further chewing. This process is popularly referred to as 'chewing the cud' or, more scientifically, as rumination.

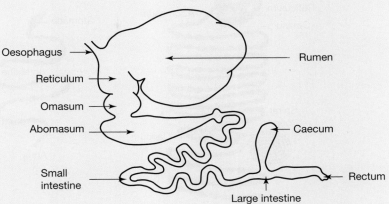

FIGURE 3.56 Line diagram showing the hollow organs of the digestive tract of a cow, including its four-chambered stomach

   - Chewing the cud grinds the hay into smaller fragments and causes the release of more saliva — up to 75 litres daily.
   - The more fibrous the cud, the more time is spent chewing it, and a cow may spend up to 10 hours per day in this activity.
   - Multiple cycles occur of the cud being chewed, returned to the rumen, regurgitated to the mouth and so on.
5. En-route to its return to the rumen, the food passes via the reticulum and any heavy food particles or other solid objects are filtered out and drop out the bottom of this compartment, while lighter material returns to the rumen.
6. On reaching the rumen again, digestion of the cud by gut microbes continues.
7. The end products of the microbial fermentation of cellulose, starch, sugars and proteins in the rumen are:
   - *volatile fatty acids*. These are used by the cow as its major source of energy; they are absorbed across the epithelium of the rumen and transferred into the cow's blood circulatory system.
   - *gases, such as methane and carbon dioxide*. About 30 to 50 litres of these gases are produced each hour and they are largely burped into the atmosphere.
   - *microbial proteins*. Nitrogen-containing compounds in the hay are broken down by microbes and are re-synthesised into microbial proteins.
8. The watery suspension of fine particles along with masses of microbial cells move from the rumen into the omasum where water is absorbed and the residue, including the microbial cells, flows into the next chamber, the abomasum.
9. In the abomasum, its lining of glandular epithelium releases hydrochloric acid and the precursor of a protein-digesting enzyme that is activated to the enzyme pepsin in the acid conditions. In the abomasum, digestion is carried out by the cow's own enzymes. Among the proteins digested here are the millions of microbial cells carried from the rumen, and these are the major source of protein for the cow. This microbial protein will be broken down to amino acids and used to synthesise cow protein.
10. From the abomasum, the digested nutrients move into the small intestine where other digestive enzymes and bile are present that break down protein to amino acids. Absorption of digested nutrients also occurs in the small intestine.
11. The large intestine (caecum and colon) is where water and minerals are absorbed. Undigested material passes in the rectum and is eliminated as faeces.

SAMPLE PROBLEM 3 Comparing digestive systems

The following diagram shows the digestive system of three different organisms: a herbivore, a carnivore and a pollen eater.

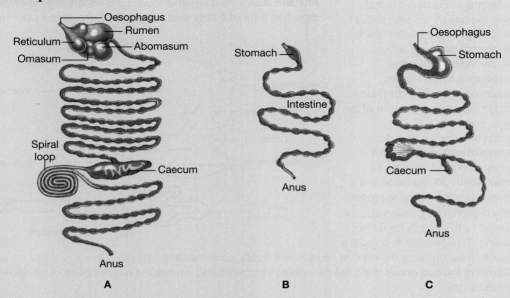

a. Identify which digestive system belongs to which organism. Justify your response. (2 marks)
b. Herbivores can be either a foregut or hindgut fermenter. Provide two clear pieces of evidence that suggest that this herbivore is foregut fermenter. (2 marks)
c. Explain the role of microbes in the digestion of herbivores. (2 marks)

THINK

a. 1. This question is asking you two things for 2 marks. Firstly, you need to identify which digestive system belongs to a herbivore, carnivore and pollen eater. Consider what you know about digestive systems:
- Herbivores often have a caecum and longer digestive system to breakdown plant matter.
- Pollen eaters have very simple digestive systems.

2. For the second mark, you are required to justify your response. Ensure you highlight the evidence that allowed you to determine your answer.

WRITE

A. Herbivore
B. Pollen eater
C. Carnivore (1 mark)

A. was determined to be the digestive system of a herbivore, due to the presence of a large caecum, spiral loop and longer, more complex digestive system.
B. was determined to belong to a pollen eater due to its short length and simplicity.
C. was determined to be a carnivore due to having a more complex digestive system than the pollen eater, but a much smaller, less pronounced caecum than the herbivore (1 mark).

b. This question requires evidence to justify the answer that was provided. Consider features that are seen in foregut fermenters and select the two that best support the diagram: • structure of the stomach • size of the caecum • presence of a spiral loop.	Two pieces of evidence that suggest this digestive system belongs to a foregut fermenter: • Foregut fermenters have larger, more complex, multi-chambered stomach, including a rumen. This is seen in the digestive system shown (1 mark). • The presence of a spiral loop and a smaller caecum (compared to a hindgut fermenter) also supports the statement that this is a foregut fermenter (1 mark).
c. This is an explain question. You need to provide a detailed response that outlines the relationship between microbes and digestion.	Herbivores eat plant material that contains cellulose. This cellulose cannot be broken down by mammals as they do not possess the appropriate enzymes (1 mark). The microbes in the gut allow for cellulose to be broken down and digested, enabling the nutrients to be obtained by the herbivore (1 mark).

Digestive system in birds

The saying 'as rare as hen's teeth' reminds us that modern birds do not have teeth. The lack of teeth means that key differences can be seen between the digestive systems of birds and of mammals. However, as in mammals, the walls of the alimentary canal are composed of four major tissue layers: mucosa, sub-mucosa, muscularis and serosa.

Figure 3.57 shows the digestive system of a typical bird. Let's look at its components, starting with the mouth and ending with the cloaca.

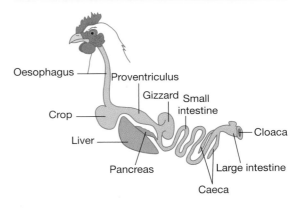

FIGURE 3.57 The components of the digestive systems of birds

Mouth

The beak consists of the upper and lower jaws covered by a layer of keratin, the same material that forms your fingernails. Beaks vary greatly in shape and size in different species. The beak size and shape may provide an indication of the diet of the bird, such as seed eaters with their short thick beaks for cracking seeds versus nectar-feeders with their long slender beaks for probing flowers.

Birds have no teeth. The absence of teeth means that no mechanical digestion occurs in a bird's mouth. Their preferred food items must be swallowed whole — OK for seeds, but not so easy for a fish (figure 3.58).

For most birds, the beak is their only means of obtaining and holding their food — the exceptions are raptors (birds of prey) and parrots that can also use their feet.

Oesophagus

The oesophagus in birds is a muscular and flexible tube that transports food from the mouth to the crop. It is relatively large in diameter compared to mammals, especially in those birds that swallow large prey whole, such as cormorants.

FIGURE 3.58 A bird with its preferred dietary item of fish. The absence of teeth means that the fish must be swallowed whole.

Crop

In birds the crop is an out-pocketing of the oesophagus located in the lower neck region and is capable of stretching. It provides temporary storage for food before it is passed to the stomach for digestion — this enables birds to consume large amounts of food quickly and then move to a safer location for digestion to occur.

Stomach

The stomach in birds consists of two parts: the proventriculus and the gizzard.

The proventriculus is the first part of the stomach and its function is enzymic digestion, especially of protein. It is lined by thick glandular epithelium that secretes mucus, hydrochloric acid and the inactive proenzyme, pepsinogen. As in the human stomach, pepsinogen is converted to active pepsin enzyme in the acid condition of the proventriculus.

The gizzard is the second part of the stomach and consists of a strong muscular sac that functions in mechanical digestion (figure 3.59).

It grinds, mashes and mixes foods, a function that is very important for animals without teeth, and creates a much larger surface area for enzyme action. It may contain grit and small stones (gastroliths) that birds swallow to assist the gizzard function.

FIGURE 3.59 Chicken gizzard. Note the strong muscular sac which may contain small stones to aid in mechanical digestion.

The gizzard has a structure that includes several layers of smooth muscle, a glandular epithelium, and a thick lining of tough non-cellular material (cuticle) that protects the underlying tissues from the acidic conditions within. As the cuticle lining is worn away by the grinding action of the gizzard, it is replaced by fluid material secreted by the glandular epithelium. This fluid moves to the surface of the cuticle where it hardens.

The gizzard is also present in the digestive tract of other animals, including crocodiles, and evidence for its existence in dinosaurs has been found. (What might that evidence be?)

Food moves back and forth between the two parts of the stomach creating cycles of digestion, grinding, more digestion, more grinding.

Small intestine

The small intestine is similar to that of mammals and is composed three segments: duodenum, jejunum and ileum. A single layer of columnar epithelium together with mucus-secreting goblet cells form the inner lining of the small intestine. Its functions are similar to those in the mammalian digestive tract.

The duodenum receives digestive enzymes from the pancreas and bicarbonate (to neutralise the acid from the proventriculus), and receives bile from the liver and pancreatic fluid from the pancreas. It is the site where most of the digestion of food by enzymes produced by a bird occurs and where digestion is completed.

The jejunum and ileum are the major sites of absorption of digested nutrients and water.

Large intestine

The large intestine consists of a pair of caeca and a colon.

The caeca are two blind pouches located at the junction of the small and large intestine. They are sites for microbial fermentation of any remaining matter that has not been digested in the duodenum, such as cellulose. Products generated through this fermentation can be used by birds as an energy source.

The colon is a short tube that ends at the cloaca. Its primary function is the absorption of water and electrolytes. The cloaca is a wide tube that is the common opening of the digestive, reproductive and urinary systems. It opens to the outside of the bird through the vent.

Resources

eWorkbook Worksheet 3.5 Different digestive systems in animals (ewbk-3494)
Weblink Health conditions of the digestive system

KEY IDEAS

- Food for an animal must contain organic matter that can be broken down and absorbed and used as chemical energy for living and the organic matter to build and repair its own structures.
- The digestive system typically breaks down macromolecules (proteins, carbohydrates, lipids (fats) to be absorbed and used for energy production, growth, maintenance and repair.
- The vertebrate digestive system includes:
  - alimentary canal (gastrointestinal tract) and the hollow organs of oesophagus, stomach and small intestine
  - accessory organs (solid organs): salivary glands, liver, gall bladder, pancreas.
- Tissues of the digestive system include:
  - muscosa (epithelial tissue)
  - sub-mucosa (connective tissue)
  - muscularis (muscle tissue)
  - serosa (connective tissue).
- Epithelial tissue has a number of types; its functions include protection, absorption and secretion.
- Connective tissues perform various functions: they bind and support, they protect and insulate, they store reserve energy as fat and they transport materials.
- Muscle tissues in the digestive system are largely smooth involuntary muscle cells — the exceptions are the muscle tissues of the throat and upper region of the oesophagus and that at the end of the anal canal.
- Coordinated contraction of muscle tissues controls peristalsis along the length of the system from oesophagus to anus.
- Regulation of movement between segments of the gut are controlled by sphincters.
- The pyloric sphincter separates the large acidic stomach from the smaller alkaline duodenum.
- Digestion of food relies on many enzymes throughout the digestive system.
- Organs of the digestive system include:
  - Mouth: teeth provide mechanical breakdown; enzymes in saliva begin chemical breakdown.
  - Oesophagus: transports food to the stomach via peristalsis
  - Stomach: temporary holding chamber for food; three layers of muscle actively churn the contents, acidic conditions and active enzymes continue digestion of fats and start digestion of proteins.
  - Liver (accessory organ): produces bile which aids in digestion
  - Pancreas (accessory organ): produces enzymes
  - Small intestine: the final stage of digestion and absorption of nutrients; mucosa distinguished by villi, crypts and microvilli.
  - Large intestine: includes the caecum, colon, rectum and anus. Functions include reabsorption of water, formation and storage of faeces, elimination of faeces and maintenance of gut bacteria.
- Animals can be classified as herbivores, carnivores and omnivores; their teeth and digestive systems reflect their diet, including length of alimentary canal, number of stomach chambers and the microbial fermentation processes.
- Digestive system of birds includes mouth (beak, no teeth), oesophagus, crop, stomach (proventriculus and gizzard), small intestine and large intestine.

3.4 Activities

To answer questions online and to receive **immediate feedback** and **sample responses** for every question, go to your learnON title at **www.jacplus.com.au**. A downloadable solutions file is also available in the resources tab.

| 3.4 Quick quiz | 3.4 Exercise | 3.4 Exam question |

3.4 Exercise

1. What is an accessory organ in the digestive system?
2. Identify the substances that contribute to the digestion process from the salivary gland and the liver.
3. Name the following:
   a. The structure that controls the flow of chyme from stomach to duodenum
   b. The enzyme that can function in both a neutral and acidic environment
   c. The least complex kind of animal that has a one-way digestive system
   d. The two major locations where mechanical digestion of food occurs
   e. The organ where most digestion of food occurs
   f. The organ where most reabsorption of water occurs
4. a. List the four tissue layers of the alimentary canal starting from the lumen of the gut.
   b. Identify the type of epithelium that is present in the mucosa of the oesophagus and small intestine, and their major function(s).
5. a. You examine two light microscope images of a cross-section of the wall of the alimentary canal. Image A shows a mucosa with many villi and crypts. Image B shows a smooth mucosa with many crypts. Identify which region the images of the alimentary canal might have come from.
   b. You are told that gastric pits are present in the surface of the organ of image B. Would you now revise your answer to part a or not?
6. Most people with the inherited disorder cystic fibrosis show various symptoms, including what is termed exocrine pancreatic insufficiency (EPI). EPI shows as a progressive loss of acinar tissues in the pancreas. Treatment includes enzyme replacement capsules.
   a. Would you expect that these enzymes would be enzyme precursors (proenzymes)?
   b. What enzymes might be present in these capsules?
7. a. What is a brush border?
   b. Identify where the brush border occurs in the alimentary canal.
   c. Identify one role of the brush border.
8. After they are absorbed, how are the following digested nutrients transported from the alimentary canal?
   a. Amino acids
   b. Fats
9. What is a benefit to the mammalian host of its gut microbiota?
10. What reasonable statements may be made about an animal that displays the following features? Justify your responses.
    a. High-crowned molars with large grinding surfaces
    b. An extremely short alimentary canal
    c. A multi-chambered stomach
    d. Carnassial teeth
    e. A gizzard
11. Which animal would be expected to have the longest alimentary canal relative to its body size: a grass-eating herbivorous mammal or a flesh-eating carnivorous mammal? Explain.

3.4 Exam questions

Question 1 (1 mark)

MC In humans, mechanical digestion of food
A. occurs in the mouth, stomach and small intestine.
B. occurs only in the mouth and stomach.
C. requires the presence of an enzyme.
D. occurs best at a temperature of 37 °C.

▶ **Question 2 (1 mark)**
MC The major type of macromolecule broken down in the stomach is
A. carbohydrate.
B. protein.
C. lipid.
D. nucleic acid.

▶ **Question 3 (1 mark)**
MC Humans do not have digestive enzymes capable of acting on
A. starch in potato.
B. sucrose in peaches.
C. polypeptides in meat.
D. cellulose in breakfast cereals.

▶ **Question 4 (1 mark)**
MC Most monomers of organic compounds are absorbed from the digestive system and into the blood stream when in the
A. oesophagus.
B. stomach.
C. small intestine.
D. large intestine.

▶ **Question 5 (2 marks)**
People with cystic fibrosis produce abnormally thick mucus that can block the ducts of the pancreas. Explain how this disease affects the human digestive system.

More exam questions are available in your learnON title.

3.5 Endocrine system in animals

KEY KNOWLEDGE

- Specialisation and organisation of animals cells into tissues, organs and systems with specific functions: endocrine

Source: Adapted from VCE Biology Study Design (2022–2026) extracts © VCAA; reproduced by permission.

3.5.1 What is the endocrine system?

The endocrine system is composed of a network of **endocrine glands** and it functions as a chemical messenger system. Among the many glands of the endocrine system are the pituitary gland, the adrenal glands, the thyroid gland and the parathyroid glands (figure 3.60).

endocrine glands ductless glands that distribute hormones via the bloodstream

The endocrine system and the nervous system are both major communicators of messages to body cells. One difference is that the nervous system uses nerves to transmit information, while the endocrine system uses blood vessels to deliver its chemical messages. Which types of message — nervous or hormonal — would be expected to be acted on more quickly? Which would be expected to be longer-lasting?

FIGURE 3.60 Diagram showing the glands of the endocrine system

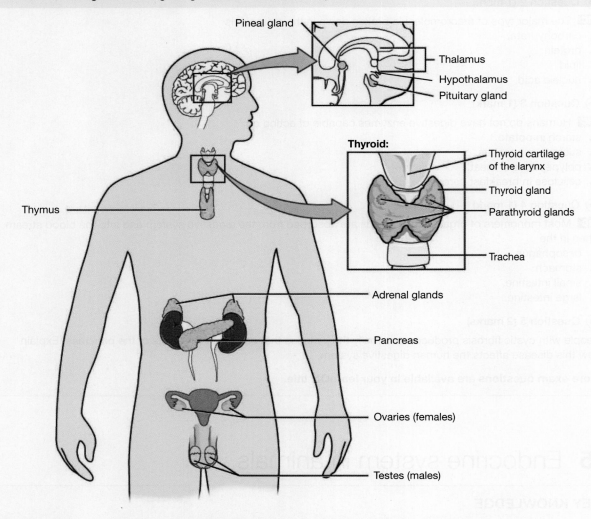

Hormones are the chemical messengers produced by the endocrine glands of the endocrine system. They can be:
- proteins, polypeptides or amino acid derivatives such as antidiuretic hormone. Protein hormones are **hydrophilic** and so cannot diffuse easily across the plasma membrane. Hence, their complementary receptors are on the plasma membrane.
- steroids, which are synthesised from cholesterol and include cortisol and aldosterone (figure 3.61). Steroid hormones are **hydrophobic** and so can easily diffuse across the plasma membrane. Hence, their complementary receptors are in the cytosol/nucleus.

Hormones regulate many body functions, such as metabolism, growth, the concentration of glucose and electrolytes in body fluids, and sexual development.

Hormones carry messages to their target tissues. For example:
- 'secrete T3 and T4 hormones' — a message to the thyroid carried by thyroid-stimulating hormone
- 'reabsorb more calcium' — a message to cells of the small intestine carried by parathyroid hormone.

The target cells have specific receptors for their respective hormones that enable them to receive and respond to the message carried by the hormone (figure 3.62). Receptors for protein hormones are located on the surfaces of their target cells, while receptors for steroid hormones are located within the target cells.

hormones chemical messengers, released by endocrine glands, that regulate the function of distant organs, each with a specific receptor for its hormone

hydrophilic substances that dissolve easily in water; also called polar

hydrophobic substances that tend to be insoluble in water; also called non-polar

FIGURE 3.61 a. Cortisol: a steroid hormone released by the adrenal glands **b.** Antidiuretic hormone (ADH): a peptide hormone released by the pituitary gland that regulates the water content of blood

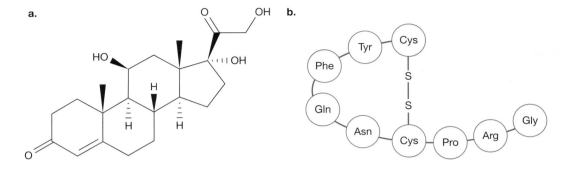

FIGURE 3.62 The delivery of a hormone via the bloodstream to its specific target cell that carries on its surface the correct receptor. Only cells with the matching receptor can accept the messenger.

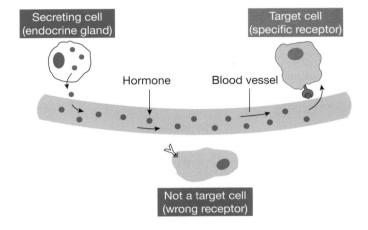

Endocrine glands:
- release hormones *directly* into the bloodstream that carries them to their distant target cells
- do not have ducts and so differ from the type of gland found in the digestive system that releases its secretions via a duct to a surface, such as the skin or the mucosal surface of the gut (figure 3.63)

FIGURE 3.63 a. Exocrine gland such as an intestinal crypt in the small intestine that releases its secretion to the surface via a duct **b.** Endocrine gland that releases its hormone directly into a blood capillary

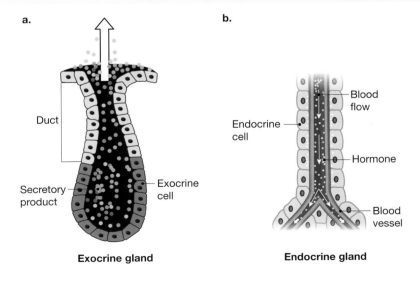

Table 3.7 shows some of the glands of the endocrine system, the hormones they release, the target organ of each hormone and the signal that it communicates.

TABLE 3.7 Summary of hormones produced by some glands of the endocrine system, their target organ and the signals or messages each hormone communicates

Endocrine gland	Hormone released	Target organ	Signal to target
Hypothalamus	Thyrotropin-releasing hormone (TRH)	Anterior pituitary gland	Start releasing thyroid-stimulating hormone (TSH)
	Growth-hormone releasing hormone (GHRH)	Anterior pituitary gland	Start releasing growth hormone (GH)
	Growth-hormone inhibitory hormone (GHIH)	Anterior pituitary gland	Stop releasing growth hormone (GH)
	Corticotitropin-releasing hormone (CRH)	Anterior pituitary gland	Start releasing ACTH (adreno-corticotropic hormone)
	Gonadotropin-releasing hormone (GnRH)	Anterior pituitary gland	Start releasing FSH or LH
Anterior pituitary gland	Growth hormone (GH)	Liver, fat Muscle and bone tissues	Release IGF-1 Build more tissue
	Adrenocorticotropic hormone (ACTH)	Adrenal glands	Secrete cortisol
	Thyroid-stimulating hormone (TSH)	Thyroid gland	Secrete T3 and T4 hormones
	Prolactin (PRL)	Mammary glands	Produce milk
	Follicle-stimulating hormone (FSH)	Ovaries/Testes	Produce gametes (eggs or sperm)
	Luteinising hormone (LH)	Ovaries/Testes	Produce sex hormones (oestrogens or testosterone)
Posterior pituitary gland	Antidiuretic hormone (ADH)	Kidneys	Raise blood pressure by increasing water reabsorption
	Oxytocin (OXT)	Uterus	Increase birth contractions
Thyroid gland	T3 and T4 hormones	Many organs	Raise the metabolic rate
	Calcitonin (CT)	Bones and kidneys	Lower blood calcium levels
Adrenal glands	Aldosterone	Kidneys	Control blood pressure by selective reabsorption of water and ions
	Cortisol	Ovaries and testes	Activate the stress response
	Androgens	Many organs	Convert precursor androgens to female or male sex hormones
	Adrendaline and noradrenaline	Many organs	Begin the 'fight or flight' response: increase heart rate and blood pressure
Parathyroid glands	Parathyroid hormone (PTH)	Bones and kidneys	Raise the blood calcium level

IGF-1 = insulin like growth factor 1, a hormone that promotes growth of muscle and bone

Note that:
- the hypothalamus produces hormones that target various cells of the anterior pituitary gland
- hormones released by the anterior pituitary gland target other endocrine glands
- the pathway to the final action of one hormone may involve earlier action by other hormones; for example, the release of the thyroid hormones, T3 and T4, requires the prior action of two hormones as shown.

Interrelationship between thyroid hormones

The release of thyroid hormones, T3 and T4, requires the prior action of other hormones as shown in the pathway provided.

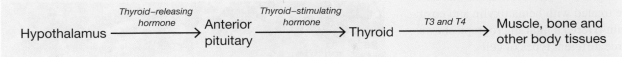

Resources

Digital document Extension: Hormones in the digestive system — without a gland (doc-35879)

3.5.2 Tissues in the endocrine system

Each of the organs of the endocrine system are composed of different tissue types which are discussed with reference to their particular organ below. The role of each of these organs will be examined in section 3.5.3.

Pituitary gland

The pituitary gland:
- is located at the base of the brain immediately below the hypothalamus
- is composed of two major lobes: the **anterior pituitary** and the **posterior pituitary**, that are separated by a thin strip of tissue.

Figure 3.64 shows the two lobes of the pituitary gland and their close proximity to the hypothalamus. The anterior and the posterior lobes differ markedly in their structure and function and are discussed separately below.

anterior pituitary anterior lobe of the pituitary gland; it is made of glandular tissue that synthesises and secretes several releasing hormones that activate other endocrine glands.

posterior pituitary posterior lobe of the pituitary gland; it is made of neural tissue that stores and releases hormones sent from the hypothalamus.

FIGURE 3.64 Simplified diagram showing the two major lobes of the pituitary gland: the anterior and the posterior

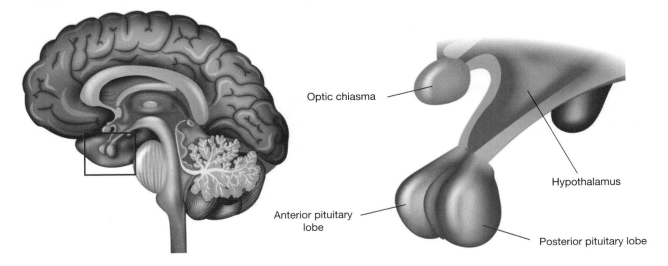

Anterior pituitary

The anterior pituitary is composed of winding cords of glandular epithelial cells with capillaries interspersed throughout. When sections of this tissue are treated with hematoxylin and eosin stain (H&E) a light micrograph shows three different types of cells:
- acidophil (acid-loving) cells with pink cytoplasm — these cells produce and secrete hormones
- basophil (base-loving) cells with bluish cytoplasm — these cells produce and secrete hormones
- chromophobes (colour-hating) — these cells are fewer in number, take up little stain and do not produce any hormone.

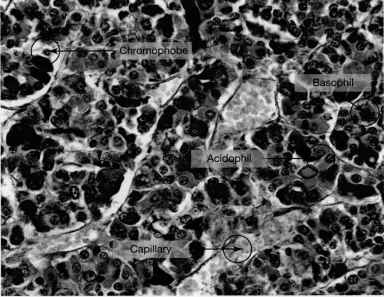

FIGURE 3.65 Light micrograph of a section through the anterior pituitary gland showing acidophil cells (bright pink cytosol) and basophils (blue cytosol). Small red blood cells can be seen in the blood vessels.

Figure 3.65 is a cross-section through the anterior pituitary gland stained with H&E that shows the hormone-producing acidophils and basophils. Note the presence of many capillaries, which transport the hormone released by each type of cell to its target tissue.

The H&E stain does *not* identify the type of hormone produced by an individual secretory cell in the anterior pituitary. However, more specialised staining techniques show that each individual acidophil and basophil produces just one hormone, as follows:
- acidophils that produce either growth hormone (GH) or prolactin (PRL)
- basophils that produce one of thyroid-stimulating hormone (TSH), follicle-stimulating hormone (FSH), luteinising hormone (LH) or adrenocorticotropic hormone (ACTH).

The hormones produced by these various cells are stored in granules in their cytosol; the size and the location of the granules differs in these various cells. The release of stored hormones is triggered by a specific hormone from the hypothalamus; for example, thyrotropin-releasing hormone (TRH) released from the hypothalamus stimulates specific basophil cells to release their stored thyroid-stimulating hormone (TSH).

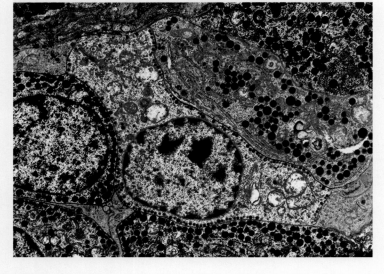

FIGURE 3.66 False colour TEM showing several endocrine cells of the anterior pituitary. The orange profiles belong to a support cell.

Figure 3.66 shows a transmission electron micrograph of several cells of the anterior pituitary. Note the variation in the size of their granules and their location within these cells. The red-coloured cell is a basophil that produces thyroid-stimulating hormone (TSH), the green-coloured cell is an acidophil that produces prolactin (PRL), and blue-coloured cells are basophils that produce growth hormone (GH). Note that the colours used in this micrograph are not related to the colours of cells in H&E-stained tissue.

Posterior pituitary

In light microscope views, the posterior pituitary looks very different from the anterior pituitary — the posterior lobe is much paler. Figure 3.67a shows a cross-section through a human pituitary gland stained with H&E. The posterior pituitary is in the lower area, the anterior pituitary is in the upper area and they are separated by a thin strip of tissue. The areas of white in the posterior pituitary are nerve fibres (axons) and nerve endings that are packed with secretory granules — no acidophils and basophils. Figure 3.67b shows a section of the posterior pituitary treated with a special stain that reveals its neurosecretory elements and shows their activity.

FIGURE 3.67 a. H&E stained posterior pituitary at lower right and the anterior pituitary at upper left **b.** Section of the posterior pituitary showing the areas of activity of neurosecretory cells (blue) and the blood-filled capillaries (orange)

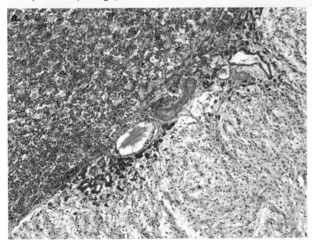

The posterior pituitary is composed of tissue that differs from the glandular epithelial tissue of the anterior pituitary. This is nervous tissue that consists of nerve fibre (axons) and nerve endings that originate in nerve cells located in the hypothalamus (figure 3.68). These cells are **neurosecretory cells** — that is, they receive nerve impulses from the brain and respond by releasing hormones. The tissue includes support cells and contains many capillaries.

The posterior pituitary does not produce hormones itself; instead it receives hormones — oxytocin (OCT) and antidiuretic hormone (ADH), which are produced by clusters of nerve cells in the hypothalamus. These hormones are transported to the posterior pituitary via nerve fibres. The hormones are stored in granules in the nerve endings.

neurosecretory cells cells that receive nerve impulses and respond by a chemical stimulus

FIGURE 3.68 Line diagram showing the nerve connection between the hypothalamus and the posterior pituitary. The OCT and ADH hormones from the hypothalamus move to the posterior pituitary via these nerve fibres.

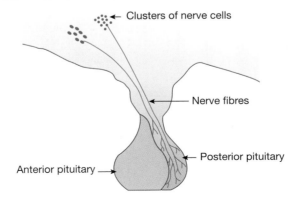

Adrenal glands

The **adrenal glands** are paired endocrine glands that are located on the top of the kidneys.

Each adrenal gland is enclosed in a fibrous connective tissue capsuleand has two distinct regions:
- an outer cortex
- an inner medulla.

Both regions are richly supplied with blood capillaries, and, in addition, the medulla has many nerve fibres and nerve endings.

Figure 3.69 is a cross-section of an adrenal gland showing the medulla with blood vessels in the centre of the gland surrounded by the outer cortex. Note, in particular, that the cortex is not uniform and shows some layering. Also visible is the fibrous capsule that encloses the gland.

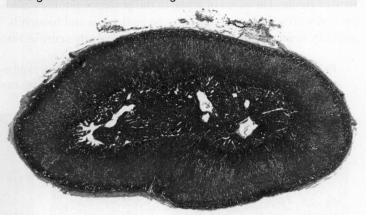

FIGURE 3.69 Light micrograph of a section of the adrenal gland showing the medulla with blood vessels in the centre of the gland and the surrounding cortex

The adrenal cortex

The adrenal cortex is composed of glandular tissue. It produces and releases various **corticoid hormones** — these are produced in the cortex of adrenal glands and are steroids.

adrenal glands small endocrine glands located on top of the kidneys that produce various hormones

corticoid hormones steroid hormones, including aldosterone and cortisol, produced by cells of the cortex of adrenal glands

FIGURE 3.70 a. The arrangement of the cells of the adrenal cortex to form three distinct zones that lie between the capsule and the medulla **b.** Light micrograph of a section of the adrenal gland showing the zonation of the adrenal cortex

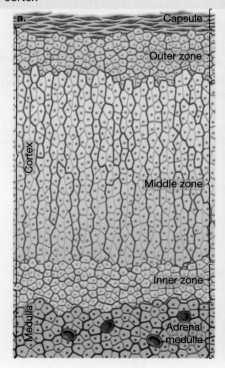
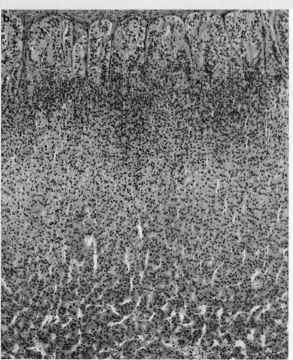

The adrenal cortex is not uniform in microscopic appearance, and three zones can be identified based on the different organisation of the cells, which each produce specific hormones:
- the outer zone: cells are cuboidal to columnar in shape and are arranged in clusters. These cells cells produce **aldosterone**, which regulates blood pressure and the salt and water balance in the body.
- the middle zone: cells produce **cortisol**, which plays a role in regulating blood sugar levels and in recovery from the **stress response**
- the inner zone: the thinnest zone, with small cells arranged in clusters that cross and intersect creating a net-like appearance. These cells secrete androgens, the precursors of sex hormones.

The adrenal medulla

The adrenal medulla forms the central part of the adrenal gland and is composed of cells called chromaffin cells that are highly modified nerve cells.

Chromaffin cells produce the hormones, adrenaline and noradrenaline, which are very rapidly released in physically and emotionally stressful situations. These hormones are released in response to impulses from nerves connected directly to them. The release of hormones in response to nerve impulses occurs much more rapidly than for hormones released in response to signals carried to their endocrine glands via the bloodstream.

The flood of adrenaline into the bloodstream starts the 'flight or fight' response that has several effects, including an increase in heart rate and the release of glucose into the bloodstream to provide fuel for cells. (You will no doubt have felt that flood of adrenaline when you received a sudden fright.)

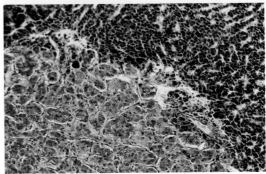

FIGURE 3.71 Light micrograph of a section of an adrenal gland showing innermost zone of the adrenal cortex (upper right) and the cells of the medulla (lower left)

Figure 3.71 is a light micrograph of a section through the adrenal gland at the junction of the medulla (at lower left) with inner zone of the cortex (at upper right). The cells of the inner zone of the cortex are dark-stained secretory cells that produce sex hormone precursors. The paler chromaffin cells of the medulla are organised into clumps of nervous tissue that secrete adrenaline and noradrenaline.

Thyroid gland

The **thyroid gland**, commonly described as a butterfly-shaped organ, is located in the neck around the trachea and below the larynx.

The thyroid gland has a rich blood supply with capillaries in close association with its secretory cells. It is composed of many structural units called **follicles**. They are lined with a single layer of cuboidal epithelial cells called follicular cells. The follicles are filled with **colloid** which consists of thyroglobulin, which is needed to produce the major thyroid hormones, T3 and T4. The thyroid gland has the only cells of the body that can absorb iodine, which is needed to produce T3 and T4.

aldosterone a steroid hormone produced by the adrenal cortex that regulates levels of salt and water and so controls blood pressure

cortisol a steroid hormone produced by the adrenal cortex that has many roles, including control of blood glucose levels during stress and the body's recovery from the stress response

stress response combined physiological reactions to stress, also known as the 'fight or flight' response, that results from release of the hormones, adrenaline and noradrenaline from the adrenal medulla

thyroid gland endocrine gland located in throat that produces and secretes the hormones including T3 and T4

follicles secretory cells that enclose the thyroid follicles and that synthesise and release the hormones, T3 and T4

colloid gel-like material inside follicles that is the source of thyroglobulin, a prohormone of the T3 and T4 thyroid hormones

The thyroid gland also contains a number of cells called **parafollicular cells** which produce and secrete calcitonin, another thyroid hormone. Parafollicular cells are typically found among groups of cells in regions between follicles.

parafollicular cells secretory cells, located in small areas between follicles, that produce the thyroid hormone, calcitonin

chief cells in the parathyroid gland, secretory cells that produce parathyroid hormone (PTH) (Note: different cells with same name are present in the gastric pits of stomach.)

parathyroid hormone a protein hormone synthesised and secreted by the parathyroid glands in response to a falling blood calcium levels

adipocytes cells with large deposits of stored fat

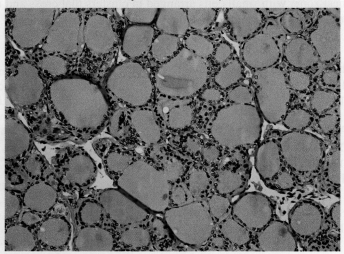

FIGURE 3.72 Light micrograph of a cross-section of thyroid gland showing its many follicles, each filled with colloid, that is used by follicle cells to produce T3 and T4

Parathyroid glands

The parathyroid glands are endocrine glands located on the rear surface of the thyroid gland. The parathyroid glands are surrounded by a thin fibrous capsule and have networks of capillary vessels throughout their tissue.

The glandular tissue of the parathyroid glands contains several types of cell, including:
- **chief cells** that produce **parathyroid hormone** (PTH), which regulates calcium levels; PTH is secreted in response to falls in blood calcium levels — these are the smallest and most numerous cells and have with intensely-stained nuclei
- fat cells (**adipocytes**), which are large cells with clear contents
- other cell types of unknown function.

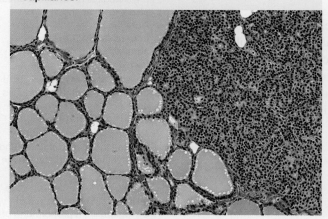

FIGURE 3.73 Light micrograph showing parathyroid gland tissue (right half) embedded within and attached to the thyroid gland tissue (left half). The thin red areas in the parathyroid tissue are blood capillaries.

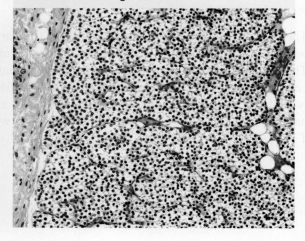

FIGURE 3.74 High power micrograph of a section through a parathyroid gland stained with H&E. Note the large fat cells and the numerous hormone-secreting chief cells.

3.5.3 Organs involved in the endocrine system

Pituitary gland

The pituitary gland is a pea-shaped endocrine gland, about 1 centimetre in diameter, that is located at the base of the brain immediately below the hypothalamus (figure 3.75). Two major components of the pituitary gland are its anterior and posterior lobes. The lobes are both formed from epithelial tissue during embryonic development: the anterior lobe is formed from surface tissue while the posterior lobe is formed from nerve tissue.

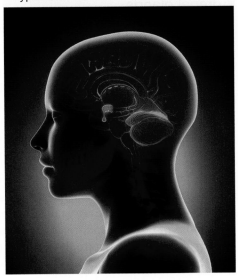

FIGURE 3.75 Image showing the pituitary gland linked to the hypothalamus

The function of the pituitary gland is to form the link between the brain and other endocrine glands through a series of hormones (see figure 3.76). As a result, it is the target of a number of hormones produced in the hypothalamus of the brain:
- some are *releasing hormones* that stimulate the anterior pituitary to produce and secrete a specific hormone
- one is an *inhibiting hormone* that stops the secretion of a specific hormone by the anterior pituitary
- some are stored in the posterior lobe for later release.

Note that in the case of the anterior pituitary, its targets include other endocrine glands (adrenal cortex and thyroid) as well as other body organs. For the posterior pituitary, its targets are the kidney and the uterus.

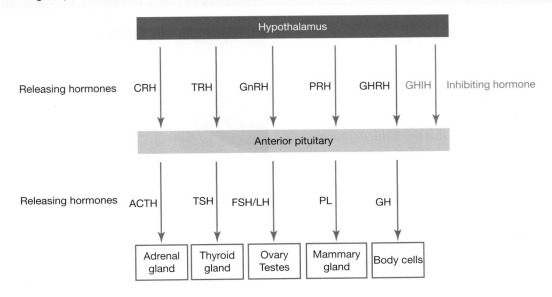

FIGURE 3.76 Diagram showing the link between the hormones of the hypothalamus, the anterior pituitary and the final targets (Hormone abbreviations are shown in table 3.7.)

The level of any hormone must be carefully regulated since an oversupply or undersupply will cause serious problems. For example:
- Oversupply: if an excess of growth hormone occurs in childhood when growth plates of the long bones are still unfused, gigantism will result, as seen in Sultan Kösen (see Case study below). If an oversupply occurs in adulthood, it will result in acromegaly, with excessive growth of hands, feet, jaw and nose.
- Undersupply: if a deficiency of growth hormone occurs, it results in growth retardation or dwarfism.

Control systems are normally in place to ensure that the levels of circulating hormone in the blood are maintained within narrow limits. For most of the hormones of pituitary gland, a key mechanism for hormone regulation is through negative feedback. In negative feedback, a hormone inhibits its own further production by blocking the glands(s) that produce it when the hormone concentration or its effects exceed or drop below the normal range. Feedback is discussed in detail in topic 4, section 4.4.3.

Adrenal glands

The adrenal glands are a pair of organs, each located on the upper margin of a kidney (figure 3.77).

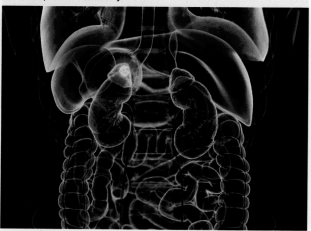

FIGURE 3.77 The location of the adrenal glands on the top of the kidneys

The adrenal glands are composed of two parts — the cortex and the medulla — which synthesise and secrete different hormones. The cortex produces the steroid hormones cortisol and aldosterone. The medulla produces the hormones adrenaline and noradrenaline, which are derived from amino acids.

The secretion of cortisol and aldosterone is regulated by a system comprising the hypothalamus, the anterior pituitary and the adrenal glands and involves the actions of corticotropin-releasing hormone (CRH) and adrenocorticotropic hormone (ACTH). The regulation of hormone levels involves negative feedback (see section 4.4.3).

Thyroid gland

The thyroid gland consists of two lobes, joined by connecting tissue, located in front of the trachea in the lower neck. The size of the thyroid varies with age and sex, but for an adult, the length is 40 to 60 millimetres and the antero-posterior diameter is 13 to 18 millimetres.

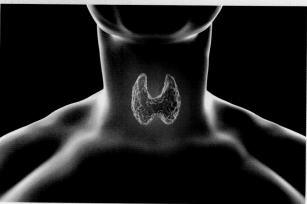

FIGURE 3.78 The thyroid gland with its two lobes near the base of the throat

The thyroid gland secretes three hormones:
- T4 (thyroxine) and T3 are produced by its follicular cells. Note that the label, thyroid hormone (TH), is a convenient term to encompasses both hormones.
- calcitonin is produced by its parafollicular cells in response to a rise in blood calcium levels.

Figure 3.79 is a high-power light micrograph image showing the relative locations of these cells. Note the blood vessels that are in close association with the thyroid cells — these will carry the hormone messengers to their target tissues.

The T3 and T4 thyroid hormones regulate metabolism, growth and development. They have an effect on almost all body systems, including:
- increasing basal metabolic rate (see section 4.4.2)
- mobilising fat from adipocytes (fat cells)
- increasing cardiac output
- increasing blood flow to the skeletal muscles, liver and kidney
- increasing glomerular filtration rates.

FIGURE 3.79 High-power light micrograph showing a section of thyroid that identifies the follicular cells that produce the hormones T4 and T3 and the parafollicular cells that produce calcitonin

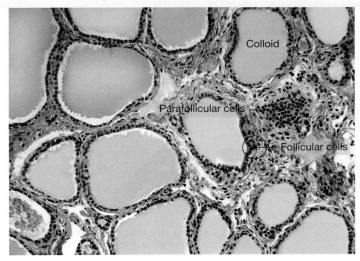

These thyroid hormones are critical for brain development in the developing foetus and during childhood. Figure 3.80 shows the structure of the amino acid tyrosine, and those of the T3 and T4 thyroid hormones. Note the presence of iodine atoms. A normal thyroid gland produces about 80 per cent T4 and about 20 per cent T3. T3 has a much stronger physiological effect than T4. T4 is enzymically converted to T3 in the thyroid and in other tissues including the liver.

FIGURE 3.80 Structural formulas of tyrosine, the amino acid that is a building block of the thyroid hormones T4 and T3. Note the presence of iodine atoms in the hormone structures.

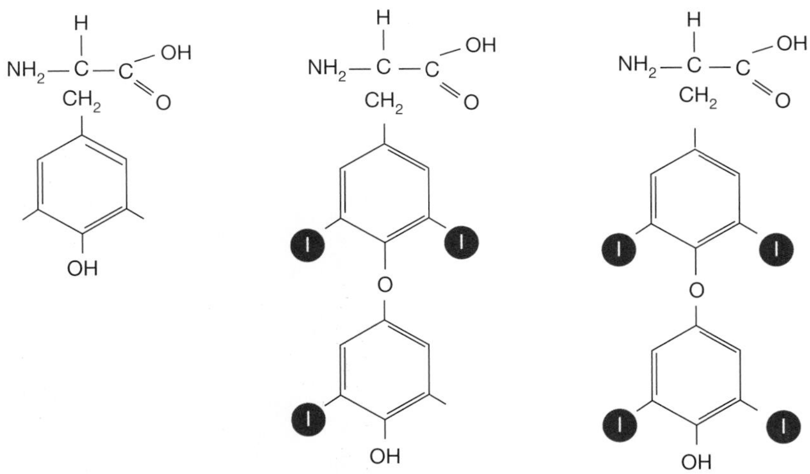

T3 and T4 hormones are produced in a series of steps by the follicular cells of the thyroid. They are the only hormones that contain iodine and this is obtained through a person's dietary intake.

Thyroid hormone production is regulated to ensure that T3 and T4 levels are neither too high nor too low. The elements in the regulation of the thyroid activity include the hypothalamus, which secretes thyrotropic-releasing hormone (TRH) and the anterior pituitary, which secretes thyroid-stimulating hormone (TSH). The mechanism of this regulation is negative feedback, which is discussed in section 4.4.2.

FIGURE 3.81 Production of T3 and T4 hormones

Thyroglobulin synthesis	Thyroglobulin, a protein, is synthesised by follicular cells and transferred into the colloid in the thyroid follicles.
Uptake and concentration of iodide (I⁻)	Iodide from dietary sources is actively taken up from the blood by follicular cells, accumulates and is then transported into the follicle adding to the colloid.
Oxidation of iodide (I⁻) to iodine (I)	In the colloid, iodide is converted to iodine.
Iodination of thyroglobulin	Iodine atoms form bonds with one particular amino acid residue (tyrosine) that is present in multiple numbers in the polypeptide chain of thyroglobulin.
Formation of MIT and DIT	MIT (mono-iodo-tyrosine) is a tyrosine residue with one iodine attached. DIT (di-iodo-tyrosine) is a tyrosine with two iodines attached. MIT + DIT = T3 and DIT + DIT = T4.
Digestion and secretion	Thyroglobulin with its iodinated tyrosine amino acids is taken from the colloid into the follicular cells where proteases cut the MITs and DIT molecules free and they join, forming T3 and T4 hormones.

Parathyroid glands

The parathyroid glands are endocrine glands on the posterior surface of the thyroid gland. Typically, a person has four parathyroid glands, each about 5 millimetres long, 3 millimetres wide and 1 to 2 millimetres thick. Two parathyroid glands lie behind each lobe of the thyroid gland (figure 3.82).

The parathyroid glands produce a hormone called parathyroid hormone (PTH), which is involved in the regulation of calcium blood levels — that is, the maintenance of calcium homeostasis. Secretion of calcitonin, a thyroid hormone, is stimulated by increases in the serum calcium concentration. Calcitonin protects against the development of hypercalcemia.

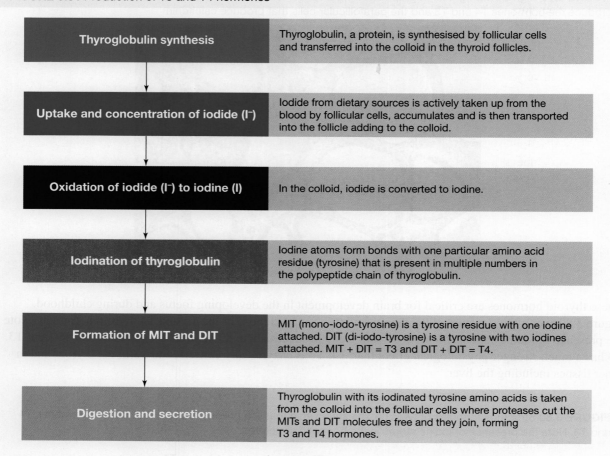

FIGURE 3.82 The parathyroid glands are located on the surface of the thyroid.

The main target organs of parathyroid hormone are bones and the kidneys. When calcium levels are low, parathyroid hormone (PTH) is released into the blood, causing the calcium levels to return to normal. Mechanisms by which this is achieved include:
- loss of calcium from bone — this is achieved when bone cells (called osteoclasts) are stimulated by the hormone to resorb bone from the surface of particular bones, releasing that calcium into the blood.

- decreased loss of calcium in urine — this is achieved through increased reabsorption of calcium by cells of the kidney tubules.
- increased conversion of inactive vitamin D in the kidney to its active form, namely 1, 25-dihydroxy vitamin D; this, in turn, increases calcium absorption in the intestine.
- increased absorption of calcium in the intestine. In the absence of vitamin D, dietary calcium is not absorbed very efficiently.

Some people suffer from excessive secretion of parathyroid hormone, which leads to structural damage to bone and excess calcium in the blood. In many cases, this condition is due to a parathyroid tumour.

Note that the actions of a thyroid hormone, calcitonin, are opposite to those of parathyroid hormone. Its targets are bone and kidneys. Calcitonin acts to decrease blood calcium levels and it does this by inhibiting the activity of osteoclasts so preventing bone being absorbed, and it inhibits the resorption of calcium in the kidney tubules.

CASE STUDY: Sultan Kösen — the world's tallest man

In 2009, the name of a Turkish man was entered the *Guinness Book of Records* as 'the world's tallest living man'. He is Sultan Kösen, born in 1982, and his recorded height is now 2.51 metres. As a comparison, the Building Code of Australia identifies 2.4 metres as the minimum ceiling height for most rooms in new houses. Figure 3.83 shows a photo of Sultan Kösen and his brother taken in 2009. Sultan's abnormal growth spurt began during childhood and was the result of a non-cancerous tumour on his pituitary gland that caused it to release excessive growth hormone, which affected his bone and muscle growth. Surgery on the tumour in 2010 stopped the excessive production of growth hormone by his pituitary gland and this prevented Sultan from becoming any taller. Reading about Sultan Kösen reminds us that the hormones that are the products of the endocrine system exert major influences on the functioning of the human body. When the production of a particular hormone is either in excess or in deficit, negative and, in some cases fatal, effects on body function follow.

FIGURE 3.83 Sultan Kösen and his brother

tlvd-1853

SAMPLE PROBLEM 4 Explaining the role of iodine

A 9-year old patient was admitted to hospital with an enlarged thyroid gland. Through various tests, they were diagnosed as having an iodine deficiency.
a. Why is iodine important for the proper functioning of the endocrine system? **(2 marks)**
b. Suggest the impact of an iodine deficiency on the production of the T3 and T4. **(2 marks)**
c. With reference to both the pituitary and thyroid gland, propose a reason why an iodine deficiency may lead to an enlarged thyroid gland. **(2 marks)**
d. Outline some of the symptoms that may result from a poorly functioning thyroid. **(3 marks)**

THINK

a. The question requires simple statements. As the question is worth 2 marks, you should highlight two key points: the function of the endocrine system and why iodine is important for this function.

WRITE

The endocrine system is a chemical messaging system that uses hormones to regulate body functions (1 mark). Iodine is required to produce some of these hormones (1 mark).

b. You are asked to suggest, which involves providing a logical explanation with reasoning. This question requires two key points.
You should consider your answer around the need for iodine to produce T3 and T4.

Iodine is found in both the T3 and T4 hormones, which are produced in the thyroid gland from the amino acid tyrosine (1 mark).
Without iodine, T3 and T4 would not be able to be produced, leading to a reduction in the production of these hormones (1 mark).

c. Carefully read the question. You need to propose a reason for an enlarged thyroid gland. The question specifically states that you need to link to both the pituitary and the thyroid glands in your response to achieve both marks.

As the hormones are not being produced, more thyroid stimulating hormone will be released from the pituitary gland (1 mark). As the thyroid gland is unable to make enough T3 and T4 hormones, it is likely that the thyroid gland will enlarge to try to increase and amplify the production of the T3 and T4 hormones (1 mark).

d. This is an outline question, so brief statements are required.
T3 and T4 act to increase basal metabolic rate, cardiac output, blood flow and glomerular filtration. They also mobilise fat and aid in brain development. Use these points to consider what a deficiency would lead to.
TIP: As the question is worth 3 marks, you should make 3 clear points. These may be written as a bulleted list. If you make 5 points, but some are incorrect, you will lose marks, so pick 3 points that you are most confident with.

T3 and T4 have an effect on metabolism, growth and development. Some of the symptoms of a deficiency in these hormones for a poorly functioning immune system may include:
- a lower than normal body temperature (due to a lower metabolic rate) (1 mark)
- low blood flow to muscles, liver and kidney (1 mark)
- reduced brain development and possible intellectual disability (1 mark).

 Resources

 eWorkbook Worksheet 3.6 The human endocrine system (ewbk-3496)

3.5.4 Endocrine systems in different animals

Communication in the form of chemical signalling between cells is present in *all* animals. Chemicals that carry a signal from one cell to other cells include hormones and nuerotransmitters.

The first cells involved in chemical communication were probably neurosecretory cells, that is, nerve cells that when stimulated release a chemical signal (hormone). The least complex means of endocrine communication is the individual cells that produce and release chemical signals that travel by diffusion to neighbouring target cells — there are no specialised endocrine glands and no involvement of a circulatory system. The most complex means of endocrine communication is an endocrine system with specialised glands, closely associated with the nervous system and the circulatory system — as seen, for example, in the endocrine system of vertebrates.

Insects are among the first animal groups in which specialised endocrine organs can be seen.

Communication through chemical signalling occurs in all animals via hormones.
The simplest form of intercellular communication is by the diffusion of a chemical signal released by one cell that diffuses neighbouring target cells, with no endocrine glands or involvement of the circulatory system.

Endocrine system of insects

The insect endocrine system produces a large number of hormones. These hormones are mainly synthesised by the glandular cells and the neurosecretory cells of endocrine glands. Many of the hormones are small peptides, but some are steroids, and they regulate growth, development, homeostasis, reproduction and behaviour.

Figure 3.84 shows a stylised representation of the major organs of the insect endocrine system. The hormones identified are those that regulate moulting and development.

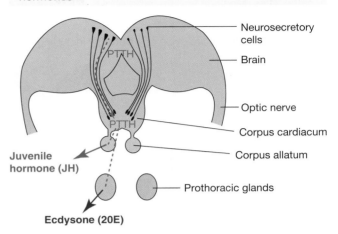

FIGURE 3.84 Stylised representation of the major organs of the insect endocrine system and some major hormones

The major organs of the insect endocrine system are:
- the brain, which has clusters of neurosecretory cells that produce hormones, including prothoracicotropic hormone (PTTH), allotropin and allostatin
- corpora cardiaca (singular = corpus cardiacum), which are paired neuroglandular bodies that produce their own neurohormones. They receive, store and release neurohormones produced in the brain, including PTTH.
- corpora allata (singular = corpus allatum), which are paired glands. Their glandular cells produce and secrete **juvenile hormone** (JH) in response to stimulation by the hormone allotropin. Another hormone, allostatin, inhibits the production of JH by the corpora allata.
- prothoracic glands, which are paired diffuse glands located partly in the thorax. Their glandular cells produce and secrete the hormone ecdysone (a steroid hormone derived from cholesterol). Ecdysone is a prohormone and is converted to its active form, 20E (20-hydroxy ecdysone) after its release into the hemolymph (body fluid).

The signal to produce ecdysone originates in neurosecretory cells in the brain that release PTTH and this hormone stimulates the prothoracic glands to secrete **ecdysone**.

> **juvenile hormone** insect hormone in the nymphal and larval stage of all insects; it is a lipid, derived from a fatty acid.
>
> **ecdysone** steroid hormone which controls moulting in insects
>
> **moult** in insects, the periodic event during development of larvae or nymphs which involves growth in epidermal cell numbers, shedding of the noncellular outermost cuticle and production of a new expanded cuticle

---

CASE STUDY: Hormones and moulting in insects

In the immature stages of their life cycle, insects (either larvae or nymphs) are voracious eaters but they do not increase in size in a smooth fashion. Instead, they grow in discrete steps that are separated by **moults**. Before moulting, a larva or a nymph enlarges its epidermis by growth in cell numbers. It then sheds the noncellular outermost layer (cuticle) and produces a new enlarged cuticle. This provides space for an increase in size (figure 3.85).

FIGURE 3.85 Black Swallowtail butterfly larva (*Papilio polyxenes*) eating the cuticle that it has shed in the process of moulting

Figure 3.86 shows the life cycle of a butterfly, including the stages egg, larva, pupa and adult. After hatching, the larva undergoes five moults. The first four moults are 'larva in–larva out', but the fifth moult is 'larva in–pupa out'. Within the pupal case, an extraordinary process of **metamorphosis** occurs, with a butterfly emerging from the pupal case. Think about how a caterpillar and a butterfly differ in their shape, diet, appendages and motor capabilities. Moulting and metamorphosis in insects are complex processes that are governed by hormones.

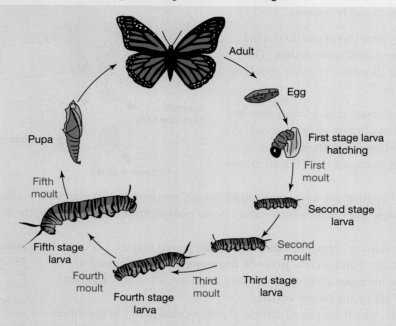

FIGURE 3.86 The life cycle of a butterfly including the five larval stages

Two of the insect hormones involved in moulting are ecdysone (20E) and juvenile hormone (JH).

The 'larva in–larva out' moults occur in the presence of both ecdysone and JH. The role of ecdysone is to:
- initiate and coordinate the moult
- initiate the changes in gene expression needed for metamorphosis.

The role of JH is to prevent these changes in gene expression.

So in these moults, the presence of JH limits the action of ecdysone. The epithelial cells of the larval integument (skin) are stimulated to detach from the overlying old cuticle and secrete to a new cuticle. By swelling up, the larva splits the old cuticle and extracts itself as a next-stage larva.

The 'larva in–pupa out' moult occurs in the presence of ecdysone only. The production of JH is blocked by the release of a neurohormone, allatostatin, from the brain to the corpora allata (CA). The role of ecdysone in this stage is to initiate a moult that inactivates larval genes and stimulates the pupa-specific genes, shifting the insect from larva to pupa.

The metamorphosis from larva to adult occurs within the pupal cuticle. Most of the larval cells are destroyed by apoptosis, and new adult tissues and organs develop from special clusters of undifferentiated cells in the larva. Ecdysone plays a major role in regulating the genes involved in these processes and in cell growth and division. Figure 3.88 shows the changes in concentration of ecdysone during the insect life cycle of the fruit fly, *Drosophila*. Far from being a 'moulting hormone' that is active during larval life, ecdysone plays a major coordinating role in the pupa-to-adult metamorphosis. Note that the release of ecdysone is not constant, but occurs in pulses, with the largest being during embryonic development, at the larva-to-pupa moult and during metamorphosis.

metamorphosis process of transformation — in insects (holometabolous) and amphibians — from an immature form to an adult that involves major changes in body structure and physiology

FIGURE 3.87 Diagram showing the hormones involved in moulting in insects. Organs are shown in boxes, hormones in ovals, excitatory hormone pathways in green and inhibitory pathways in red.

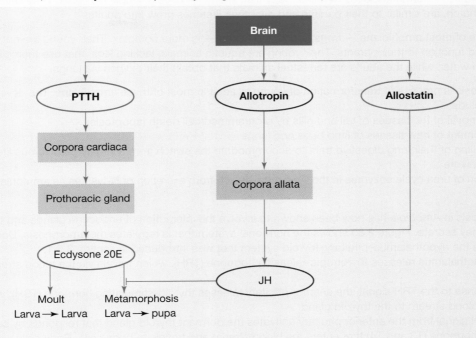

FIGURE 3.88 a. Graph showing the release of pulses of ecdysone during development of the fruit fly *Drosophila*, over its embryonic, larval and pupal stages **b.** *Drosophila melanogaster*, the common fruit fly

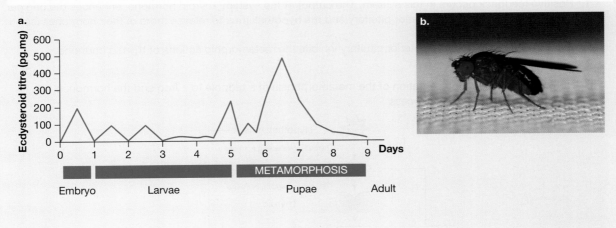

INVESTIGATION 3.6

Observing the life cycle of a butterfly

Aim

To observe the life cycle of a butterfly including the various moulting stages

CASE STUDY: Thyroid hormones in amphibians

Most vertebrates — mammals, birds, reptiles, bony fish, sharks — have a simple life cycle. They have young that are born or hatch, are similar to their parents and over their lifetimes grow into adults.

The life cycle of most amphibians — frogs, toads and newts — is more complex. Their young are not similar in structure and function to their parents. Their young are aquatic animals, lacking legs, that use their gills to obtain oxygen from water, while the adults are terrestrial animals that obtain their oxygen via lungs.

Metamorphosis is process of transformation of young to adult in most of these amphibians. The developmental changes include:
- total removal of the tissues of tail and gills by programmed cell death (apoptosis)
- development of new tissues of limb buds and lungs
- remodelling of their long digestive tract to accommodate the switch from a largely herbivorous lifestyle to that of a carnivore
- induction of urea cycle enzymes in the liver in the switch from excretion of N-wastes as ammonia to excretion as urea.

Metamorphosis in *Amphibia* has now been shown to involve the interaction of endocrine glands and the hormones they secrete. Figure 3.89 shows the hormonal system that is regulates metamorphosis. Does it look familiar? It is the hypothalamus–pituitary–thyroid system that was introduced in section 3.5.1.
- The hypothalamus releases thyrotropic-releasing hormone (TRH), which travels via the blood stream to the pituitary.
- In response to the TRH signal, the anterior pituitary releases thyroid-stimulating hormone (TSH), which travels via the blood stream to the thyroid gland.
- The TSH signal from the anterior pituitary activates the dormant thyroid gland that responds by secreting thyroid hormone (T3 and T4) that enters the blood stream and travels to body cells. Body cells with receptors for the thyroid hormone respond by initiating the particular biochemical and physiological changes of the metamorphic process for their tissue.
- Note that:
  1. Positive feedback occurs in this system. The output of the system, thyroid hormone, enhances the original signals by stimulating the anterior pituitary and the hypothalamus to release more of their hormones (see topic 4).
  2. Prolactin released from the anterior pituitary inhibits the metamorphic actions of thyroid hormone.

FIGURE 3.89 Schematic representation of the metamorphosis of a tadpole to a frog and the hormones involved in the regulation of this process

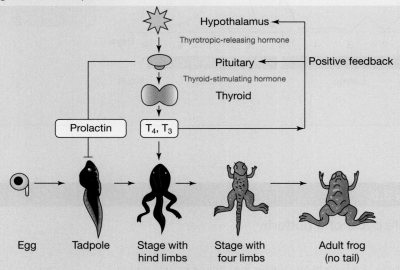

Resources

eWorkbook Worksheet 3.7 Endocrine system in other animals (ewbk-7379)
Weblink What is endocrinology?

KEY IDEAS

- The endocrine system is a network of endocrine glands that deliver chemical messages through hormones in the blood stream to target tissues.
- Hormones regulate body functions, including metabolism, growth, the concentration of glucose and electrolytes in body fluids, and sexual development.
- Target cells have specific receptors for their respective hormones to receive and respond to messages.
- The hypothalamus produces various releasing hormones that target various cells of the anterior pituitary gland.
- The pituitary gland is composed of the anterior pituitary and the posterior pituitary.
- The glandular tissue of the anterior pituitary contains acidophil and basophil cells that release various stimulating hormones that target other endocrine glands.
- The posterior pituitary gland is composed of nervous (neural) tissues and does not produce hormones but stores and releases hormones produced by the hypothalamus.
- Adrenals are paired endocrine glands on the top surface of the kidneys.
- The cortex of the adrenal gland has three zones, the cells of each one producing different hormones.
- The adrenal medulla produces the hormones adrenaline and noradrenaline that are involved in the stress response, also known as the fight-or-flight response.
  - Colloid-containing follicles, enclosed by follicular cells, form the structure of the thyroid gland.
  - Follicular cells of the thyroid gland synthesise the hormones T3 and T4 and its parafollicular cells produce the hormone calcitonin.
- Parathyroid glands are located on the rear surface of thyroid gland and secrete the parathyroid hormone.
- Regulation of hormone action typically involves negative feedback.
- Communication between cells through chemical signalling occurs in all animals.
- The simplest form of intercellular communication is via a chemical signal produced by one cell that moves by diffusion to a nearby cell.
- Insects were among the first animal groups to develop specialised endocrine systems.
- The main organs of insect endocrine system are the corpora cardiaca, corpora allata and prothoracic glands.
- Important hormones in moulting and metamorphosis in insects are PTTH, JH and ecdysone (20E).

3.5 Activities

learn on

To answer questions online and to receive **immediate feedback** and **sample responses** for every question go to your learnON title at **www.jacplus.com.au**. A **downloadable solutions** file is also available in the resources tab.

| 3.5 Quiz quiz | 3.5 Exercise | 3.5 Exam questions |

3.5 Exercise

1. **MC** The thyroid gland
   A. requires iodine to produce the hormones T3 and T4.
   B. releases calcitonin to increase calcium levels when they fall too low.
   C. is located just above the kidneys.
   D. releases growth hormones during development.
2. What is an endocrine gland?
3. Briefly identify:
   a. the chemical nature of hormones
   b. the general function of hormones
   c. their mode of delivery to their target organs.
4. Describe how a hormone finds its correct target cells.
5. Consider the hormones that are produced and released by the anterior pituitary gland. Which of these hormones has as its target another endocrine gland that it stimulates to release its hormones?

6. Compare and contrast the anterior pituitary and the posterior pituitary in terms of:
   a. their ability to synthesise hormones
   b. the composition of their tissue.
7. Briefly identify the role of follicular and parafollicular cells of the thyroid gland.
8. Where in the endocrine system would you find the following?
   a. Chief cells
   b. Cells that secrete aldosterone
   c. Cells that synthesise thyroglobulin
9. List three actions by which parathyroid hormone acts to increase calcium levels in the blood when the concentration falls below the normal range.
10. The hormones ecdysone (20E) and juvenile hormone (JH) are both involved in the insect life cycle.
    a. Which are involves in larva in–larva out moults?
    b. Which are involves in the larva in–pupa out moult?
    c. At which stage of the insect life cycle are 20E levels at their highest?

3.5 Exam questions

Question 1 (1 mark)

MC Certain tumors of the adrenal gland cause high production of the hormone aldosterone. This change in production of aldosterone will most likely affect blood

A. carbon dioxide concentration.
B. cell count.
C. glucose concentration.
D. pressure.

Question 2 (2 marks)

Source: VCAA 2013 Biology Section B, Q3a, 3b

A signalling molecule, epinephrine (adrenaline), is released from the adrenal gland when a human feels threatened. The molecule is transported in the bloodstream and initiates responses in cells in other parts of the human body.

a. To which group of signalling molecules does epinephrine belong? **1 mark**
b. Receptors for epinephrine are found on the exterior surface of the plasma membrane of cells. What does this suggest about the nature of the epinephrine molecule? **1 mark**

Question 3 (3 marks)

Source: VCAA 2007 Biology Exam 1, Section B, Q1b, Q1ci-ii

A hormone was produced in one cell, entered the bloodstream and travelled to two groups of cells adjacent to each other. One group of cells responded to the hormone but the neighbouring group did not.

a. What is the most likely reason for this difference in response by cells to the same hormone? **1 mark**
b. Consider one hormone you have studied this year that is transported through the blood to one or more types of cells.
   i. Name the hormone. **1 mark**
   ii. Name the tissue or gland that produces the hormone. **1 mark**

Question 4 (1 mark)

Source: VCAA 2008 Biology Exam 1, Section B, Q6c

Hormones are found in all multicellular organisms.

A hormone is sometimes defined as 'a chemical that is produced in one organ and transported by the blood to other cells where it causes a specific change'.

We now understand that this definition fails to account for all hormones found in multicellular organisms.

Write a new definition for a hormone, covering the majority of situations in which we know hormones are involved.

Question 5 (4 marks)

Compare and contrast the location and hormones released by the thyroid gland and the parathyroid gland. **4 marks**

More exam questions are available in your learnON title.

3.6 Excretory system in animals

> **KEY KNOWLEDGE**
>
> - Specialisation and organisation of animals cells into tissues, organs and systems with specific functions: excretory
>
> **Source:** Adapted from VCE Biology Study Design (2022–2026) extracts © VCAA; reproduced by permission

3.6.1 What is the excretory system?

Living organisms carry out life-sustaining metabolic reactions in their cells all the time. Some of these reactions generate left-over products that are toxic, or substances that disturb the homeostatic balance of the body if their concentration build to too high levels, or substances that cannot be stored but are in excess of immediate needs.

Leftover products that must be removed include harmful nitrogenous wastes (N-wastes) such as:
- ammonia and urea from the metabolism of protein
- guanine from the breakdown on nucleic acids
- creatine and creatinine from the metabolic activities of skeletal muscle.

Substances that must be removed when they are in excess include water and inorganic ions, such as Na^+, K^+ and Cl^-.

These various types of wastes are removed in a process known as **excretion**. In vertebrates, the key organ of the excretory system is the kidney. (Organs in other systems also contribute to excretion of wastes — lungs excrete gaseous carbon dioxide, skin excretes excess salt and water in sweat, the liver excretes a product of the breakdown of red blood cells in bile — but that is not their primary function.)

Figure 3.90 is a simple diagram of the human excretory system showing its four organs — kidney, ureters, urinary bladder and urethra — and its major blood vessels. (We will explore these organs and their functions in section 3.6.3.) Note that the kidney is a solid organ, while the **urinary tract** is composed of the three hollow organs: ureters, urinary bladder and urethra. The urinary tract is also known as the excretory passage.

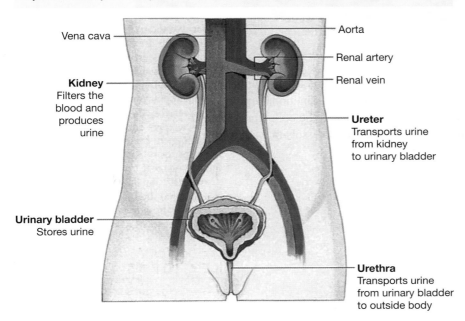

FIGURE 3.90 A diagram showing the four organs of the excretory (urinary) system: kidneys, urinary bladder, ureter, and urethra

excretion process of removal from the body of various types of waste material arising from its metabolic activities

urinary tract a series of hollow organs comprising ureters, bladder and urethra that transport urine to the outside of the body

3.6.2 Tissues in the excretory systems

Epithelial tissues

Epithelial tissue is important in the mammalian excretory system where it forms the linings of the hollow organs of the urinary tract — ureters, bladder and urethra — and the linings of various tubules within the kidney.

Linings of the hollow organs

The linings of the hollow organs are formed by **transitional epithelium**, which occurs nowhere else in the human body. All the cell layers in transitional epithelium are living cells.

The apical cells, that is, the cells that face into the lumen, have distinctive rounded outer surfaces that can be seen in light micrographs.

Transitional epithelium provides a waterproof barrier so that no component of the urine can pass out of the urinary tract into surrounding tissues, nor can anything enter. It can also expand and contract — this is particularly important in the urinary bladder that, when full, can expand to hold 400 to 600 mL. When empty the epithelial lining forms folds and the bladder shrinks.

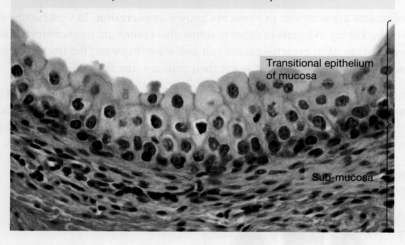

FIGURE 3.91 High-power light micrograph of a section of the mucosa of the human bladder. The transitional epithelial tissue is underlaid by dense connective tissue. Note the rounded outer surfaces of the apical cells of the transitional epithelium.

Linings of the tubules

The lining of various tubules of the kidney are composed of simple epithelial tissue. These various tubules in the kidney form part of the basic functional unit of the kidney, the **nephron**. Figure 3.92 shows these tubules and the collecting duct into which they empty. What is not shown here are the blood vessels that are intimately associated with the tubules.

transitional epithelium type of stratified epithelium present only in the hollow organs of the excretory system

nephron functional unit of the kidney

FIGURE 3.92 Line diagram showing the structures (Bowmans's capsule and tubules) that form part of the nephron. Arrows indicate the direction of movement of the filtrate through tubule. The star ★ shows the location of the cluster of capillaries from where the process of excretion starts.

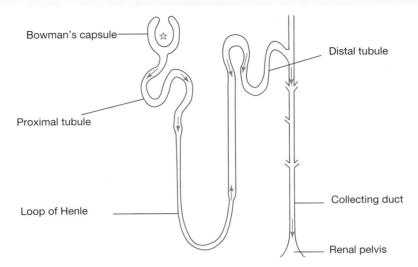

The linings of all the Bowman's capsule, kidney tubules and the collecting duct are various types of simple epithelium (single layer of cells).
- The Bowman's capsule (figure 3.93) encloses a space with an outer wall composed of simple squamous epithelium. The inner wall consists of a layer of special epithelial cells called podocytes that are closely associated with the cluster of capillaries (glomerulus) enclosed within Bowman's capsule. The simple squamous epithelium of the outer wall of the capsule is very thin, but the nuclei of some of its cells are visible (arrowed).
- Proximal tubules have a lining of simple low columnar epithelium, with each cell having a brush border of microvilli (figure 3.94).
- The loop of Henle has a lining of simple squamous epithelium, while distal tubules have a lining of simple cuboidal cells, with no brush borders.
- The lining of the collecting duct is simple cuboidal to columnar epithelium.

FIGURE 3.93 Micrograph of a cross-section of the kidney cortex. The Bowman's capsule encloses a cluster of capillaries (glomerulus). Surrounding the Bowman's capsule are cross-sections of tubules.

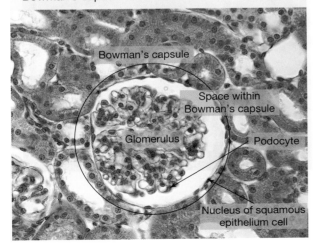

FIGURE 3.94 Micrograph of a cross-section of proximal tubules of the kidney. The specific stain used in this preparation distinguishes the brush border from the rest of the proximal tubule cells.

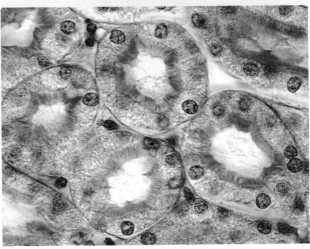

Connective tissue

Fibrous connective tissue is present in the hollow organs of the excretory system where it underlies and supports the transitional epithelium.

Muscle tissue

Muscle tissue is present in all the hollow organs of the urinary tract. Typically two layers are present, an inner layer of circular smooth muscle and an outer layer of longitudinal smooth muscle. Smooth muscle is not under voluntary control.

In the ureter walls, smooth muscle layers continually contract and relax in a peristaltic movement that forces urine away from the kidneys and towards the bladder. In the bladder wall, relaxation of the muscle layers enables the bladder to expand as the volume of stored urine increases. When the bladder empties, its walls contract and flatten, emptying urine into the urethra. At the junction of the bladder and the urethra are sphincter muscles, which are circular muscles that keep urine from leaking by closing tightly like a rubber band around the opening of the bladder.

3.6.3 Organs involved in the excretory system

Kidneys

All vertebrates depend on an excretory system to remove various wastes. In humans, each kidney weighs only approximately 150 grams and is about 11 centimetres long, 5 to 7.5 centimetres wide and about 2.5 centimetres thick (figure 3.95a) but they receive about 25 per cent of the blood output by the heart. This very rich blood supply highlights their importance as organs that remove wastes, and regulate the water and electrolyte balance of body fluids. The cortex is the location of Bowman's capsules and glomeruli and they give it a granulated appearance. The renal pelvis is the intersection of all the collecting ducts. Figure 3.95b shows the blood vessels within the kidney tissue, two renal arteries (purple), branching from the aorta, that supply blood to the kidney. The network of blood vessels within the kidney includes clusters of capillaries that form the glomeruli and the capillaries that surround the rest of the tubules.

FIGURE 3.95 a. The basic features of the anatomy of the human kidney **b.** A resin cast showing the rich blood supply of the kidney. The urine produced by the kidney leaves via the two ureters (yellow) and travels to the bladder.

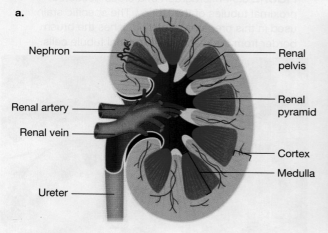

The functional unit of the kidney is the nephron. Each kidney has a large number of nephrons — an average normal human kidney contains about 1 million nephrons. Figure 3.96a shows a simplified diagram of a kidney nephron that starts at Bowman's capsule and ends where the tubule joins the collecting duct. The blood flow from the renal artery to the renal vein is indicated.

Figure 3.96b shows a more detailed picture of a nephron. Each tubule consists of five parts: Bowman's capsule, the proximal tubule, the loop of Henle, the distal tubule and the collecting duct. In this diagram, trace the tubule from its start at Bowman's capsule to its end where it joins the collecting duct. The intimate contact between the tubule and the surrounding peritubular capillaries enables material to be moved from within the tubule to the blood, and vice versa. In both diagrams note the associated blood vessels: the cluster of capillaries of the glomerulus in Bowman's capsule and the peritubular capillaries that coil around the rest of the kidney tubule.

The basic functions performed by the nephrons are:
- filtration
- reabsorption
- secretion
- excretion.

It is through these processes that the kidneys:
- remove metabolic wastes from the blood to the outside while conserving useful substances (glucose and other nutrients) — this is its excretory function.
- conserve salts and water so that the balance of water and ions in the body and the pH of the blood is maintained within narrow limits — this is its osmoregulatory function.

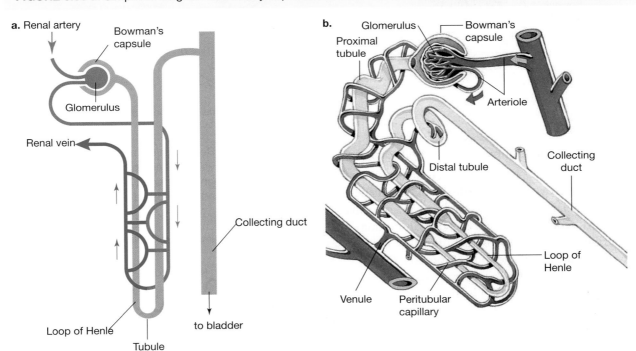

FIGURE 3.96 a. Simplified diagram of a kidney nephron b. Detailed structure of a nephron

Functions of the kidney: filtration

The site of filtration is Bowman's capsule and the cluster of capillaries (glomerulus) that it encloses. Filtration is the mass movement of the fluid component of the blood from the glomerular capillaries into the space within the Bowman's capsule (see figure 3.96b).

Blood cells and macromolecules (molecular weight > 60 000) cannot cross the barrier that consists of the lining of the capillary walls, the underlying basement membrane and the podocytes that surround the capillaries. The image on the opening page of this topic is a false-coloured scanning electron micrograph of some glomerular capillaries (shown in green) enveloped by podocytes with their multi-branched cytoplasmic extensions (shown in brown).

Plasma is the liquid portion of the blood, minus blood cells and large plasma proteins, and, once inside the space within the Bowman's capsule, it is called the **filtrate**. About 180 litres of plasma is filtered from the blood into the kidney each 24 hours, that is, about 125 mL of plasma per minute. Since the plasma volume of the blood is about 3 litres, this means that the entire plasma of the blood is filtered 60 times per day.

The force that drives the movement of plasma into the space in the Bowman's capsule is blood pressure created in the glomerulus because the exit arteriole is smaller in diameter than the arteriole that enters glomerulus (figure 3.97a). Because filtration only stops blood cells and macromolecules from moving into the space in Bowman's capsule, this means that the filtrate contains useful molecules as well as wastes including glucose and amino acids.

filtrate fluid composed of blood plasma minus large proteins that is filtered into Bowman's capsule from the glomerulus

FIGURE 3.97 a. The glomerular capillaries within Bowman's capsule; red arrows show blood flow, green arrows show filtrate forced from glomerulus. **b.** The close association of the inner wall of Bowman's capsule and the wall of a glomerular capillary. This is the path the plasma is filtered across.

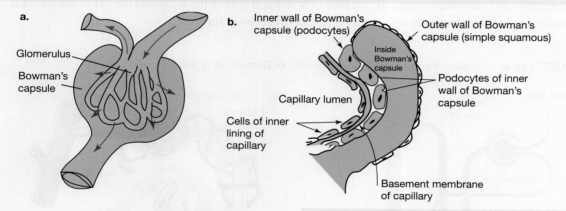

Functions of the kidney: tubular reabsorption

Reabsorption is the process by which water and useful solutes are removed from the filtrate and returned to the blood. The major site of reabsorption is the proximal tubule, with more than 70 percent of reabsorption occurring here. The brush border of microvilli increases the surface area available for reabsorption. Other segments of the tubule are involved to a lesser extent in reabsorption of water and inorganic ions.

All nutrients, such as glucose and amino acids, and most of the water and inorganic ions, such as sodium, potassium, phosphate and calcium are reabsorbed across the wall of the proximal tubule and enter the peritubular capillaries of the blood stream (figure 3.98). Reabsorption occurs by diffusion, facilitated diffusion and active transport (see topic 1).

FIGURE 3.98 Reabsorption in a proximal tubule cell. Substances are taken up from the filtrate in the tubule lumen, transferred across the tubule cell and are returned to the blood in the peritubular capillary.

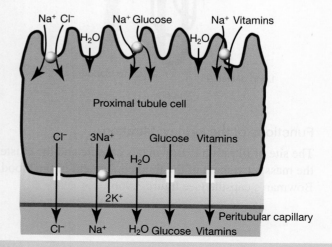

Reabsorption from the proximal tubule to the blood is enhanced by the presence of microvilli on the epithelium that greatly increase the surface area available for reabsorption.

Functions of the kidney: tubular secretion

Secretion is the process of transporting specific compounds, typically waste products, out of the blood of the peritubular capillaries into the tubular filtrate that will eventually become urine. Secretion occurs mainly in the proximal tubule, but some also occurs in other regions of the tubules.

Substances secreted from the blood into the tubule filtrate include urea, potassium ions, ammonia (NH_3), creatinine, and foreign substances taken into the body, such as various pharmaceutical drugs. Secretion occurs mainly by active transport, but some also occurs by passive diffusion.

Functions of the kidney: excretion

Of the 180 litres of filtrate that enters Bowman's capsule each daily, the excretory output of the kidney of an average person on a healthy diet is only 1 to 2 litres of urine daily. Urine is a watery solution containing organic wastes: urea, uric acid, ammonia and creatinine, as well as sodium and chloride ions. Urine is transferred out of the kidneys via the collecting ducts in the ureters.

FIGURE 3.99 Summary of urine formation in mammals

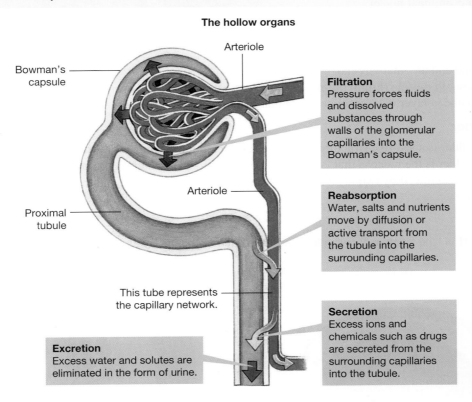

INVESTIGATION 3.7 online only

Kidney dissection

Aim

To dissect a kidney and observe its structure

The hollow organs

The hollow organs of the excretory system — ureters, urinary bladder and urethra — function in transporting urine from the kidney to the outside after storage in the bladder. No change in the composition of the urine occurs during its passage through the urinary tract.

Ureter

Two ureters form the first part of the urinary tract. They are tubes and their sole function is the transport of the urine made in the kidney to the bladder. Figure 3.100 shows a photomicrograph of the cross-section of the ureter of a mammal. Moving out from the central lumen, the following layers may be seen:

- the mucosa that includes a wide layer of transitional epithelium that is folded (light purple layers) of cells with prominent nuclei
- a thin layer of connective tissue that under lies the epithelium
- a layer of circular muscle, an inner longitudinal and an outer circular layer of smooth muscle (bright purple)
- on the outside, a layer of connective tissue that include fat tissue and blood vessels.

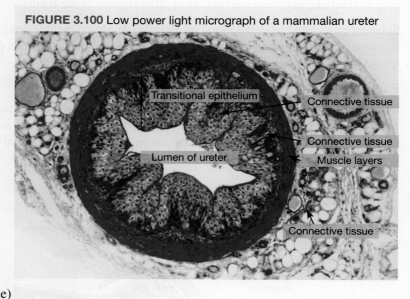

FIGURE 3.100 Low power light micrograph of a mammalian ureter

Urinary bladder

The urinary bladder is a sac-like structure that serves as a temporary storage for urine (figure 3.101a). The inner lining of the bladder is the mucosa that is composed of transitional epithelium. Below this is a thin layer of connective tissue, followed by two layers of smooth muscle, one longitudinal and the next circular (figure 3.101b).

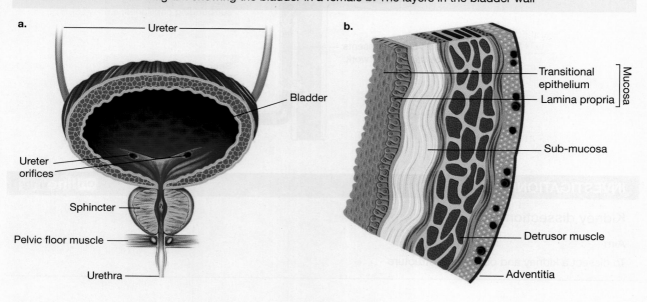

FIGURE 3.101 a. Line diagram showing the bladder in a female b. The layers in the bladder wall

Urethra

The urethra is a tube that allows urine to pass outside the body. Because it passes through the penis, the urethra is longer in human males (18–20 cm) than in human females (3 cm).

Urination occurs when a signal from the brain causes the bladder muscle layer to squeeze urine from the bladder. At the same time, a signal goes to the sphincter muscles at the junction of the bladder and the urethra to relax and enable urine to flow from the bladder to the exterior.

EXTENSION: Kidney disease in humans

Kidney disease occurs in humans when the filtration rate of the kidneys falls, resulting in the kidneys becoming less efficient. While diet can help with early stage kidney disease, more advanced kidney disease requires haemodialysis. Without haemodialysis, nitrogenous wastes from the breakdown of proteins build up in the patient's blood, causing tiredness and nausea. The excess fluid in the blood causes swelling and raises blood pressure. High concentrations of sodium ions in the blood further elevate blood pressure and excess potassium ions can result in a decrease in heart muscle activity.

For more information on the results of kidney failure, and how haemodialysis works, please see the digital document.

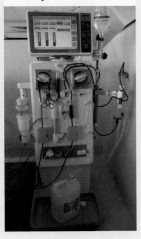

FIGURE 3.102 A dialyser used to treat kidney disease

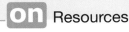 Resources

Digital document Extension: Kidney disease in humans (doc-35877)

INVESTIGATION 3.8 — online only

Modelling kidney filtration and dialysis

Aim
To model kidney filtration and investigate how the process of dialysis works

---

SAMPLE PROBLEM 5 Describing excretion

a. The kidneys are a vital component in the excretory system. Identify the functional unit of the kidney. (1 mark)
b. Outline the four main steps involved in the production of urine. (4 marks)
c. Two individuals have consumed vastly different amounts of water. Explain what differences you would expect in the urine of these individuals. (2 marks)
d. Some species such as the kangaroo rat have structural adaptations in the loop of Henle that allow them to better survive in dry conditions and reduce water loss through urine. Identify and explain what adaptation they would likely have in regard to the loop of Henle. (2 marks)

THINK	WRITE
a. This is an identify question, and worth only 1 mark. It requires only a simple one word answer.	Nephron (1 mark)
b. This question requires an outline of each step, which is a brief statement. You should make sure that you name each of the four steps of: • filtration • reabsorption • secretion • excretion.	The four main steps are as follows: • Filtration: the movement of the fluid component (plasma) of the blood within the from the glomerulus at the Bowman's capsule to form the filtrate (1 mark). • Reabsorption: the process in which water and useful solutes from the filtrate in the proximal tubule (1 mark). • Secretion: the secretion of waste products, excess ions and chemicals from the surrounding capillaries into the tubule (1 mark). • Excretion: The removal of excess water and solutes in the form of urine (1 mark).
c. This is an explain question. You need to provide a detailed response that describes the differences between the urine of two individuals — one who drinks less water and one who drinks more water. Urine would likely contain much more water in individuals who drink high volumes of it.	An individual who drinks large volumes of water will have more water removed from the blood during secretion and will therefore have very dilute urine (lighter in colour, lower urea concentration) (1 mark). An individual who drinks less water will have less water removed from the blood (and likely water will be reabsorbed from the proximal tubule) resulting in more concentrated and darker urine (1 mark).
d. This question requires you to identify the feature of the loop of Henle for one mark and explain the adaptation for the second mark. Remember that part of the loop of Henle is used to reabsorb water.	The loop of Henle is associated with the reabsorption of water. An animal in dry conditions would likely have a long loop of Henle (1 mark). This would allow for more water to be reabsorbed from the urine, reducing water loss, therefore allowing it to survive in drier conditions (1 mark).

 Resources

 eWorkbook Worksheet 3.8 The excretory system in mammals (ewbk-7381)

3.6.4 Excretory systems in different animals

Getting rid of the nitrogenous wastes (N-wastes) produced in various metabolic reaction is a problem for animals. The means by which animals excrete N-waste and the form in which it is excreted — ammonia, urea or uric acid — reflect an animal's evolutionary history, its body plan structure and its habitat. All vertebrates have kidneys similar to those of mammals. Let's now consider excretory systems in invertebrate animals.

Excretory systems in invertebrates

Ammonia (NH_3) is the initial N-waste produced from the breakdown of protein and other N-containing compounds, such as nucleic acids. Some animals excrete their N-wastes in the form of ammonia. Ammonia can be converted into urea (NH_2CONH_2) or uric acid ($C_5H_4N_4O_3$) but these conversions require an investment of energy. Many animals carry out these conversions and excrete their N-wastes as either urea or as uric acid. Why not save energy and excrete ammonia? See the Extension box below for a clue.

Diffusion of ammonia

Diffusion of ammonia is the least complicated means of excreting N-wastes. This does not require any specialised tissues or organs — all that is needed is lots of water to dilute the ammonia and carry it away. However, diffusion of any substance across cells or tissues can only operate over short distances, and when a high surface area to volume ratio exists.

A few animal groups can excrete ammonia by diffusion, and they are the sponges of phylum *Porifera* and all members of phylum *Cnidaria* that includes jellyfish, corals, and sea anemones (see section 3.3.2). Neither of these animal groups has distinct excretory cells or tissues.

Sponges, the least complex animals, have a two-layered body structure — this means that no sponge cell is far from the external environment. N-wastes in the form of ammonia simply diffuse across the cell surfaces of the sponge into the external environment, which is, for almost all sponges, the sea.

Jellyfish and other members of the phylum *Cnidaria* with their two-layered body structure also rely on diffusion of ammonia for excretion of their N-wastes into their marine environment.

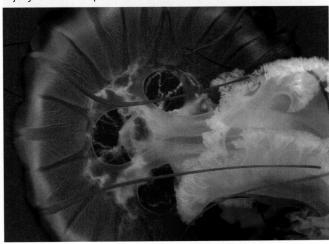

FIGURE 3.103 The simple two-layered body structure of jellyfish relies upon ammonia diffusion for excretion.

As the body size of animals increases, diffusion becomes inefficient. In more complex multicellular animals with a three-layer body structure, most of their cells are too far removed from the external environment for these animals to rely on diffusion as an excretory mechanism. These multicellular animals have special tissues and organs that enable them to collect and excrete N-wastes efficiently and rapidly as ammonia, urea or uric acid. The first animal group with excretory organs are the flatworms (planarians) and the other members of phylum *Platyhelminthes*. Other groups of invertebrates developed different systems for excreting N-wastes but they include similar features such as tubules and cells with cilia.

EXTENSION: Forms of N-wastes

Nitrogenous wastes (N-wastes) are a by-product mainly of the metabolism of proteins. Some animals can excrete their N-waste as ammonia. Others convert ammonia to urea, yet other animals convert ammonia to uric acid. Both of these conversions are energy-requiring reactions. Why use this energy?

To access more information on this extension concept please download the digital document.

FIGURE 3.104 A colony of cormorants (*Phalacrocorax aristotelis*). The extensive white areas are the uric acid droppings of these birds.

 **Resources**

    Digital document Extension: Forms of N-wastes (doc-35880)

CASE STUDY: Flame cells of planarians

Planarians are free-living flatworms that live in fresh water. Their ancestors were the first animals to have an excretory system.

The planarian excretory system consists of a large number of units (protonephridia), each containing excretory cells called **flame cells** and connecting tubules that join to a main excretory canal that opens to the exterior through excretory pores. Two excretory canals are located on either side of the planarian (figure 3.105).

The flame cells are cup-shaped and have long strands of cilia on their inner surface. When viewed with a microscope the moving cilia look like a flickering flame, and hence the cells are called flame cells.

Beating of the cilia draws tissue fluid into the flame cells where water and waste materials are removed through filtration. The beating cilia then drives waste matter into the excretory tubules that terminate at excretory pores where wastes pass to the outside. The pores open at the body surface along both sides of the flatworm's body.

FIGURE 3.105 Dugesia planarian are one of the simplest animals to have an excretory system. Inset shows a stylised representation of one unit of their excretory system.

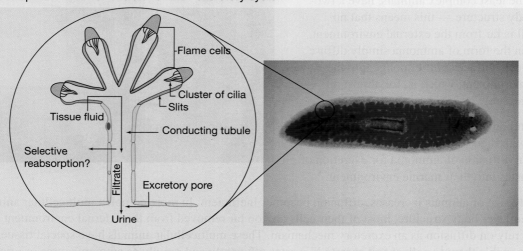

CASE STUDY: Malpighian tubules of insects

The Malpighian tubule system is the excretory system present in insects, and also in millipedes, centipedes, spiders and scorpions.

The Malpighian excretory system consists of a number of thin blind-ending tubules (Malphighian tubules) located in the abdomen of the insect. They are bathed in **hemolymph** of the insect.

The open ends of the tubules join the digestive system at the junction between the midgut and the hindgut.

The tubules are lined with a single layer of epithelial cells. The number of tubules varies between species and range from a few to more than 100. Tubule cells remove N-wastes in the form of uric acid and other solutes from the hemolymph.

flame cells excretory cells with flagella present in members of phylum *Platyhelminthes*

hemolymph the internal fluid of insects, analogous to blood in vertebrates; mostly water; it also contains ions, carbohydrates, lipids, glycerol, amino acids, hormones and some cells.

Urine is formed by a process of secretion. In this process, tubule cells secrete Na$^+$ ions into the fluid of the lumen of the tubule by active transport and this inflow of ions increases the osmotic concentration of the enclosed fluid. This change causes water, uric acid and other solutes to flow into the tubule lumen, forming urine. In contrast, in the mammalian kidney, fluid enters the nephron by a process of filtration.

The urine passes from the Malpighian tubules into the rectum of the hindgut where water and ions are reabsorbed while the almost solid uric acid is expelled along with faeces. Water is reabsorbed to a level to maintain the organism's water balance. However, larvae of some insect species are aquatic, such as mosquito larvae ('wrigglers'), and they reabsorb useful solutes in the hindgut, but do not need to reabsorb water.

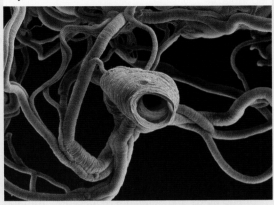

FIGURE 3.106 SEM image of a mosquito (*Anopheles* sp.) midgut (at centre) and, immediately behind it, the five Malpighian tubules. Together these form the excretory system of this insect.

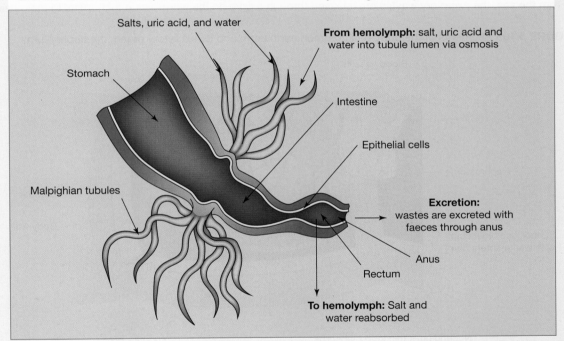

FIGURE 3.107 A diagrammatic representation of the excretory system in insects and other animals. In these animals, excretion also involves action by the digestive system.

CASE STUDY: Nephridia of the earthworm

The common earthworm (*Lumbricus terrestris*) is a segmented worm belonging to the phylum *Annelida* that also includes polychaete worms and leeches. The main form in which N-waste excreted by earthworms is urea. The unit of excretion is the nephridium and each segment of the earthworm, except for the first two, has a pair of nephridia.

In the process of excretion (see figure 3.107):
1. Body fluid with wastes from one segment is taken into a ciliated funnel (nephrostome) by the beating of its cilia, from where the fluid enters a long narrow ciliated tube
2. Fluid is carried in this tube through the dividing wall into the next segment where the tube forms a series of loops.
3. The loops of the tube are enveloped in blood capillaries and more wastes are transferred from the blood into the tube.
4. Useful substances, including water, glucose, amino acids and salts, are reabsorbed from the fluid in the tube and returned to the blood.
5. The residual fluid containing the N-wastes passes into a thick portion of the tube that forms a bladder-like sac.
6. From the sac, the excretory wastes are released through a pore on the worm's side that opens to the outside.

FIGURE 3.108 The common earthworm (*Lumbricus terrestris*)

FIGURE 3.109 Segments of the body of a common earthworm and its excretory organ, the nephridium

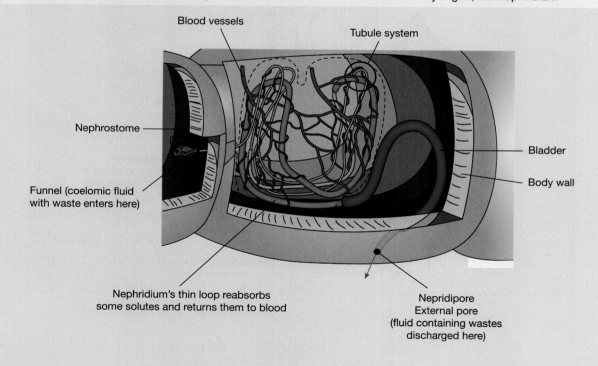

Resources

- **eWorkbook** Worksheet 3.9 Excretion in other animals (ewbk-7383)
- **Weblink** Keeping kidneys healthy

KEY IDEAS

- In unicellular microbes, protists and the simplest animals, wastes are removed by a process of simple diffusion across the plasma membrane.
- Animals with complex multicellular structures use specialised tissues, organs and systems for waste removal.
- In mammals, the excretory system is responsible for removing metabolic and other wastes from the blood and plays a role in water balance.
- N-wastes include:
  - ammonia and urea from the metabolism of protein
  - guanine from the breakdown of nucleic acids
  - creatine and creatinine from the metabolic activities of skeletal muscle.
- In vertebrates, the excretory (urinary) system is the kidneys and urinary tract.
- The urinary tract comprises the ureters, urinary bladder and urethra.
- Tissues in the excretory system include:
  - epithelial tissue in the linings of the hollow organs of the urinary tract and in the various tubules within the kidney
  - connective tissue
  - muscle tissue.
- Organs of the excretory system include the kidneys and the hollow organs of the ureters, urinary bladder and urethra.
- The main organ is the kidney; the functional unit of the kidney is the nephron.
- The functions of the nephron are filtration, reabsorption, secretion and excretion.
- Diffusion of ammonia is the simplest way to get rid of N-wastes. It does not require specialised organs but is limited by availability of water and animal size.
- End stage kidney disease is a serious and life-threatening condition.
- Haemodialysis is a medical intervention that can replace some of the defective kidney functions.
- The planarian excretory system relies on flame cells and excretory canals.
- The Malpighian excretory system in insects consists of thin blind-ending tubules which remove uric acid and solutes from the hemolymph; urine is formed by changes on osmotic pressure, and is then excreted with other wastes.
- Earthworms excrete N-wastes using the nephridia, which absorb useful substances into the body. The residual fluids are excreted.
- N-wastes may be excreted as ammonia, urea or uric acid that differ in their solubility, toxicity and energy cost of production.
- Aquatic invertebrates, bony fish and amphibian larvae excrete their N-wastes as ammonia.
- Birds, reptiles and some terrestrial invertebrates, including insects and spiders, excrete their N-waste as uric acid.

3.6 Activities

learn on

To answer questions online and to receive **immediate feedback** and **sample responses** for every question, go to your learnON title at **www.jacplus.com.au**. A **downloadable solutions** file is also available in the resources tab.

3.6 Quick quiz on	3.6 Exercise	3.6 Exam questions

3.6 Exercise

1. a. Identify the key organs of the mammalian excretory system.
   b. What is the functional unit of the kidney?
2. In the human kidney:
   a. What produces the pressure that forces filtrate from the glomerulus into the space within Bowman's capsule?
   b. What type of tissue lines:
      i. the proximal tubules of the kidney
      ii. the urethra of the urinary tract?

3. Identify the key difference between the members of the following pairs.
   a. Secretion and reabsorption
   b. Filtrate and urine
4. Think about the passage of filtrate through the kidney tubules and that of urine through the urinary tract. Does either the filtrate or the urine undergo a change in composition as it moves along its particular route? Briefly justify your decisions.
5. Explain the following observations:
   a. The tubular filtrate in a healthy kidney contains no red blood cells.
   b. Healthy people have glucose in their blood, but no glucose in their urine.
6. Identify two sources of nitrogenous waste in mammals.
7. By what means to the following animals excrete their nitrogenous and other wastes?
   a. Jellyfish of phylum *Cnidaria*
   b. Flatworms of phylum *Platyhelminthes*
   c. Insects of phylum *Arthropoda*
   d. Earthworms of phylum *Annelia*
8. Jellyfish and all other members of the phylum *Cnidaria* excrete their nitrogenous waste in the form of ammonia. Is it reasonable to predict that the next species in that phylum to be discovered might excrete its waste as uric acid rather than ammonia?

3.6 Exam questions

Question 1 (1 mark)

MC A major role of the kidneys is
A. the removal of carbon dioxide.
B. maintaining blood glucose levels.
C. eliminating nitrogenous wastes.
D. balancing fat-soluble vitamin blood levels.

Question 2 (1 mark)

MC Antidiuretic hormone (ADH) acts by increasing the reabsorption of water from the collecting duct. In the presence of ADH, what would be expected?
A. Urine would contain more glucose.
B. Urine would be more diluted.
C. Urine production would cease.
D. Urine solutes would be more concentrated.

Question 3 (1 mark)

MC Which of the following molecules would not be found in the filtrate of a nephron?
A. Urea
B. Glucose
C. Amino acids
D. Large plasma proteins

Question 4 (2 marks)

Inflammation of the glomeruli in the kidneys can lead to the presence of blood proteins and blood cells in the urine. Explain why blood protein and blood cells are not normally found in urine.

Question 5 (2 marks)

Proximal tubule cells lining the beginning of a renal nephron in the kidneys have a folded plasma membrane and contain many mitochondria. Explain the significance of each of these structures.

More exam questions are available in your learnON title.

3.7 Review

3.7.1 Topic summary

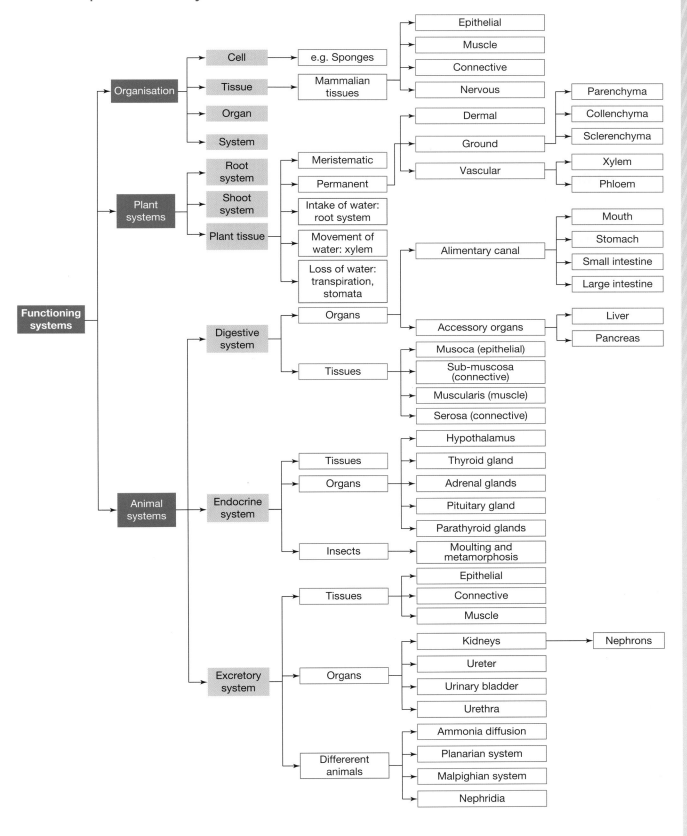

3.7 Exercises

To answer questions online and to receive **immediate feedback** and **sample responses** for every question go to your learnON title at **www.jacplus.com.au**. A **downloadable solutions** file is also available in the resources tab.

3.7 Exercise 1: Review questions

1. Comment on or briefly explain the following observations.
   a. The release of a hormone from the anterior pituitary gland in response to a stimulus detected in the brain occurs more slowly than the release of a hormone from the posterior pituitary gland.
   b. Only very few animals can excrete ammonia by simple diffusion.
   c. Of the 180 litres of filtrate that enters Bowman's capsule every day, the excretory output of the kidney of an average person eating a healthy diet is only 1 to 2 litres each day.

2. Comment on, or briefly explain, the following observations:
   a. When a plant is transplanted, root hairs are often damaged, so before transplanting a gardener usually removes many of the leaves.
   b. When florists shorten the stems of a bunch of flowers, they take care to cut each stem under water and not in the air.

3. Consider a tiny living organism, known as a hydra — just visible to the unaided eye — that lives in freshwater. The figure shows a diagram of its body plan: it has a two-layer body wall separated by mesoglea. Several different cell types are present in this organism including cells with flagella that line the gut cavity and several kinds of stinging cells (cnidocytes) with a mini-harpoon that fires, penetrating and paralysing the prey on which this organism feeds.
   a. What kind of organism is a hydra?
   b. On what information did you base your decision?
   c. Does a hydra show evidence of cell specialisation?
   d. What is the mesoglea?
   e. Would you predict that organs are present in a hydra? Explain.
   f. What level of organisation does a hydra show (cellular, tissue, organ or system)?

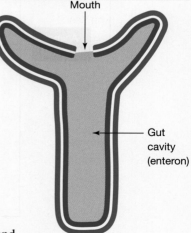

FIGURE Simplified body plan of a hydra with its two main cell layers: an outer ectoderm and an inner endoderm

4. What is a key difference between members of the following pairs?
   a. Cohesion and adhesion of water molecules
   b. Parenchyma and sclerenchyma tissues in vascular plants
   c. Calcitonin and calcitriol
   d. Tissue of the anterior pituitary gland and that of the posterior pituitary gland
   e. Follicular cells and parafollicular cells of the thyroid gland
   f. A releasing hormone and a stimulating hormone

5. The figure shows the alimentary canal of a mammal.
   a. Based on this diagram, what conclusion might reasonably be drawn about the preferred diet of this mammal. Explain your choice.
   b. Suggest two features (either presence or absence) that might be expected to be seen in the teeth of this mammal. Explain your decision.
6. Identify two key differences between ammonia and uric acid and relate them to the types of animal that have ammonia as their excretory product and those that have uric acid.

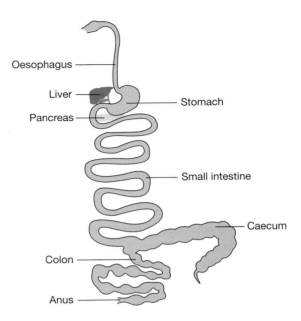

3.7 Exercise 2: Exam questions

on Resources

▶ **Teacher-led videos** Teacher-led videos for every exam question

Section A — Multiple choice questions

All correct answers are worth 1 mark each; an incorrect answer is worth 0.

▶ **Question 1**

An example of permanent tissue in plants is

A. collenchyma tissue.
B. xylem tissue.
C. dermal tissue.
D. sclerenchyma tissue.

▶ **Question 2**

The internal lining of the oesophagus and the anal canal is composed of

A. simple squamous epithelial cells.
B. stratified columnar epithelial cells.
C. stratified squamous epithelial cells.
D. multi-layered transitional epithelial cells.

▶ **Question 3**

Processes that occur during formation of urine by the mammalian kidney include

A. blood from the glomerular capillaries enters Bowman's capsule and becomes the tubular filtrate.
B. reabsorption of water and useful solutes occurs mainly in the proximal tubules.
C. secretion of drugs into the collecting ducts.
D. about half the filtrate that enters Bowman's capsule is finally excreted as urine.

Question 4

Which of the following events take place in the human digestive system?

A. Protein digestion is completed in the stomach; digestion of fats and carbohydrates occurs later in other regions.
B. The digestive enzyme trypsin is release by acinar cells of the pancreas.
C. Secretion of hydrochloric acid by the pancreas provides the optimal pH for enzymes that act in the duodenum.
D. End products of fat digestion are carried from the alimentary canal in lacteals.

Question 5

A calcium-sensing receptor in the parathyroid gland detects a fall in the calcium concentration in the blood. This would likely lead to

A. the release of calcitonin from the parathyroid gland.
B. a signal being sent to the adrenal glands to stimulate the release of cortisol.
C. the inactivation of the T3 and T4 hormones.
D. the release of the parathyroid hormone (PTH) from the parathyroid gland.

Question 6

In vascular plants, identify the best description of the parenchyma tissue.

A. It forms the bulk of the cells in xylem tissue.
B. It is composed of living cells with thickened cell walls.
C. It acts as a major storage site for starch in roots and stems.
D. Its primary function is as support and protection in leaves.

Question 7

Identify the generation of the major force that pulls water up the xylem tissue.

A. The loss of water by transpiration in the leaves
B. The closing of the stomata of the leaves
C. The pressure by soil particles on the root hairs
D. The lateral movement of water through pits in the walls of xylem vessels

Question 8

All animals are multicellular, but various animals show different levels of organisation of their cells. It is reasonable to state that

A. animals at the cellular level of organisation could have tissues but not organs.
B. animals at the organ level of organisation include members of the phylum *Cnidaria*.
C. animals at the system level of organisation would be expected to have organs and tissues.
D. animals at the cellular level of organisation include flatworms of phylum *Platyhelminthes*.

Question 9

Enzymes involves in digestion include

A. amylases that are secreted by the salivary glands and the stomach chief cells.
B. maltase that is embedded in the microvilli of enterocytes of the small intestine.
C. pepsin that is secreted as a proenzyme pepsinogen by the pancreas into the duodenum.
D. lipases that are secreted by the liver and pancreas.

Question 10

As occurs in vertebrates, excretion in invertebrate animals is also concerned with the removal of waste products from within their cells. It is reasonable to state that

A. all invertebrates excrete their nitrogenous waste as ammonia.
B. flame cells are the functional excretory cells in flatworms.
C. the Malpighian tubules of insects open directly to the external environment.
D. the nephridia of earthworms lack the ability to secrete substances.

SECTION B — Short answer questions

Question 11 (10 marks)

a. Identify the first two enzymes to which ingested food is exposed. 2 marks
b. After being swallowed, food travels to the stomach.
   i. Descibe the mechanism how food moved. 1 mark
   ii. Identify the organ and kind of tissue involved in bringing about this movement of food to the stomach. 2 marks
c. Describe the difference between a foregut fermenter and a hind gut fermenter. 2 marks
d. Identify one principal site of protein digestion in the alimentary canal. 1 mark
e. Suggest, giving a reason, which amylase enzyme — from the salivary glands or the pancreas — would be expected to be more effective in digesting carbohydrates? 2 marks

Question 12 (8 marks)

a. Consider the structure of the wall of the digestive tract.
   i. Identify the four layers, starting from the gut lumen, that are typically visible in a cross-section of the digestive tract. 2 marks
   ii. Identify the specific type of tissue that is present in the layer in immediate contact with the gut contents. 2 marks
   iii. List the two major functions that are served by this type of tissue in the small intestine. 2 marks
b. Identify a structural feature of the lining of the small intestine that contributes to or facilitates its function. 1 mark
c. Explain why protease digestive enzymes are released into the gut as inactive proenzymes. 1 mark

Question 13 (13 marks)

a. Identify two key functions of the excretory system. 2 marks
b. What is the functional unit of the mammalian kidney? 1 mark
c. Briefly describe the process by which the kidney filtrate is formed. 2 marks
d. i. What is the key difference between the tubular processes of reabsorption and secretion? 2 marks
   ii. Explain why the process is termed *reabsorption* rather than *absorption*? 1 mark
e. Briefly describe the structure of the lining of the proximal tubules that are in contact with the filtrate and link this structure of its function. 2 marks
f. Identify the excretory organ of insects, their major excretory product and the means by which it is formed. 3 marks

Question 14 (10 marks)

a. In addition to secreting hormones, what is a distinctive feature of the endocrine glands? **1 mark**
b. What kind of tissue forms the bulk of the endocrine glands? **1 mark**
c. You examine a light microscope view of an endocrine gland that has been stained with standard hematoxylin and eosin stain (H&E). You are told that it is a cross-section of either the thyroid gland or the anterior pituitary gland. For each possibility, identify two distinguishing features that you would expect to see that would allow you to identify the gland. **4 marks**
d. Identify the change in the body (or the stimulus) that would lead the release of the following hormones:
   i. The thyroid hormones T3 and T4 **1 mark**
   ii. Parathyroid hormone (PTH) **1 mark**
e. Identify and provide an example of a way in which the production of a hormone by an endocrine gland can be blocked. **2 marks**

Question 15 (9 marks)

a. The blood (plasma) concentration of calcium is regulated and in a healthy person is maintained within a narrow range.
   i. Which hormones are involved in maintaining calcium levels of the blood within normal limits? **2 marks**
   ii. For each hormone, identify one target tissue and one action on this tissue. **4 marks**
b. Alcohol consumption interferes with the normal secretion of antidiuretic hormone (ADH) by inhibiting its release.
   i. Identify the gland that releases ADH. **1 mark**
   ii. ADH acts on the kidneys, increasing water reabsorption from the collecting tubules back into the blood. Explain how alcohol consumption would affect the volume of urine. **2 marks**

3.7 Exercise 3: Biochallenge online only

Resources

 eWorkbook Biochallenge — Topic 3 (ewbk-8073)

 Solutions Solutions — Topic 3 (sol-0648)

teach on

Test maker
Create unique tests and exams from our extensive range of questions, including practice exam questions.
Access the assignments section in learnON to begin creating and assigning assessments to students.

Online Resources

Below is a full list of **rich resources** available online for this topic. These resources are designed to bring ideas to life, to promote deep and lasting learning and to support the different learning needs of each individual.

eWorkbook

- 3.1 eWorkbook — Topic 3 (ewbk-3158)
- 3.2 Worksheet 3.1 Systems, organs and tissues in plants (ewbk-3486)
- 3.3 Worksheet 3.2 Systems, organs and tissues in animals (ewbk-3488)
- 3.4 Worksheet 3.3 Structure and function of the digestive system (ewbk-3490)
  Worksheet 3.4 The ins and outs of digestion (ewbk-3492)
  Worksheet 3.5 Different digestive systems in animals (ewbk-3494)
- 3.5 Worksheet 3.6 The human endocrine system (ewbk-3496)
  Worksheet 3.7 Endocrine system in other animals (ewbk-7379)
- 3.6 Worksheet 3.8 The excretory system in mammals (ewbk-7381)
  Worksheet 3.9 Excretion in other animals (ewbk-7383)
- 3.7 Worksheet 3.10 Reflection — Topic 3 (ewbk-7387)
  Biochallenge — Topic 3 (ewbk-8073)

Solutions

- 3.7 Solutions — Topic 3 (sol-0648)

Practical investigation eLogbook

- 3.1 Practical investigation eLogbook — Topic 3 (elog-0160)
- 3.2 Investigation 3.1 A closer look at vascular tissue (elog-0254)
  Investigation 3.2 Root hairs (elog-0256)
  Investigation 3.3 Leaf epidermal layer (elog-0258)
  Investigation 3.4 Water movement in the plant (elog-0260)
- 3.4 Investigation 3.5 Digestive systems (elog-0796)
- 3.5 Investigation 3.6 Observing the life cycle of a butterfly (elog-0798)
- 3.6 Investigation 3.7 Kidney dissection (elog-0800)
  Investigation 3.8 Modelling kidney filtration and dialysis (elog-0802)

Digital documents

- 3.1 Key science skills — VCE Biology Units 1–4 (doc-34648)
  Key terms glossary — Topic 3 (doc-34651)
  Key ideas summary — Topic 3 (doc-34662)
- 3.2 Extension: Growing plants without soil (doc-35878)
- 3.5 Extension: Hormones in the digestive system — without a gland (doc-35879)
- 3.6 Extension: Kidney disease in humans (doc-35877)
  Extension: Forms of N-wastes (doc-35880)

Teacher-led videos

- Exam practice questions — Topic 3
- 3.2 Sample problem 1 Comparing and contrasting tracheids and vessels (tlvd-1850)
- 3.3 Sample problem 2 Tissue types in mammals (tlvd-1851)
- 3.4 Sample problem 3 Comparing digestive systems (tlvd-1852)
- 3.5 Sample problem 4 Explaining the role of iodine (tlvd-1853)
- 3.6 Sample problem 5 Describing excretion (tlvd-1855)

Video eLesson

- 3.4 Digestive system (eles-2643)
- 3.6 Urine formation in the kidney (eles-2644)

Interactivities

- 3.4 Digestive system (int-3030)
  The digestive system (int-3398)
- 3.5 Endocrine glands (int-5766)
- 3.6 Labelling the kidneys (int-8234)

Weblinks

- 3.4 Health conditions of the digestive system
- 3.5 What is endocrinology?
- 3.6 Keeping kidneys healthy

Teacher resources

There are many resources available exclusively for teachers online.

To access these online resources, log on to **www.jacplus.com.au**.

AREA OF STUDY 2 HOW DO PLANT AND ANIMAL SYSTEMS FUNCTION?

4 Regulation of systems

KEY KNOWLEDGE

In this topic you will investigate:

Regulation of systems
- regulation of water balance in vascular plants
- regulation of body temperature, blood glucose and water balance in animals by homeostatic mechanisms, including stimulus–response models, feedback loops and associated organ structures
- malfunctions in homeostatic mechanisms: type 1 diabetes, hypoglycaemia, hyperthyroidism.

Source: VCE Biology Study Design (2022–2026) extracts © VCAA; reproduced by permission.

PRACTICAL WORK AND INVESTIGATIONS

Practical work is a central component of learning and assessment. Experiments and investigations, supported by a **practical investigation eLogbook** and **teacher-led videos**, are included in this topic to provide opportunities to undertake investigations and communicate findings.

4.1 Overview

Numerous **videos** and **interactivities** are available just where you need them, at the point of learning, in your digital formats, learnON and eBookPLUS, and at **www.jacplus.com.au**.

4.1.1 Introduction

Plants and animals are continually sensing and responding to changes in their external and internal environments as part of maintaining their living states and a relatively stable internal environment known as homeostasis. They do this through the process of regulation, which allows plants to regulate water balance and animals to regulate everything from body temperature to blood glucose. For example, a marathon runner needs to regulate their body temperature to prevent heat stroke.

FIGURE 4.1 Marathon runners face challenges in regulating body temperature to prevent heat stroke.

Some conditions, such as hyperthyroidism, which includes excessively high body temperature, result from malfunctions in homeostatic mechanisms — in this case, the thyroid is overactive and produces too much thyroxine. The image that opens this topic is of the thyroid gland, showing large follicles (top left and lower right) that secrete T3 and T4 hormones. The thyroid follicles are filled with a liquid called thyroglobulin or thyroid colloid.

In this topic, we will explore the regulation of systems, including water balance in plants and animals, and the regulation of body temperature and blood glucose levels in humans. We will then turn to malfunctions in homeostatic mechanisms that give rise to conditions such as type 1 diabetes, hypoglycaemia and hyperthyroidism, as mentioned above.

LEARNING SEQUENCE

4.1	Overview	252
4.2	Regulation of water balance in vascular plants	253
4.3	Homeostatic mechanisms and stimulus–response models	263
4.4	Regulation of body temperature in animals	278
4.5	Regulation of blood glucose in animals	295
4.6	Regulation of water balance in animals	305
4.7	Malfunctions in homeostatic mechanisms	314
4.8	Review	326

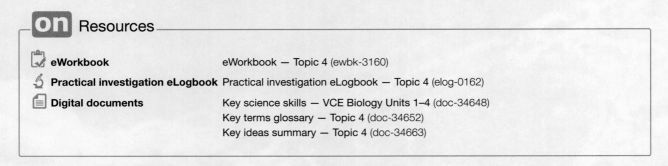

4.2 Regulation of water balance in vascular plants

KEY KNOWLEDGE
- Regulation of water balance in vascular plants

Source: VCE Biology Study Design (2022–2026) extracts © VCAA; reproduced by permission.

4.2.1 Water balance in plants

Water makes up about 90 to 95 per cent of the living tissues of plants by weight. Water is involved in many life-sustaining processes in plants, including as:
- an input to the essential processes of photosynthesis
- the solvent for minerals transported in the **xylem** tissue and for sugars transported in phloem tissue
- the aqueous medium in which all biochemical reactions in plant cells occur
- a means of cooling plants by evaporation of water from leaves during transpiration
- the source of **turgor pressure** that maintains plant cell shape and enables non-woody plants to stand erect.

Given the importance of water in the life of plants, plants have regulatory mechanisms and structural adaptations that contribute to maintaining a water balance, such that overall:

Water IN (by roots) = Water OUT (at leaves)

However, an imbalance can occur in plants in hot, dry and windy conditions or in drought, when the rate of water uptake at the roots does not keep pace with water loss at the leaves.

Maintaining water balance is the key challenge for terrestrial plants and involves:
- keeping water loss within manageable limits, so that plant tissues are hydrated and able to function
- keeping water uptake at a level to compensate for water loss.

Plant organs that are critical in water balance have both regulatory mechanisms and structural adaptations.
- Regulatory mechanisms are active responses by plants that are induced by changes in water balance.
- Structural adaptations are always there, regardless of the state of water balance. These adaptations are heritable physical traits that have evolved in plant populations and equip plants to survive and reproduce under their particular environmental conditions.

4.2.2 Controlling water uptake

As described in Topic 3, roots are the plant organs that are specialised for locating and absorbing water. As such, roots supply the water on the input side of the water balance equation in plants

In terrestrial plants, roots have several functions:
- anchoring the plant in the soil
- storing products of photosynthesis, for example starch-containing potato tubers and modified tap roots in plants such as carrots, parsnips and radishes
- importantly, in terms of water balance, locating and absorbing water (and essential mineral nutrients) in the soil.

xylem the part of vascular tissue that transports water and minerals throughout a plant and provides a plant with support

turgor pressure force or pressure potential within a fully hydrated plant cell that pushes the plasma membrane against the cell wall

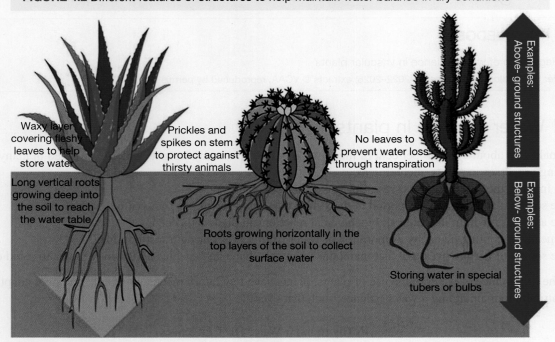

FIGURE 4.2 Different features of structures to help maintain water balance in dry conditions

Root systems: finding water

Root systems are of two types:
- In grasses and other monocot plants, the root system consists of many fine fibrous roots that arise from the base of stems (see figure 4.3a). Lateral roots can have multiple levels of branching and are usually restricted to the shallow levels of soil. Fibrous root systems provide a very large surface area for water absorption.
- In dicot plants, the root system includes a primary (tap) root that grows vertically downwards with lateral roots originating from the primary root (see figure 4.3b). Tap root systems can penetrate the soil to reach water at deeper levels. In a study of maximum rooting depths, the maximum depth observed was 68 metres for a tree, *Boscia albitrunca*, that grows in hot dry areas of southern Africa.

FIGURE 4.3 a. The fibrous root system of a grass; the roots originate from the base of the stems. b. Tap root system of a dicot plant showing main roots from which lateral roots arise

Roots do not just grow to a point and stay fixed. Rather, roots are actively growing explorers that can sense features of the soil around them, such as variations in water potential, soil compaction, and the distribution of mineral nutrients. Plants respond to this information by modifying their root systems, for example by developing lateral roots or by changing the degree of lengthening and the direction of growth. These responses are discussed in Topic 10 with reference to plant adaptations.

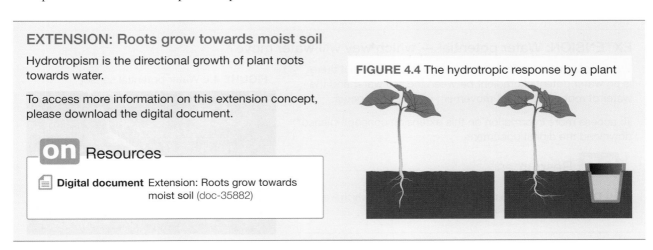

EXTENSION: Roots grow towards moist soil

Hydrotropism is the directional growth of plant roots towards water.

To access more information on this extension concept, please download the digital document.

Resources

Digital document Extension: Roots grow towards moist soil (doc-35882)

FIGURE 4.4 The hydrotropic response by a plant

Root systems: absorbing water

The entry point of water in plants is through the **root hairs**. Root hairs are fine outgrowths of the specialised epidermal cells in the maturation zone of roots. Root hairs greatly increase the surface area available for absorption of soil water. In general, apart from root tips, the remainder of a root cannot absorb water because the epidermis is covered in a waterproof layer of **suberin**.

Figure 4.5a is a scanning electron micrograph (SEM) of a cross-section through the maturation zone of a plant root. Note the root hairs (white) projecting from the epidermal cells of the root and the central vascular bundle with its large xylem vessels. Once absorbed, water must travel from the root hairs to the xylem.

root hairs extensions of cells of the epidermal tissue that forms the outer cellular covering of the root, responsible for absorption and uptake of liquid water

suberin a non-cellular waxy substance that forms a waterproof barrier found on the cell walls of some plant cells, including root epidermal cells (except for the root tip) and root endodermal cells

FIGURE 4.5 a. SEM of a cross-section of a root, showing many fine root hairs (white). b. Cross-section of a root tip at the zone of maturation. Note how some epidermal cells have root hair extensions.

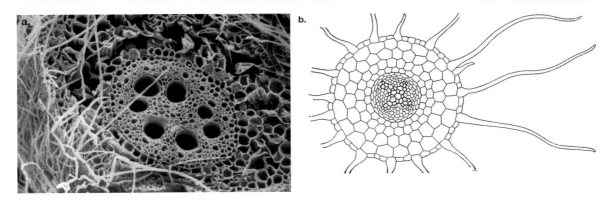

Absorption of water by root hairs mainly occurs passively, through osmosis and facilitated diffusion, with neither process requiring an input of energy. These passive forms of water uptake — osmosis and facilitated diffusion — can only occur in the presence of a water potential gradient between the soil water (higher potential)

and the root cell water (lower potential). Cell water normally has more dissolved solutes than soil water and thus has a lower water potential that that of the soil water. This difference means that water can move down a water potential gradient from the soil into root cells. The soil water also contains various dissolved solutes that are picked up from the soil.

For mineral nutrients, the plasma membranes of root hairs provide some control over their uptake.

EXTENSION: Water potential — which way will water move?

Water potential is an important variable in biology, as if there is no water potential gradient between the soil water and the water of root cells, no net movement of water can occur.

To access more information on this extension concept, please download the digital document.

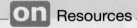 **Resources**

📄 **Digital document** Extension: Water potential — which way will water move? (doc-35883)

FIGURE 4.6 Water potential sensor

Roots: regulating water entry to the xylem

After being absorbed by the root hairs, water must reach the xylem, from where it is transported to all living cells of a plant.

The soil water that enters the root hairs brings valuable mineral nutrients with it, but it also brings other solutes that may be toxic to plants, such as pesticide residues, as well as solutes that are either in excess or unwanted. Roots have a mechanism to prevent the unregulated passive movement of soil water and its solutes into the xylem. The root cells that play a major role in this regulation are those of the **endoderm**.

Water first moves radially across the root cortex towards the xylem. In its journey across the cortex, some water may follow a pathway from within one cell to the next. Other water may not enter cells but instead travel along cell walls and through intercellular spaces. Figure 4.7 shows two pathways for soil water from root hairs across the root cortex: a cell-to-cell pathway (shown in red) and the cell wall pathway (shown in blue).

Before reaching the xylem, all soil water is forced to cross the plasma membranes of the cells of the endoderm. This forced route is due to the presence of a ring of waterproof material that encircles the cell wall of endodermal cells — the **Casparian strip**.

Endodermal cells are active gatekeepers that control the quality and amount of water that enters the xylem. They also selectively control the entry of mineral nutrients and other solutes:

- Endodermal cells control the number and the activity of water channels (**aquaporins**) in their plasma membranes. Because of this, the endoderm can regulate the rate of water transport into the xylem, keeping it in balance with the needs of the plant. If water loss by the leaves of a plant increases, the water uptake by roots and its transport into the xylem can also be increased.
- Endodermal cells have transport proteins in their plasma membranes that selectively control the entry and exit of solutes — this means that toxic or excessive solutes can be differentially excluded and prevented from reaching the xylem, while required mineral nutrients can continue to the xylem.

> **endoderm** a layer of cells that forms the innermost part of the cortex and encircles the vascular bundle that includes the xylem
>
> **Casparian strip** a ring of waterproof material, composed of lignin, deposited on the walls of endodermal cells that forces all soil water to move into the cells of the endoderm before it can reach the xylem
>
> **aquaporins** protein channels in the plasma membrane that allow the rapid flow of water into and out of cells

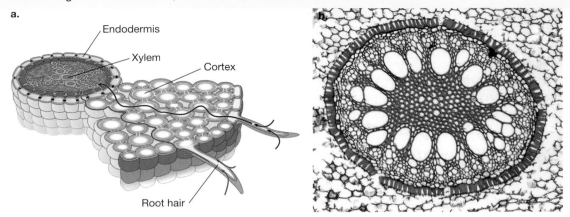

FIGURE 4.7 a. Two pathways for water from its uptake at root hairs into the xylem. Casparian strips are denoted by black dots between adjacent endodermal cells. **b.** Photomicrograph of a cross-section of part of a plant root. Note the red ring of the endodermis, which surrounds the central vascular bundle.

Lack of water uptake causes water stress

Water stress is the negative physiological effects on plants from a lack of water that is available for uptake. This upset in the water balance (water IN is less than water OUT) can be seen in wilting.

When rainfall is low or absent, roots can sense decreasing soil water. When water deficit is signalled, the roots respond. Some of these responses can take effect rapidly, while others are longer-term growth responses. Responses of plants in times of water stress from low rainfall include:

- a greater investment into the growth of roots rather than shoots. The hormone abscisic acid (ABA) plays a key role in this since it promotes root elongation and inhibits shoot growth.
- penetration of roots into deeper layers of soil where there is less danger of drying out
- increased deposition of suberin, a waterproofing material, on most root cells. This measure is an attempt to protect against water loss from root cells to the soil through a process of **reverse osmosis**
- **osmotic adjustment** (OA), an important and rapid response that also prevents the loss of water from root cells to the drying soil, enabling normal osmosis to continue.

reverse osmosis a process in which water moves by osmosis from root cells to the soil; occurs when an increase in the solute concentration in the soil water lowers its water potential to below that of the cell water

osmotic adjustment a control mechanism for preventing water loss due to reverse osmosis in plants under water or salt stress; involves root cells accumulating additional solutes that lower the cell water potential to below that of the soil water

Root systems: responses in waterlogging

Too little water creates stress for plants, but so does too much! Figure 4.8 shows a soy bean field where persistent heavy rainfall and/or poor drainage have produced waterlogged soil with obvious surface water. The soybeans show the signs of water stress through wilting and the yellowing of their leaves (leaf chlorosis). How can these plants show wilting, a classic sign of water deficit in plants, in all this water?

In waterlogged soil, the air spaces in the soil become filled with water. Within about 24 hours, cellular respiration by root cells can deplete the oxygen in the soil water to near zero. As a result, the oxygen consumed by plant roots cannot be replaced at

FIGURE 4.8 Waterlogged soy bean field with plants showing signs of water stress

rates sufficient to maintain their cellular respiration. The roots of plants in waterlogged soil become starved of oxygen and are damaged so that they cannot absorb water. Death of the plants eventually follows if waterlogging continues.

Features that enable plants to tolerate waterlogging include:
- inhibition of growth of main roots and inhibition of formation of new lateral roots
- outgrowth of adventitious roots, into the soil, along the water surface or into the floodwater. Adventitious roots originate from the stem nodes and are commonly seen in wetland plants. These roots can compensate in part for the death of other roots and replace their water uptake function.
- presence of aerenchyma, a spongy plant tissue that contains spaces or air channels, found in roots and stem. Aerenchyma tissue provides an internal channel for the diffusion of gaseous oxygen from shoots to roots.

4.2.3 Controlling water loss

Leaves are the source of water loss from plants; thus, they are on the output side of the water balance equation. The key function of leaves is to capture the energy of sunlight and use that energy indirectly to produce glucose from carbon dioxide. The carbon dioxide that is required for photosynthesis comes from the atmosphere and diffuses into leaves through open pores in the leaf epidermis (figure 4.9). These pores are called **stomata** (singular = stoma). Each pore (or stoma) is surrounded by a pair of guard cells.

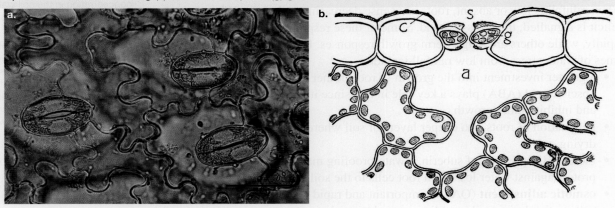

FIGURE 4.9 a. Three stomata on a leaf surface. Note the paired guard cells. **b.** Cross-section of part of the epidermis of a leaf showing (s) the stomatal pore, (g) guard cell, (c) cuticle and (a) airspace or sub-stomatal cavity

As carbon dioxide molecules enter leaves through the stomatal pores, water molecules are simultaneously being lost from the leaves through the same pores. The cost of importing carbon dioxide for photosynthesis from the air is very high in terms of water loss — for every molecule of carbon dioxide that enters a leaf through a stomatal pore, several hundred water molecules exit at the same time. However, this is a necessary cost to the plant. Plant stomata are the site of water loss in leaves.

Transpiration: the major source of water loss

The loss of water by diffusion through leaf stomata is termed **transpiration**. Transpiration is not simply a water loss process, it is an essential process in plants. Transpiration pulls the water required for photosynthesis and other biochemical processes up to the leaves, brings the mineral nutrients required for growth into leaf cells, and is a source of evaporative cooling of leaves, helping to prevent heat stress — as water evaporates, heat energy is taken with it.

stomata pores, each surrounded by two guard cells that regulate the opening and closing of the pores

transpiration loss of water from the surfaces of a plant

The rate of transpiration is affected by environmental factors:
- humidity: the greater the content of water vapour in the air, the lower the rate, such that at 100 per cent humidity in the air, net water loss by transpiration stops
- wind speed: transpiration is least in still air and, all other things being equal, increases as wind speed increases, as moving air removes water vapour from around the leaves
- temperature: as temperatures increase, the rate of evaporation of water increases
- light intensity: light stimulates the opening of the leaf pores (stomata).

Transpiration through open stomata is the major source of water loss by plants.

Studies show that only about one per cent of the water taken up by plant roots is used by the plant. Of the reminder, 90 per cent or more is lost by transpiration. For a large leafy tree, this is a considerable amount. For example, the water loss by transpiration from the leaves of one large oak tree has been estimated at about 150 000 litres per year. On a global scale, transpiration is significant — studies have shown that about 10 per cent of the moisture in the atmosphere comes from transpiration by plants.

Water imbalance, when water input is less than output, results in dehydration of plants. This leads to a loss of turgor pressure in plant cells and is expressed by the wilting of a plant. Wilted plants usually regain their turgor pressure when they are re-watered, unless they have reached the 'permanent wilting point' from which there is no recovery.

FIGURE 4.10 At left: a fully hydrated plant with its water in balance; at right: a dehydrated plant that has lost turgor pressure and wilted because of a water imbalance

INVESTIGATION 4.1

Investigating transpiration in plants

Aim

To investigate factors that affect the process of transpiration in plants

Closing stomata to minimise water loss

Stopping or slowing transpiration by closing stomata is the major means of minimising water loss, but at the same time, this affects the supply of carbon dioxide that is needed for photosynthesis.

The opening and closing of stomata must be finely regulated. Surrounding guard cells are critical to the operation of the pore as a regulated gateway that can open and close as needed.

The opening and the closing of a stomatal pore depends on the state of hydration of the guard cells around the pore:
- When fully hydrated, the contents of the guard cells exert pressure on their cell walls. This turgor pressure causes the thinner outer walls of the guard cells to move outward, pulling apart the thicker inner walls and opening the pore.
- When guard cells lose water, this turgor pressure is lost and the pores close (refer back to figure 4.9a).

SAMPLE PROBLEM 1 Exploring the role of stomata and guard cells in water balance

tlvd-1747

Describe what would happen to stomata and guard cells when water is low and what would happen when water is high. **(2 marks)**

THINK

1. Identify what the question is asking you to do. Here you are being asked to describe. The question is worth two marks. One mark is for the description when water is low. Ensure you mention both the stomata and guard cells.

2. The second mark is for the description when water is high. Ensure you mention both the stomata and guard cells.

WRITE

When water levels are low, water and ions leave the guard cells, causing them to become flaccid. This causes the stomata to close (1 mark).

When water levels are high, water and ions enter the guard cells, causing them to become turgid. This causes the stomata to open (1 mark).

The plant hormone abscisic acid (ABA) plays an important role in regulating the water balance of plants. When plants become water-deficient, a regulatory response occurs that closes the leaf stomata and, as a result, water loss by transpiration is stopped. ABA is a key player in regulating the closing of stomatal pores. ABA also inhibits the opening of stomata, so while it is present in guard cells, stomatal pores will remain closed. The following extension box provides a simplified outline of the regulatory role of ABA on stomatal pores.

EXTENSION: ABA hormone and stomatal pore closing

When plants experience water stress, the hormone ABA is synthesised, which triggers a series of responses in the guard cells of the stoma.

To access more information on this extension concept, please download the digital document.

 Resources

Digital document Extension: ABA hormone and stomatal pore closing (doc-35884)

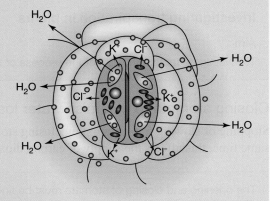

FIGURE 4.11 Plant hormones can trigger responses in stoma.

Stoma closing (cells flaccid)

260 Jacaranda Nature of Biology 1 VCE Units 1 & 2 Sixth Edition

INVESTIGATION 4.2

Distribution of stomata and guard cells

Aim

To investigate stomata and guard cells in plants adapted to different environments

Other signals for stomatal action

The regulation of water balance of a plant does not occur in response to water stress only. Water stress is just one signal to a plant to close its stomatal pores. In addition to soil moisture content, other variables such as high temperatures and the darkness of night are also signals to close the pores. Signals to open stomatal pores include the blue light of early morning and falls in carbon dioxide within leaves. These various signals are highly coordinated so that the regulation of stomatal pores is finely balanced between the need for carbon dioxide for photosynthesis by a plant's leaves and the need for adequate hydration of all living plant cells.

Some plants, such as succulents, also regulate water balance by altering when they close their stomatal pores. This is described as water-conserving photosynthesis. In many plants, water loss is high during the day when they are photosynthesising, due to higher temperatures and humidity. Succulents open their stomata during the cooler night times, take in the carbon dioxide needed for photosynthesis, and store it as an organic acid. At sunrise, the stomata close and the carbon dioxide is released from the store to be used for photosynthesis. This helps them reduce water loss during the day.

FIGURE 4.12 Plants such as succulents undertake water-conserving photosynthesis.

In addition to stomata, plants have a number of other adaptations that minimise water loss, including leaf rolling, sunken stomata and waxy cuticles. These are discussed in detail in Topic 10.

 Resources

 eWorkbook Worksheet 4.1 Controlling water loss and uptake in plants (ewbk-7491)

KEY IDEAS

- Water is vital for the survival of plants.
- Water is in balance when water uptake by roots equals water loss from leaves.
- Vascular plants have control strategies that can be activated to maintain their water balance.
- Water potential is an important variable that identifies the direction of water movement between soil and roots.
- Roots can sense and actively move towards moist patches of soil.
- In response to environmental signals, plants can initiate regulatory mechanisms that contribute to the maintenance of water balance.
- Plants also have various structural adaptations that assist in maintaining their water balance.
- Root hairs, outgrowths of epidermal cells on root tips, are the sites of water absorption by roots.
- The major loss of water in plants occurs when water vapour is lost by transpiration from the leaf stomata.
- Plants can actively regulate water loss by the opening and closing of their stomata.
- Water loss occurs when guard cells are turgid, causing the stomata to open; when guard cells lose turgor, stomata close, preventing water loss.
- Plants can also actively regulate water loss by other methods, including osmotic adjustment.

4.2 Activities

To answer questions online and to receive **immediate feedback** and **sample responses** for every question, go to your learnON title at **www.jacplus.com.au**. A **downloadable solutions** file is also available in the resources tab.

4.2 Quick quiz	4.2 Exercise	4.2 Exam questions

4.2 Exercise

1. What is the major avenue of water loss in plants?
2. Identify an example of:
   a. a plant response that reduces water loss from leaves
   b. a root response that enables roots to absorb water even though the plant is drying out.
3. Identify one example of:
   a. reducing water loss from cells at the leaf surface
   b. reducing water loss from the leaf stomata.
4. Describe how regulation of guard cells contributes to water balance.
5. Identify one important role of each of the following.
   a. Root endodermis
   b. Aerenchyma in shoots and roots
6. Measurements in a field showed that the water potential (WP) of the root cells of plants was −0.6 MPa and that of the soil was −0.3 MPa.
   a. Which has the higher water potential: the soil water or the cell water?
   b. In which direction will water move?
7. a. What is succulence?
   b. Identify how this growth form contributes to water balance during periods of water shortage.

4.2 Exam questions

Question 1 (1 mark)

MC Water is absorbed from the soil and into the vascular plant root system through

A. tracheids.
B. xylem vessel elements.
C. sieve tube cells.
D. root hair cells.

Question 2 (1 mark)

MC Four populations of a particular plant species are found in four locations with different environmental conditions.

Location	Daily average temperature (°C)	Daily average humidity (%)
1	18	20
2	25	60
3	18	60
4	25	20

The average transpiration rate of the plants was calculated. At which location would the plants with the highest average transpiration rate be found?

A. Location 1
B. Location 2
C. Location 3
D. Location 4

Question 3 (1 mark)

MC Root hair cells increase the efficiency of water absorption by

A. having a large surface area.
B. being spherical in shape.
C. containing many mitochondria.
D. possessing a large nucleus.

▶ **Question 4** (2 marks)

Endodermal cells control the quality and amount of water that enters the xylem, and the entry of mineral nutrients and other solutes.
  a. How do endodermal cells control the amount of water that enters a plant?   **1 mark**
  b. How do endodermal cells control the entry and exit of solutes to the plant?   **1 mark**

▶ **Question 5** (3 marks)

An experiment was set up to investigate the rate at which water vapour evaporated from a plant.

Three identical plants were enclosed in airtight containers and the rate of water vapour evaporation from the plant was recorded. The environment of each of the three plants was slightly different. Each plant was given the same amount of water in the soil. The diagram below illustrates how the experiment was set up.

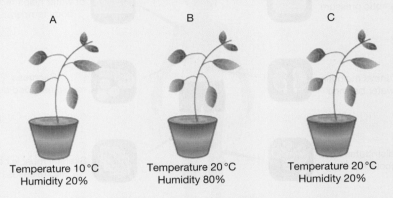

A — Temperature 10 °C, Humidity 20%
B — Temperature 20 °C, Humidity 80%
C — Temperature 20 °C, Humidity 20%

In which plant would you expect the greatest rate of water vapour evaporation? Explain.

More exam questions are available in your learnON title.

4.3 Homeostatic mechanisms and stimulus–response models

KEY KNOWLEDGE

- Regulation by homeostatic mechanisms, including stimulus–response models, feedback loops and associated organ structures

Source: Adapted from VCE Biology Study Design (2022–2026) extracts © VCAA; reproduced by permission.

4.3.1 What is homeostasis?

Homeostasis is the body's attempt to maintain a constant and balanced internal environment. Homeostasis can be defined as the outcome of processes that maintain a steady state (constant internal environment) for certain physiological variables and for the chemical compositions of body fluids. Homeostasis requires the continual monitoring of the levels of key variables and the making of adjustments as conditions change.

homeostasis condition of a relatively stable internal environment maintained within narrow limits

TOPIC 4 Regulation of systems 263

Monitoring the level of variables involves comparing the current level against the value of its **set point**. A set point is not a single value. Rather, it is a narrow range of values that are seen as 'normal'. Refer to column 2 of table 4.1 to identify the ranges around set points for some important variables. When the value of a variable moves outside its normal range, homeostatic mechanisms come into play to counteract the change and restore the variable to within the narrow range around its set point.

set point midpoint of a narrow range of values around which a physiological variable fluctuates in a healthy person

FIGURE 4.13 Homeostasis is an incredibly important process in regulation of body systems.

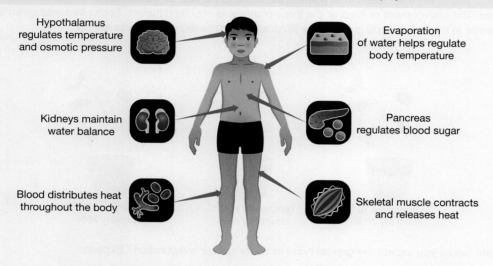

TABLE 4.1 Summary of major variables that are subject to homeostasis in humans

Variable	Normal range	Comments
Temperature	36.1–37.2 °C	The temperature of internal cells of the body is called the core temperature.
Blood glucose	3.6–6.8 mmol per L	Blood glucose is typically maintained within narrow limits regardless of diet.
Water balance: blood volume blood osmolality (osmotic concentration)	approx. 70 mL/kg 275–295 mOsm/kg	Blood volume and osmolality are a sign of whether or not water intake and water loss are in balance. If water intake is in deficit, blood volume falls and blood osmolality rises.
Ions, e.g. plasma Ca^2	2.3–2.4 mmol per L	Specific ions are required by some tissues.
pH of arterial blood	7.4	Regulation of pH is necessary for enzyme action and nerve cells.
Arterial blood pressure: diastolic (relaxed) systolic (contracted)	13.3 kPa (100 mmHg) 5.33 kPa (40 mmHg)	Transport of blood depends on maintenance of an adequate blood volume and pressure.

The tolerance ranges for various vertebrate animals differ. Some vertebrates have much higher or much lower tolerance ranges compared to humans. For example, the normal arterial blood pressure of a giraffe is 300/180 mmHg, and the normal body temperature of many birds is 40.5 °C. Other vertebrates are unable to regulate certain variables. For example, reptiles do not maintain a relatively stable core temperature and do not have the homeostatic mechanisms that would enable them to do this.

Body systems in homeostasis

In mammals (including humans), all body systems contribute in various ways to the maintenance of homeostasis. The most important body systems involved are:
- the nervous system
- the **endocrine system** (figure 4.14).

The nervous system and the endocrine system are both vital for maintaining homeostasis at the level of the whole organism. Their critical role is in the communication of signals. Communication in the nervous system occurs through electrochemical signals that are transmitted by neurons. The endocrine system communicates through hormones that are released by endocrine glands into the blood stream and travel to target organs. Table 4.2 is a summary of the contrasting features of these two systems.

> **endocrine system** system of ductless glands that produce hormones and release them directly into the bloodstream

FIGURE 4.14 The endocrine system in the human body and various hormones that it releases

Hypothalamus
- **Growth-hormone-releasing hormone:** stimulates pituitary gland to release GH
- **Corticotropin-releasing hormone (CRH):** stimulates pituitary gland to release ACTH
- **Thyroid-releasing hormone:** stimulates thyroid gland to release TSH
- **Gonadotropin-releasing hormone (GnRH):** stimulates pituitary gland to release FSH and LH
- **Antidiuretic hormone (ADH):** promotes reabsorption of H_2O by kidneys
- **Oxytocin:** induces labour and milk release from mammary glands in females

Pineal gland
- **Melatonin:** regulates sleep-wake cycles

Thyroid and parathyroid glands
- **Thyroxine:** increases metabolic rate and heart rate; promotes growth
- **Parathyroid hormone (PTH):** increases blood concentration of calcium ions

Kidneys
- **Erythropoietin (EPO):** increases synthesis of red blood cells
- **Vitamin D:** decreases blood concentration of calcium ions

Ovaries (in females)
- **Estradiol:** regulates maintenance and development of secondary sex characteristics
- **Progesterone:** prepares uterus for pregnancy

Placenta (during pregnancy)
- **Chorionic gonadotrophin:** supports early pregnancy

Testes (in males)
- **Testosterone:** regulates development and maintenance of secondary sex characteristics

Anterior pituitary gland
- **Growth hormone (GH):** stimulates growth
- **Adrenocorticotropic hormone (ACTH):** stimulates adrenal glands to secrete glucocorticoids
- **Thyroid-stimulating hormone (TSH):** stimulates thyroid gland to secrete thyroxine
- **Follicle-stimulating hormone (FSH) and luteinizing hormone (LH):** regulates menstrual cycle in females; production of sex hormones
- **Prolactin (PRL):** stimulates mammary gland growth and milk production in females

Thymus
- **Thymosin:** triggers T cell formation

Pancreas (islets of Langerhans)
- **Insulin:** lowers levels of blood glucose
- **Glucagon:** raises levels of blood glucose

Adrenal glands
- **Epinephrine:** produces short-term stress response
- **Cortisol:** produces short-term and long-term stress responses
- **Aldosterone:** increases reabsorption of sodium ions by kidneys

- ■ Polypeptides/proteins
- ■ Amino acid derivatives
- ■ Steroids (lipid-derived)

 Resources

📄 **Digital document** Background knowledge: The nervous system (doc-35885)

TABLE 4.2 Comparing the nervous and endocrine systems

Feature	Nervous system	Endocrine system
Speed of message	Fast	Slow
Speed of response	Immediate	Slow
Duration of response	Short	Long
Type of message	Electrical and chemical (neurotransmitters) along nerves	Chemical (hormones) through the bloodstream

4.3.2 Stimulus–response models

Homeostatic regulation involves the monitoring of the value of a variable, such as body temperature; detecting if it starts to move outside the normal range; and making adjustments to correct the situation. This process can be represented by a **stimulus–response model** with feedback (figure 4.18).

The model starts with a stimulus and ends with a response that feeds back to and, typically, counteracts the original stimulus. Stimulus–response feedback models can be used to show how homeostatic mechanisms act in the body and maintain a fairly constant state.

There are two types of stimulus–response models: open stimulus–response models and closed homeostatic stimulus–response loop models. Each of these models involves five main components.

> **The components of a stimulus–response model**
>
> 1. *Stimulus*: a change, either an increase or a decrease, in the level of an internal variable
> 2. *Receptor*: the structure that detects the change and sends signals to the control centre
> 3. *Control centre*: the structure (**central nervous system**) that evaluates the change against the set point for that variable and sends signals to the effector about the correction needed
> 4. *Effector*: the structure that adjusts its output to make the required correction
> 5. *Response*: the corrective action taken
>
> An additional process in the closed homeostatic stimulus response loop model is feedback:
>
> 6. *Negative feedback*: the counteracting or negating effect of the response on the stimulus.

What are receptors?

Receptors are structures that detect stimuli and, except for reflex arcs, transmit signals via sensory neurons to various control centres in the brain.

- Some receptors sense changes that are stimuli for open stimulus–response actions.
- Other receptors sense changes that are stimuli for homeostatic stimulus–response loops.

Some examples of different types of receptors include:

- *chemoreceptors*: receptors that detect changes in dissolved chemicals, such as the carotid body in the carotid artery, which detects changes in pH and carbon dioxide concentration in the blood
- *mechanoreceptors*: receptors that detect physical forces; examples include receptors for touch, pressure, stretch and vibration in your skin (figure 4.15b), pressure receptors in blood vessels and stretch receptors in joints
- *photoreceptors*: receptors that detect changes in light, such as the rods and cones in the retinas of your eyes, which respond to visible light
- *thermoreceptors*: receptors that detect changes in temperature, such as heat-sensitive and cold-sensitive nerve endings in your skin (figure 4.15b) and groups of temperature-sensitive nerve cells in some body organs, including the hypothalamus of the brain
- *osmoreceptors*: receptors that respond to changes in osmotic concentration (osmolality) of body fluids. These receptors are groups of cells located in the hypothalamus.

stimulus–response model a representation of an action that starts with a stimulus and ends with the response to that stimulus
central nervous system the part of the nervous system composed of the brain and spinal cord

FIGURE 4.15 a. Diagrammatic representations of different receptors **b.** Photomicrograph showing two Pacinian corpuscles in the dermis of the skin — the nerve endings of these receptors are enclosed in a capsule.

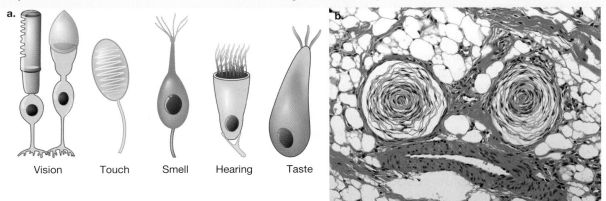

The open stimulus–response model

Consider a runner on the starting line who hears the start signal and pushes off the blocks (figure 4.16).

This scenario can be represented by a stimulus–response model of the action of the nervous system, as shown in figure 4.17. The model is an open pathway with three structural components — receptor, control centre and effector — plus an input (stimulus) and an output (response). The response has no effect on the stimulus. The neurons involved in communicating signals between the components belong to the **peripheral nervous system**.

FIGURE 4.16 The starting gun provides the stimulus for the runners.

FIGURE 4.17 Simplified representation of an open stimulus–response model. Nerve connections within the control centre, such as interneurons in the brain, are not shown.

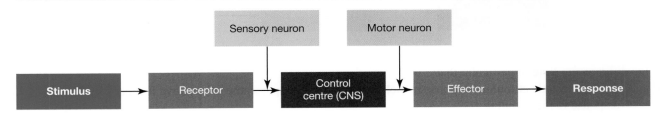

peripheral nervous system
the part of the nervous system containing nerves that connect to the central nervous system

Consider each component of the model for the runner:
- *Stimulus:* the sound of the start signal
- *Receptor:* the ears (specifically, hair cells of the cochlea) detect the vibration of the sound of the start signal.
- *Control centre:* the brain cortex receives and processes the nerve signal from the ears and transmits a message about the required response.
- *Effector:* the skeletal muscles of their arms and legs receive the signal from the control centre.
- *Response:* the contractions of the muscles of their arms and legs push them off the starting block.

In an open stimulus–response model, there is no feedback provided after the response occurs. We will compare this to homeostatic stimulus–response loop models, which contain feedback and are a mechanism of homeostasis.

The homeostatic stimulus–response loop model

Many physiological variables, for example core temperature and blood glucose level, are regulated by homeostatic stimulus–response loops. A homeostatic stimulus–response loop includes the same components as the open stimulus–response model, but the stimulus and response are connected by feedback to form a loop. Figure 4.18 shows the components of a homeostatic stimulus–response loop (homeostatic control loop).

The key additional element in a homeostatic control loop is negative feedback from the response that counteracts the stimulus.

FIGURE 4.18 Diagram of a stimulus–response model with feedback

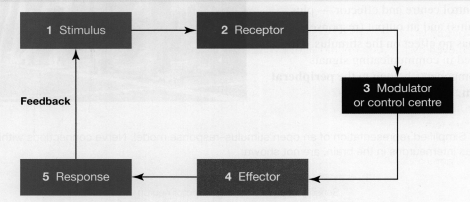

Homeostatic control loops differ significantly from open stimulus–response models. Table 4.3 identifies some of the differences. Note in particular that:
- the homeostatic stimulus–response model forms a closed loop because of negative feedback by the response to the stimulus
- the sensing of a stimulus by a receptor and the response by an effector are not one-off events; instead, they occur continually as long as the change in the internal variable (stimulus) means that it is outside its normal range
- the response counteracts the initial stimulus and is not under conscious control.

In this topic, different homeostatic stimulus–response loop models will be examined for temperature regulation (subtopic 4.4), blood glucose regulation (subtopic 4.5) and water balance (subtopic 4.6).

TABLE 4.3 Comparison of an open stimulus–response model and a closed homeostatic stimulus–response loop

Feature	Stimulus–response model	Homeostatic loop model
Outcome of operation	Specific motor actions performed	Steady level (homeostatis) of internal variable
Stimulus	Change in external or internal environment	Change in level of a variable outside its narrow range
Receptor	One-off sensing of the change	Continual sensing until change is corrected
Control centre	Cerebral cortex of the brain	Hypothalamus and hindbrain
Effector	Skeletal muscles	Smooth muscles, glands and organs
Response	Mainly voluntary responses but also involuntary (reflex) responses	Only involuntary physiological responses, but behavioural responses can occur
Feedback from response to stimulus	No	Yes, negative feedback
Pathway	Open	Closed
Communication: receptor to control centre	Sensory nerve pathway of the somatic nervous system[1]	Sensory nerve pathway of the autonomic nervous system[2]
Communication: control centre to effector	Motor nerve pathway of the somatic nervous system	Signal sent by nerve or by hormone carried in blood

[1] The somatic nervous system is the part of the peripheral nervous system associated with voluntary control of movement.
[2] The autonomic nervous system is the part of the peripheral nervous system associated with involuntary control of bodily functions.

Two homeostatic loops in action

The regulation of key variables typically depends on more than one homeostatic control loop. In many cases, regulation is achieved by the joint operation of two homeostatic loops.

The continual monitoring by the receptors, combined with the corrective responses of the effectors, ensures that the variable is tightly regulated within its set point range. Figure 4.19 shows the operation of two homeostatic control loops concerned with the regulation of a key variable. If the response by one loop overcompensates for the change in core temperature, the second loop makes the needed adjustment.

FIGURE 4.19 Diagram showing only the stimuli and the responses of two homeostatic loops involved in the regulation of a variable. (Not shown here are the associated receptor, control centre and effector of each loop.)

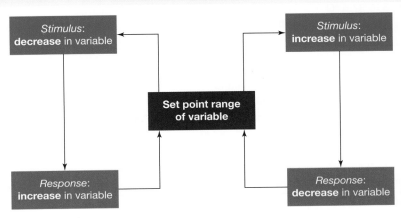

Homeostatic control loops do not simply run through a cycle once. Receptors are continually monitoring the levels of variables, and effectors are constantly producing corrections when levels move outside the narrow ranges around set points, as shown in figure 4.20. This process is known as negative feedback, which is explored in section 4.3.3.

FIGURE 4.20 Diagram showing the actions of the receptors and effectors of two homeostatic control loops involved in regulating the level of a variable. The receptor and effector of one loop are shown in pink, and those of the second loop are shown in green.

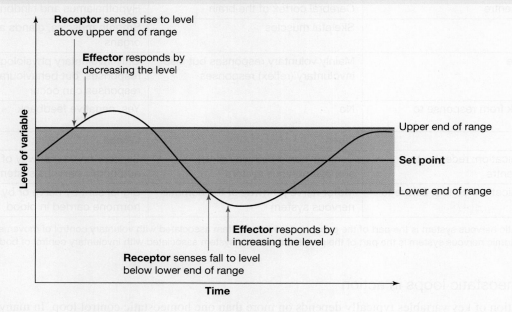

What variables are under homeostatic control loops?

Many variables in our body are regulated through the use of homeostatic control loops. Some of these are shown in table 4.4.

TABLE 4.4 Examples of regulation by homeostatic control loops

Regulated variable	Normal range	Receptors and location	Control centre	Effectors
Core body temperature	36.5 to 37.5 °C	Thermoreceptors in skin and in hypothalamus	Hypothalamus	Blood vessels, sweat glands and skeletal muscles
Blood glucose	3.6 to 6.8 mmol/L	*When fasting:* chemoreceptors in pancreas and hypothalamus	Hypothalamus and pancreas	Liver and adipose tissue
		Non-fasting: chemoreceptors in pancreas	Pancreas	
Plasma Ca^{2+}	2.1 to 2.4 mmol/L	Chemoreceptors in parathyroid gland	Parathyroid gland	Bone, kidney and intestine
Plasma K^+	3.5 to 5.0 mmol/L	Chemoreceptors in adrenal cortex	Adrenal cortex	Kidney

(continued)

TABLE 4.4 Examples of regulation by homeostatic control loops *(continued)*

Regulated variable	Normal range	Receptors and location	Control centre	Effectors
H^+ concentration in arterial blood (pH)	7.35 to 7.45	Chemoreceptors in carotid, aorta and brain	Brain stem (medulla)	Lungs and diaphragm muscles
		Chemoreceptors in kidney	Kidney	Muscles and kidney
Mean arterial blood pressure	70 to 100 mmHg	Mechanoreceptors in aorta and carotid artery	Brain stem	Heart and blood vessels
		Mechanoreceptors and chemoreceptors in kidney	Kidney	Kidney and adrenal cortex
Osmolality of blood	275 to 295 mmol/kg	Osmoreceptors in hypothalamus	Hypothalamus and posterior pituitary	Kidney
Volume of blood (circulating)	Approx. 5 L	Mechanoreceptors in carotid artery and heart	Blood vessels	Blood vessels
		Mechanoreceptors in kidney	Kidney	Kidney

on Resources

eWorkbook Worksheet 4.2 Maintaining the balance (ewbk-7493)

Weblink Sensory receptors

4.3.3 Feedback loops

Feedback can be variously defined as:
- the response of a system that influences its ongoing activity
- a mechanism in which the output of a process is directed back to and affects the input to the process.

The effect of feedback may either amplify the input or counteract it.

This process of stimulus–response and feedback is referred to as a **feedback loop**. Figure 4.21 shows a representation of a feedback loop. Here, the system includes a detector to receive the input, a control centre to evaluate the input and identify a response, and an effector to generate the output. The arrows denote communication between the parts of a feedback loop.

feedback loop a process in which the response (output) in a stimulus–response action affects the stimulus (input), either increasing or decreasing it

FIGURE 4.21 Diagram of a feedback loop

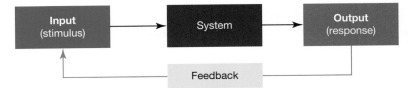

Result: Input altered (either amplified or counteracted)

There are two main types of feedback:
- **negative feedback**: when the response feeds back to and counteracts the change in the variable. This is a homeostatic mechanism.
- **positive feedback**: when a change in a variable produces a response that further amplifies the stimulus and increases its effect.

> **TIP:** Be very careful of a common misconception when talking about negative and positive feedback. Negative feedback is not always a decrease of a variable. It is about returning a variable back to the narrow range around its set point. Sometimes, negative feedback may involve increasing a variable to bring it back to normal.

Negative feedback loops in homeostasis

A negative feedback loop is a process in which the body senses a change in a variable and activates mechanisms to reverse the change. Negative feedback is a key component of homeostatic control loops that regulate many body variables and maintain the body's internal conditions within narrow limits.

> A key feature of a negative feedback loop is that the response is opposite in direction to that of the original stimulus.

So:
- if the stimulus is an increase in a variable, then the response is a decrease in the same variable.
  For example, if the blood glucose level is too high, negative feedback results in it decreasing to normal.
- if the stimulus is a decrease in a variable, then the response is an increase in the same variable.
  For example, if the blood glucose level is too low, negative feedback results in it increasing to normal.

Negative feedback is an important control mechanism in almost all processes of homeostatic regulation. In subtopics 4.4, 4.5 and 4.6, we will explore specific examples of negative feedback loops in body temperature, blood glucose and water balance.

Figure 4.22 shows the negative feedback process. In each example, the input is the stimulus and the output is the response. The response counteracts the change in stimulus. For simplicity, these figures focus on the interaction between the response and the stimulus — they do not show the other components of a homeostatic feedback loop, namely the receptor, control centre and effector.

negative feedback a regulatory mechanism in which a stimulus results in a response that is opposite in direction to that of the stimulus

positive feedback a process in which the response to a stimulus causes an amplification of the initial change, increasing its level further away from the starting value

FIGURE 4.22 During a negative feedback loop, if a stimulus is **a.** too high or **b.** too low, a response acts to bring it back to normal.

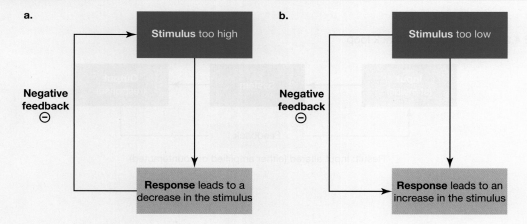

CASE STUDY: Negative feedback loops in the endocrine system

Case 1: Negative feedback in thyroid hormone secretion

The thyroid gland synthesises and releases the hormones T3 (triiodothyronine) and T4 (thyroxine). The image that opens this topic shows large follicles (top left and lower right) in the thyroid gland, lined by cuboidal secretory epithelial cells (red), that are secreting T3 and T4. The thyroid follicles are filled with a liquid called thyroglobulin or thyroid colloid, which is a storage form of the thyroid hormones.

After release into the blood, much of the T4 hormone is converted into the more active T3 hormone, mainly in the liver but also in other tissues.

Thyroid hormones influence development, growth and metabolism of tissues, organs and systems throughout the body, beginning in foetal life. Almost all body cells have receptors for these hormones. The hormone–receptor combination binds to DNA in target cells and so acts directly on genes, either stimulating or inhibiting gene expression.

The release of the T3 and T4 hormones involves the following:
- Specific cells in the hypothalamus are stimulated by signals such as stress, cold and nutritional status to release thyrotropin-releasing hormone (TRH).
- TRH travels via the blood to the anterior pituitary, which responds by releasing thyroid-stimulating hormone (TSH).
- TSH is carried by the blood to the thyroid gland, which is stimulated to release its T3 and T4 hormones (the *response*).

When the circulating thyroid hormones increase above a particular level, negative feedback mechanisms come into play. They inhibit the production of TSH by the anterior pituitary and TRH by the hypothalamus, stopping the release of T3 and T4. When the blood levels of these hormones fall, the thyroid gland again releases T3 and T4. Figure 4.23 shows a summary of these processes.

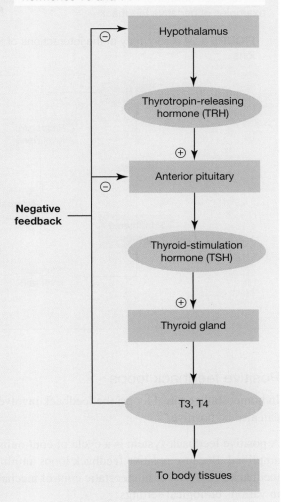

FIGURE 4.23 Feedback loops in the regulation of the release of the thyroid hormones T3 and T4

Case 2: Negative feedback in hormonal control of blood calcium level

Calcium ions in the extracellular fluids of the body are involved in many cell activities, including nerve transmission, bone and cartilage formation, blood clotting and muscle contraction. The level of calcium ions (Ca^{2+}) in the blood is tightly controlled by a homeostatic mechanism involving two hormones and negative feedback.

There are two hormones involved in the homeostatic control of the level of calcium ions in the blood:
- Parathyroid hormone (PTH) is released by the parathyroid glands:
  - Calcium sensors in the parathyroid glands detect falls in blood calcium levels below the normal range.
  - PTH is released in response to these falls.
  - PTH acts on bone to release some of its calcium store, and also activates the kidneys to convert inactive vitamin D to an active form that increases calcium reabsorption by the kidneys and calcium absorption from the gut.

- Calcitonin hormone is released by the thyroid gland:
  - Calcium sensors in the thyroid detect rises of blood calcium levels above the normal range.
  - Calcitonin is released in response to these rises.
  - Calcitonin inhibits reabsorption of calcium from bone, decreases calcium reabsorption by the kidney and decreases the calcium absorption from the gut.

Figure 4.24 shows the interaction of PTH and calcitonin in the homeostatic control of blood calcium levels. Note that for each hormone, the response that it produces is opposite in direction to that of the stimulus. These are examples of negative feedback.

FIGURE 4.24 A summary of the interactions of PTH and calcitonin in regulating the blood levels of calcium ions

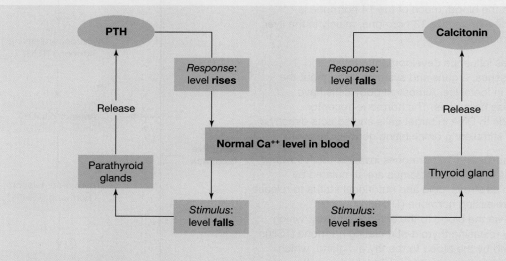

Positive feedback loops

In homeostatic control loops, the feedback involved is negative and counteracts the stimulus. However, feedback can also be positive.

A positive feedback system is a cycle of continuing change in which an original change is increasingly amplified. Because positive feedback loops amplify an initial change further from the set point, positive feedback is not part of homeostatic control mechanisms. However, positive feedback can be seen in operation in various biological settings.

Positive feedback acts to increase the magnitude of a stimulus, moving it further from a set point.

Childbirth

Positive feedback is involved in the process of childbirth.

During labour, as a baby starts its journey from its mother's uterus, the hormone **oxytocin** is released from the posterior pituitary gland in response to a signal from the hypothalamus. The action of oxytocin casuses contractions of the wall of the uterus.

oxytocin a hormone that induces labour and milk release from mammary glands in females

The cervix at the exit of the uterus has pressure receptors that are stimulated by pressure from the baby's head. As the pressure from the baby's head increases, more oxytocin is released, and this stimulates even stronger and more rapid contractions of the wall of the uterus. The increased rate of contractions causes the release of more oxytocin, creating a positive feedback cycle that leads to even stronger contractions (figure 4.25). This cycle continues, producing more and stronger contractions that continue until the baby is born.

Two other examples of positive feedback are:
- *breastfeeding:* a baby suckles at the breast, leading to the release of oxytocin and prolactin, which causes the ejection and release of breast milk. As the baby continues to suckle, positive feedback occurs and more oxytocin and prolactin are released, resulting in the production of more breast milk.
- *fruit ripening:* Do all the fruit on a tree change quickly from all being unripe to all being ripe? The trigger for this event is the ripening of the first fruit. As it ripens, this single fruit produces ethylene, releasing it through its skin. Ethylene is a gaseous plant hormone that stimulates fruit ripening. Release of ethylene from the first fruit to ripen causes its near neighbours to ripen, so they also produce ethylene that will stimulate even more fruit to ripen, and so on. Very shortly, all the fruit on the tree will have ripened in response to the ethylene hormone.

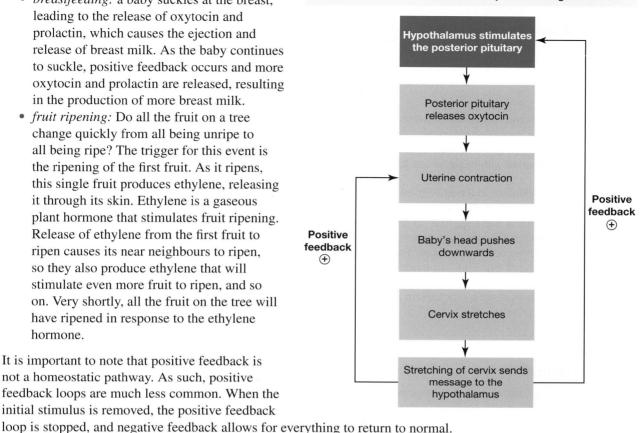

FIGURE 4.25 Positive feedback system during childbirth

It is important to note that positive feedback is not a homeostatic pathway. As such, positive feedback loops are much less common. When the initial stimulus is removed, the positive feedback loop is stopped, and negative feedback allows for everything to return to normal.

SAMPLE PROBLEM 2 Comparing feedback loops

Compare and contrast the processes of negative and positive feedback. (2 marks)

THINK

1. In compare and contrast questions, you need to provide both similarities and differences between the processes. As this question is worth 2 marks, 1 mark is for comparing and 1 mark is for contrasting.

2. Compare the processes by considering similarities between negative and positive feedback loops.

3. Contrast the processes and outline differences between them. You must ensure you address both processes in your response.

WRITE

Both negative and positive feedback loops use the stimulus–response model, in which a stimulus is detected by a receptor, messages are sent to control centres and effectors, and a response occurs (1 mark).

Negative feedback is when a response counteracts the change in a variable, allowing it to be returned to normal. This can be contrasted with positive feedback, in which a response causes an amplification of a variable, moving it further away from a set point. Negative feedback, unlike positive feedback, is a mechanism of homeostasis (1 mark).

Resources

- **eWorkbook** Worksheet 4.3 Positive and negative feedback loops (ewbk-7495)
- **Video eLesson** Maintaining a stable internal environment (eles-3884)

KEY IDEAS

- Homeostasis is the maintenance within a narrow range of conditions in the internal environment.
- Many variables, including blood glucose and core body temperature, are kept within a narrow range around their set points by homeostatic control loops.
- Stimulus–response with negative feedback is a key mechanism for the homeostatic regulation of physiological variables.
- The first step involved in a homeostatic stimulus–response loop is the detection of a stimulus (such as a change in variable) by a receptor.
- A receptor sends a message to a control centre (such as the hypothalamus in the brain) that evaluates the signal, determines the response needed, and sends this message to an effector.
- An effector carries out the response that counteracts the stimulus, returning the variable to within the set range.
- Negative feedback loops are key mechanisms of homeostasis that allow for the return of a variable to its required range.
- Positive feedback is not a mechanism of homeostasis. It causes an amplification of a stimulus further from a set point.

4.3 Activities

To answer questions online and to receive **immediate feedback** and **sample responses** for every question, go to your learnON title at **www.jacplus.com.au**. A **downloadable solutions** file is also available in the resources tab.

| 4.3 Quick quiz | 4.3 Exercise | 4.3 Exam questions |

4.3 Exercise

1. Identify why homeostasis is important for sustaining life.
2. Identify four physiological variables in humans that are under homeostatic control.
3. Would you expect the normal range for blood pressure to be the same across all animals? Justify your response.
4. Identify the role of the control centre in the homeostatic stimulus–response model.
5. Compare and contrast chemoreceptors and mechanoreceptors.
6. Identify a difference between an open stimulus–response model and a homeostatic closed loop model in terms of:
   a. any relationship between the stimulus and the response
   b. the nature of the response
   c. the means of transmission of a signal from a receptor to the control centre.
7. Describe the difference between
   a. a receptor and an effector
   b. negative and positive feedback.
8. Imagine you are walking in a garden. Your bare arm is suddenly punctured by a thorn on a rose bush and you immediately pull your arm back.
   a. Draw a flowchart outlining the stimulus–response model for this event.
   b. Explain if this is an example of an open or a closed loop.
   c. Is this an example of homeostatic regulation?

9. For each of the following processes, determine if it includes a negative or a positive feedback loop.
   a. When you cut your finger, your cells produce thrombin, an enzyme that helps with blood clotting. The presence of the thrombin enzyme causes more thrombin to be produced.
   b. As part of the immune response, a special protein known as complement is activated. This complement activates more complement molecules, causing a complement cascade.
   c. The human body maintains a stable pH of around 7.40 (with the exception of areas of digestion). When the pH rises, a condition caused alkalosis can occur. One response to alkalosis is the automatic depression of breathing rate. This causes a rise in the carbon dioxide level in the blood, which lowers the pH. Once the pH has returned to normal, the depression of breathing rate is lifted.
10. Why is positive feedback not identified as a homeostatic mechanism?

4.3 Exam questions

Question 1 (1 mark)
Source: VCAA 2011 Biology Exam 1, Section A, Q21

MC An example of homeostasis is when
A. root hairs of a pumpkin plant grow towards a source of water.
B. evaporation of water from the body surface after swimming has a cooling effect.
C. cabbage plants grown in a phosphorus-deficient soil mobilise phosphorus from tissues and release it into the phloem.
D. the body surface colour of the chameleon lizard changes to match the colour of the foliage on which it is resting.

Question 2 (1 mark)
Source: VCAA 2007 Biology Exam 1, Section A, Q16

MC Homeostatic systems comprise components such as sensors, effectors and variables. In such systems, the component being kept relatively constant is
A. the variable.
B. input to the sensor.
C. input to the effectors.
D. output from the effectors.

Question 3 (1 mark)
Source: VCAA 2012 Biology Exam 1, Section A, Q13

MC The following flow chart shows a feedback mechanism related to parathyroid hormone. Parathyroid hormone acts on various parts of the body and stimulates an increase in the concentration of blood calcium.

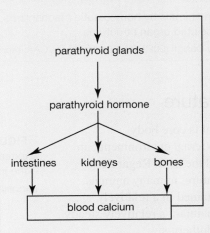

At any fall in the level of calcium in the blood, action by the parathyroid hormone would result in
A. a rise in the level of calcium in urine.
B. an increase in absorption of calcium by bones.
C. a reduction in the activity of the parathyroid glands.
D. increased absorption of calcium from the intestines.

Question 4 (1 mark)
Source: VCAA 2008 Biology Exam 1, Section A, Q23

MC In mammals the parathyroid gland secretes parathyroid hormone (PTH). PTH is involved in regulating the concentration of calcium in blood plasma. Parathyroid hormone increases the amount of calcium in plasma by causing calcium to move from bone to the plasma, and by assisting the uptake of calcium from the alimentary canal. PTH also stimulates the kidney to activate vitamin D.

The concentration of calcium in plasma acts directly, in negative feedback, to regulate the output of parathyroid hormone. From this information it would be expected that
- **A.** increased production of PTH results in reduction of vitamin D activation.
- **B.** reduced production of PTH results in increased calcium in the faeces.
- **C.** sustained overproduction of PTH results in strengthened bones.
- **D.** high levels of blood calcium stimulate release of PTH.

Question 5 (3 marks)
Source: VCAA 2004 Biology Exam 1, Section B, Q4b, 4c

Human blood-calcium concentrations are under homeostatic control. When the concentration of calcium in the blood begins to fall the parathyroid gland releases parathyroid hormone. This hormone stimulates bone cells called osteoclasts to break down bone and release calcium into the blood.

When the concentration of calcium rises in the blood, specialised cells associated with the thyroid gland release the hormone calcitonin. Calcitonin acts on bone and increases the amount of calcium that is deposited into bone.

a. What is the general name given to structures that detect the changes in blood-calcium concentrations? **1 mark**

b. The control of blood-calcium concentrations involves negative feedback. What is negative feedback? **2 marks**

More exam questions are available in your learnON title.

4.4 Regulation of body temperature in animals

KEY KNOWLEDGE

- Regulation of body temperature in animals by homeostatic mechanisms, including stimulus–response models, feedback loops and associated organ structures

Source: Adapted from VCE Biology Study Design (2022–2026) extracts © VCAA; reproduced by permission.

4.4.1 Core body temperature

One variable under homeostatic control is core body temperature. The temperature of the external environment can vary — it may be hot one day and cold the next. Regardless of the variation in the external temperature, in many animals, the core temperature needs to be maintained in a relatively narrow range. These animals, which maintain a relatively steady internal temperature, are known as endotherms.

Different species have different narrow ranges for varying factors that they must keep within. For example, the core body temperature of the human body and most mammals at rest is about 37 °C, with a typical narrow range of 36–39 °C. For birds, typical core body temperature is much higher, 39–42 °C. Other animals, such as reptiles, cannot regulate their temperature as closely.

FIGURE 4.26 The temperatures in the external environment may be different, but the internal environment stays the same.

Core body temperature relates to the temperature in organs and deep tissues within the core of the body.

Core temperature must be distinguished from **peripheral surface temperatures**, which can be many degrees cooler. For example, when the ambient temperature is 23 °C, the core temperature in the deep body tissue of a person is about 37°C, but the temperature of that person's hands will be less, perhaps only 30 °C, and the temperature of the feet even cooler, perhaps 25 °C. It is the core temperature that is important for maintaining body functions. A person's core body temperature can be measured at several sites, including the mouth (oral measurement), the rectum (rectal measurement) and the ear, where the temperature of the ear's tympanic membrane or eardrum indicates the temperature of the brain stem and the hypothalamus in the brain.

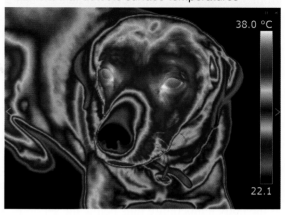

FIGURE 4.27 Heat signature of Lettie, a Labrador dog, obtained with an infra-red camera that detects surface temperatures

The difference between the core temperature of a mammal and its peripheral surface temperature can be revealed using an infra-red camera that produces an image called a **thermograph**. A thermograph shows the surface temperatures or heat signature of an object as seen in the heat signature of Lettie, a Labrador dog, in figure 4.27. The bar at the right shows colour-coded temperatures in order. In spite of all this variation in her surface temperatures, Lettie's core body temperature is maintained within the narrow range of 37.7 °C to 39.2 °C.

4.4.2 Why do animals need to regulate body temperature?

In order for functions in the body to be maintained, it is important that a relatively stable internal core temperature is maintained. If the core temperature is too high, proteins in the body can begin to denature, including enzymes that are vital for metabolic processes. As well as this, water is lost during to excessive perspiration, and cellular death occurs. If the core body temperature is too low, cellular processes slow. As temperatures continue to decrease, these cellular processes can no longer sustain life. Therefore, the ability to regulate core body temperature is incredibly important.

core body temperature temperature of internal cells of the body; in humans, core temperature is around 37 °C

peripheral surface temperatures temperature of cells on the outside of the body; may be many degrees cooler than core temperature

thermograph an instrument that shows the surface temperature or heat signature of an object

---

EXTENSION: How do ectotherms 'regulate' temperature?

We have been examining the control of body temperature in animals that produce their own body heat. But how do animals that rely on an external source of heat (ectotherms) regulate their body temperature?

To access more information on this extension concept, please download the digital document.

FIGURE 4.28 Snakes rely on external heat.

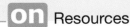 Resources

📄 **Digital document** Extension: How do ectotherms 'regulate' temperature? (doc-35886)

4.4.3 How do animals regulate body temperature?

The core body temperature of a healthy person averages about 37 °C and varies only within a narrow range, from 36 °C to 39 °C, depending on factors such as air temperature, level of physical activity and food intake. For example, during vigorous exercise, the core temperature may temporarily rise to 40 °C. If the body temperature rises, mechanisms come into action to lower the temperature and return it to within the narrow range. If the body temperature falls, other mechanisms come into action to raise the temperature and return it to within the narrow range. All other mammals and birds are also able to regulate their body temperature and maintain it within specific limits.

Thermoregulation is the regulation of body temperature. It occurs through homeostatic mechanisms, allowing for a response to either increase or decrease the core body temperature. Many of these responses are physiological.

thermoregulation the maintenance of core body temperature

Physiological processes are not under a person's conscious control; instead, they occur automatically — you do not have to think about starting them. These processes are initiated by centres in the hypothalamus of the brain.

If the body temperature falls below the set point, physiological processes increase heat production within the body and reduce heat loss. As a result, the body temperature rises. (This is a bit like switching the heater on in a room and closing the windows.) If, however, the body temperature rises, other physiological processes produce an increase in heat loss from the body so that the body temperature falls. (This is a bit like opening the windows in a hot room to let cooler air in.) These processes are shown in figure 4.29.

FIGURE 4.29 Diagram showing physiological mechanisms that lead to the production or loss of heat

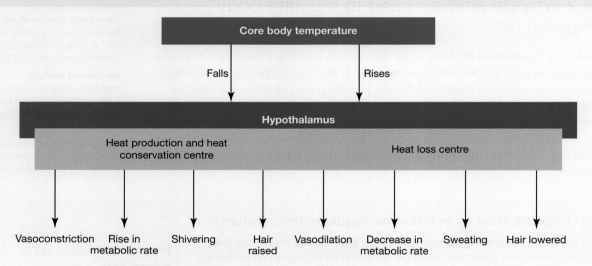

The components involved in homeostatic control of core body temperature are receptors, control centre and effectors that form a negative feedback loop. Their functions are as follows:
- *receptors sense (detect) changes in the external and internal temperatures:* this is carried out by free nerve endings that are temperature-sensitive receptors for heat and cold. These receptors are located in the skin, liver, skeletal muscles and hypothalamus.
- *the control centre identifies the response to change:* groups of nerve cells in the hypothalamus can receive either 'warm' or 'cold' signals from the thermoreceptors, compare them with the desired set point, identify the response needed and signal this to effectors.
- *effectors respond to change:* blood vessels in skin, brown adipose tissue (BAT), skeletal muscles and sweat glands defend against heat or cold through automatic (involuntary) responses (see table 4.5). Effectors and their responses are covered in detail in sections 4.4.4 and 4.4.5.

TABLE 4.5 Effectors and their automatic (involuntary) responses involved in thermoregulation

Effectors	Responses in cold	Responses in heat
Blood vessels in skin	Reduced blood flow through skin surface vessels (vasoconstriction)	Increased blood flow through skin surface vessels (vasodilation)
Brown adipose tissue (BAT)	Heat-producing BAT metabolism activated	BAT metabolism inhibited
Skeletal muscles of limbs and trunk	Shivering occurs	Shivering inhibited
Sweat glands	Sweating inhibited	Sweating occurs

Means of losing body heat

The body can lose heat in a variety of ways. This is important in regulating body temperature. When our body temperature is too high, we want to maximise heat loss. When our body temperature is low, we do not want to lose heat, so our bodies work to minimise heat loss.

Body heat can be lost by:
- *convection:* occurs when colder air or water moves past exposed areas of a warmer body — the faster the wind speed or the water current and the colder the air or the water, the greater the rate of heat loss
- *conduction:* occurs when heat is lost to a thin layer of still air or water (or a solid) that is in direct contact with a warm body. Heat is conducted away from the body about 25 times faster in cold water than in air at the same temperature.
- *radiation:* all heated objects radiate heat as infra-red electromagnetic radiation — no air or water is required and no contact is needed
- *evaporation:* occurs when water from a person's skin or airways changes state from liquid to vapour, a process that requires an input of heat energy that is taken from the body.

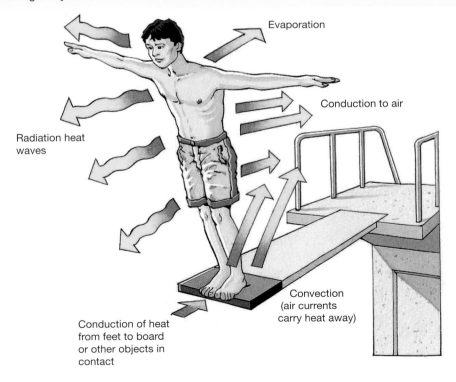

FIGURE 4.30 Losing body heat from skin

4.4.4 When body temperature is too low

Usually, homeostasis ensures that body temperature is regulated so that when body temperature decreases, mechanisms allow it to increase back to normal levels. People wearing light clothing who are exposed for an extended period to cold still air, cold wind or rain, or are immersed in cold water are at risk of **hypothermia**. If a person's core body temperature falls below 35 °C, a state of hypothermia is said to exist. In this situation, heat gain and heat loss are out of balance, with heat loss in excess of heat gain. Table 4.6 shows indicative stages of hypothermia. Prolonged severe hypothermia with further falls in core temperature can result in death.

FIGURE 4.31 Ice water rescue training

TABLE 4.6 Stages of hypothermia

Stage	Core temperature	Symptoms include:
Mild	35–32 °C	Shivering; increased heart rate and breathing; cold pale skin; loss of manual dexterity
Moderate	32–28 °C	Shivering stops; decreased heart rate and breathing; pupils dilated; confused and irrational thinking
Severe	Less than 28 °C	Slow, weak or absent pulse; fluid gathers in lungs; loss of consciousness; coma; heart failure

Several forms of hypothermia are recognised:
- **Acute hypothermia** occurs when a person is suddenly exposed to extreme cold, such as immersion in cold water. This was the situation for hundreds of passengers on board the ill-fated *RMS Titanic* when it sank in the North Atlantic Ocean on 15 April 1912. The water temperature on that morning is thought to have been just above −2 °C. Those passengers who were immersed in the ocean and could not get into life boats died quickly from hypothermia.
- **Exhaustion hypothermia** occurs when a person is exposed to a cold environment, is exhausted and does not have sufficient food. As a result, such persons cannot generate sufficient metabolic heat to compensate for their loss of heat, and their core body temperature falls. Antarctic explorers and mountaineers climbing for days at high altitude on Earth's highest mountains are at risk of exhaustion hypothermia. Many of the deaths that have occurred on Mount Everest are from exhaustion hypothermia.

Figure 4.32 summarises the homeostatic mechanisms that correct a fall in body temperature. This is a closed stimulus–response model with negative feedback by the response that counteracts the stimulus. The responses seen in various mammals are of two types:
- a heat-conserving mechanism to reduce heat loss from the body, that is vasoconstriction of cutaneous (skin) blood vessels
- heat-producing mechanisms to produce internal heat, such as shivering and brown adipose tissue (BAT) metabolism.

hypothermia condition in which an individual has an extremely low body temperature and is at risk of death

acute hypothermia occurs when a person is suddenly exposed to extreme cold

exhaustion hypothermia occurs when a person is exposed to a cold environment and cannot generate sufficient metabolic heat to maintain their core body temperature due to exhaustion or lack of food

FIGURE 4.32 The stimulus–response model when body temperature decreases

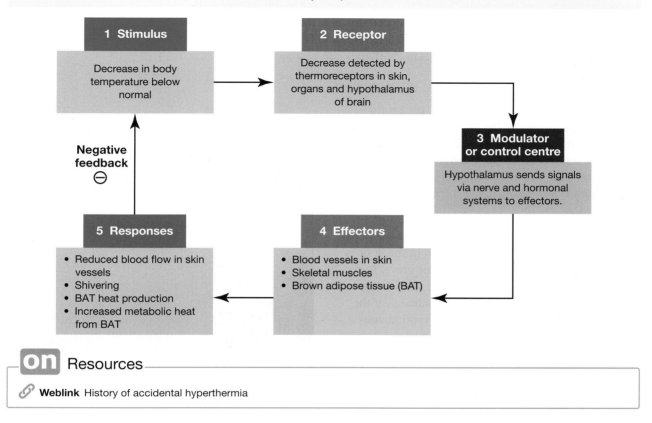

Heat-conserving mechanisms: vasoconstriction

Because of the large surface area of the skin — about 2 square metres in a typical adult — the cutaneous (skin) blood vessels can be a major source of heat loss from the body — about 90 per cent. For this reason, regulation of the cutaneous blood vessels plays an important role in thermoregulation.

Blood vessels are routinely under a low background level of **vasoconstriction**. When core body temperature falls below normal, nerve signals from the hypothalamus are sent via sympathetic nerve fibres to cutaneous blood vessels. These signals increase the constriction of cutaneous blood vessels to limit blood flow to the skin, resulting in less heat loss across the skin and more heat retained within the body.

vasoconstriction narrowing of the diameter of blood vessels

FIGURE 4.33 During vasoconstriction, a pre-capillary sphincter narrows, preventing blood flow into the capillary bed and minimising heat loss across the skin surface.

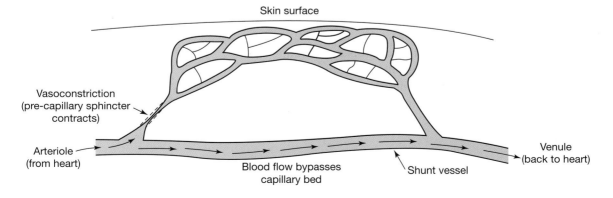

Vasoconstriction also affects small rings of smooth muscle (pre-capillary sphincters) around entries to capillary beds close to the skin surface. When this happens, the blood flow to these capillaries is greatly reduced. Instead, the blood flows through **shunt vessels** that are directly connected to arterioles and venules (figure 4.33). In extreme cold, blood flow through the skin capillaries can be almost zero.

shunt vessels direct connections between arterioles and venules that bypass capillary beds

Heat-conserving mechanisms: piloerection

A distinguishing feature of mammals, except for whales and porpoises, is the presence of hair or fur over all or most of their bodies. Piloerection means 'hair standing on end', and it occurs in many mammals. In humans it is easily seen as 'goosebumps' around each hair follicle (see figure 4.34).

FIGURE 4.34 A human arm in cold conditions that shows raised hairs and goose bumps. This insulation will not prevent heat loss.

Piloerection is an involuntary response that can occur when a mammal is fearful, aggressive, in shock or is in a cold environment. Mammalian hair grows from a cell-lined cavity (follicle) within the skin. A tiny bundle of smooth muscle fibres is attached near the base of each hair follicle. When the hypothalamus receives input from cold sensors, it sends impulses via sympathetic nerve fibres to these muscles, causing them to contract and raise the hair or fur.

Although raised hair is not important for heat conservation for humans, the piloerector response is an important heat conservation mechanism in almost all other mammals. Raised hairs create a layer of still air between the skin and the external environment that acts as an insulator and reduces heat loss from the body. Fluffed-up feathers play a similar insulating role for birds (figure 4.35).

FIGURE 4.35 A mourning dove (*Zenaida macroura*) whose fluffed-up feathers act as an insulator, restricting heat loss

Heat-producing responses: shivering

Shivering is the alternate contraction and relaxation in rapid succession of many small skeletal muscle groups in the upper limbs and body trunk. It is an involuntary action that occurs in persons exposed for an extended period to cold air or water and who are not suitably protected from the cold. Shivering usually starts when the core temperature has fallen to 35 °C. Figure 4.36 shows a young girl shivering after swimming in cold water. She also shows a voluntary behavioural response by wrapping her arms around her body — this reduces her surface area that is exposed to the cold air and reduces her heat loss.

The hypothalamus contains a centre that controls the shivering reflex. The signal for shivering is transmitted from the hypothalamus to skeletal muscles via motor neurons of the somatic nervous system. (Typically, skeletal muscle stimulation involves voluntary actions, but this is not the case for shivering — you cannot control it.)

Shivering requires energy. To power their shivering, muscles release energy from their cellular store, and heat energy is released in this process. Shivering increases the body's metabolic rate and raises the associated heat production to about five times that of the resting rate (basal metabolic rate). Thus, shivering produces significant amounts of additional heat for the body. However, because of its high rate of energy use, shivering cannot be sustained for long periods because the energy stores of muscle tissue become depleted — they run out of fuel!

FIGURE 4.36 A girl shivering after swimming in a cold lake

Heat-producing responses: brown adipose tissue (BAT) metabolism

All mammals have two types of adipose (fat) tissue: one type formed of white fat cells that serve as energy stores, and a second type composed of brown adipose cells. Newborn babies have a large supply of brown adipose tissue (BAT). The amount of BAT declines after birth, but some BAT is present in most adult humans for most of their lives. In adults, BAT is located in the neck, shoulders and armpits, and alongside the vertebral column.

All cells break down glucose and produce ATP, which is the energy source needed for life functions. This process of cellular respiration captures about 40 per cent of the energy from the glucose as ATP molecules, and about 60 per cent is released as heat.

However, when a person's core temperature falls below normal, BAT cells can switch to another form of metabolism. These cells break down fatty acids, rapidly releasing almost 100 per cent of the energy directly as heat, *with no production of ATP*. Heat production by BAT cells can be sustained in the longer term.

This heat-producing metabolic response is activated by a signal sent from the hypothalamus via sympathetic nerve fibres to BAT. This heat source is important when the core body temperature can no longer be regulated by vasoconstriction of cutaneous blood vessels and when shivering can no longer continue.

EXTENSION: Metabolic rate

Another mechanism that affects body temperature is changes in the **basal metabolic rate**. Metabolic processes in the body produce heat. The minimum amount of heat generated internally is the so-called basal metabolic rate. This rate is linked with how quickly fuels are broken down through cellular respiration to produce ATP.

Changes in metabolic rate are not often due to a fall in temperature; rather, they are responses to an increased metabolic demand, such as during exercise. Although such change is not a direct mechanism of homeostasis and thermoregulation, it is important to consider the ways in which metabolic rate affects body temperature, which as outlined above is controlled by various effectors and responses that restore it to a normal range.

Changes in metabolic rate are initiated by a centre in the hypothalamus. When metabolic demand increases, the hypothalamus releases a hormone known as thyrotropin-releasing hormone (TRH). This causes the release of thyroid-stimulating hormone (TSH) from the pituitary gland. This activates the thyroid gland, causing the hormones triiodothyronine (T3) and thyroxine (T4) to be released (see the Case study in section 4.3.3). These hormones act to increase the metabolic rate of body cells, leading to increased heat production (see figure 4.37).

basal metabolic rate the minimum amount of heat generated in the body by metabolic processes

FIGURE 4.37 The hypothalamus, pituitary gland and thyroid gland all release hormones vital for the regulation of body temperature.

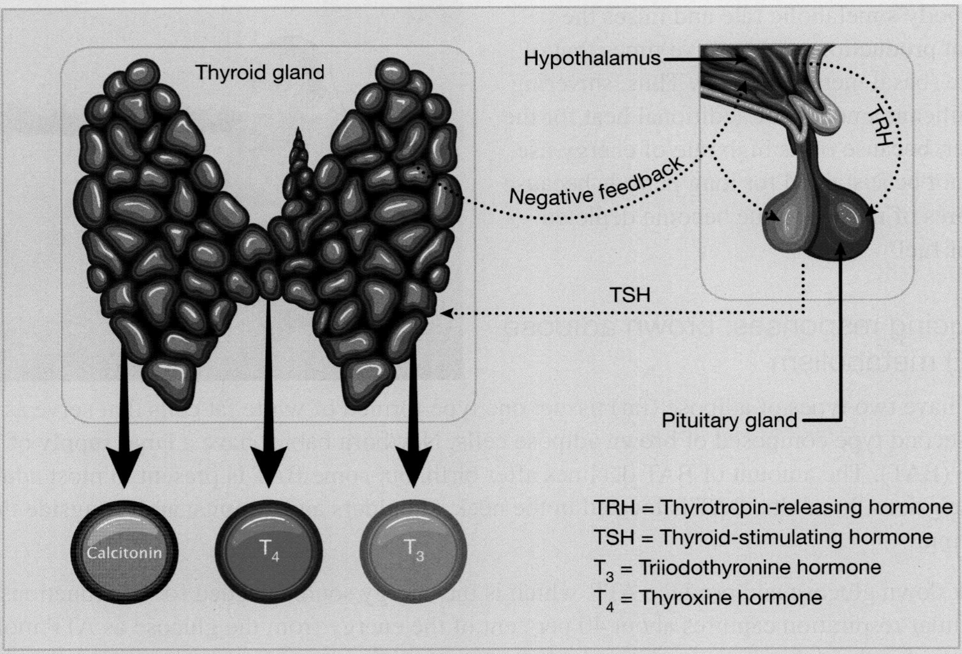

TRH = Thyrotropin-releasing hormone
TSH = Thyroid-stimulating hormone
T_3 = Triiodothyronine hormone
T_4 = Thyroxine hormone

Behavioural changes

In this subtopic, you have been introduced to many involuntary responses to a fall in core body temperature that are controlled by the hypothalamus of the brain. In addition to these involuntary responses, mammals carry out many voluntary behaviours that produce an increase their body temperature (see table 4.7).

These voluntary actions, such as rubbing your hands together when they become cold, are voluntary motor responses under the control of the brain cortex. In some cases, voluntary actions may occur *before* the cold receptors in the skin have been activated. For example, you might put on a woollen cap *before* going outside on a cold day because you are aware from other sensory inputs that it will be cold outside — maybe a news report you heard or the grey sky with rain clouds you saw through a window.

TABLE 4.7 Examples of behaviours to reduce heat loss humans

Behaviours for gaining and conserving heat
Vigorously exercising
Putting on another layer of clothing
Soaking in a hot bath
Having a hot drink
Rubbing your hands together
Reducing exposed body surface area
Wearing a hat and gloves
Standing in front of a heater

Changing our clothing is one behaviour in which we commonly engage to regulate body heat. Wearing clothing causes a significant reduction in the loss of body heat from a person. This is particular important for us humans as we do not have structures that some animals have (such as fur and blubber).

A naked person radiates body heat to the environment to varying degrees over their entire body surface (see figure 4.38a). When the person adds grade 3 thermal clothing (see figure 4.38b), heat losses from the trunk and the arms that are covered by the clothing are greatly reduced. If grade 5 thermal clothing is worn (see figure 4.38c), there is a further reduction in heat loss from the trunk. In addition, because the trunk is warmer, the heat loss from the legs is also reduced.

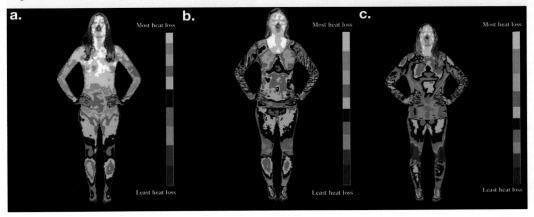

FIGURE 4.38 Infra-red images of a female subject showing the heat loss under various conditions **a.** Naked subject **b.** Subject wearing grade 3 thermal clothing **c.** Subject wearing grade 5 thermal clothing.

4.4.5 When body temperature is too high

The core body temperature of a person can increase above the upper end of the normal range. If the core body temperature rises to 38 °C or higher, a state of **hyperthermia** has started.

Hyperthermia is of two different types:
- Non-fever hyperthermia: caused by external factors, and the set point for core body temperature stays within its normal narrow range, typically 36.5 to 37.5 °C. This type can result in heat stroke.
- Fever hyperthermia: the set point for core body temperature is reset to a higher value, anything from 38 °C to greater than 40 °C. This type produces fever of differing severity depending on the reset level. This is discussed further in the Extension box at the end of this subtopic.

Non-fever hyperthermia

Core temperature rises above the normal narrow range when a person is exposed to a hot environment for an extended period or when a person engages in strenuous physical activity for a lengthy period. These situations involve an imbalance between heat gain and heat loss, with heat gain being in excess of heat loss. As a result, heat builds up in the deep body tissues, raising the core body temperature.

Hyperthermia may be mild at first, but it can become more serious if the core body temperature continues to rise. As the core temperature increases, a person will show signs of **heat exhaustion**. This may then develop into **heat stroke**. Heat stroke is a serious and life-threatening condition in which the core temperature rises above 40 °C and a person's brain function becomes affected, as seen in the appearance of symptoms of delirium, convulsions or coma (table 4.8). If untreated, the excessive heat load in heat stroke eventually overwhelms the body's homeostatic mechanisms. (Further details of heat stroke are covered in the Case study at end of this section.)

hyperthermia condition in which core body temperature exceeds the upper end of the normal range without any change in the temperature set point

heat exhaustion an increase in core body temperature; symptoms include poor coordination, slower pulse and excessive sweating; may develop into heat stroke

heat stroke a critical and life-threatening condition where brain function is affected; symptoms include high core body temperature in excess of 40 °C, slurred speech, hallucinations and multiple organ damage

Two types of heat stroke are recognised:
- *exertional:* can occur in persons engaged in extended periods of strenuous physical activity in hot and humid conditions, such as when running a marathon
- *situational:* also termed non-exertional; can occur, for example, during heat waves in older persons living alone in poorly ventilated housing without air conditioning, in young children left in locked cars with closed windows, or in people in hot tubs when the water temperature is above 40 °C.

TABLE 4.8 Symptoms of heat exhaustion and heat stroke

Heat exhaustion	Heat stroke
Headaches, nausea and vomiting	Headaches, nausea and vomiting
Excessive sweating	Minimal sweat
Fatigue and exhaustion	Possible seizures or coma
Poor coordination	Dizziness and delirium, with slurred speech
Slow pulse	Fast pulse
Thirst	Excessive thirst
Increase in body temperature	Skin may feel cool compared to internal body temperature

An increase in core body temperature is the stimulus that starts the homeostatic mechanisms that will lower the temperature to within the normal narrow range (figure 4.39).
- The stimulus is detected by heat receptors in the skin and various organs, and is signalled to the hypothalamus.
- The hypothalamus signals a range of automatic 'cooling down' instructions for effectors.
- The effectors are cutaneous blood vessels and sweat glands in the skin.
- The responses by the effectors are an increase in blood flow in the skin blood vessels and the production of sweat by the sweat glands.

FIGURE 4.39 Diagram showing a stimulus–response model for an increase in core body temperature.

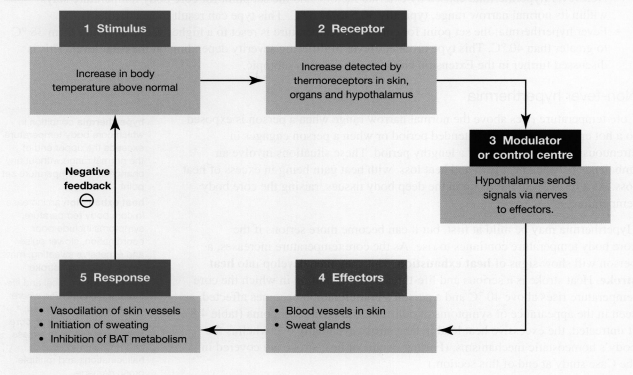

Heat loss mechanisms: vasodilation

Blood vessels are routinely under a low background level of vasoconstriction. When the core body temperature rises above the upper limit of the set point, the hypothalamus sends a nerve signal that inhibits vasoconstriction of cutaneous (skin) blood vessels. Removing the vasoconstriction results in the dilation of these blood vessels, increased blood flow through capillary beds in the skin and increased heat loss across the skin (figure 4.40). **Vasodilation** is the first response to a rise in body temperature above its normal range. In heat stress, the blood flow through the skin capillaries can reach up to 8 litres per minute.

vasodilation widening of blood vessels to increase blood flow

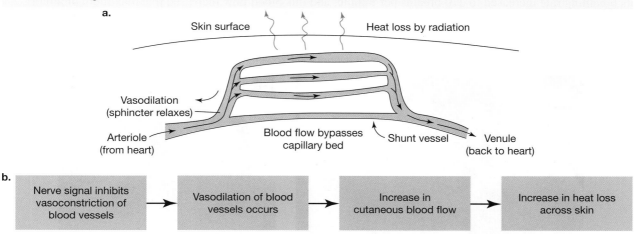

FIGURE 4.40 a. During vasodilation, a pre-capillary sphincter relaxes, allowing blood flow into the capillary bed and maximising heat loss across the skin surface. **b.** Flowchart of vasodilation

In some mammals, particular regions of their bodies — thin, with little or no hair, and with a rich blood supply — are major sites of heat loss from their bodies when core body temperature rises. One example is the disproportionately large ears of the fennec fox, *Vulpes zerva* (figure 4.41), which is native to the Sahara Desert. Warm blood from the body core travels to the fox's ears, where heat can be lost from dilated blood vessels to a relatively cooler environment — the higher the core temperature, the greater the blood flow through the ears.

FIGURE 4.41 Photo of a fennec fox showing its very thin-skinned ears with their blood supply. How might its ear size be expected to compare with that of the Arctic fox?

Heat loss mechanisms: sweating

Sweating is a physiological response that specifically allows for a decrease in body temperature. Like the other physiological processes concerned with thermoregulation, sweating is controlled by a centre in the hypothalamus.
- Nerve impulses from the hypothalamus activate sweat glands.
- Liquid sweat on the skin evaporates, forming a vapour.
- When liquid water evaporates, energy is needed to change its state from liquid to gas. The evaporation of sweat requires heat energy, and this is taken from blood vessels close to the skin, thus cooling the body. Cooling achieved through sweating is called evaporative cooling (see figure 4.42a). You can observe the effect of evaporative cooling by placing your hand in front of a fan and comparing the cooling effect when your hand is dry with the effect when your hand is wet. However, evaporative cooling becomes less and less efficient as a cooling mechanism in conditions of increasing humidity in the air. Because sweating involves water loss from the body, it is not the first response when the core body temperature increases above the normal range – sweating can put a person at risk of dehydration.

Humans have sweat glands over almost all their body surface. In contrast, members of the dog and cat families have sweat glands on only a few small areas that are not fur-covered, such as noses and paw pads. The primary means of heat loss to the environment for these mammals is by **panting**, which removes body heat through evaporation from the tongue and moist surfaces of the mouth. The panting dog in figure 4.42b is carrying out some canine thermoregulation. Cooling is achieved by evaporation of saliva from the mucous membranes of the oral cavity, in particular, from the tongue. During panting, a dog's breathing rate is greatly increased, and its tongue with its rich supply of *dilated* blood vessels flattens to produce an increase in surface area. As warm air passes over the tongue during panting, heat from the blood vessels in the tongue produces evaporation of saliva and cooling. In one study, it was found that in cool conditions (20 °C), the blood flow through a dog's tongue was 11 mL/min. In hot conditions (38 °C,) the dog was panting, its breathing rate increased to 270 breaths per minute, and this blood flow increased to a maximum of almost 75 mL/min.

> **panting** an increase in breathing rate that acts to reduce body temperature through evaporation

FIGURE 4.42 a. Evaporation of sweat is a major means of heat loss from the human body. **b.** Panting in dogs is another example of evaporative cooling that reduces core temperature.

INVESTIGATION 4.3

The regulation of body temperature

Aim

To investigate how skin blood flow and sweat production change in response to exercise

CASE STUDY: Hyperthermia and heat stroke — extreme heat can kill

Australia is well known for its extreme climate. This case study examines a case of two farm workers who became stranded in the Simpson Desert. It describes the physiological response of extreme heat stroke, which led to the death of one of them. To access more information on this Case study, please download the digital document.

FIGURE 4.43 The Simpson Desert

Resources

Digital document Case study: Hyperthermia and heat stroke — extreme heat can kill (doc-35881)

Behavioural changes

There are many behavioural changes that occur in order to reduce body temperature (see table 4.9).

Specialised clothing has also been designed to assist in cooling the body. Body cooling vests have been developed to prevent overheating or to assist in reducing the heat load of a body after exercise. The *Arctic Heat Body Cooling Vest* is an Australian-made lightweight garment designed specifically for body cooling. The vest is two-layered, with an internal layer of merino wool and an outer micromesh layer made of 'wicking' material. This is a fabric that is woven in a particular way in order to draw moisture from the skin to the exterior from where it can evaporate, so that the skin is kept dry and cool.

TABLE 4.9 Different behaviours that allow for heat loss

Behaviours for losing heat
Removing a layer of clothing
Having a cold shower
Resting in the shade
Using an ice pack
Removing your hat and gloves
Sitting in front of a fan
Maximising the body surface area exposed to a cooling wind
Soaking your feet in cold water

EXTENSION: What happens when you have a fever?

The development of a fever is a common and useful response when a person has a bacterial or viral infection. The fever develops as the result of an initial stimulus of the presence of viruses or bacteria that release fever-inducing substances into the bloodstream. This initiates a multi-step process that is described in the downloadable digital document.

FIGURE 4.44 Core body temperature from fever onset to break

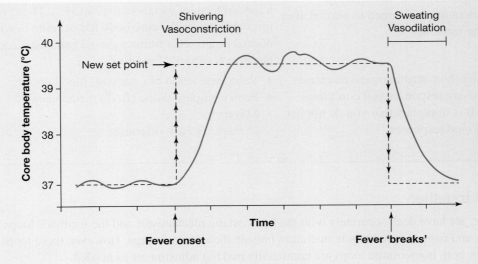

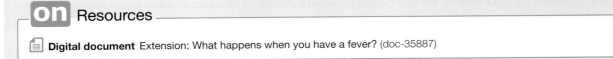

Digital document Extension: What happens when you have a fever? (doc-35887)

SAMPLE PROBLEM 3 Understanding body temperature regulation

An individual was outside in a very cold region for an extended period and entered a state of hypothermia.
a. What is meant by the term hypothermia? (1 mark)
b. Using the stimulus–response model, clearly explain how homeostatic mechanisms act to return body temperature to normal. (3 marks)
c. Describe three physiological responses that will allow for the regulation of body temperature to normal levels. (3 marks)

THINK

a. 1. The question is asking you to explain the meaning of a term. As it is a one-mark question, you only need to provide a simple definition.

2. Examine the stimulus material and the term used. The stimulus material described an individual being in the cold. The term hypothermia can be broken up into two parts — *hypo* meaning under and *thermia* meaning heat.

b. The question is worth three marks, so you should consider the three main things the question is asking you to include:
- a reference to the stimulus–response model
- an explanation of homeostatic mechanisms
- an explanation of how to body temperature is returned to normal after being too low.

c. Consider three physiological responses — these are responses you don't think about. It is important that you do not list behavioural responses.

WRITE

Hypothermia is a condition in which an individual has a lower than normal body temperature (1 mark).

Thermoreceptors throughout the body and within the hypothalamus detect that the body temperature is low (the stimulus). These receptors sends nerve signals to the hypothalamus. The hypothalamus sends nerve signals to activate various effectors. It signals blood vessels near the skin surface to constrict, reducing body heat loss; it signals skeletal muscles to shiver, producing metabolic heat through the release of thyroxine; and it signals brown adipose tissue to stimulate heat generation though a special form of metabolism (2 marks). These various responses return the core body temperature to within the normal range. This forms a closed homeostatic loop with negative feedback (1 mark).

- Vasoconstriction of cutaneous blood vessels
- Brown adipose tissue (BAT) metabolism
- Shivering
(1 mark for each response)

Putting it all together

In this subtopic, we have dealt separately with the homeostatic mechanisms and the feedback loops involved in regulating falls and rises in core body temperature outside their normal range. However, these loops do not act in isolation; rather, both homeostatic loops are continually making adjustments as needed.

Figure 4.45 shows the operation of the two closed homeostatic control loops involved in the regulation of core body temperature.

FIGURE 4.45 Diagram showing the different stimuli and responses of two closed homeostatic loops involved in thermoregulation

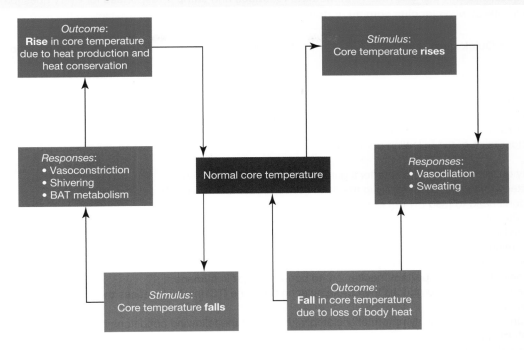

Resources

eWorkbook Worksheet 4.4 Thermoregulation (ewbk-7497)

Interactivity Homeostasis and body temperature regulation (int-8081)

Weblinks Fevers
Tragedies are a reminder of the fury of Australia's outback heat

KEY IDEAS

- Body temperature is a highly regulated variable in mammals.
- Heat loss can occur by evaporation of excreted water including sweat, from the pores and from the airways and lungs.
- Thermoreceptors detect changes in body temperature and send messages to the hypothalamus.
- The nervous system is an essential component of thermoregulation. It conveys signals from sensors to the control centre in the hypothalamus and from the hypothalamus to various effectors to produce relevant responses.
- The thyroid gland produces the hormones triiodothyronine (T3) and thyroxine (T4), which act to regulate basal metabolism, which affects body temperature.
- When the core body temperature rises above normal, homeostatic mechanisms operate to increase heat loss and inhibit heat gain mechanisms.
- Other responses that decrease body temperature include sweating, vasodilation of cutaneous blood vessels and behavioural responses.
- When the core body temperature drops below normal, homeostatic mechanisms operate to increase heat production and decrease heat loss (heat conservation).
- Other responses that increase body temperature include shivering, vasoconstriction of cutaneous blood vessels and behavioural responses.
- Homeostatic mechanisms can be overwhelmed in extremely hot and extremely cold conditions.

4.4 Activities

To answer questions online and to receive **immediate feedback** and **sample responses** for every question, go to your learnON title at **www.jacplus.com.au**. A **downloadable solutions** file is also available in the resources tab.

| 4.4 Quick quiz | 4.4 Exercise | 4.4 Exam questions |

4.4 Exercise

1. Some homes have central heating. When the air reaches a certain temperature, a thermostat turns the heat off. When the temperature drops below a certain level, a thermostat turns the heater on. Explain whether you think this is similar to or different from the control of internal body temperature.
2. Consider each of the following comments and indicate, giving a brief explanation, whether it is biologically valid.
   a. 'My quick response to the starter's gun is a reflex action.'
   b. 'After working out strenuously in the gym, I towel myself down so as to avoid getting a chill.'
   c. 'I don't need to feed my pet lizard as often as my pet budgerigar.'
3. All of the following statements are false. Rewrite each statement to make it true.
   a. Core body temperature is a highly regulated variable in all animals.
   b. Warming mechanisms that come into action when the core body temperature falls include shivering and a decrease in metabolic rate.
   c. Piloerection is an important mechanism to conserve heat in humans
   d. The hypothalamus produces thyroid-stimulating hormone (TSH), which causes the thyroid to produce thyroxine.
4. Give an example of one symptom associated with each of the following conditions and briefly explain why this symptom appears.
   a. Hyperthermia   b. Hypothermia
5. Briefly explain a key difference between fever resulting from a microbial infection and hyperthermia arising from prolonged strenuous physical activity in hot conditions.
6. State the difference between a receptor and an effector in the homeostatic regulation of core body temperature.
7. Give an example of each of the following.
   a. A behavioural factor that can increase heat loss
   b. A physiological factor that results in heat gain
   c. A behaviour not seen in humans that produces heat loss
8. Explain the interrelationship between vasodilation and vasocontriction and how this is used to regulate body temperature.

4.4 Exam questions

Question 1 (1 mark)

Source: Adapted from VCAA 2005 Biology Exam 1, Section A, Q23

MC The body temperature of two different mammals was recorded over 24 hours. The average daytime temperature was 40 °C and average night-time temperature was 20 °C. The temperatures of the mammals over the 24 hours are shown in the graph.

From the information given it would be reasonable to conclude that
A. between midnight and 6.00 am, mammal H would be gaining heat by conduction.
B. at 12.00 noon, mammal H would be gaining heat by radiation.
C. at 12.00 noon, mammal G is losing heat by radiation.
D. between midnight and 6.00 am, mammal G would be losing heat through evaporation of sweat.

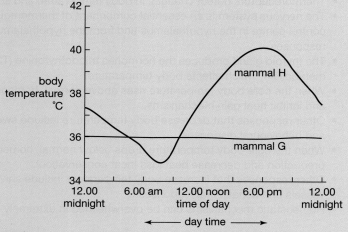

Question 2 (1 mark)
MC A response to an increase in body temperature includes
A. vasodilation of arterioles in the skin.
B. increased release of thyroxine from the thyroid gland.
C. an increase in skeletal muscle activity.
D. putting on extra clothes.

Question 3 (1 mark)
MC In a person suffering from hypothermia
A. sweat would be produced to lower body temperature.
B. cell function would be unaffected.
C. their body temperature would be lower than 35 °C.
D. effectors would act to decrease heat production.

Question 4 (2 marks)
A person with a low body temperature often has pale skin. Explain this observation.

Question 5 (4 marks)
Explain the role of the endocrine system in the regulation of body temperature.

More exam questions are available in your learnON title.

4.5 Regulation of blood glucose in animals

KEY KNOWLEDGE
- Regulation of blood glucose in animals by homeostatic mechanisms, including stimulus–response models, feedback loops and associated organ structures

Source: Adapted from VCE Biology Study Design (2022–2026) extracts © VCAA; reproduced by permission.

4.5.1 Why do animals need to regulate blood glucose?

Glucose is the primary energy source for the functioning of cells. Through the process of cellular respiration, glucose is broken down to carbon dioxide and water, and its chemical energy is transferred to ATP.

ATP provides the energy that drives all the energy-requiring biological reactions in cells. Some cells, such as brain cells and red blood cells, rely exclusively on glucose to generate their ATP. Many other body cells, including those of the liver, heart and skeletal muscles, can use fatty acids to produce ATP. Fatty acids are produced from the breakdown of fats (triglycerides) in adipose tissue.

Sources of glucose for body cells

Almost all your glucose comes from carbohydrates (starches and sugars) in the food that you eat. After digestion and absorption, this dietary carbohydrate appears in solution in the blood or is stored in liver and skeletal muscle in the form of **glycogen**, a polymer built of glucose subunits. Under normal conditions, the total amount of glucose in solution in the plasma of your blood is just a teaspoon full — about 4 grams. Your body cells are constantly drawing on the glucose in the blood for their energy needs. As a result, blood glucose is constantly being topped up by glucose released from the glycogen store (about 100 grams in the liver of a well-nourished person).

A second source of glucose is known as 'new' glucose. This provides a supply of glucose to the brain during periods of fasting. This involves **gluconeogenesis** (gluco = glucose, neo = new and genesis = creation), which is the synthesis of new glucose from non-carbohydrate precursors.

glycogen a polysaccharide storage carbohydrate built from glucose; found mainly in liver and muscle tissue

gluconeogenesis cellular production of glucose using non-carbohydrate precursors; occurs mainly in liver cells

The blood glucose level in a healthy person who is fasting is typically between 4.0 and 5.5 millimoles per litre (mmol/L). Regardless of what a person eats, blood glucose levels are normally back under 5.5 mmol/L within about two hours of eating a meal. Blood glucose levels can be monitored using a digital mini-glucometer (figure 4.46).

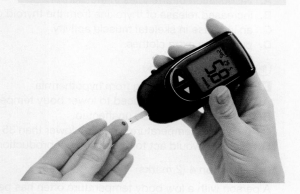

FIGURE 4.46 A digital glucometer can be used to measure the blood glucose level using a tiny drop of blood from a finger prick.

Why regulate blood glucose levels?

In everyday life, the glucose uptake by cells to supply energy can vary — glucose uptake ranges from high when we are active to basal when we are resting. At the same time, the glucose input from our diet also varies from high, soon after we have eaten, to zero when we are fasting. *Without regulation, our blood glucose levels would fluctuate widely* — they would become dangerously low when we are active while fasting, and dangerously high when we rest after eating a carbohydrate-rich meal.

It is important that your blood glucose is maintained within narrow limits, neither too high (hyperglycaemia) nor too low (hypoglycaemia), because both conditions, if prolonged, can result in a range of symptoms as seen in figure 4.47. Persistent hyperglycaemia can produce serious effects, including damage to nerves, kidneys and the blood vessels of the retina, and foot ulcers that may result in lower limb amputation. Severe hypoglycaemia can occur suddenly; if blood glucose levels fall to 2.2 mmol/L or lower, unconsciousness, coma and brain damage and even death can occur. However, healthy people with normal diets are protected from these extremes because homeostatic mechanisms are rapidly initiated when blood glucose levels either rise above or fall below the normal range. Both hypoglycaemia and hyperglycaemia are explored further in section 4.7.3.

FIGURE 4.47 When blood glucose falls too low or goes too high, it can cause health issues. Therefore, it is important that blood glucose is maintained in a relatively steady range.

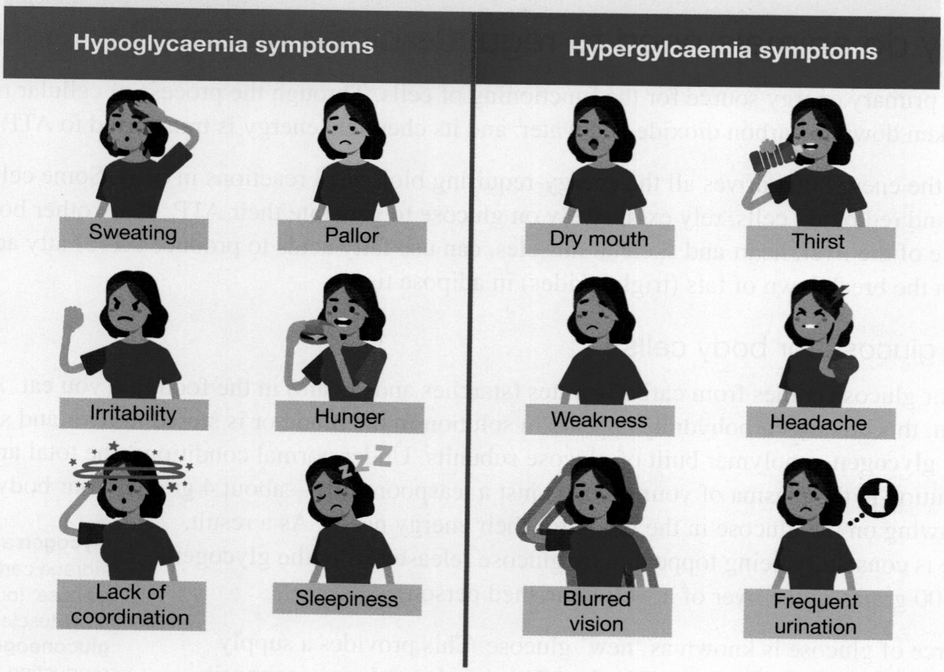

Hypoglycaemia symptoms		Hypergylcaemia symptoms	
Sweating	Pallor	Dry mouth	Thirst
Irritability	Hunger	Weakness	Headache
Lack of coordination	Sleepiness	Blurred vision	Frequent urination

4.5.2 How do animals regulate blood glucose?

Two hormones are important in the homeostatic regulation of blood glucose levels: **insulin** and **glucagon**. Both hormones are produced in the **pancreas** by special cells within the pancreas called the islets of Langerhans. These islets are surrounded by tissue composed of acinar cells, which produce pancreatic digestive enzymes (figure 4.48a). The central core of each islet is composed of large numbers of beta cells, which produce insulin. Alpha cells, which produce glucagon, are less abundant and are located around the margin of the islet (figure 4.48b).

Even when your blood glucose level is within the normal range, say 5.0 mmol/L, both glucagon and insulin are constantly being secreted at basal (very low) levels. Glucagon and insulin counterbalance each other's effects, resulting in a stable blood glucose level.

> **insulin** hormone produced by beta cells of the pancreas that acts to increase the uptake of glucose from the blood by body cells
>
> **glucagon** hormone produced by alpha cells of the pancreas that acts on liver cells resulting in increased release of glucose from the liver cells into the bloodstream
>
> **pancreas** organ that secretes digestive enzymes into the duodenum and hormones into the bloodstream

When blood glucose extends beyond normal range

It is only when the blood glucose level rise above or fall below the normal range that the balance between the two hormones, glucagon and insulin, is altered:
- When the blood glucose level falls below normal, secretion of glucagon increases and that of insulin stops.
- When the blood glucose level rises above normal, secretion of insulin increases and that of glucagon stops.

Glucagon is active in the liver. Insulin is active in the liver, skeletal muscle and adipose tissue.

FIGURE 4.48 a. Section through pancreatic tissue showing one islet of Langerhans (the pale circle, lower left). **b.** Groups of cells in the islet of Langerhans (yellow) surrounded by acini cells (pink). Note the large numbers of beta cells (yellow with red nucleus), which secrete insulin, and alpha cells (red with red nucleus), which secrete glucagon.

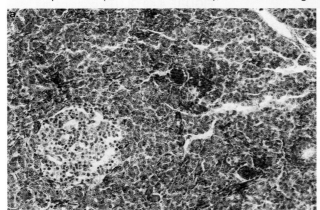
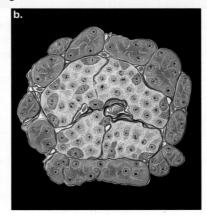

TABLE 4.10 Properties and actions of the pancreatic hormones insulin and glucagon

Feature	Insulin	Glucagon
Signal for release from islet cells	Rise in blood glucose Hyperglycemia: > 5.5 mmol/L	Fall in blood glucose Hypoglycemia: < 4.0 mmol/L
Produced by	Beta cells of pancreatic islets	Alpha cells of pancreatic islets
Type of molecule	Peptide hormone (51 amino acids)	Peptide hormone (29 amino acids)
Mode of transport	Dissolved in blood plasma	Dissolved in blood plasma

(continued)

TABLE 4.10 Properties and actions of the pancreatic hormones insulin and glucagon *(continued)*

Feature	Insulin	Glucagon
Primary targets	Skeletal muscle, adipose tissue and liver	Liver
Actions	Increases uptake of glucose by skeletal muscle and fat cells Converts glucose to glycogen for storage in liver and muscle cells	Increases breakdown of glycogen to glucose in liver Synthesises 'new' glucose in liver cells
Final result	Fall in blood glucose	Rise in blood glucose
Negative feedback	Fall in blood glucose stops release of insulin by beta cells	Rise in blood glucose stops release of glycogen by alpha cells

Insulin: the glucose-storing hormone

After we eat a carbohydrate-rich meal, our blood glucose can rise to excessively high levels. Insulin is released when glucose-sensing beta cells in the pancreatic islets detect an above-normal rise in the level of blood. The excess glucose could be excreted via our kidneys, but glucose is an energy source and is far too valuable to waste as sweetened urine. Instead, excess glucose is moved out of the blood in response to insulin.

Insulin lowers high blood glucose levels through several actions:
- stimulating the movement of glucose from the blood into skeletal muscle and adipose tissues — this is its major action, because these two tissues form about 60 per cent of the body mass
- activating enzymes that build glucose into glycogen in liver and skeletal muscle cells
- inhibiting the breakdown of fats in adipose tissue — this reduces the supply of fatty acids in the blood and causes some body cells to make more use of glucose for their energy needs, in particular, those cells that mainly use fatty acids for their energy supply.

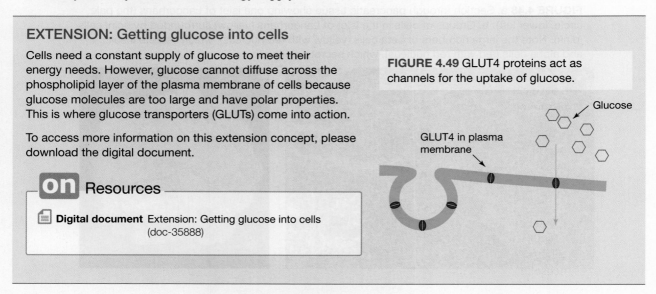

EXTENSION: Getting glucose into cells

Cells need a constant supply of glucose to meet their energy needs. However, glucose cannot diffuse across the phospholipid layer of the plasma membrane of cells because glucose molecules are too large and have polar properties. This is where glucose transporters (GLUTs) come into action.

To access more information on this extension concept, please download the digital document.

Resources

Digital document Extension: Getting glucose into cells (doc-35888)

FIGURE 4.49 GLUT4 proteins act as channels for the uptake of glucose.

Glucagon: the glucose-releasing hormone

Glucagon is stored as granules in the alpha cells of pancreatic islets (figure 4.50). During fasting or periods of activity, blood glucose levels can fall below normal. A fall below 4.0 mmol/L is the trigger for the release of the hormone glucagon by the alpha cells of the pancreatic islets (figure 4.51).

The main target of the glucagon hormone is the liver. Glucagon binds to receptors on liver cells and its action causes the release of glucose, raising the blood glucose level by breaking down glycogen and simulating the production of new glucose. These actions restore the blood glucose level to within the normal range and ensure that body cells have a continual adequate supply of glucose for their needs.

In addition, glucagon acts on cells of adipose tissue and activates the enzyme lipase, which breaks down fats to form glycerol and fatty acids that are released into the bloodstream. Glycerol is taken up by liver cells and used to form 'new' glucose. Fatty acids are used by many body cells as a source of energy, and their use of fatty acids reduces their demand on blood glucose for this purpose. Figure 4.51 shows the main actions of glucagon on blood glucose levels.

Glycogen: the key to blood glucose homeostasis

- Depending on blood glucose levels, glycogen in liver cells can store or release glucose.
- This allows regulation of blood glucose and enables us to endure periods of fasting.

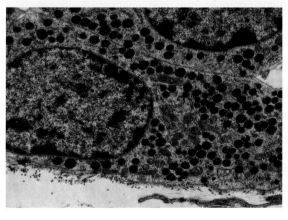

FIGURE 4.50 Coloured TEM of two alpha cells in the pancreas, with glucagon granules shown in blue and the nucleus shown in purple

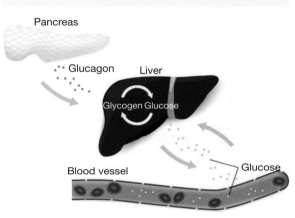

FIGURE 4.51 Glucagon acts on the liver, stimulating glycogen to break down into glucose.

Storing glucose as glycogen

Soon after you have eaten, blood glucose levels rise. Insulin stimulates liver cells to build glucose into glycogen as a short-term store of glucose.

When the store of glycogen in liver cells is at capacity, insulin stimulates the production of fatty acids from glucose, and these are transported to other tissues, in particular adipose tissues. In cells of adipose tissue, fatty acids are combined with glycerol to form fat (triglycerides).

Releasing glucose

If you have not eaten for some time, your blood glucose level begins to fall. When this occurs, the glycogen in liver cells is rapidly mobilised and glucose is released into the blood. The hormone glucagon stimulates the release of glucose from its glycogen store. The liver is the only internal organ that can release significant amounts of glucose into the blood.

Skeletal muscle cells can also release glucose from their glycogen store, but this glucose can be used only by the muscle cells themselves. For example, marathon runners depend to a large degree on the release of glucose from the glycogen in the skeletal muscles to supply their energy needs. Unlike liver cells, skeletal muscle cells cannot release glucose into the blood for use by other body cells because they lack a critical enzyme.

TIP: Although you are not always required to have perfect spelling in biology, there are some cases where correct spelling is vital to ensure your communication is clear. This is particularly important in the use of glucose, glycogen and glucagon. These three words are very similar but have very different meanings. For example, if you misspell a word as glycagon, it is very hard to know if you are talking about glucagon or glycogen.

4.5.3 When blood glucose levels are too low

When blood glucose level falls below normal (less than 4.0 mmol/L), negative feedback and homeostasis occurs to restore this to normal. This involves the release of glucagon from alpha cells in the pancreas, which works to increase blood glucose levels by travelling through the bloodstream binding to receptors on the liver (see figure 4.52).

The binding of glucagon activates key liver enzymes involved in two pathways:
- the breakdown of stored glycogen to glucose — this pathway is usually the major supplier of glucose to the blood
- the production of 'new' glucose — usually a minor supplier of glucose, except when glycogen stores are depleted.

Glucagon also inhibits the formation of glycogen by blocking the activity of key enzymes.

> The overall response of the actions of glucagon is an increase in blood glucose level. This is through the release of glucose from liver cells into the blood.

FIGURE 4.52 The stimulus–response model showing the response initiated by a fall in blood glucose level.

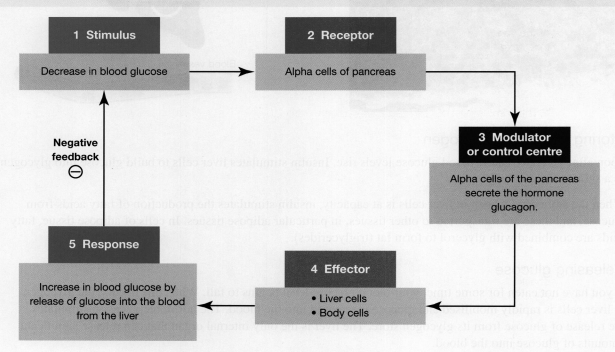

4.5.4 When blood glucose levels are too high

Intake and digestion of dietary carbohydrates can produce an above-normal increase in the circulating blood glucose level after glucose is absorbed from the intestine. When blood glucose levels increase above the normal range (greater than 5.5 mmol/L), a negative feedback loop and homeostasis occurs to restore this to normal. This involves the release of insulin from beta cells in the pancreas, which works to decrease blood glucose levels, travelling to target cells through the bloodstream (see figure 4.53).

The main responses (and the locations of these responses) that occur as a result of the release of insulin are as follows:
- *Skeletal muscle and adipose tissue*: Insulin binds to its receptors on the plasma membranes of these cells. This leads to special glucose transporters moving to the membrane so more glucose can move into skeletal muscle and adipose tissue.
  - In skeletal muscle, this glucose is converted to glycogen.
  - In adipose cells, this glucose is converted to fats (triglycerides). Insulin also acts to prevent these fats from breaking down and being used as an energy source.
- *Liver*: Insulin plays no role in the uptake of glucose by liver cells, but it activates key enzymes in liver cells that are needed to build glucose into glycogen, thus preventing release of glucose into the blood.

Note: The hypothalamus of the brain also plays a coordinating role in blood glucose homeostasis. It contains both glucose-sensing neurons and insulin-sensing neurons that respond to changes in the level of blood glucose. This can send signals to the cells in the pancreas to release insulin or glucagon.

The results of the various responses to the actions of insulin is a decrease in the level of blood glucose. This is mainly due to uptake of glucose from the blood into cells of skeletal muscle and adipose tissue.

FIGURE 4.53 The stimulus–response model showing the response initiated by an increase in blood glucose level

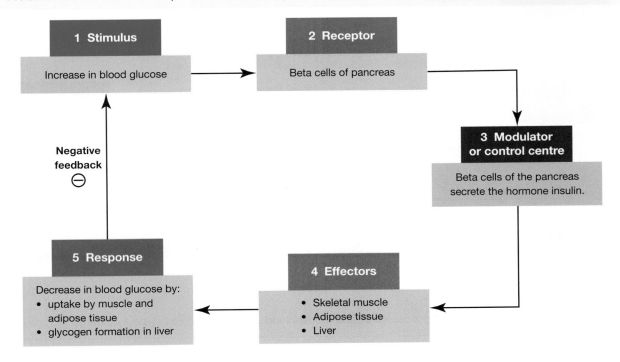

Feedback loops working together

The homeostatic mechanisms that increase and decrease blood glucose levels are acting all the time, making the necessary adjustments to these levels as they rise and fall (see figure 4.54).

As a result of negative feedback mechanisms involving both insulin and glucagon, a steady state is achieved in blood glucose levels, with small fluctuations.

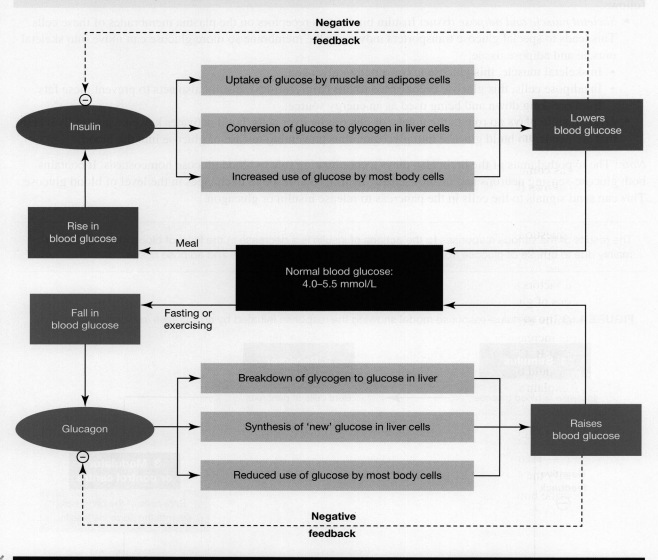

FIGURE 4.54 A summary of major events that maintain blood glucose levels with a narrow range

INVESTIGATION 4.4

Testing glucose levels

Aim

To test blood glucose levels using glucose strips and model blood glucose samples in type 1 diabetes

SAMPLE PROBLEM 4 Understanding the regulation of blood glucose

A student has not eaten for a few hours and has low blood glucose levels. Their blood glucose levels over time are shown below.

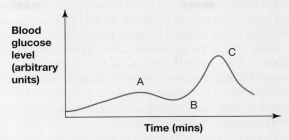

a. What caused the initial increase of blood glucose at point A despite the student not consuming food? (2 marks)
b. Explain what probably happened at point B that allowed for a rapid spike in blood glucose levels. (2 marks)
c. Insulin has come into play at point C. Identify where insulin is produced and describe how this allows the decrease in blood glucose levels. (2 marks)

THINK

a. The question is worth two marks, so you need to outline two key factors that allow for an increase in blood glucose. You should consider what factors cause glucose to rise, that is, the actions of glucagon.

b. Carefully examine the graph and compare the initial increase from glucagon to the sudden spike at B. Consider what would cause a spike. You should be able to identify what caused it and explain why the results on the graph were seen.

c. 1. Carefully read what the question is asking you to do — identify a site and describe a process. Identify the site in which insulin is processed.

   2. Describe how insulin allows for blood glucose levels to decrease.

WRITE

Glucagon has been released from alpha cells in the pancreas (1 mark). This causes glycogen in the liver to be broken down into glucose, which diffuses into the blood stream (1 mark).

It is likely that an individual consumed something high in glucose (a sugary drink or a lolly for example) (1 mark). This would cause the sudden spike in blood glucose, which is much more significant than the small rise seen in the release of glucagon (1 mark).

Insulin is produced in the beta cells in the pancreas (1 mark).

Insulin allows for glucose to be taken up by cells. By allowing for glucose to move into cells, the amount of glucose in the bloodstream decreases (1 mark).

Resources

- **eWorkbook** Worksheet 4.5 Glucose ups and downs (ewbk-7499)
- **Video eLesson** Blood glucose regulation (eles-4735)
- **Interactivity** Homeostasis and blood glucose (int-8082)

KEY IDEAS

- Blood glucose levels are normally highly regulated.
- Insulin and glucagon are two hormones involved in the homeostatic regulation of blood glucose levels.
- Insulin is produced in beta cells of the pancreatic islets. Its secretion is triggered by a rise in blood glucose above the normal level. It allows blood glucose levels to decrease.
- Glucagon is produced in alpha cells of the pancreatic islets. Its secretion is triggered by a fall in blood glucose below the normal level. It allows blood glucose levels to increase.
- Insulin stimulates the uptake of blood glucose by cells of skeletal muscle and adipose tissue.
- Insulin activates key enzymes needed to convert glucose to glycogen in liver and muscle cells.
- Insulin inhibits the breakdown of fats, reducing the supply of fatty acids. This results in an increased use of blood glucose as the energy source of many body cells.
- Glucagon stimulates the breakdown of liver glycogen to glucose, which is released into the blood.
- Glucagon stimulates production of 'new' glucose in liver cells, an important source when glycogen stores are depleted.
- As well as glucose, fatty acids can be used by many body cells as an energy source. Glucagon stimulates fat breakdown, increasing the supply of fatty acids, leading to less glucose being used. This results in a decrease in the use of blood glucose as the energy source for many body cells.

4.5 Activities

To answer questions online and to receive **immediate feedback** and **sample responses** for every question, go to your learnON title at **www.jacplus.com.au**. A **downloadable solutions** file is also available in the resources tab.

| 4.5 Quick quiz | 4.5 Exercise | 4.5 Exam questions |

4.5 Exercise

1. Identify whether each of the following statements is true or false. Justify your response.
   a. Insulin promotes the breakdown of glycogen in the liver.
   b. The hormone glucagon causes the breakdown of glycogen to its glucose subunits.
   c. 'New' glucose results from the digestion and absorption of dietary carbohydrates.
   d. In addition to glucose, fatty acids in the blood are used by many body cells as their source of energy.
2. What is the difference between hyperglycaemia and hypoglycaemia?
3. Which hormone is more likely to be produced soon after eating a meal?
4. Identify which cells in the pancreas are the sites of secretion of
   a. insulin.
   b. glucagon.
5. Compare and contrast the regulation of body temperature and the regulation of blood glucose.
6. Explain the effect of each of the following on blood glucose levels. In each case, assume the concentration of the other paired hormone is unchanged.
   a. An increase in insulin
   b. An increase in glucagon
   c. A decrease in insulin
   d. A decrease in glucagon

4.5 Exam questions

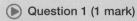 Question 1 (1 mark)

MC The hormone insulin
A. acts to increase blood glucose levels.
B. is produced by alpha cells in the pancreas.
C. increases uptake of glucose into the body cells.
D. is a steroid molecule.

▶ **Question 2 (1 mark)**
MC Glucagon
A. is a carbohydrate.
B. acts in liver cells to increase the conversion of glycogen to glucose.
C. is produced by beta cells in the pancreas.
D. is released when blood glucose levels are above normal levels.

▶ **Questions 3 (2 marks)**
Discuss two responses that occur within the human body when blood glucose levels are above the normal range.

▶ **Question 4 (2 marks)**
Explain why it is important to maintain a constant level of glucose in the internal environment of human body cells.

▶ **Question 5 (3 marks)**
Source: VCAA 2010 Biology Exam 1, Section B, Q1bi, ii
Blood glucose levels are controlled by a homeostatic mechanism.

Two females of the same age and similar body structure were each given an identical meal. The following graph shows the level of blood glucose in each female for the five-hour period after eating the meal.

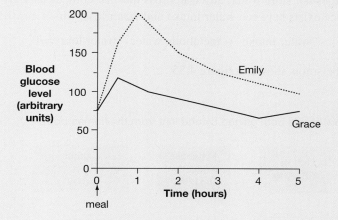

i. Explain whether Emily or Grace had a defect in the blood-glucose homeostatic mechanism. Refer to at least two parts of the graph to support your answer. **2 marks**
ii. Explain the small rise in her level of glucose between four and five hours after the meal. **1 mark**

More exam questions are available in your learnON title.

4.6 Regulation of water balance in animals

KEY KNOWLEDGE

- Regulation of water in animals by homeostatic mechanisms, including stimulus–response models, feedback loops and associated organ structures

Source: Adapted from VCE Biology Study Design (2022–2026) extracts © VCAA; reproduced by permission.

4.6.1 Why do animals need to regulate water?

Water is the most common compound present in all living organisms, including the human body. Overall, a typical lean human body is about 60 per cent water, that is about 40 litres. Individual organs differ in their water content — muscles are about 80 per cent water, the brain is about 73 per cent water, and bones are about 30 per cent water. Water in the body is present mainly as the intracellular fluid that consists of the cytosol of all cells, and the interstitial fluid that surrounds cells and the plasma of the blood. These fluids contain many solutes, such as sodium ions (Na^+) in the extracellular fluid and potassium ions (K^+) in the intracellular fluid.

In the human body, the metabolic reactions essential for living occur in the aqueous medium of cells. In addition, water is essential for many functions:
- The absorption of nutrients from the alimentary canal into the bloodstream depends on their being water-soluble.
- The watery plasma of the bloodstream transports digested nutrients to cells and also circulates the red blood cells that carry oxygen to cells.
- Wastes are excreted from the body via the kidney in solution in aqueous urine.
- Sweating helps cool the body during periods of exercise and in hot environments.
- Water acts as a cushioning fluid around the brain (cerebrospinal fluid) and in joints (synovial fluid).
- Water is a major component of the mucus that acts as the lubricant for the mucous membranes that line the airways and the passages of the alimentary canal.
- Water is a major component of the interstitial fluid that surrounds cells and forms their external environment.

Because the human body does not store water, any water loss must be compensated by water gain to ensure balance. Normally, a balance exists between water intake and water loss, which may be shown as follows:

$$\text{water intake} + \text{metabolic water} - \text{water loss} = 0.$$

The balance for a healthy person is shown in figure 4.55.

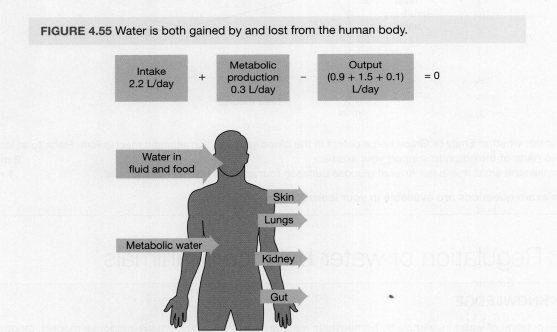

FIGURE 4.55 Water is both gained by and lost from the human body.

Average values for water intake and output differ between various animals. Some animals, for example the kangaroo rat, have adaptations that minimise water loss, so their intake of water is much less compared to humans. These adaptations are investigated further in Topic 10.

Like other key variables, water needs to be maintained within a set range. Water levels can fluctuate for a variety of reasons, as shown in table 4.11.

TABLE 4.11 The sources of water gain and water loss

Sources of water gain	Sources of water loss
• Fluids from drinks (60%) • Fluid content of foods (30%) • Internally produced metabolic water (10%)	• From skin and lungs (28%) • From sweat glands (as fluid) (8%) • From the gut (as fluid in faeces) (4%) • From the kidneys (as urine) (60%)

The level of water may become too low because of:
- an excessive loss of water (such as the result of sweating in hot conditions)
- an inadequate intake of fluids
- an abnormal loss of body fluids (such as may occur with severe diarrhoea, prolonged vomiting, significant haemorrhage or serious burn injuries).

Likewise, the water levels in the body can become too high because of:
- impaired kidney function that produces insufficient urine
- drinking excessive amounts of water.
- medical conditions, such as SIADH (syndrome of inappropriate secretion of antidiuretic hormone).

Effects of water imbalance:
- If water intake is less than water output, this can lead to a toxic increase in excess ions and waste products in the blood. This can cause disruption of muscle and nerve function, as well as shrinkage of cells, including those of the brain. Possible symptoms include confusion and seizures.
- If water intake is in excess of water output, a person can become over-hydrated, a situation sometimes called 'water intoxication'. This can cause cells to swell. The adverse effects of unregulated overhydration include swelling of the legs and feet, as well as swelling of the brain (cerebral edema), with symptoms of confusion, lethargy, headache, and drowsiness.

4.6.2 How do animals regulate water?

Water balance is closely associated with the concentration of mineral salts (electrolytes) such as sodium and potassium in the blood. When water levels in the body are unbalanced, electrolyte levels are also unbalanced.

osmoregulation process by which the volume of body fluids and their solutes concentrations are controlled

The concentration of fluids in the human body is regulated (that is, kept within a narrow range) by homeostatic mechanisms.

> The regulation of body fluids is called **osmoregulation**. Osmoregulation is the process of controlling the water content of the human body and its solute concentration.

Regulation to maintain water balance occurs through two means:
1. controlling the *volume* of body fluids by regulating the amount of water excreted in the urine, and by initiating the thirst response when needed
2. controlling the *osmolality* (concentration of dissolved solutes) of body fluids by regulating the water-to-sodium balance of extracellular fluids, so that any gains or losses of sodium are matched by corresponding gains or losses of water. Sodium is by far the major solute in extracellular fluids.

Osmolality is an indicator of water level. If the blood plasma osmolality is higher than normal, this means that the water level is too low. If osmolality is below normal, this means that the water level is too high. Osmoreceptors can detect the change in osmolality from passing blood. Osmolality of the blood plasma is tightly regulated and is largely due to the presence of sodium ions.

TABLE 4.12 Regulation actions to restore normal water balance in terms of volume and osmolality

Event	Effect on extracellular fluid (ECF) and plasma	Message	Control response
A person drinks excessive bottled water within a short time frame.	ECF volume increased Plasma osmolality decreased	Water out of balance Water level too high	Increased output of dilute urine
A person fails to drink water during a 24-hour period.	ECF volume increased Plasma osmolality increased	Water out of balance Water level too low	Thirst sensation activated Increased water reabsorption by kidneys
Bouts of severe diarrhoea with major loss of water and salts	ECF volume increased Plasma osmolality unchanged	Water level reduced Both water and salt are needed	Increased reabsorption of sodium and water by kidneys Fluid intake, such as soup or juices
A person receives an excessive infusion of isotonic saline.	ECF volume increased Plasma osmolality unchanged	Water level raised	Excess fluid removed in urine

Key organs and hormones in osmoregulation

The brain

One area of the brain, the hypothalamus, is the location of osmoreceptors, which are specialised nerve cells able to detect changes in blood plasma osmolality. These osmoreceptors stimulate the release of a hormone known as **antidiuretic hormone** (ADH) or vasopressin.

Another area of the brain, the lamina terminalis, stimulates the *thirst response*, so that a person has a strong desire to have a glass of water. This occurs when water levels drop too low (and dissolved salts are detected at a higher level).

The kidneys

The kidneys are the organs that are essential for regulating both the volume and the osmolality (composition) of body fluids.
- When plasma osmolality falls, the kidneys respond by producing larger volumes of more dilute urine.
- When plasma osmolality rises, the kidneys respond by conserving water and producing smaller volumes of more concentrated urine. This response is stimulated by the hormone ADH.

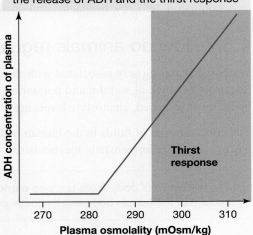

FIGURE 4.56 Graph of the plasma osmolality levels required to stimulate the release of ADH and the thirst response

Figure 4.57 shows urine of increasing concentration; the darker the urine, the higher its osmolality. Note that on its own, conservation of water by the kidneys cannot correct a fall in plasma volume. To do this, a person must have a net intake of fluid by drinking water.

Water balance is controlled in two main parts within the nephrons of kidneys: the descending limb of the loop of Henle and the collecting tubule (figure 4.58). When water levels are high, more water is excreted in urine, and a higher volume of urine is produced. When water levels are low, more water is reabsorbed in the nephron and retained in the bloodstream.

> **antidiuretic hormone** hormone produced by neurosecretory cells in the hypothalamus; increases reabsorption of water into the blood from distal tubules and collecting ducts of nephrons in the kidney

FIGURE 4.57 One of the main ways water balance is maintained is through the dilution of urine. More water has been excreted in the dilute urine in the left. The urine shown on the far right has much higher concentrations of solutes and urea.

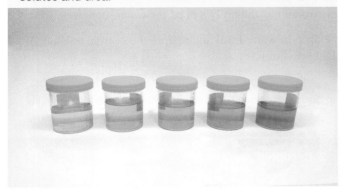

FIGURE 4.58 The structure of the nephron including the collecting tubule and descending limb of the loop of Henle

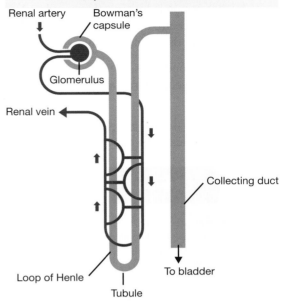

Antidiuretic hormone

Antidiuretic hormone (ADH) is a peptide hormone that is released from the posterior pituitary in response to a signal from the hypothalamus. ADH travels though the bloodstream to the kidney, where it binds mainly to receptors on the cells that line the collecting ducts in the kidney, making these cells more permeable to water. ADH stimulates the insertion of specific channel proteins called aquaporins into the plasma membrane of these cells. Aquaporins enable the more rapid movement of water from fluid in the collecting ducts back into the bloodstream. In the presence of ADH, smaller volumes of more concentrated urine are produced. In the absence of ADH, larger volumes of more dilute urine are produced.

- When water levels are low and solute levels are high, ADH is released from the hypothalamus and water is reabsorbed in the kidneys.
- When water levels are high and solute levels are low, ADH levels are decreased and excess water is excreted in the kidneys, resulting in lots of dilute urine being produced

4.6.3 When water levels are too low

When water levels in the body fall below the normal level, the concentration of dissolved compounds (solutes) in body fluids rises. In order to restore the balance, homeostatic mechanisms are activated.

Figure 4.59 shows an outline of events that occur when the water levels in the body are too low. The drop in water level and rise in osmolality of blood plasma (increase in solutes) is the stimulus that is detected by osmoregulator cells in the hypothalamus as blood circulates through the brain.

These cells monitor the osmolality, or concentration of solutes, of the blood plasma against a set point. In this case, the water levels are low, so the concentration of dissolved solutes is too high and is above the set point.

The hypothalamus identifies the corrective actions required and sends signals to effectors that take action as follows:

1. The hypothalamus sends a signal to the posterior pituitary gland to release antidiuretic hormone (ADH).

2. The collecting ducts of the kidney become more permeable to water through the action of ADH, so the reabsorption of water into the blood is increased.
3. As a result, the volume of urine produced falls. When a person is extremely dehydrated, the urine produced by that person is low in volume and dark yellow in colour.
4. The thirst centre of the hypothalamus sends a signal that stimulates the sensation of thirst, which motivates a person to drink fluids. The urge to drink becomes stronger as the body's need for water increases.

Under normal conditions, these responses counteract the drop in water level and restore the normal balance of water level (and concentration of dissolved solutes in the body fluids).

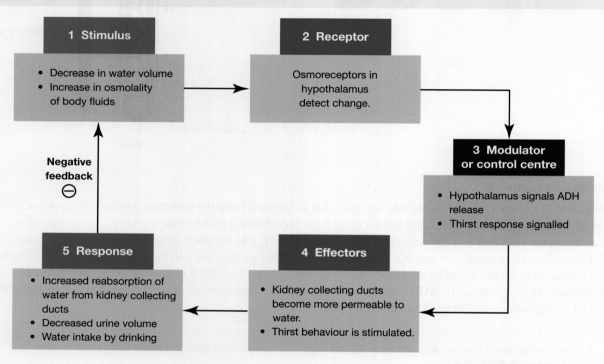

FIGURE 4.59 The stimulus–response model for a drop in body water levels that causes the concentration of dissolved solutes in the body fluid to increase

4.6.4 When water levels are too high

When water levels in the body increase above normal, the ECF increases in volume and the osmolality of the blood decreases (reduction in solutes). In order to restore the balance, homeostatic mechanisms are activated.

Figure 4.60 shows an outline of events that occur when the water levels in the body are too high. Water levels that are too high dilute the body fluids, causing the decrease in osmolality. The osmolality decrease is the stimulus that initiates homeostatic mechanisms to correct this situation and restore the normal balance.

This stimulus is detected by osmoreceptors in the hypothalamus as blood circulates through the brain.
1. The hypothalamus identifies that corrective actions are required, namely that water levels in the body must decrease (and osmolality must increase), and signals this to effectors.
2. A signal from the hypothalamus to the pituitary gland results in the inhibition of the release of ADH. When ADH is not present, the collecting ducts of the kidneys become impermeable to water so that water reabsorption from the kidneys back into the blood is reduced and greater volumes of urine are produced.
3. The sensation of thirst from the thirst centre in the hypothalamus is suppressed.

These responses counteract the increase in water volume in the body. This decrease in volume raises the osmolality of the blood to normal level and water balance is restored.

FIGURE 4.60 The stimulus–response model for an increase in body water levels that causes the concentration of dissolved solutes in the body fluid to decrease

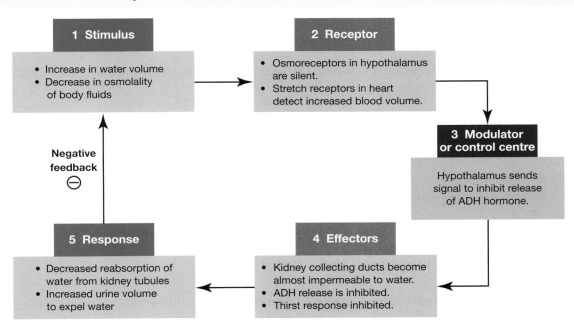

SAMPLE PROBLEM 5 Understanding water balance in animals

tlvd-1751

An individual has had a throat infection. As a result, they had copious amounts of water to ease their sore throat.

a. Would you expect the production of ADH from the hypothalamus to increase or decrease? Justify your response. **(2 marks)**
b. Explain the role of the kidneys in maintaining water balance in this individual. **(2 marks)**

THINK	WRITE
a. You are being asked to identify if an increase or decrease in ADH occurred and to justify your response. It is important to address both factors to be awarded marks. Carefully answer both parts of the question	ADH levels would be expected to decrease (1 mark). As the amount of water has increased, more water will be excreted. As ADH allows for the reabsorption of water in times of low water, a decrease in ADH would be required when water levels are high (1 mark).
b. The question is worth two marks, so ensure you outline two points — the role of the kidney in maintaining water balance in general, and how this would specifically occur in the individual who had consumed large quantities of water.	ADH acts on the kidneys, specifically on the nephrons and the collecting tubules. Water is either excreted and left in urine or reabsorbed from the tubule (and the loop of Henle) back into the bloodstream (1 mark). In this case, the kidneys would produce very dilute urine with large volumes of water in order to return water balance back to normal (1 mark).

Resources

- **eWorkbook** Worksheet 4.6 The regulation of water (ewbk-7501)
- **Video eLesson** Urine formation in the kidney (eles-2644)
- **Interactivity** Homeostasis and water balance (int-8083)

TOPIC 4 Regulation of systems **311**

4.6.5 Other methods of water balance in different animals

There are many other methods that help animals with water balance. Adaptations to maintain water balance are discussed in more detail in Topic 10.

Water balance in fish

Freshwater fish are exposed to constantly high levels of water. In order to ensure they do not have extremely low levels of salt and high levels of water, they:
- do not drink
- have nephrons that produce large amounts of very dilute urine
- lose water through their gills
- actively absorb ions through their gills

Saltwater fish, on the other hand, are exposed to water that is concentrated with ions and less concentrated with water. In order to ensure they do not lose water to the environment and uptake a high level of salts, they:
- drink sea water
- produce minimal urine (and what urine is produced has a high ion concentration)
- have gills that are impermeable to water
- secrete excess ions such as sodium and chloride through their gills.

FIGURE 4.61 Freshwater and saltwater fish have very different homeostatic mechanisms to maintain water balance.

Water balance in birds

Birds do not excrete urine. Rather, their nitrogenous waste is excreted as uric acid, which allows for better water conservation. Water is also reabsorbed in the cloaca, causing uric acid to be highly concentrated with urine but with minimal water loss. This allows for a greater volume of water to be retained.

Water balance in amphibians

Amphibians, such as frogs, live near and in fresh water. Their skin is highly permeable to water, which results in continual movement of water via osmosis. In order to counteract this, frogs produce very dilute urine that removes excessive water.

Amphibians also often live on land. When in these conditions, they are not gaining water through osmosis, so instead they produce highly concentrated urine with very little water.

KEY IDEAS

- Water is essential for life and is a major component of the human body.
- For survival, water loss must be balanced by water gain.
- Water loss from the human body occurs through several channels, the main one being via the kidneys.
- Water gain is either from food or drink, or from internal metabolic processes.
- Water levels and solute concentrations of body fluids in the human body are under homeostatic regulation.
- When water levels fall, the solute concentrations of body fluids increase.
- Osmoreceptors work to detect changes in water balance
- When water levels decrease, a message is sent to the hypothalamus, causing the activation of thirst centres in the brain and an increase in ADH, leading to greater reabsorption of water in the kidneys.
- When water levels increase, a message is sent to the hypothalamus, causing the suppression of thirst centres in the brain and a decrease in ADH, leading to less reabsorption of water in the kidneys and an increase in water excreted in urine.

4.6 Activities

To answer questions online and to receive **immediate feedback** and **sample responses** for every question, go to your learnON title at **www.jacplus.com.au**. A **downloadable solutions** file is also available in the resources tab.

4.6 Quick quiz on	4.6 Exercise	4.6 Exam questions

4.6 Exercise

1. Describe how ADH is used to regulate water balance. Ensure you identify the site of action of ADH in your response.
2. Describe three ways in which humans are able to gain water.
3. The concentration of salt and other ions in the blood is detected to have increased by a receptor.
   a. Would you expect ADH in the plasma to increase or decrease? Justify your response.
   b. Explain what would be expected to happen in the kidneys.
4. The collecting duct is an important feature in the kidneys for water balance. Explain why the collecting duct can assist with water balance.
5. Describe what would happen if
   a. the concentration of solutes in body fluids decreased.
   b. the water gain by the body increased.
6. What causes the thirst centre to be stimulated?
7. Diuretics, which have the opposite effect of the antidiuretic hormone, are banned in sport.
   a. What effect on water balance would you expect diuretics to have?
   b. Justify three reasons why these might be banned in sport.

4.6 Exam questions

Question 1 (1 mark)

MC Which of the following human hormones is involved in the regulation of water balance?
A. Insulin
B. Thyroxine
C. Glucagon
D. Antidiuretic hormone

Question 2 (1 mark)

MC A large volume of urine would be produced by a person who
A. ate very salty food.
B. exercised at high intensity for over an hour.
C. drank many cups of coffee in a short period of time.
D. ate food that contained very little water.

Question 3 (1 mark)

MC Hormones the help regulate the amount of water in the internal environment of human cells include
A. antidiuretic hormone and renin.
B. insulin and glucagon.
C. thyroxine and adrenaline.
D. oestrogen and testosterone.

Question 4 (1 mark)

MC Water balance in the human body is associated with the maintenance of
A. oxygen concentration.
B. heart rate.
C. osmolality.
D. glucose concentration.

Question 5 (3 marks)

Explain how the consumption of alcohol affects water balance in humans.

More exam questions are available in your learnON title.

4.7 Malfunctions in homeostatic mechanisms

KEY KNOWLEDGE

- Malfunctions in homeostatic mechanisms: type 1 diabetes, hypoglycaemia, hyperthyroidism

Source: Adapted from VCE Biology Study Design (2022–2026) extracts © VCAA; reproduced by permission.

4.7.1 Malfunctions in homeostatic mechanisms

Sometimes, homeostatic mechanisms fail to regulate variables that usually need to be in a set range. This can lead to a various of illnesses and in extreme cases can result in death. This shows the importance of homeostatic mechanisms to ensure the survival and functioning of organisms.

A malfunction of homeostatic mechanisms may be due to various factors, including disease or environment.

Examples of these can be seen in the regulation of temperature, blood glucose and water balance. Some of these are outlined in table 4.13.

TABLE 4.13 Some examples of malfunctions in homeostasis

Variable	Examples of malfunctions	
Basal metabolic rate	• Hyperthyroidism	• The thyroid is overactive, producing excessive thyroxine, leading to an increase in basal metabolic rate. Symptoms of this include elevated body temperature.
	• Hypothyroidism	• The thyroid is underactive, resulting in a reduced basal metabolic rate. Symptoms include decreased body temperature.
Temperature	• Hypothermia	• Exposure to extreme cold can cause homeostatic mechanisms to fail.
	• Heat stroke	• Exposure to high temperatures can cause homeostatic mechanisms to fail.

Variable	Examples of malfunctions	
Blood glucose	• Type I diabetes	• Insulin is not produced, causing high blood glucose.
	• Type II diabetes	• Cells are resistant to insulin, causing high blood glucose.
	• Hypoglycaemia	• Blood glucose becomes too low.
Water balance	• Dehydration	• Extreme conditions cause water levels to drop to excessively low amounts.
	• Kidney failure	• Kidneys are unable to regulate water balance, leading to either low or high levels of water in the bloodstream.

4.7.2 Type 1 diabetes

In subtopic 4.5, we explored how blood glucose level is normally regulated by homeostatic mechanisms that maintain glucose levels within a narrow range. The beta cells of the pancreas are responsible for the production of the hormone insulin. This hormone plays a key role in stopping blood glucose from increasing above the normal level by facilitating the uptake of glucose by body cells. This process is summarised in figure 4.62.

For some individuals, this process is not properly regulated, leading to the blood glucose levels moving beyond normal ranges without external treatment.

What is type 1 diabetes?

Type 1 diabetes is a chronic autoimmune disease in which the beta cells of the pancreas produce little or no insulin, greatly affecting the regulation of blood glucose levels (see figure 4.63).

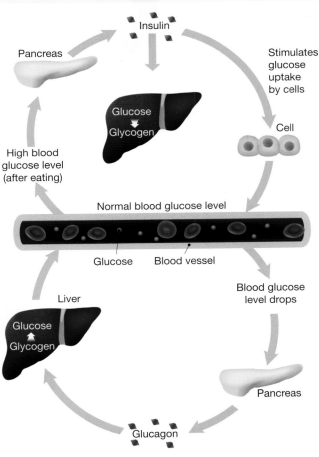

FIGURE 4.62 Reviewing the relationship between blood glucose, insulin and glucagon

When insulin is not produced, blood glucose levels remain high and cannot be returned to normal through negative feedback and homeostatic mechanisms.

This condition most commonly appears in childhood but may also appear later in life (see figure 4.64). The cause of this condition is not certain, but one view is that type 1 diabetes is an autoimmune disorder in which the immune system specifically turns against the body's own beta cells in the pancreas, attacking them as if they were foreign cells. It is thought that over 100 000 individuals in Australia are currently living with type 1 diabetes.

Risk factors for type 1 diabetes include a family history of the disorder and the presence of certain genes in a person's genetic make-up.

type 1 diabetes a condition that results when the homeostatic mechanisms that regulate blood glucose levels fail when insulin production fails, characterised by a blood glucose level that is higher than normal

TOPIC 4 Regulation of systems 315

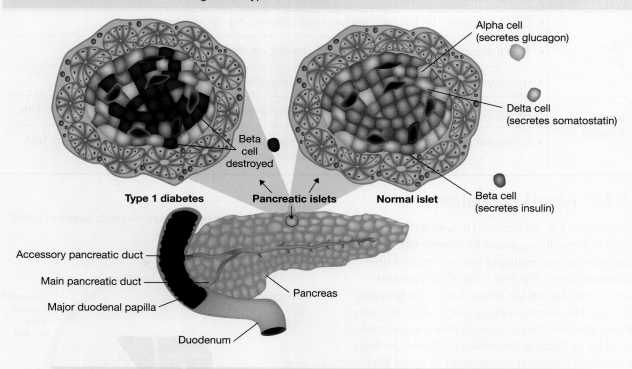

FIGURE 4.63 The beta cells are targeted in type 1 diabetes.

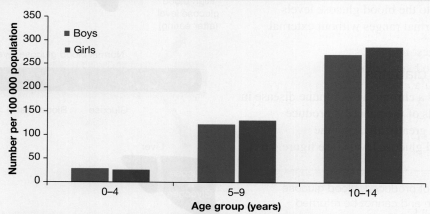

FIGURE 4.64 The prevalence of type 1 diabetes in Australia in children

Source: AIHW, https://www.aihw.gov.au/reports/diabetes/diabetes-snapshot/contents/how-many-australians-have-diabetes/type-1-diabetes

Glucose levels out of control

The homeostatic mechanisms that regulate blood glucose levels can fail when insulin production fails.

Type 1 diabetes, therefore, is characterised by a blood glucose level that is higher than normal.

Why?

- Glucose is the main source of energy for body cells, but glucose is too large to diffuse across the plasma membrane and must be actively transported into cells. Insulin facilitates this transportation.
- Because insulin production is defective in type 1 diabetes, the body cells of a person affected by this condition cannot take up glucose from the bloodstream.
- As a result, the glucose levels in the blood rise above normal, a condition termed **hyperglycaemia** (*hyper* = above; *glykys* = sweet; *haima* = blood).

hyperglycaemia a condition where glucose levels in the blood rise above normal

FIGURE 4.65 Glucose cannot cross the plasma membrane without insulin.

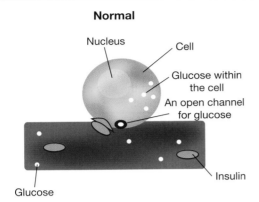
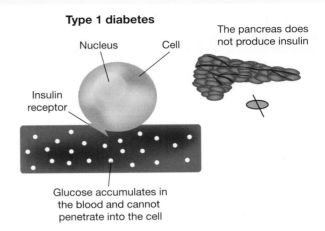

This inability to regulate through homeostatic mechanisms can be seen in figure 4.66. At the start of the test, both individuals ingested a standard quantity of glucose, administered in solution.
- Person A, who is not affected with diabetes, had a normal homeostatic process in which insulin was released and glucose in the blood was moved into cells, causing a decrease in blood glucose levels. Their levels fell back within the normal range an hour after ingesting glucose.
- Person B, who has type 1 diabetes, was unable to regulate their blood glucose and have it fall back to normal levels. As insulin was not released, the blood glucose level in person B is still very high two hours after ingesting glucose.

FIGURE 4.66 Blood glucose levels after ingestion of a standard quantity of glucose

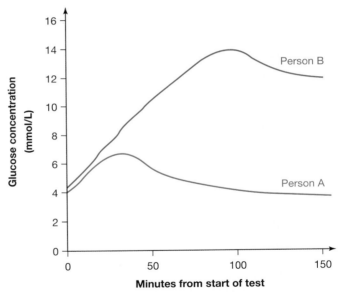

Glucose is also present in the urine of an affected person. This idea has been understood since ancient times, when the Chinese recognised people as having diabetes by the fact that their urine attracted ants due to the high glucose levels.

Why does this increase of glucose in blood urine occur?
- Normally, any glucose that enters the kidney tubules from the bloodstream is reabsorbed back into the blood through a process of active transport.
- However, the carrier proteins involved in returning glucose from the fluid in the kidney tubules back into the bloodstream are not able to deal with the high level of glucose filtered from the blood of a person with diabetes.
- Therefore, much of the glucose from the bloodstream is lost in the urine instead of being reabsorbed.

Symptoms of type 1 diabetes

TABLE 4.14 Symptoms of diabetes that individuals may experience

Symptom	Details
Increased thirst and frequent urination	These symptoms arise in response to the higher than normal levels of glucose in the blood. The body increases its output of urine in an attempt to remove the excess glucose from the blood. The large volume of urine excreted means that the loss of water from the body via this route is excessive, and this in turn stimulates thirst.
Low energy levels, fatigue Extreme hunger and possible weight loss	These symptoms arise in diabetes sufferers because the body cells are starved of glucose and hence are starved of energy. Without an adequate supply of glucose in body cells, they cannot generate sufficient ATP through cellular respiration to meet their needs. In addition, the loss of glucose from the body in urine can lead to weight loss.
Blurred vision	Untreated elevated glucose blood levels can, over time, cause damage to the capillaries in various organs and tissues, including the eyes, kidneys and nerves. For example, blurred vision is a consequence of damage to capillaries in the retina of the eye.
Diabetic ketoacidosis	This symptom arises in type 1 diabetes because the body cells cannot use their normal fuel, glucose, for energy production. Instead, the body cells metabolise fat for energy production. The breakdown products of fat metabolism include acids called **ketones**. If ketones reach high concentrations in the blood, as may happen with persons whose diabetes is not under control, the blood becomes more acidic, producing a serious condition known as ketoacidosis. One of the symptoms of diabetic ketoacidosis is a strong fruity smell to the breath.

Diagnosis of diabetes

Diabetes is typically recognised by a distinctive set of symptoms (see table 4.14) and is confirmed by blood tests:

ketones acid; breakdown product of fat metabolism

- Fasting plasma glucose test: This test is done after a person has fasted overnight. The blood glucose levels in a healthy person typically fall in the range of 3.9–5.5 mmol/L. If this value exceeds 7.0 mmol/L, a diagnosis of diabetes is identified.
- Oral glucose tolerance test (OGTT): In this test, a subject who has not eaten for 12 hours is given a standard dose of 75 mg of glucose in solution. Blood samples are taken from the person immediately before drinking the glucose solution, and then at 30-minute intervals up to 120 minutes.

Treatment of diabetes

There is currently no cure for type 1 diabetes, similar to many other chronic autoimmune diseases.

People with type 1 diabetes are treated with insulin replacement, as they are unable to produce any insulin of their own. This typically occurs by injection or through an insulin pump (see figure 4.67). There are many types of insulin that may be used. They often differ in how long they take to act and the time they are active for. Some examples of different types of insulin are shown in figure 4.68.

The insulin used for the treatment of type 1 diabetes has changed over time. Before the 1980s, insulin used was from the pancreases of cattle and pigs obtained from abattoirs. However, insulin from pigs and cattle is not a perfect match to human insulin. There is also the risk of allergic reactions occurring in some people.

Now, recombinant human insulin, made through a process known as gene cloning, is used instead. The development of recombinant human insulin was a major breakthrough in the treatment of type 1 diabetes.

The amount of insulin injected must be carefully controlled. People with type 1 diabetes need to constantly monitor their glucose levels. This used to be done only through finger-prick tests, in which blood from the finger is drawn and tested. However, improved technologies to easily detect glucose levels are becoming increasingly available. Some of these that have been approved in Australia include smart phone apps that can sync up to sensors that can monitor blood glucose levels (figure 4.69). AI technologies are also constantly being explored to make the daily lives of those with insulin easier.

If a person inadvertently injects too much insulin, a dangerous condition of too little glucose in the blood occurs, termed **hypoglycaemia**. Hypoglycaemia is explored further in section 4.7.3.

hypoglycaemia glucose levels in the blood drop below normal

FIGURE 4.67 a. A portable insulin pump being used by a young boy with diabetes injects insulin slowly and continuously into the bloodstream. **b.** An insulin pen that can be used to administer insulin

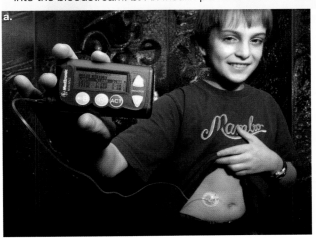
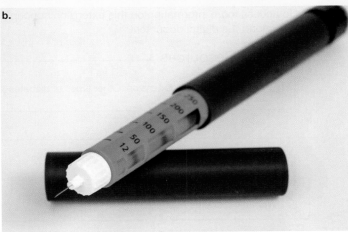

FIGURE 4.68 Time course of action of various forms of recombinant insulin on blood glucose levels

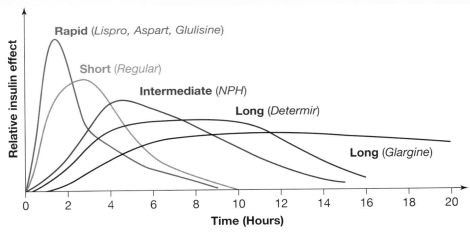

FIGURE 4.69 Monitoring blood glucose: **a.** using a finger-prick test; **b.** sensors with results shown through a phone app

> **EXTENSION: Other types of diabetes**
>
> Type 1 diabetes is an autoimmune disease. However, there are two other types of diabetes that affect the homeostatic regulation of blood glucose in humans: type 2 diabetes and gestational diabetes. Type 2 diabetes is mostly caused by lifestyle factors, whereas gestational diabetes develops during pregnancy.
>
> To access more information on this extension concept, please download the digital document.
>
> 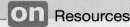 Resources
>
> **Digital document** Extension: Other types of diabetes (doc-35889)

FIGURE 4.70 A pregnant woman undergoing blood glucose testing

 Resources

Weblink Diabetes Australia

4.7.3 Hypoglycaemia

Type 1 diabetes leads to an increase in blood glucose levels (hyperglycaemia) due to insulin not being produced.

Hypoglycaemia (*hypo* = under or below; *glykys* = sweet; *haima* = blood), on the other hand, is a condition in which there is too little glucose in the blood.

The most common time hypoglycaemia occurs is during diabetes treatment. An important part in the treatment of type 1 diabetes is through the injection of insulin. However, the amount of insulin needs to be carefully monitored. If too much insulin is injected, an excessive amount of glucose leaves the bloodstream, leading to the effects of hypoglycaemia (commonly referred to as 'a hypo').

Hypoglycaemia may lead to fainting or even coma. It is treated by giving the affected person a source of glucose, such as glucose tablets, honey or a sweet. This is why individuals with type 1 diabetes often carry a small bag of jellybeans with them — to save their lives in the case of hypoglycaemia. Glucose in this form allows for a quick increase in their blood glucose levels. (Other sources of glucose such as carbohydrates are too slow in giving this glucose spike.)

It is important to recognise the signs of hypoglycaemia, especially in those with type 1 diabetes.

FIGURE 4.71 Jellybeans, which can often by purchased from the chemist, are important in the treatment of hypoglycaemia.

Signs of hypoglycaemia include:
- excessive hunger
- trembling
- sweating
- nausea.

If you notice someone with these symptoms (particularly if they have type 1 diabetes), it is important to encourage them to have a source of glucose, to prevent further symptoms of hypoglycaemia developing.

Another treatment for severe hypoglycaemia is through the use of glucagon. In type 1 diabetes, due to the issues with producing insulin, natural glucagon does not work as expected, as the use of insulin and glucagon in the negative feedback loops of glucose regulation is interrupted. Therefore, administered glucagon can counteract the effects of hypoglycaemia by triggering the liver to break down stored glycogen and release this as glucose into the bloodstream. Glucagon medication is usually only used when an individual is not responsive or unconscious, or in cases of severe hypoglycaemia.

Although hypoglycaemia is more common in diabetic individuals, it can also occur in other individuals. Although this is rare, there are some triggers or risk factors that can cause this, such as:
- the production of too much insulin following a meal: this causes blood glucose to drop more than expected
- drinking too much alcohol: this can cause damage to liver functionality. Usually, when blood glucose decreases, glucagon from the pancreas acts on the liver to break down stored glycogen into glucose. If the liver is damaged, it can be difficult for the liver to break down glycogen and release glucose.
- hepatitis: hepatitis is inflammation of the liver. Similar to drinking too much alcohol, this can prevent the liver from functioning correctly, and not release glucose from glycogen.
- some medications: different medications can cause side effects such as hypoglycaemia, such as medication for malaria.
- anorexia: blood glucose levels can drop extremely low due to the lack of food consumption
- tumours and pancreatic cancer: tumours in the pancreas and pancreatic cancer can affect the production of insulin and glucagon. More particularly, they can cause the overproduction of insulin. Too much insulin can cause blood glucose levels to fall, resulting in hypoglycaemia.

tlvd-1752

SAMPLE PROBLEM 6 Exploring type 1 diabetes and hypoglycaemia

A 17-year-old male with type 1 diabetes has his lunch — a sandwich and a small block of chocolate. His friend, who does not have diabetes, consumes the exact same lunch.
a. What differences would you exact to see in the blood glucose and blood insulin concentrations between the two individuals? (2 marks)
b. The diabetic student monitors his blood glucose level and takes insulin. However, soon after taken insulin he becomes very dizzy and faints. Explain what may have caused these symptoms and what this is called. (2 marks)
c. Describe two possible treatments for this scenario. (2 marks)

THINK	WRITE
a. 1. Carefully read the question. You need to outline both blood insulin and blood glucose. Consider the blood insulin levels in the diabetic male, who cannot produce insulin, and his friend, who can produce insulin.	The friend without diabetes would have an increase in insulin production and blood insulin in response to the meal. The diabetic male cannot produce insulin as they have non-functioning beta cells (1 mark).
2. Consider the blood glucose levels in the diabetic male, who is unable to regulate blood glucose due to malfunctions in homeostatic mechanisms, and his friend, who can regulate their glucose levels.	The non-diabetic friend would have an initial increase in blood glucose before it returns to normal levels. The diabetic individual would have a continual rise in blood glucose levels (1 mark).
b. 1. Identify what the question is asking you to do — explain a set of symptoms and identify the name. It is easier to identify the name first to help guide your answer.	The diabetic student is in a state of hypoglycaemia (1 mark).

2. Read the question carefully to identify the situation and the symptoms — the individual had taken insulin, which acts to lower blood glucose, but they soon felt dizzy and faint.

It is likely that the student has injected too much insulin, causing his blood glucose to drop too low, leading to symptoms such as dizziness and causing the student to faint (1 mark).

c. The question is asking for two treatments for two marks.

- Have the individual consume something high in glucose, such as lollies or a fruit juice (1 mark).
- If the individual is unresponsive or the situation become worse, glucagon infusions can be used to increase blood glucose levels (1 mark).

4.7.4 Hyperthyroidism

The thyroid gland produces the hormone thyroxine in response to the release of thyroid-stimulating hormone (TSH) from the pituitary gland. Thyroxine directly affects the metabolic rate and the cardiovascular function of a person. If thyroxine levels in the blood are too high, the metabolic rate increases, but if they are too low, the metabolic rate falls. Severe metabolic disorders arise when the thyroid gland is overactive, producing excessive amounts of thyroxine, or when it is underactive, producing insufficient amounts of the hormone.

An overactive thyroid gland results in an increase in the basal metabolic rate of an affected person — a condition known as **hyperthyroidism**.

> In hyperthyroidism, excessive production of thyroid hormones drives up the basal metabolic rate, affecting many functions.

A person with hyperthyroidism will show many symptoms, including:
- an increase in the resting heart rate
- an elevated body temperature
- an increase in appetite
- unexplained weight loss
- sensitivity to and sweating in warm conditions
- relative insensitivity to cold conditions.

Hyperthyroidism may be the result of the growth of nodules on the thyroid or local inflammation. Genetic factors can also predispose a person to hyperthyroidism.

One disease that causes hyperthyroidism is Graves' disease. Graves' disease is an autoimmune disease that is more commonly seen in women under the age of 40. In Graves' disease, the immune cells target cells in the thyroid gland. These antibodies act like the pituitary hormone, causing the thyroid to produce excess thyroxine rather than producing the amount that allows for homeostatic mechanisms.

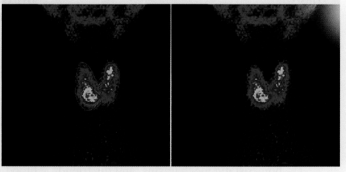

FIGURE 4.72 Hyperthyroidism shown in a gamma camera scan, in which the uptake of iodine by the thyroid is observed. Some parts of the thyroid are much brighter, indicating an overactive thyroid.

hyperthyroidism condition in which there is an overabundance in thyroid hormone production

Management of hyperthyroidism may involve surgical removal of part of the thyroid gland or the administration of anti-thyroid medication that interferes with the ability of the thyroid gland to take up iodine from the blood. Iodine is an essential component of thyroxine (figure 4.74). Carbimazole is one such anti-thyroid medication. Carbimazole is converted to an active form in the body that prevents the precursor of thyroxine from binding the iodine atoms that are needed to produce active thyroxine.

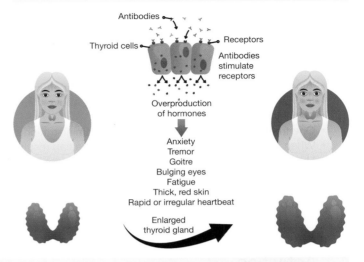

FIGURE 4.73 Graves' disease leads to excessive thyroxine being produced and an enlarged thyroid.

FIGURE 4.74 a. Molecular structure of the hormone thyroxine. Note the four iodine atoms that form part of its structure. b. Carbimazole, an anti-thyroid medicine

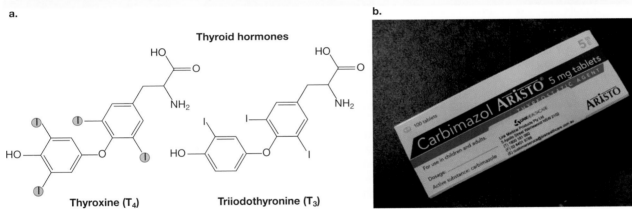

EXTENSION: Hypothyroidism

As opposed to an overactive thyroid as seen in hyperthyroidism, hypothyroidism can occur due to an underactive thyroid. It can also occur in people when their long-term diet is chronically deficient in iodine. Symptoms include an enlarged thyroid gland (goitre).

In hypothyroidism, the metabolic rate of the body falls below normal. It can be treated with hormone replacement by administration of thyroxine tablets.

To access more information on this extension concept, please download the digital document.

 Resources

 Digital document Extension: Hypothyroidism (doc-35890)

FIGURE 4.75 A severe case of goitre in a woman. The goitre most probably resulted from an iodine-deficient diet over the course of the woman's life.

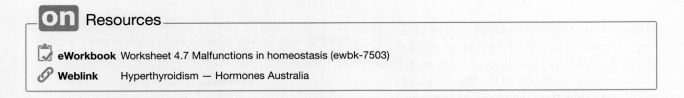

KEY IDEAS

- Type 1 diabetes in an autoimmune disease that affects the beta cells in the pancreas.
- In type 1 diabetes, blood glucose levels are above the normal range (hyperglycaemia) due to the individual's inability to produce insulin.
- One treatment for type 1 diabetes is the administration of insulin to reduce blood glucose levels.
- When blood glucose levels drop too low, this can result in hypoglycaemia.
- Hypoglycaemia is often the result of excessive insulin production.
- Hyperthyroidism causes an elevation in basal metabolic rate that affects the function of many body organs.
- Hyperthyroidism results from an overactive thyroid, leading to an increased production of the thyroid hormones T3 and T4 (thyroxine).

4.7 Activities

To answer questions online and to receive **immediate feedback** and **sample responses** for every question, go to your learnON title at **www.jacplus.com.au**. A **downloadable solutions** file is also available in the resources tab.

| 4.7 Quick quiz | 4.7 Exercise | 4.7 Exam questions |

4.7 Exercise

1. Identify whether the following statement is true or false: 'Treatment of type 1 diabetes uses hormone replacement' Justify your response.
2. Why might a person with diabetes keep a sweet in their pocket?
3. Describe how the treatment of type 1 diabetes can lead to hypoglycaemia.
4. Give three symptoms of hyperthyroidism, and explain why these symptoms might occur.
5. Two individuals each ingested 60 g of glucose in solution. Their blood plasma glucose levels were measured over the next few hours and the results are shown below.

    TABLE Blood plasma glucose levels over time

Individual	Time after ingestion				
	0 mins	20 mins	40 mins	60 mins	80 mins
Individual 1	4.9 mmol/L	7.8 mmol/L	12.2 mmol/L	13.1 mmol/L	15.6 mmol/L
Individual 2	4.9 mmol/L	7.8 mmol/L	9.1 mmol/L	7.4 mmol/L	6.1 mmol/L

    a. Draw a clear graph to show this data.
    b. Describe why the blood glucose levels did not decrease in individual 1.
    c. Both individuals are given an injection of insulin at 90 minutes. What would happen to their blood glucose concentration?
6. Insulin for the treatment of type 1 diabetes comes in different forms. These are shown in figure 4.68.
    a. Explain an advantage that a rapid-acting insulin such as Lispro might have over a longer-acting insulin such as glargine.
    b. Explain a disadvantage that a rapid-acting insulin such as Lispro might have over a longer-acting insulin such as glargine.
7. Hyperthyroidism is commonly caused by diseases such as Graves' disease, an autoimmune disease that attacks the thyroid and causes it to be overactive. Suggest a possible treatment that may be used to combat the hyperthyroidism caused by Graves' disease.

4.7 Exam questions

Question 1 (1 mark)

MC Which of the following diseases is associated with hyperthyroidism?
A. Diabetes mellitus
B. Graves' disease
C. Diabetes insipidus
D. Nephritis

Question 2 (1 mark)

MC Diabetes mellitus may occur when
A. the pancreas fails to produce enough insulin.
B. your brain is not producing enough antidiuretic hormone.
C. the pancreas produces high levels of glucagon.
D. your body cells increase uptake of glucose from their internal environment.

Question 3 (1 mark)

MC Goitre is an abnormal enlargement of the thyroid gland. This is likely to affect homeostatic mechanisms that help regulate
A. water balance.
B. blood glucose.
C. metabolic rate.
D. oxygen concentration.

Question 4 (3 marks)

Source: VCAA 2003 Biology Exam 1, Section B, Q5a,b

The table below shows the mean levels of glucose and insulin in two groups of people sampled one hour after the ingestion of 75 g of glucose. One of the experimental groups consisted of people with diabetes; the other acted as a control.

	group X		group Y	
time after glucose ingestion	0 min	60 min	0 min	60 min
plasma glucose (mmol/L)	5.3	13.0	5.3	7.8
plasma insulin (mmol/L)	68	66	69	380

a. What is the purpose of the control group? **1 mark**
b. Using the above data, give two reasons to explain which group (X or Y) included the people with diabetes. **2 marks**

Question 5 (3 marks)

Source: Adapted from VCAA 2004 Biology Exam 1, Section B, Q6b, 6c

Regulation of blood-glucose concentration relies upon functioning β cells in the Islets of Langerhans in the pancreas. These cells produce a hormone. Diabetes mellitus develops when β cells are destroyed.

a. Currently, individuals with diabetes mellitus are treated with injections of the missing hormone. These injections are usually given several times each day.
Recently, a pump capable of delivering the hormone continuously in response to changing blood-glucose concentration has been developed. The pump turns on and off automatically when the blood glucose changes.
What would be the signal for the pump to turn on and begin delivering the hormone? **1 mark**

b. Episodes of abnormally low concentration of blood glucose are reported to be less frequent in individuals when an automatic hormone pump is used rather than hormone injections.
Explain why this would be the case. **2 marks**

More exam questions are available in your learnON title.

4.8 Review
4.8.1 Topic summary

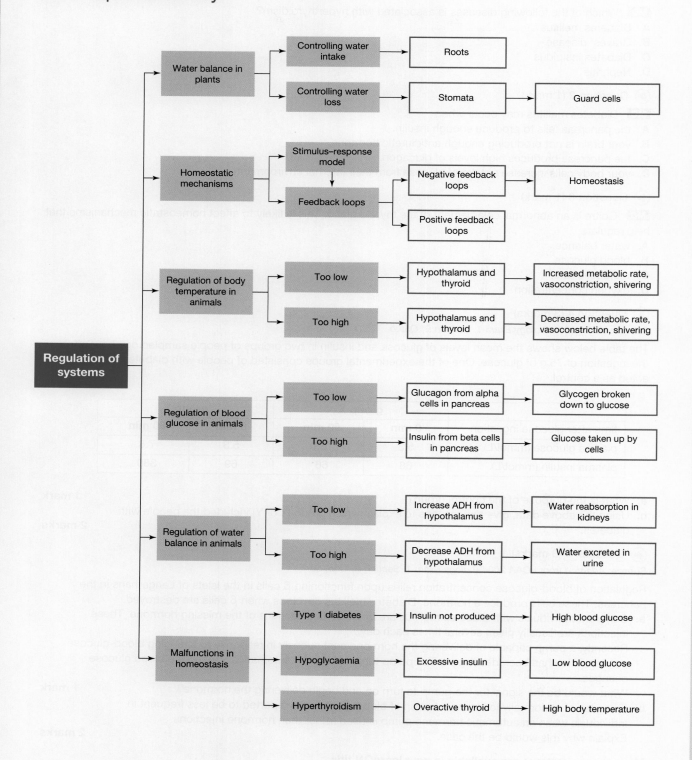

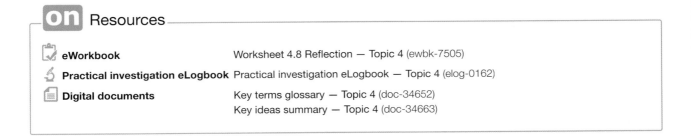

4.8 Exercises

To answer questions online and to receive **immediate feedback** and **sample responses** for every question, go to your learnON title at **www.jacplus.com.au**. A **downloadable solutions** file is also available in the resources tab.

4.8 Exercise 1: Review questions

1. Explain the difference between the members of the following pairs.
   a. Ectotherm and endotherm
   b. Vasoconstriction and vasodilation
   c. Thermoregulation and osmoregulation
   d. Core temperature and peripheral temperature

2. Give an explanation in biological terms for each of the following observations.
   a. High temperatures in a humid climate are more uncomfortable than the same temperatures in a dry climate.
   b. More heat is lost from the human body when the air surrounding the body is cool and moving than when the air at the same temperature is still.
   c. The cooling effect of sweat is less when the surrounding air is humid than when it is dry.
   d. Immersion in cold water can produce greater and faster heat loss than exposure to air at the same temperature.

3. The stimulus–response model is an important homeostatic mechanism.
   a. Explain how the stimulus–response model allows for homeostasis.
   b. To maintain homeostasis, would the feedback loop need to be negative or positive? Justify your response.
   c. Describe what is meant by each of the following terms.
   d. Draw a clear diagram showing the relationship between the terms you described in part **c**.
      i. Response
      ii. Control centre
      iii. Effector
      iv. Stimulus
      v. Receptors

4. Draw a single flowchart showing how water balance occurs in animals. On your flowchart, you should show what happens when water is both low and high.

5. An 28-year-old female had been feeling unwell, with swollen glands, an increased body temperature, unexplained weight loss, an increased appetite and a loss of sensitivity to the cold. She told the doctor that no matter what she tried, she always seemed to have an increased body temperature.
   a. What do you think the doctor may diagnose this female with?
   b. Why would he have made this decision?
   c. What hormone would likely be elevated to cause these symptoms in the 28-year old female?
   d. Outline how homeostatic mechanisms failed in this individual.

6. When the walls of a small blood vessel are damaged, platelets in the blood cling to the damaged site and start clot formation. These platelets release chemicals that attract more platelets to the site of the damage, and this process continues until a blood clot is formed. What type of feedback system is this? Briefly explain your decision.

7. A person has fallen into the cold, still water of a lake and will not be rescued immediately. In these circumstances, what might the person do to avoid loss of body heat while waiting for rescue? (The person is not in danger of drowning as he is wearing a life jacket.)
    - Student A suggested that the person should keep active by continually swimming around.
    - Student B suggested that the person should stay as still as possible and keep their legs close together and arms close to the body.

   Carefully consider the two alternatives and decide, giving reasons, which is likely to be the better strategy for minimising heat loss.

8. The body fluid of some animals, such as octopuses, squids, sharks and rays, has the same osmotic pressure as sea water. Identify a possible advantage of this situation in terms of water balance.

9. An experiment was carried out on mice to determine whether the pituitary gland controlled growth. Ninety mice were divided into three equal groups. Treatments and results are shown in the following table.

Treatment	Group A Pituitary gland removed	Group B Pituitary gland removed and daily injections of pituitary gland hormone given	Group C No treatment
Average mass at start	218 g	221 g	214 g
Average mass after one month	200 g	530 g	527 g

   a. Explain which you consider to be the control group in this experiment.
   b. What hypothesis does the data support? Explain your answer.

10. Bony fish (all fish excluding sharks and rays) use homeostatic mechanisms to maintain their water balance. Water balance also involves maintaining a balance of the salts in body fluids. Fish that live in fresh water and fish that live in sea water face different challenges in achieving a water balance.
    Here are some facts:
    - The scale-covered skin of most fish is relatively impermeable to water and salts.
    - However, both freshwater and saltwater (marine) fish must have permeable surfaces across which oxygen can be taken into the body and carbon dioxide can be excreted from the body. These permeable surfaces are their gill surfaces.
    - The gill surfaces of both kinds of fish allow not only the passage of oxygen and carbon dioxide, but also the movement of water and salts.
    - The body fluids of freshwater fish have a higher solute concentration than fresh water, so are hypertonic to the water in which freshwater fish live.

- In freshwater fish, water tends to move into the body across the gill surfaces and salts tend to be lost from the body via the same surfaces.
- The body fluids of marine fish have a lower solute concentration than sea water, so are hypotonic to the water in which marine fish live.
- In marine fish, water tends to be lost from the body across the gill surfaces and salts tend to be gained by the body via the same surfaces.

Complete the table below by placing the following entries into the correct cells. Your final result should show some of the homeostatic mechanisms involved in maintaining water and salt balance in these fish.

- Does not drink
- Produces large volumes of urine
- Drinks large amounts of water
- Takes in salts across gills
- Produces very small volumes of urine
- Secrete salts out across gills

Variable	Freshwater fish	Saltwater fish
Volume of urine produced		
Volume of water drunk		
Salt movement across gills		

4.8 Exercise 2: Exam questions

Resources

Teacher-led videos Teacher-led videos for every exam question

Section A — Multiple choice questions

All correct answers are worth 1 mark each; an incorrect answer is worth 0.

Question 1

Source: VCAA 2006 Biology Exam 1, Section A, Q16

The following diagram is a summary of a homeostatic mechanism for compound X.

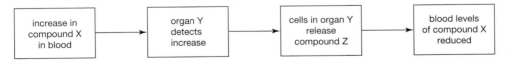

In organ Y
- A. the cells must act as exocrine glands.
- B. there must be sensors for compound X levels.
- C. compound Z must act to increase blood levels of compound X.
- D. compound X sensors and compound X effectors are part of the nervous system.

▶ **Question 2**

The hormone responsible for the control of water balance in animals is

A. thyroxine.
B. ADH.
C. FSH.
D. insulin.

▶ **Question 3**

When a plant is trying to prevent water loss

A. guard cells become flaccid and close stomata.
B. guard cells become turgid and close stomata.
C. guard cells become flaccid and open stomata.
D. guard cells become turgid and open stomata.

▶ **Question 4**

Insulin is produced by

A. alpha cells in the pancreas.
B. beta cells in the pancreas.
C. the hypothalamus.
D. the pituitary gland.

▶ **Question 5**

Which of the following is NOT an example of a negative feedback loop?

A. When body temperature increases, homeostatic mechanisms bring it back to normal.
B. During childbirth, the baby pushing down onto the cervix leads to more oxytocin being released.
C. When ion levels in the blood are high, the kidney produces highly concentrated urine.
D. Upon eating a meal, insulin is produced by the pancreas.

▶ **Question 6**

In hyperthyroidism

A. individuals have a lower than normal body temperature.
B. the body does not produce thyroxine.
C. individuals are more susceptible to the cold.
D. individuals have a higher than normal heart rate.

▶ **Question 7**

Source: VCAA 2008 Biology Exam 1, Section A, Q23

In mammals the parathyroid gland secretes parathyroid hormone (PTH). PTH is involved in regulating the concentration of calcium in blood plasma. Parathyroid hormone increases the amount of calcium in plasma by causing calcium to move from bone to the plasma, and by assisting the uptake of calcium from the alimentary canal. PTH also stimulates the kidney to activate vitamin D.

The concentration of calcium in plasma acts directly, in negative feedback, to regulate the output of parathyroid hormone. From this information it would be expected that

A. increased production of PTH results in reduction of vitamin D activation.
B. reduced production of PTH results in increased calcium in the faeces.
C. sustained overproduction of PTH results in strengthened bones.
D. high levels of blood calcium stimulate release of PTH.

Question 8

Source: VCAA 2003 Biology Exam 1, Section A, Q14

The following diagram shows the regulation of thyroxine secretion.

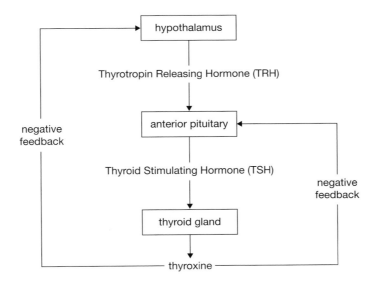

A person who lacks sufficient iodine in the diet is unable to manufacture thyroxine. This is because iodine is required by the thyroid gland to manufacture thyroxine.

It is reasonable to conclude that when a person lacks sufficient iodine in the diet

A. TSH production would increase.
B. TRH production would decrease.
C. the negative feedbacks would cease to operate.
D. the metabolic rate of cells in the body would increase.

Question 9

The hypothalamus produces antidiuretic hormone (ADH), which is stored in the pituitary gland. ADH released by the pituitary acts on the tubules in the kidney to absorb more water. It would be expected that less ADH would be released:

A. on a hot day.
B. during exercise.
C. after drinking lots of water.
D. when high ions levels are present in the bloodstream.

Question 10

An individual with type 1 diabetes injects insulin after a meal. However, they inject too much insulin, leading to a state of hypoglycaemia. It is important that they:

A. immediately consume glucose.
B. administer more insulin to restore glucose levels to normal.
C. eat a meal high in protein.
D. wait two hours as homeostatic mechanisms will take over and increase blood glucose levels.

Section B — Short answer questions

Question 11 (7 marks)

A plant in the desert has very minimal access to water.

a. Explain how the structure of the roots allows for water balance to be achieved in vascular plants. **2 marks**
b. How is water obtained through the roots able to be supplied through the entire plant? **1 mark**
c. What is the main process by which plants lose water? **1 mark**
d. One of the main ways that plants regulate water is through the use of stomata. Describe how stomata are able to be opened or closed in different conditions, and draw a clear diagram to show this. **3 marks**

Question 12 (5 marks)

Source: VCAA 2012 Biology Exam 1, Section B, Q6

a. Homeostasis is essential for the survival of any organism.
   i. What is homeostasis? **1 mark**
   ii. State one variable, other than body temperature and blood calcium levels, that is under homeostatic control in a mammal. Explain why the homeostatic control of this variable is essential for survival. **2 marks**
b. The flow chart below summarises a sequence of events occurring inside a rod cell in the retina of the human eye.

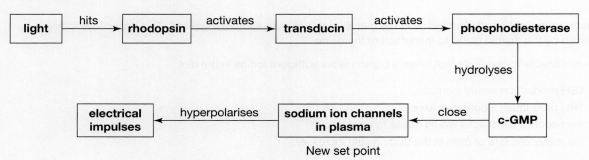

This sequence of events is part of a stimulus-response system.

i. Explain why a rod cell is regarded as a receptor in this system. **1 mark**
ii. Many of a human's regulating systems are based on negative feedback.
Does the series of events taking place in a rod cell form part of a negative feedback system?
Explain your answer. **1 mark**

Question 13 (10 marks)

The regulation of blood glucose is important for the survival of animals. Glucose allows for our cells to produce energy to sustain various processes that enable our survival.

a. What is the name of the process that allows for the regulation of blood glucose? **1 mark**
b. The two hormones involved in blood glucose regulation are the antagonistic hormones insulin and glucagon. Describe how insulin and glucagon work alongside each other to control blood glucose levels. **2 marks**

An individual consumes a meal at point A, as shown in the graph below. The change in blood glucose levels over a period of time is shown.

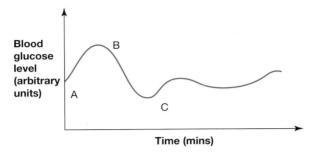

c. Explain how the blood glucose levels decreases at point B. **2 marks**

d. The individual did not consume another meal over this time period. Despite that, their blood glucose level increases at point C. Explain what has caused this. **2 marks**

Type 1 diabetes is a common autoimmune disease, in which the immune cells attack the beta cells in the pancreas.

e. What is the role of beta cells in healthy individuals? **2 marks**

f. On a copy of the graph shown above, outline the expected change in blood glucose levels of an individual with Type 1 diabetes. **1 mark**

▶ Question 14 (11 marks)

a. The body temperature of a human is kept constant at 37 °C. Why is it important to keep the body temperature constant at 37 °C? **1 mark**

b. Regulation of body temperature involves negative feedback. Explain what is meant by negative feedback. **2 marks**

c. The temperature of an individual was recorded over a 20-minute period and the data was displayed in a graph.

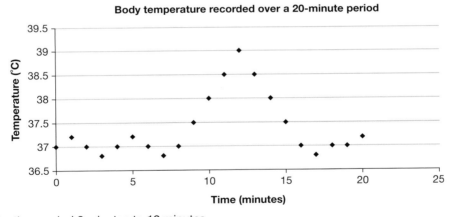

Consider the time period 8 minutes to 12 minutes.

  i. What change can be seen in the body temperature of the individual during this time? **1 mark**

  ii. What may the individual be doing during this time to cause the change in temperature seen in this 4-minute time period? **1 mark**

  iii. Name two physiological responses in the individual that may be occurring after 12 minutes to bring the body temperature back to 37 °C by 16 minutes. **2 marks**

d. Both the nervous system and endocrine system are involved in the regulation of body temperature. Describe the role played by the endocrine system. **4 marks**

Question 15 (7 marks)

A group of students had no food to eat over an 8-hour period.

Each of their blood glucose levels were measured and recorded. The data is shown in the table below.

Student	Blood glucose level (mmol/L)
1	5.6
2	5.0
3	5.4
4	5.5
5	8.0

If an individual has not eaten for this length of time, a blood glucose level of less than 5.6 mmol/L is considered normal.

Consider the hormones associated with the maintenance of blood glucose levels.

a. Which hormone would you expect to be at a relatively high concentration when the students' blood glucose levels were measured? Explain your choice. **3 marks**

b. i. Name the disease that student 5 may be suffering from. **1 mark**
   ii. What medication could the student be prescribed to help maintain blood glucose levels? **1 mark**
   iii. Explain how this medication works within the body to help maintain blood glucose levels. **2 marks**

4.8 Exercise 3: Biochallenge online only

Resources

eWorkbook   Biochallenge — Topic 4 (ewbk-8074)

Solutions   Solutions — Topic 4 (sol-0649)

teachon

Test maker
Create unique tests and exams from our extensive range of questions, including practice exam questions.
Access the assignments section in learnON to begin creating and assigning assessments to students.

Online Resources

Below is a full list of **rich resources** available online for this this topic. These resources are designed to bring ideas to life, to promote deep and lasting learning and to support the different learning needs of each individual.

eWorkbook

- 4.1 eWorkbook — Topic 4 (ewbk-3160)
- 4.2 Worksheet 4.1 Controlling water loss and uptake in plants (ewbk-7491)
- 4.3 Worksheet 4.2 Maintaining the balance (ewbk-7493)
  Worksheet 4.3 Positive and negative feedback loops (ewbk-7495)
- 4.4 Worksheet 4.4 Thermoregulation (ewbk-7497)
- 4.5 Worksheet 4.5 Glucose ups and downs (ewbk-7499)
- 4.6 Worksheet 4.6 The regulation of water (ewbk-7501)
- 4.7 Worksheet 4.7 Malfunctions in homeostasis (ewbk-7503)
- 4.8 Worksheet 4.8 Reflection — Topic 4 (ewbk-7505)
  Biochallenge — Topic 4 (ewbk-8074)

Solutions

- 4.8 Solutions — Topic 4 (sol-0649)

Practical investigation eLogbook

- 4.1 Practical investigation eLogbook — Topic 4 (elog-0162)
- 4.2 Investigation 4.1 Investigating transpiration in plants (elog-0808)
  Investigation 4.2 Distribution of stomata and guard cells (elog-0810)
- 4.4 Investigation 4.3 The regulation of body temperature (elog-0812)
- 4.5 Investigation 4.4 Testing glucose levels (elog-0814)

Digital documents

- 4.1 Key science skills — VCE Biology Units 1–4 (doc-34648)
  Key terms glossary — Topic 4 (doc-34652)
  Key ideas summary — Topic 4 (doc-34663)
- 4.2 Extension: Roots grow towards moist soil (doc-35882)
  Extension: Water potential — which way will water move? (doc-35883)
  Extension: ABA hormone and stomatal pore closing (doc-35884)
- 4.3 Background knowledge: The nervous system (doc-35885)
- 4.4 Extension: How do ectotherms 'regulate' temperature? (doc-35886)
  Case study: Hyperthermia and heat stroke — extreme heat can kill (doc-35881)
  Extension: What happens when you have a fever? (doc-35887)
- 4.5 Extension: Getting glucose into cells (doc-35888)
- 4.7 Extension: Other types of diabetes (doc-35889)
  Extension: Hypothyroidism (doc-35890)

Teacher-led videos

- Exam questions — Topic 4
- 4.2 Sample problem 1 Exploring the role of stomata and guard cells in water balance (tlvd-1747)
- 4.3 Sample problem 2 Comparing feedback loops (tlvd-1748)
- 4.4 Sample problem 3 Understanding body temperature regulation (tlvd-1749)
- 4.5 Sample problem 4 Understanding the regulation of blood glucose (tlvd-1750)
- 4.6 Sample problem 5 Understanding water balance in animals (tlvd-1751)
- 4.7 Sample problem 6 Exploring type 1 diabetes and hypoglycaemia (tlvd-1752)

Video eLessons

- 4.3 Maintaining a stable internal environment (eles-3884)
- 4.5 Blood glucose regulation (eles-4735)
- 4.6 Urine formation in the kidney (eles-2644)

Interactivities

- 4.4 Homeostasis and body temperature regulation (int-8081)
- 4.5 Homeostasis and blood glucose (int-8082)
- 4.6 Homeostasis and water balance (int-8083)

Weblinks

- 4.3 Sensory receptors
- 4.4 History of accidental hyperthermia
  Fevers
  Tragedies are a reminder of the fury of Australia's outback heat
- 4.7 Diabetes Australia
  Hyperthyroidism — Hormones Australia

Teacher resources

There are many resources available exclusively for teachers online.

To access these online resources, log on to **www.jacplus.com.au**

UNIT 1 | AREA OF STUDY 2 REVIEW

AREA OF STUDY 2 How do plant and animal systems function?

OUTCOME 2
Explain and compare how cells are specialised and organised in plants and animals, and analyse how specific systems in plants and animals are regulated.

PRACTICE EXAMINATION

STRUCTURE OF PRACTICE EXAMINATION		
Section	Number of questions	Number of marks
A	20	20
B	5	30
	Total	50

Duration: 50 minutes

Information:
- This practice examination consists of two parts. You must answer all question sections.
- Pens, pencils, highlighters, erasers, rulers and a scientific calculator are permitted.

SECTION A — Multiple choice questions

All correct answers are worth 1 mark each; an incorrect answer is worth 0.

1. What is an example of an organ in a plant?
   A. Flower
   B. Root hair
   C. Stomata
   D. Epidermis

2. The specialised tissue responsible for the movement of sucrose within a plant is the
   A. collenchyma.
   B. parenchyma.
   C. phloem.
   D. xylem.

3. The function of dermal tissue in the plant is to
   A. provide cells for photosynthesis.
   B. protect and minimise water loss.
   C. support tissue of elongating stems.
   D. transport water and dissolved minerals.

4. Water can continue to move up the stem of a vascular plant as a continuous column without breaking due to
   A. osmosis.
   B. diffusion.
   C. facilitated diffusion.
   D. cohesion of water molecules.

5. What is the function of connective tissue in mammals?
   A. To cover external and internal surfaces
   B. To contract and enable movement
   C. To provide structural support
   D. To detect changes in the internal environment

6. The small intestine in a human is classified as
   A. a tissue.
   B. an organ.
   C. a group of cells.
   D. the digestive system.

7. The liver is an accessory organ of the digestive system. What is the role of the liver in this system?
   A. To produce bile
   B. To produce enzymes
   C. To digest macromolecules
   D. To move food through the small intestine

8. What is the tissue responsible for the absorption of digested food in humans?
   A. Muscle tissue
   B. Nervous tissue
   C. Epithelial tissue
   D. Connective tissue

9. The diagram shows a hormone being carried in the blood inside a blood vessel. Three different cell types are shown outside of the blood vessel.

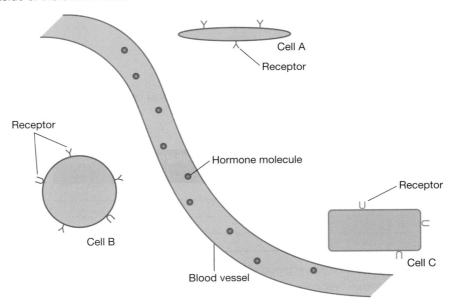

Which of the cells shown would most likely respond to the hormone?
A. Cell A
B. Cell B
C. Cell C
D. Both cell A and cell B

10. Consider the endocrine gland labelled X in the diagram provided.

The endocrine gland would
A. have a poor vascular supply.
B. release hormones from cells into the surrounding fluid tissue.
C. have ducts to carry hormones away from the gland.
D. play a major role in maintaining blood glucose levels.

11. Hormones may be grouped as either protein hormones or steroid hormones. A difference in the cell producing a protein hormone as opposed to a steroid hormone would be
A. the size of the nucleus.
B. the molecules in the plasma membrane.
C. more extensive rough endoplasmic reticulum.
D. a higher number of mitochondria.

12. A person was given a drink of glucose at time zero and then the level of glucose in their blood was recorded for the next 90 minutes. The results are shown in the table.

TABLE Blood glucose levels

Time (mins)	Blood glucose concentration (mg/dL)
0	70
30	115
60	100
90	85

What is the likely cause of the increase in blood glucose levels in the first 30 minutes?
A. The release of glucose from liver cells
B. Glucose being absorbed from the small intestine
C. Reduced levels of glucagon being released from the pancreas
D. Increased levels of insulin being released from the pancreas

13. When is insulin released into the blood?
    A. When sleeping
    B. After eight hours of fasting
    C. When blood glucose levels are high
    D. During periods of intense physical exercise

14. The alpha cells of the pancreas release which hormone?
    A. Insulin
    B. Glucagon
    C. Thyroxine
    D. Antidiuretic hormone

15. Consider a human liver cell. What is the effect of insulin on a liver cell?
    A. It increases the metabolic rate of the cell.
    B. It increases the release of glucose from the cell.
    C. It inhibits synthesis of fatty acids.
    D. It increases glycogen stores.

16. Hypoglycaemia can be experienced by a diabetic person after which of the following events?
    A. Eating a carbohydrate meal
    B. An insulin injection
    C. A glucagon injection
    D. Drinking a sports drink

17. Which of the following statements is true of type 1 diabetes?
    A. It is an autoimmune disease.
    B. It can be cured by taking medication.
    C. It occurs when the pancreas is producing too much insulin.
    D. It prevents a person from participating in regular exercise.

18. Birds do not produce urine but instead produce a semi-solid white paste. What is an advantage to birds of producing this semi-solid paste?
    A. It allows them to concentrate urea.
    B. It allows them to produce dilute urine.
    C. It removes the need for a kidney.
    D. It allows embryos to develop inside eggs.

19. Antidiuretic hormone (vasopressin) acts on cells within the tubules of the nephron to
    A. increase secretion of urea from the tubules.
    B. remove nitrogenous wastes.
    C. increase the return of water to the blood.
    D. cause dehydration.

20. Antidiuretic hormone is important in water regulation for humans. Another hormone that helps regulate water balance in humans is
    A. insulin.
    B. renin.
    C. thyroxine.
    D. thyroid-stimulating hormone.

SECTION B — Short answer questions

Question 21 (8 marks)

Consider the human excretory (urinary) system.

a. Describe the role played by the system. — 2 marks

b. i. Name two organs that are part of the human excretory system. — 2 marks

   ii. Consider one of the organs that you have named in part i. Would this organ be made from one tissue type? Justify your answer. — 2 marks

c. Organisms such as the single-celled *Paramecium caudatum* shown in the image do not have an excretory system.

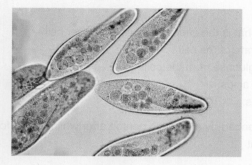

Explain why an excretory system is not necessary. — 2 marks

Question 22 (3 marks)

Consider this cross-section of a sunflower stem.

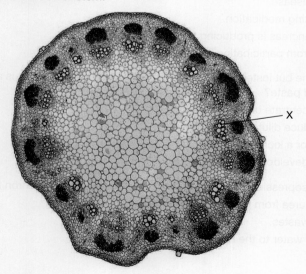

a. What is the name of the tissue labelled X? — 1 mark

b. Explain the importance of tissue X to the plant. — 2 marks

Question 23 (8 marks)

Consider the diagram showing plants growing up the branch of a tree.

a. Identify two ways that these plants may lose water to their environment.   **2 marks**

b. Describe a response that plants may produce to reduce water loss on a hot day.   **2 marks**

c. These plants grow in regions with high humidity and high rainfall, and they rarely wilt.
   i. What causes a plant to wilt?   **2 marks**
   ii. Explain why the plants that grow in these regions with high humidity and high rainfall rarely wilt.   **2 marks**

Question 24 (7 marks)

Consider the diagram of a stimulus–response model with feedback.

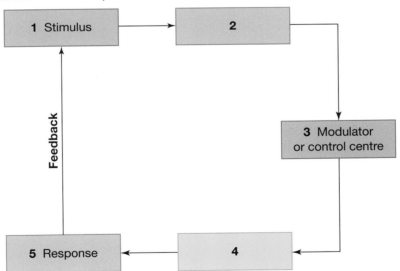

a. The labels from rectangles 2 and 4 are missing.
Name the components that should have been written in rectangles 2 and 4.   **2 marks**

b. Consider the regulation of human body temperature. If the stimulus was an increase in body temperature, name the modulator or control centre and then identify two suitable responses that would result in a lowering of body temperature.   **3 marks**

c. Explain the significance of the tongue to temperature regulation in dogs. **2 marks**

Question 25 (4 marks)

In humans, thyroxine is a hormone produced by the thyroid gland.
a. What is the role of thyroxine? **2 marks**
b. Explain how medication that interferes with the ability of the thyroid gland to take up iodine from the blood results in the reduction of symptoms in a person with hyperthyroidism. **2 marks**

END OF EXAMINATION

UNIT 1 | AREA OF STUDY 2

PRACTICE SCHOOL-ASSESSED COURSEWORK

ASSESSMENT TASK — A case study analysis on regulations of and malfunctions in homeostatic mechanisms

In this task, you will focus on the regulation of body temperature, blood glucose and water balance in animals by homeostatic mechanisms, including stimulus–response models, feedback loops and associated organ structures. You will also focus on malfunctions in homeostatic mechanisms: type 1 diabetes, hypoglycaemia and hyperthyroidism.
- This practice SAC comprises of 16 questions; you must complete ALL question sections.
- Pens, pencils, highlighters, erasers, rulers and a scientific calculator are permitted.
- Mobile phones and/or any other unauthorised electronic devices including wrist devices are NOT permitted.

Total time: 50 minutes (5 minutes reading time, 45 minutes writing time)
Total marks: 40 marks

PART A: HOMEOSTATIC CONTROL (7 marks)

Answer the following questions about homeostatic control.
1. Define homeostasis. **2 marks**
2. How are homeostatic mechanisms controlled? **2 marks**
3. Give three examples of things kept under homeostatic control. **3 marks**

PART B: BLOOD SUGAR CONTROL (13 marks)

Blood sugar is kept under tight control.
4. Using the stimulus–response model, outline how an increase in blood sugar would be dealt with in a healthy person. You may choose to draw a flow chart to answer. **6 marks**

A person has a blood glucose level of 80 mg/dL when they consume a meal high in carbohydrates. An hour after eating, their blood glucose level starts to rise, reaching a peak of 150 mg/dL 2.5 hours after the meal. Their blood glucose is lowest at 7.5 hours after the meal, at a level of 75 mg/dL, before returning to the original level after 8 hours.

5. On the following axis, show what you would expect the blood glucose levels of the person to look likefollowing the meal. Include title and axis labels where appropriate. **5 marks**

6. Provide an explanation for the blood glucose dip at 7.5 hours. **2 marks**

PART C: MALFUNCTIONS IN HOMEOSTATIC MECHANISMS (20 marks)

Diabetes is a group of diseases that result in problems with regulating the amount of sugar in the blood. In type 1 diabetes, insulin is not produced by the pancreas. This means that there is no signal for cells to take up glucose out of the blood.

7. Add a line to your graph in question 5 to show the blood glucose of someone with diabetes. **2 marks**

In a person with diabetes, the body has to deal with the high blood glucose level in other ways, rather than cell uptake and conversion to glucagon. One of these methods is to balance glucose and water.

8. Describe how an increase in water in the blood would impact the concentration of glucose. **1 mark**

Urination is another way to remove glucose. In the kidneys the fluid component of blood is pushed out of a series of capillaries, called a glomerulus, into a structure called the renal capsule of the nephron (see figure). From here, substances such as water, salts and glucose are reabsorbed by the blood. Any excesses are not reabsorbed and removed through urine.

Part of a nephron (glomerulus, renal capsule and renal tubule

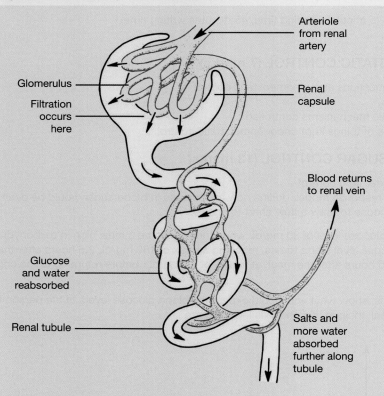

9. Name the process responsible for glucose reuptake in the renal tubule. **1 mark**
10. Name the process responsible for the reuptake of water in the renal tubule. **1 mark**
11. Both processes above are passive. Explain why they do not require energy. **2 marks**
12. A diabetic can become easily dehydrated. Explain why this symptom occurs as a result of high blood sugar. **4 marks**

Diabetes has many known complications. One of these is kidney nephropathy (kidney damage).

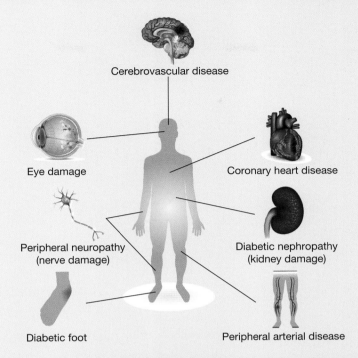

13. A decrease in kidney function can lead to a condition known as oedema. Symptoms include: bloating; swollen legs, feet and ankles; puffiness of the face, hips and abdomen; stiff joints; and weight fluctuations. Suggest a reason for this swelling. **3 marks**

The increased volume of blood places a strain on the cardiovascular system. This in turn can damage the vessels in both the lymphatic and vascular systems.

These vessels return fluid to the heart with the aid of muscular contractions. Valves, such as those seen in the normal vein in the figure, prevent the backflow of this fluid

14. Using the image, explain why damaged blood vessels such as varicose veins would contribute to oedema. **3 marks**

High blood pressure due to the increased volume of blood can also lead to problems with heart function, making the heart less efficient at pumping blood around the body.

15. What effect do you think this would have on a diabetic's ability to heal from a wound? **1 mark**

The treatment of a person with type 1 diabetes is injections of insulin. In a healthy person, insulin is released in an inactive form called a hexamer (a unit of six insulin molecules); however, diabetes patients inject themselves with insulin monomers (units of one molecule), which are immediately active in bringing about glucose uptake.

16. Suggest why non-sufferers have inactive forms of insulin in their blood. **2 marks**

A normal vein and a varicose vein

Normal vein

Varicose vein

Resources

 Digital document Unit 1 Area of Study 2 School-assessed coursework (doc-35085)

AREA OF STUDY 3 HOW DO SCIENTIFIC INVESTIGATIONS DEVELOP UNDERSTANDING OF HOW ORGANISMS REGULATE THEIR FUNCTIONS?

5 Scientific investigations

KEY KNOWLEDGE

online only

In this topic you will investigate:

Investigation design

- biological science concepts specific to the selected scientific investigation and their significance, including the definition of key terms
- scientific methodology relevant to the selected scientific investigation, selected from: classification and identification; controlled experiment; correlational study; fieldwork; modelling; product, process or system development; or simulation
- techniques of primary qualitative and quantitative data generation relevant to the investigation
- accuracy, precision, reproducibility, repeatability and validity of measurements in relation to the investigation
- health, safety and ethical guidelines relevant to the selected scientific investigation

Scientific evidence

- the distinction between an aim, a hypothesis, a model, a theory and a law
- observations and investigations that are consistent with, or challenge, current scientific models or theories
- the characteristics of primary data
- ways of organising, analysing and evaluating generated primary data to identify patterns and relationships including sources of error
- use of a logbook to authenticate generated primary data
- the limitations of investigation methodologies and methods, and of data generation and/or analysis

Science communication

- the conventions of scientific report writing including scientific terminology and representations, standard abbreviations and units of measurement
- ways of presenting key findings and implications of the selected scientific investigation.

Source: VCE Biology Study Design (2022–2026) extracts © VCAA; reproduced by permission.

This topic is available online at www.jacplus.com.au.

Online Resources

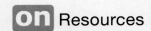

Below is a full list of **rich resources** available online for this this topic. These resources are designed to bring ideas to life, to promote deep and lasting learning and to support the different learning needs of each individual.

eWorkbook

- 5.1 eWorkbook — Topic 5 (ewbk-3162)
- 5.2 Worksheet 5.1 Controlled, dependent and independent variables (ewbk-7511)
  Worksheet 5.2 Writing an aim and forming a hypothesis (ewbk-7513)
- 5.6 Worksheet 5.3 Displaying primary data (ewbk-7515)
  Worksheet 5.4 Analysing primary data (ewbk-7517)
  Worksheet 5.5 Identifying errors (ewbk-7519)
- 5.7 Worksheet 5.6 Identifying strong and weak evidence (ewbk-7521)
- 5.10 Worksheet 5.7 Scientific posters (ewbk-7523)
- 5.11 Worksheet 5.8 Reflection – Topic 5 (ewbk-7525)
  Biochallenge — Topic 5 (ewbk-8075)

Solutions

- 5.11 Solutions — Topic 5 (sol-0650)

Digital documents

- 5.1 Key science skills — VCE Biology Units 1–4 (doc-34648)
  Key terms glossary — Topic 5 (doc-34653)
  Key ideas summary — Topic 5 (doc-34664)

Teacher-led videos

Exam questions — Topic 5

- 5.2 Sample problem 1 Identifying types of variables (tlvd-1820)
  Sample problem 2 Developing questions, making aims and formulating hypotheses (tlvd-1821)
- 5.5 Sample problem 3 Comparing accuracy and precision (tlvd-1822)
- 5.6 Sample problem 4 Drawing a graph (tlvd-1823)
- 5.10 Sample problem 5 Writing numbers in scientific notation (tlvd-1824)
  Sample problem 6 Converting units (tlvd-1825)

Video eLessons

- 5.2 SkillBuilder — Controlled, dependent and independent variables (eles-4156)
  SkillBuilder — Writing an aim and forming a hypothesis (eles-4155)

Interactivities

- 5.2 SkillBuilder — Controlled, dependent and independent variables (int-8090)
  SkillBuilder — Writing an aim and forming a hypothesis (int-8089)
- 5.6 Selecting graphs (int-7733)

Weblink

- 5.4 Animal ethics

Teacher resources

There are many resources available exclusively for teachers online

To access these online resources, log on to **www.jacplus.com.au**

UNIT 2 How does inheritance impact on diversity?

AREA OF STUDY 1

How is inheritance explained?

OUTCOME 1

Explain and compare chromosomes, genomes, genotypes and phenotypes, and analyse and predict patterns of inheritance.

- **6** From chromosomes to genomes 351
- **7** Genotypes and phenotypes 411
- **8** Patterns of inheritance 445

AREA OF STUDY 2

How do inherited adaptations impact on diversity?

OUTCOME 2

Analyse advantages and disadvantages of reproductive strategies, and evaluate how adaptations and interdependencies enhance survival of species within an ecosystem.

- **9** Reproductive strategies 509
- **10** Adaptations and diversity 561

AREA OF STUDY 3

How do humans use science to explore and communicate contemporary bioethical issues?

OUTCOME 3

Identify, analyse and evaluate a bioethical issue in genetics, reproductive science or adaptations beneficial for survival.

- **11** Exploring and communicating bioethical issues 655

Source: VCE Biology Study Design (2022–2026) extracts © VCAA; reproduced by permission.

AREA OF STUDY 1 HOW IS INHERITANCE EXPLAINED?

6 From chromosomes to genomes

KEY KNOWLEDGE

In this topic, you will investigate:

From chromosomes to genomes
- the distinction between genes, alleles and a genome
- the nature of a pair of homologous chromosomes carrying the same gene loci and the distinction between autosomes and sex chromosomes
- variability of chromosomes in terms of size and number in different organisms
- karyotypes as a visual representation that can be used to identify chromosome abnormalities
- the production of haploid gametes from diploid cells by meiosis, including the significance of crossing over of chromatids and independent assortment for genetic diversity

Source: VCE Biology Study Design (2022–2026) extracts © VCAA; reproduced by permission.

PRACTICAL WORK AND INVESTIGATIONS

Practical work is a central component of learning and assessment. Experiments and investigations, supported by a **practical investigation eLogbook** and **teacher-led videos**, are included in this topic to provide opportunities to undertake investigations and communicate findings.

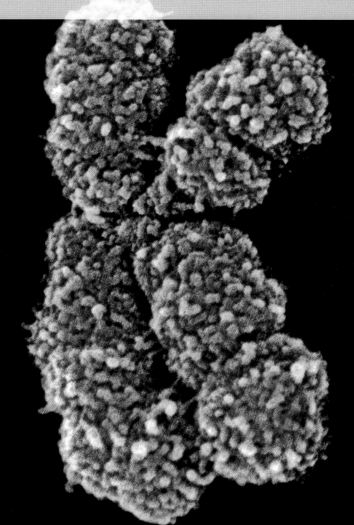

6.1 Overview

Numerous **videos** and **interactivities** are available just where you need them, at the point of learning, in your digital formats, learnON and eBookPLUS at **www.jacplus.com.au**.

6.1.1 Introduction

Do you share any characteristics with people around you at the moment? If so, what are they? Do you share any characteristics with your parents or grandparents? Do you look like any siblings you might have?

Answering these questions can be achieved through simple observation. Seeking an explanation to these questions, however, requires a knowledge of chromosomes, DNA, genes, alleles and genomes. An intimate knowledge of these concepts was not achieved until relatively recently and there is still much research being conducted by scientists today.

In this topic, you will learn about the importance of genes, their various forms and their contribution towards the genome of an organism. You will learn how variability in chromosome number, size and shape contribute towards genotypic and phenotypic differences. You will also understand how autosomes are different from sex chromosomes and the concept of homologous chromosomes. You will examine the importance of karyotyping in identifying genetic abnormalities and understand how meiosis produces variation in haploid daughter cells.

FIGURE 6.1 Chromosomes were discovered to carry genes — the basis of heredity — by American geneticist Thomas Hunt Morgan in 1910.

LEARNING SEQUENCE

6.1 Overview	352
6.2 BACKGROUND KNOWLEDGE The role of chromosomes as structures that package DNA	353
6.3 The distinction between genes, alleles and a genome	357
6.4 Homologous chromosomes, autosomes and sex chromosomes	369
6.5 Variability of chromosomes	375
6.6 Karyotypes	382
6.7 Production of haploid gametes	392
6.8 Review	402

on Resources

- **eWorkbook** — eWorkbook — Topic 6 (ewbk-3164)
- **Practical investigation eLogbook** — Practical investigation eLogbook — Topic 6 (elog-0166)
- **Digital documents** — Key science skills — VCE Biology Units 1–4 (doc-34648)
  - Key terms glossary — Topic 6 (doc-34654)
  - Key ideas summary — Topic 6 (doc-34665)

6.2 BACKGROUND KNOWLEDGE The role of chromosomes as structures that package DNA

BACKGROUND KNOWLEDGE

This subtopic will review concepts from the Biological sciences of Levels 9 and 10 to help you understand key knowledge points of the Study Design for Units 1 & 2.
- The relationship between DNA and chromosomes

6.2.1 Why are chromosomes important?

Located in the nucleus in almost all cells are **chromosomes**. These thread-like structures are composed of a single molecule of **deoxyribonucleic acid (DNA)** and associated proteins (figure 6.2). Chromosomes are not only found in human cells, but in almost all eukaryotic and prokaryotic cells. Their exact chemical composition, size and shape can differ between cells but their presence in so many cell types indicates their importance for life.

The question then is, why are chromosomes so important? To answer this we need to examine their structure more carefully. This structure differs slightly depending on whether the DNA is present in eukaryotic or prokaryotic cells.

chromosome a thread-like structure composed of DNA and protein

deoxyribonucleic acid (DNA) nucleic acid containing the four bases — adenine, guanine, cytosine and thymine — which forms the major component of chromosomes and contains coded genetic instructions

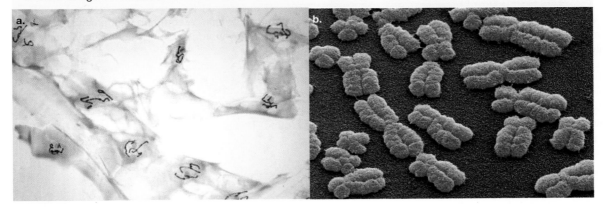

FIGURE 6.2 a. Stained chromosomes are visible in salivary gland cells undergoing cell division. **b.** SEM image of human chromosomes

6.2.2 Eukaryotic chromosome structure

There is a huge quantity of DNA inside cells; in fact, a typical human cell contains approximately 2 metres of DNA! This volume of DNA needs to be carefully packaged into a cell of an average size of only 30 μm (0.00003 m). This packaging must be done in such a way that the **genes** retain their chemical make-up and physical position.

Eukaryotic chromosomes are composed of two main ingredients — DNA and proteins called **histones**. A small section of DNA coils tightly around a core of eight histones to form a **nucleosome**, which in turn combines with other nucleosomes to form supercoils of tightly compacted DNA (figure 6.3).

gene a section of a chromosome that codes for a protein through the order of the nucleotide base sequence it possesses

histone a protein found in eukaryotic chromosomes that assists in packaging the DNA

nucleosome a section of supercoiled DNA around histones

FIGURE 6.3 Coiling and supercoiling of DNA, forming chromosomes

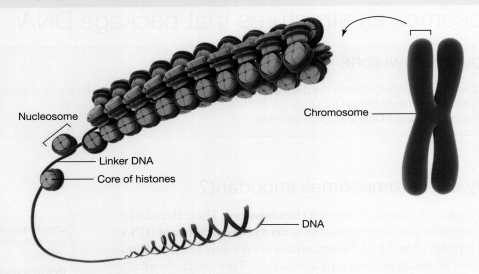

The coiling of DNA around histones to form nucleosomes enables the huge quantity of DNA to condense. This condensed DNA is now known as **chromatin**. Specific enzymes can control how tightly the chromatin is packaged, which in turn can regulate how easily certain genes within these sections are expressed. Ultimately, this coiling around the histones means the physical space into which the DNA can fit is dramatically reduced. The integrity of the DNA is also maintained, meaning its chemical composition has not been altered, and so the information encoded by the DNA is preserved in this process. The chromosomes formed can also be more easily manoeuvred to the poles of the cell during cell division.

TIP: Be careful to ensure that you refer to the condensed DNA as *chromatin*. A common exam mistake is instead writing *chromatid*, which is one half of a replicated chromosome. Many words in biology sound very similar but have very different meanings.

If you have ever had to coil a long length of hose, you will know that it is much easier to do if you have something to coil the hose around. This is a useful model for how DNA is coiled around the histones.

Resources

▶ **Video eLesson** Coiling and supercoiling of DNA (eles-4140)

> **chromatin** a mass of genetic material composed of DNA and proteins that condense to form chromosomes during eukaryotic cell division

6.2.3 Prokaryotic chromosome structure

Prokaryotic cells also have their DNA packaged into chromosomes. However, their DNA is usually in the form of a single circular chromosome rather than the several linear chromosomes we find in eukaryotes (see topic 1). The DNA that makes up the single chromosome is also less condensed compared to DNA in eukaryotes because histones are not used in the supercoiling process. Nevertheless, the DNA in prokaryotes is still coiled very tightly, forming the distinctive loops and twists characteristic of supercoiled DNA (figure 6.4). This looping and twisting is the result of supercoiling of the DNA double helix.

FIGURE 6.4 a. Looping and twisting of DNA of the single circular chromosome in bacteria compared to human DNA **b.** The bacterium *E. coli* surrounded by its supercoiled DNA, which has been released from the cell

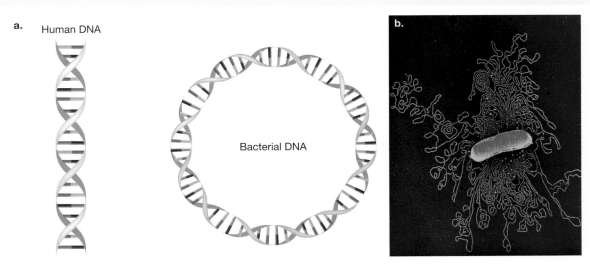

For both eukaryotic cells and prokaryotic cells, if the DNA is not packaged properly:
- it will not physically fit into the cell, resulting in the loss of genetic information
- the cell will not be able to distribute the DNA during cell division and so daughter cells will not receive the correct genetic information.

CASE STUDY: Mitochondrial DNA

Mitochondria contain their own DNA, which comprises of approximately 16 500 base pairs that code for 13 proteins. When mitochondrial chromosomes are examined, they are found to be very different from eukaryotic chromosomes. Mitochondrial chromosomes do not contain any proteins and are circular, meaning they are very similar to prokaryotic chromosomes. This is important evidence in support of endosymbiotic theory (see topic 1).

FIGURE 6.5 Mitochondria contain their own DNA.

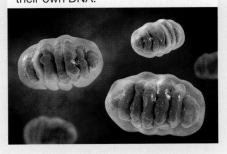

mitochondria in eukaryotic cells, organelles that are the major site of ATP production; singular = mitochondrion

SAMPLE PROBLEM 1 Comparing and contrasting eukaryotic and prokaryotic chromosomes

Compare and contrast the features of chromosomes present in eukaryotes and prokaryotes. (3 marks)

THINK

1. Identify what the question is asking you to do. This question asks you to both *compare* (find the similarities) and *contrast* (find the differences). Start by describing the similarities.

2. Contrast the two differences. When you contrast, ensure that you address *both* cell types, not just one.

3. Bring both aspects of your answer together.

WRITE

Both involve supercoiling to condense the DNA molecule, reducing the physical space that the DNA occupies.

Eukaryotic DNA is coiled around histones to produce nucleosomes. This condensed DNA is called chromatin. However, prokaryotic chromosomes do not contain histones and therefore do not contain nucleosomes or chromatin. Eukaryotic chromosomes are linear whereas prokaryotic chromosomes are circular.

The DNA in both cell types is highly twisted, producing supercoiled DNA, which reduces the physical space into which the DNA can fit (1 mark). However, eukaryotic DNA is coiled around histones to produce nucleosomes. This condensed DNA is called chromatin. However, prokaryotic chromosomes do not contain histones and therefore do not contain nucleosomes or chromatin (1 mark). Eukaryotic chromosomes are linear, however prokaryotic chromosomes are circular (1 mark).

INVESTIGATION 6.1

online only

Extraction of DNA from kiwi fruit

Aim
To extract DNA from within the nucleus of cells in a kiwi fruit

KEY IDEAS

- Chromosomes are condensed single molecules of DNA with associated proteins.
- Condensing of DNA allows a great deal of genetic information to be stored in a cell.
- Prokaryotic chromosomes do not contain proteins within their structure.
- Chromosomes that are highly condensed can be safely moved around the cell during cell division.

6.2 Activities

To answer questions online and to receive **immediate feedback** and **sample responses** for every question, go to your learnON title at **www.jacplus.com.au**. A **downloadable solutions** file is also available in the resources tab.

6.2 Quick quiz	6.2 Exercise

6.2 Exercise

1. State the two major components of eukaryotic chromosomes.
2. DNA is negatively charged but histones are positively charged. Explain why this is an advantage to the cell.
3. The human cell contains approximately 2 metres in length of DNA. Explain why *E. coli* cells cannot contain such a large volume of DNA.
4. A mutation in the gene that codes for histone protein production has been found in a cell. Explain a consequence of this mutation in the ability of a cell to package its DNA.
5. Suggest why cells undergoing replication with uncondensed DNA are more at risk of DNA damage occurring during the cell division process.
6. Cells can control access to their DNA by modifying their chromatin so that some sections are more tightly packed than others. For cells in the retina of the eye, explain whether the insulin gene is tightly packaged or not.

6.3 The distinction between genes, alleles and a genome

KEY KNOWLEDGE

- The distinction between genes, alleles and a genome

Source: VCE Biology Study Design (2022–2026) extracts © VCAA; reproduced by permission.

6.3.1 DNA and its bases

In the previous subtopic we learned about the importance of chromosomes in storing huge quantities of DNA. We also know that the DNA has specific sections, called genes, which possess the information to code for proteins.

CASE STUDY: DNA and its bases

The genetic material of an organism is its DNA. DNA is a complex molecule built of many basic building blocks called **nucleotides**, which contain a phosphate group, a nitrogenous base and a sugar (deoxyribose). There are four bases — **adenine (A)**, **thymine (T)**, **cytosine (C)** and **guanine (G)**. The nucleotides form long chains, with bases on one chain pairing in a specific manner with bases on a second chain. This pairing occurs according to **Chargaff's rules** — adenine pairs with thymine and cytosine pairs with guanine to form a complete DNA molecule with the famous double-helix shape (figure 6.6).

Further information can be found in the digital document.

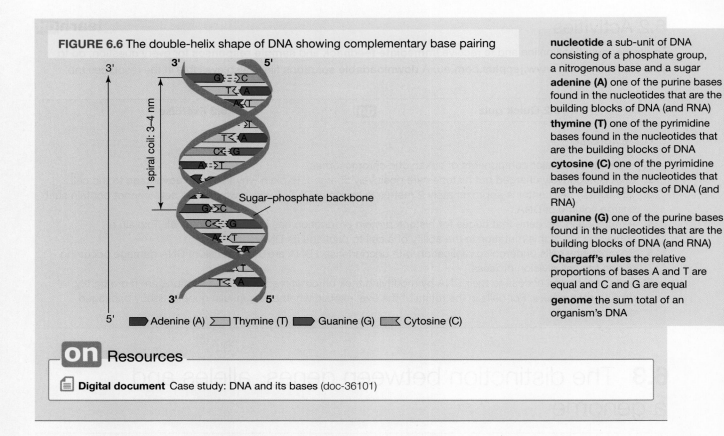

FIGURE 6.6 The double-helix shape of DNA showing complementary base pairing

nucleotide a sub-unit of DNA consisting of a phosphate group, a nitrogenous base and a sugar
adenine (A) one of the purine bases found in the nucleotides that are the building blocks of DNA (and RNA)
thymine (T) one of the pyrimidine bases found in the nucleotides that are the building blocks of DNA
cytosine (C) one of the pyrimidine bases found in the nucleotides that are the building blocks of DNA (and RNA)
guanine (G) one of the purine bases found in the nucleotides that are the building blocks of DNA (and RNA)
Chargaff's rules the relative proportions of bases A and T are equal and C and G are equal
genome the sum total of an organism's DNA

Resources

Digital document Case study: DNA and its bases (doc-36101)

6.3.2 How do genes differ?

Genes are sequences of bases located on chromosomes that code for specific proteins. These proteins control a particular characteristic or trait of the organism.

Every genetic instruction can be shown as a sequence of bases in the nucleotides that form the DNA of the gene. The genetic material of all organisms is DNA and the structure of DNA is identical, regardless of whether it comes from wheat, jellyfish, wombats, bacteria, insects or humans. In all organisms, genes are built of the same alphabet of four letters, namely the A, T, C and G of the bases in the nucleotide sub-units of DNA. So the genetic instruction kit of the great white shark, that of an oak tree and that of a human consists of thousands of different instructions, each made up of DNA with different base sequences.

These proteins control nearly every aspect of cellular function. Some genes code for enzymes that catalyse important chemical reactions such as those involved in cellular respiration or photosynthesis. Others produce fibrous proteins such as keratin, which is the structural material present in hair, skin, nails, feather and horns. Alternatively, regulatory genes code for proteins that are involved in controlling how other genes are expressed. There are an estimated 20 000 to 25 000 genes in the human **genome**, but only around 1 per cent of our DNA is classified as containing genes. The other 99 per cent of bases do not code for a protein and scientists are still working to understand the specific function of the so-called non-coding regions.

How much DNA is in a gene?

An average gene consists of about 3000 base pairs (bp). Genes, however, differ markedly in size (figure 6.7). The longest human gene is the *DMD* gene that encodes the muscle protein dystrophin and is 2 220 223 nucleotides long. An error in this gene is the cause of the inherited disorder Duchenne muscular dystrophy. Among the shortest human genes is a gene that encodes a histone protein and consists of about 500 nucleotides.

FIGURE 6.7 a. Human male sex chromosomes. The larger of the two chromosomes shown is the X chromosome. The *DMD* gene is found here along with approximately another 800 genes across 153 million base pairs.
b. Space-filling molecular model of the DMD protein. During protein production, each sequence of three bases codes for a specific amino acid. In the DMD protein there are 3685 amino acids!

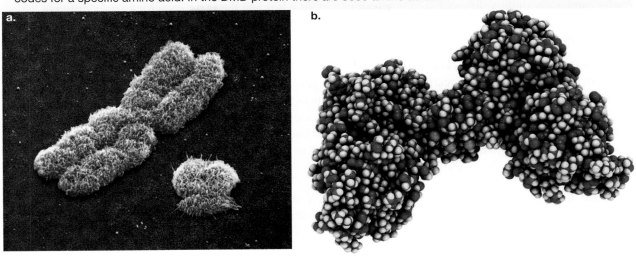

EXTENSION: How does DNA produce a protein?

The information to produce a protein is in the specific order and sequence of the four nitrogenous bases. A copy of the sequence of DNA bases is made in the form of another information molecule called **messenger RNA (mRNA)**. The base thymine (T) is replaced by uracil (U) in the strand of mRNA. This mRNA molecule leaves the nucleus, and each sequence of three bases codes for one of 20 amino acids. The order of the amino acids determines the shape, and in turn, the specific function of the protein (figure 6.8).

messenger RNA (mRNA) single-stranded RNA formed by transcription of a DNA template strand in the nucleus; mRNA carries a copy of the genetic information into the cytoplasm

FIGURE 6.8 Overview of protein synthesis. The DNA base sequence of each gene differs and therefore so does the sequence of amino acids in a protein.

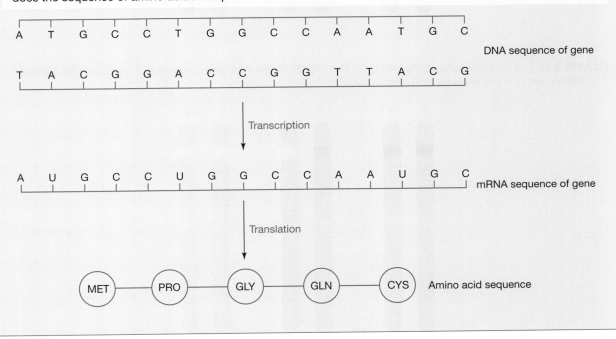

TOPIC 6 From chromosomes to genomes

6.3.3 Alleles — forms of a gene

In many organisms, pairs of chromosomes exist with the same genes at the same locations. One chromosome is inherited from each parent and so each individual has two copies of every gene. Less than 1 per cent of these copies can have a slightly different sequence of nitrogenous bases and therefore code for slightly different proteins. These alternative forms of genes are called **alleles** (figure 6.9). One gene can have several alleles, and each is identified in terms of its specific action.

The effect of alleles of a single gene can be demonstrated by the ability to taste a bitter chemical called phenylthiocarbamide (PTC). This gene (*TAS2R38*) is located on chromosome 7. One allele codes for a protein that acts as a taste receptor for bitterness. The other allele also codes for a receptor but it cannot detect the bitterness of PTC. Having one copy of the PTC-tasting allele (*T*) on each of the chromosomes in the pair (*TT*), or only on one (*Tt*), enables an individual to taste PTC (figure 6.10). Individuals who cannot detect PTC have the **genotype** *tt*.

alleles the different forms of a particular gene

genotype both the double set of genetic instructions present in a diploid organism and the genetic make-up of an organism at one particular gene locus

FIGURE 6.9 Three pairs of chromosomes, each pair from a different individual and showing a different combination of alleles. *A* and *a* are different versions of the same gene and therefore code for slightly different proteins.

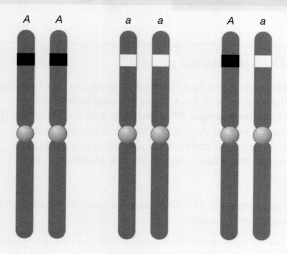

FIGURE 6.10 The combinations of the alleles for the *TAS2R38* gene on chromosome 7. *T* indicates the allele for PTC tasting and *t* is the allele for non-PTC tasting.

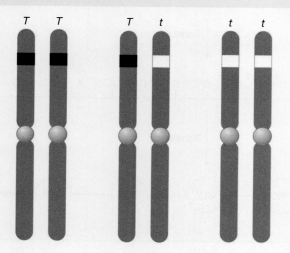

Multiple alleles for each gene

Within any population there is variation between individual members and this variation is due to the combinations of alleles they possess. Almost all characteristics in humans are controlled by several genes, each with multiple alleles, giving rise to a great deal of variation in traits such as eye colour, hair colour, skin colour, the presence of freckles and height (figure 6.11).

When three or more alleles exist for a gene, it is said to have multiple alleles. Some regions of a chromosome are hypervariable, meaning between 10 and 40 alleles can exist (table 6.1).

FIGURE 6.11 Variations in eye colour, skin colour and hair shape are considerable due to multiple genes controlling each trait. Each gene has multiple different versions or alleles.

TABLE 6.1 Multiple alleles of selected genes in various organisms. Not all alleles for each gene are shown; for example, there are 12 alleles for eye colour in *Drosophilia*.

Gene function	Multiple alleles and their action	
Controls human ABO blood type	I^A	Antigen A present
	I^B	Antigen B present
	i	Neither antigen present
Controls white spotting in dogs	S	White spots absent
	s^i	Irish spotting (as in collies) (see figure 6.12a)
	s^p	Piebald spotting (as in fox terriers) (see figure 6.12b)
	s^e	Produces extreme spotting (as in Samoyeds and Maltese terriers)
Controls pigment levels in cats	C	Intense pigment (as in a black cat, see figure 6.13a)
	c^b	Burmese dilution (see figure 6.13b)
	c^s	Siamese dilution (see figure 6.13c)
Controls colour intensity in rabbits	C	Intense pigment (as in solid black)
	c^{ch}	Chinchilla colouring (white fur with black tips)
	c^h	Himalayan colouring (colour on ears, nose, feet and tail only)
	c	Albino (white fur and pink eyes)
Controls white markings in cattle	S	White band around middle (as in Galloways)
	s^h	Hereford-type spotting
	s^c	Solid colour with no spots (as in Belmont Reds)
	s	Friesian-type spotting
Controls eye colour in fruit fly (*Drosophila* sp.)	w^+	Red eye
	w^a	Apricot
	w	White

TIP: Allele notation varies across scientific literature. They are commonly shown in italics, however, bold notation can also be used.

FIGURE 6.12 Dogs showing **a.** Irish spotting phenotype and **b.** piebald spotting phenotype

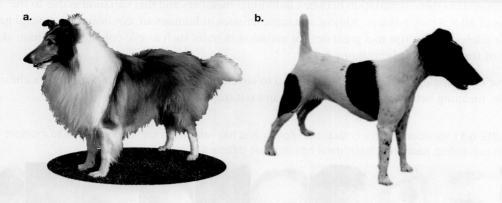

FIGURE 6.13 Cats showing **a.** intense black pigment; **b.** Burmese dilution, in which the black pigment is reduced to brown; and **c.** Siamese pigment, in which the pigment is further reduced and restricted to the ears, face, feet and tail. The order of dominance is $C > c^b = c^s$.

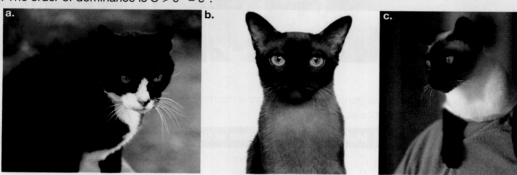

Allele variation in plants

Alleles are also responsible for the variation in plants (table 6.2 and figure 6.14). Note that the gene controls a general function, such as flower colour, but its alleles produce specific expressions of that function, such as purple and white. Figure 6.14a shows the smooth and wrinkled kernel textures in corn (*Zea mays*). These differences in texture are due to the difference in sugar levels in the kernels. Kernels with a high sugar content take up more water and swell more than kernels with a high starch content. As the kernels dry out, the greater loss of water from the sugary kernels causes them to become wrinkled. Figure 6.14b shows the purple and yellow kernels in corn, while figure 6.15 shows some of the mature fruit colours in capsicum.

TABLE 6.2 Alleles of some genes in plants

Gene function	Alleles and their action
Flower colour in delphinium	*P* purple *p* white
Kernel texture in corn (figure 6.14a)	*Su* smooth (starchy) *su* wrinkled (sugary)
Kernel colour in corn (figure 6.14b)	*Pr* purple *pr* yellow
Mature fruit colour in capsicum (figure 6.15)	*R* red *r* yellow

FIGURE 6.14 Corn showing **a.** wrinkled (sugary) and smooth (starchy) kernels, and **b.** purple and yellow kernels

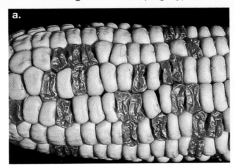

FIGURE 6.15 Various colours in mature fruit of capsicum *(Capsicum annum)*

6.3.4 The genome

A genome is the complete set of genetic instructions for an organism; it is the total DNA of an organism. The field of study of genomes is called **genomics**.

Genomes of eukaryotes

During **fertilisation** of an egg, two **gametes** fuse to form a zygote. The human zygote consists of 46 DNA molecules (chromosomes) arranged into 23 pairs, where one in each pair is inherited from each parent. The zygote cell is **diploid (2n)** and will divide by mitosis to form the cells of the human. Since the pairs of chromosomes are essentially copies and almost identical, the information in a somatic cell is in duplicate.

To determine the sum total of an organism's DNA — its genome — the **haploid (n)** number of base pairs is measured. This ensures only the number of base pairs from one of each of the nearly identical chromosomes is measured. Even so, in humans this number is still astonishing large — 3 234 830 000 base pairs — and includes the relatively small quantity of DNA in mitochondria. If you stretched the DNA out in one of your diploid somatic cells it would be about 2 metres long! If you did the same for all your cells and put them end to end, it would be approximately the same distance as twice the diameter of the solar system — a total of 575 billion kilometres!

Similarly, the genomes of other eukaryotes (animals, plants, fungi and protists) are the DNA of the haploid sets of their chromosomes. When we refer to the genome of a eukaryotic organism, such as the chimpanzee genome or the rice genome, we are speaking about the nuclear DNA. We can also talk about the genomes of those cell organelles that contain DNA, such as the mitochondrial genome or the chloroplast genome.

genomics the study of the entire genetic make-up or genome of a species

fertilisation the union of egg and sperm to form a zygote

gamete an egg (ovum) or a sperm cell

diploid (2n) having two copies of each specific chromosome in each set

haploid (n) having one copy of each specific chromosome in each set

Genomes of prokaryotes

The genomes of prokaryotes (bacteria and archaea) comprise the DNA of their single circular chromosome that carries the genetic instructions of each species. The genomes of viruses consist of their entire genetic instructions encoded in one DNA molecule, or, in the case of retroviruses, in one RNA molecule.

> Each genome is the sum total of an organism's DNA and is expressed as the base sequence of the haploid set of chromosomes.

CASE STUDY: The Human Genome Project

In 1990, an international team of scientists embarked on a huge task — to find the sequence of bases (adenine, cytosine, guanine and thymine) present in the human genome. It was not until 2003 that the Human Genome Project finally ended, with scientists announcing they had sequenced the genome. The project cost around US$1 billion. Nowadays, with more sophisticated computer hardware and software, we can sequence a human genome in only one to two days for a cost of around US$3000. Sequencing the genome has allowed scientists to find the specific chromosomal locations of important genes, which may offer some help in seeking cures for diseases. The genomes of other organisms have also been sequenced (see table 6.3).

TABLE 6.3 Dates of sequencing of genomes of a virus and various organisms

Organism	Date published	Size of genome (base pairs, bp)	Estimated number of coding genes	Comment
Virus phiX174	Apr. 1993	5 386	11	First genome sequenced
Haemophilus influenzae (bacterium)	July 1995	1 830 000	1 850	First bacterium
Saccharomyces cerevisiae (brewer's yeast)	Apr. 1996	12 069 000	6 294	First eukaryote and first fungus
Methanococcus jannaschii (archaean found at hydrothermal vents)	Aug. 1998	1 700 000	1 738	First archaean
Caenorhabditis elegans (nematode worm)	Dec. 1998	97 000 000	19 099	First animal
Arabidopsis thaliana (thale cress)	Dec. 2000	115 000 000	25 498	First plant
Equus caballus (horse)	Nov. 2009	2 689 000	20 322	Thoroughbred
Yersinia pestis (bacterium)	Aug. 2011	4 553		Cause of Black Death plague
Brassica rapa (Chinese cabbage)	Aug. 2011	485 000		
Anolis carolinensis (green anole lizard)	Aug. 2011	2 200 000	16 533	
Anopheles gambiae (mosquito)	2002	278 268 413	12 843	Main vector of malaria in sub-Saharan Africa
Gallus gallus (chicken)	2004	1 072 544 763	15 508	
Canis familiaris (dog)	2005	2 392 715 236	19 856	
Felis catus (domestic cat)	2007	2 365 745 914	19 493	

(continued)

TABLE 6.3 Dates of sequencing of genomes of a virus and various organisms *(continued)*

Organism	Date published	Size of genome (base pairs, bp)	Estimated number of coding genes	Comment
Oryctolagus cuniculus (rabbit)	Nov. 2009	2 604 023 284	19 293	
Pongo pygmaeus (orang-utan)	2011	3 109 347 532	20 424	One of the five great apes
Macropus eugenii (tammar wallaby)	Aug. 2011	2 549 429 531	15 290	
Sarcophilus harrisii (Tasmanian devil)	Feb. 2011	2 931 556 433	18 788	
Escherichia coli 0111 (bacterium)	2011	5 766 081	5 407	10 000th genome in GOLD database
Falco peregrinus (Peregrine falcon)	2013	1 200 000 000	16 263	A top predator in some ecosystems
Panthera tigris (Amur (Siberian) tiger)	Sep. 2013	~2 440 000 000	20 226	First entire genome of the endangered Amur tiger
Phascolarctos cinereus (koala)	Apr. 2013	~3 000 000 000	Approx. 15 000	

DISCUSSION

The following statement appeared in November 2014:

> Ultimately, the goal is for all of us to have our genomes sequenced and available as a medical reference for our clinical care.
>
> William Biggs, 'iCommunity Newsletter', November 2014

Another statement is as follows:

> It may be decades before interactions between genes, behavior and environment are understood well enough to provide substantial utility to warrant individualized recommendations based on genomic profiles. Furthermore, behavior change interventions that take advantage of some of the more unique aspects of genetic risk information are in their infancy.
>
> Kurt D Christensen and Robert C Green, 'How could disclosing incidental information from whole-genome sequencing affect patient behavior?', vol. 10, no. 4, 10.2217/pme.13.24

Discuss with your classmates your thoughts about these statements. Identify any positive perspectives in favour of the statements and any negative standpoints against the statements.

Resources

🔗 **Weblink** The Human Genome Project

There is considerable variation in the sizes of genomes, as shown in table 6.3. The human genome has around three billion base pairs, but the largest genome belongs to the flower *Paris japonica* (figure 6.16). It is an incredible 50 times larger than the human genome, containing around 130 billion base pairs. If all the DNA from a *P. japonica* cell was stretched out end to end it would be 91 metres long!

FIGURE 6.16 The largest known genome belongs to the plant *Paris japonica*.

SAMPLE PROBLEM 2 Calculating percentages in genomes

tlvd-1754

The genome of a blue whale (*Balaenoptera musculus*) was sequenced. Cytosine made up 35% of the bases present. Calculate the percentage of thymine present in the genome. (2 marks)

THINK	WRITE
1. Identify what the question is asking you to do. Use previous detailed scientific understanding as a basis for your answer, remembering Chargaff's rules for base-pairing.	Cytosine always pairs with guanine (Chargaff's rules). Therefore, there must be 35% guanine present.
2. When being asked to calculate, you need to show your working.	35% cytosine + 35% guanine = 70% (1 mark) Therefore, the remaining 30% must be divided evenly between the remaining bases of adenine and thymine. $$\frac{30}{2} = 15\% \text{ adenine, } 15\% \text{ thymine (1 mark)}$$

Resources

eWorkbook Worksheet 6.1 Exploring genes and alleles (ewbk-2898)
Worksheet 6.2 Genomes and the Human Genome Project (ewbk-2899)

KEY IDEAS

- A genome is the sum total of an organism's DNA measured in the haploid number of base pairs.
- A gene is a section of DNA, which is composed of the bases A, T, C and G.
- A gene codes for a protein.
- Alleles are alternative forms of genes.
- Variation in populations exists due to individuals possessing different genes and alleles.

6.3 Activities

To answer questions online and to receive **immediate feedback** and **sample responses** for every question, go to your learnON title at **www.jacplus.com.au**. A **downloadable solutions** file is also available in the resources tab.

| 6.3 Quick quiz | 6.3 Exercise | 6.3 Exam questions |

6.3 Exercise

1. Distinguish between a *gene* and an *allele*.
2. If you saw the base sequence of part of a gene, could you identify if it came from a dog or from a flea? Briefly explain.
3. State whether each of the following entries refers to a gene, or to the particular alleles of a gene:
    a. The ... that control(s) eye colour in humans
    b. The ... that produce(s) blue eye colour in humans
    c. The ... that produce(s) non-blue eye colour in humans.
4. Using the information in table 6.2, explain what colour a capsicum would be if it had the alleles *r* on one chromosome and *r* on the other.
5. Explain why the genome of the eastern grey kangaroo (*Macropus giganteus*) must contain 12 per cent cytosine if 38 per cent of the genome contains adenine.
6. A cob of corn consists of many individual cobs that are the offspring of a pair of parents. In one particular cob it is seen that some of the cobs are smooth and swollen but a smaller number are wrinkled and shrunken. This variation is due to the action of a single gene with two alleles. Using table 6.2, suggest which alleles of this gene might give rise to these two phenotypes.
7. Explain why plants with very large genomes, such as *Paris japonica,* grow very slowly.
8. Sequencing the human genome began in 1990 and took 13 years to complete. Sequencing the genome allowed scientists to obtain the specific base sequences present in a human and identify the locations of specific genes. Today, an individual's complete genome can be sequenced in just two days.
    a. Suggest why a person may want their genome to be sequenced.
    b. Despite possible benefits to sequencing a genome, some people were opposed to sequencing the genome in the 1990s. Suggest two reasons why.

6.3 Exam questions

Question 1 (1 mark)

Source: VCAA 2008 Biology Exam 2, Section A, Q17

MC Biologists have sequenced the genomes of many organisms. The number of genes found in organisms varies greatly. Some examples are listed in the table below.

Species of organism	Size of genome (approximate number of millions of base pairs)	Number of genes (approximate)
Escherichia coli (bacterium)	4.6	3 000
C. elegans (nematode worm)	100	20 621
Fugu rubripes (puffer fish)	365	38 000
Mus musculus (mouse)	3 000	22 000
Homo sapiens (human)	3 300	22 000
Psilotum nudum (whisk fern, a fern that grows in cracks in rocks)	250 000	unknown
Arabidopsis thaliana (flowering mustard plant)	100	28 000

From this data it can be concluded that
A. larger organisms have larger genomes.
B. puffer fish show greater genetic variety than *E. coli*.
C. a nematode and flowering mustard plant have the same number of chromosomes.
D. the larger the genome of an organism, the greater the number of proteins it produces.

▶ Question 2 (1 mark)
Source: VCAA 2006 Biology Exam 2, Section A, Q1

MC In eukaryotic organisms genes are
A. composed of DNA.
B. alternative forms of an allele.
C. composed of DNA and protein.
D. the same length as a chromosome.

▶ Question 3 (1 mark)
Source: VCAA 2012 Biology Exam 2, Section A, Q2

MC The genome of the woodland strawberry *Fragaria vesca* has been recently sequenced to show a relatively small genome of just 206 million base pairs. *F. vesca* is an ancestor of the garden strawberry and is a relative of apples and peaches.

The genome of *F. vesca*
A. is found only in the stem cells of the woodland strawberry.
B. includes all of the proteins made by *F. vesca*.
C. comprises all of the genes of *F. vesca*.
D. is the same as the genome of the apple.

Use the following information to answer Questions 4 and 5.

The complete genomes of many species have been sequenced. The genome size and the number of genes present in the genome for several species are shown in the table.

Common name	Species name	Number of genes in genome (estimate)	Number of base pairs in genome (bp)	Number of haploid chromosomes (n)
Bacterium	*Haemophilus influenzae*	11	5 386	1
Yeast	*Saccharomyces cerevisiae*	1 850	1 830 000	16
Rice	*Oryza sativa*	40 000	450 000 000	12
Capsicum	*Capsicum annum*	35 000	3 480 000 000	12
Horse	*Equus cabullus*	20 322	2 689 000	32
Human	*Homo sapiens*	20 000	3 000 000	23

▶ Question 4 (2 marks)
Which organism listed in the table has the largest genome? Explain your choice.

▶ Question 5 (2 marks)
A biology student stated that 'The greater the number of chromosomes present in an organism, the larger its genome'. From the information in the table, explain whether or not you agree with this statement and to what extent.

More exam questions are available in your learnON title.

6.4 Homologous chromosomes, autosomes and sex chromosomes

KEY KNOWLEDGE

- The nature of a pair of homologous chromosomes carrying the same gene loci and the distinction between autosomes and sex chromosomes

Source: VCE Biology Study Design (2022–2026) extracts © VCAA; reproduced by permission.

6.4.1 Homologous chromosomes

In section 6.3.3, we learned that in many organisms — including humans — the chromosomes are found in pairs. One of each pair is inherited from each parent. Therefore, of the 46 chromosomes in a human cell, 23 are inherited from the father and 23 from the mother. In females all 46 chromosomes can be matched up with another chomosome, and in males 44 chromosomes can be matched up with another chromosome. Each of these matched pairs are termed **homologous**, which means that the same genes are found in the same locations or **loci** (figure 6.17a). Chromosomesthat do not have the same gene loci are termed **non-homologous** (figure 6.17b). To help scientists assign chromosomes to their homologous partner, they can be stained with a chemical called Giemsa. Parts of the chromosomes that have a high concentration of adenine and thymine produce dark bands, whilethe light bands have high concentrations of cytosine and guanine. Figure 6.17 is an **ideogram** showing the stylised representation of chromosomes that demonstrates their relative sizes and their distinctive banding patterns.

homologous matching pairs of chromosomes that have the same genes at the same positions

locus the position of a gene on a chromosome; plural = loci

non-homologous non-matching chromosomes

ideogram a stylised representation of a haploid set of chromosomes arranged by decreasing size

FIGURE 6.17 a. Homologous chromosomes and **b.** non-homologous chromosomes stained to show banding patterns

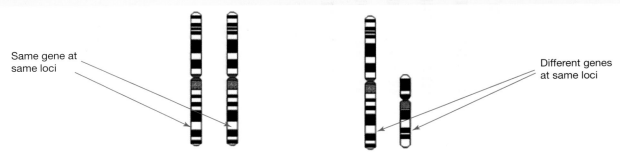

Even though homologous chromosomes share the same gene loci, it does not necessarily mean the alleles are the same. Remember that each chromosome has been inherited from a different parent and so the forms of genes on those chromosomes may be the same or different (section 6.3.3).

6.4.2 Autosomes and sex chromosomes

Chromosomes can also be divided up into two groups depending on whether they are involved in determining the sex of the organisms. The **sex chromosomes**, or allosomes, determine the sex, whereas the **autosomes** do not. In mammals, a pair of chromosomes — X and Y in males, and X and X in females— are the sex chromosomes or allosomes.

Autosomes

As discussed above, the 46 human chromosomes can be arranged into 23 pairs of chromosomes, consisting of 22 matched pairs and one 'odd' pair. The 22 matched pairs of chromosomes present in both males and females are termed autosomes. These different autosomes can be distinguished by:
- their relative size
- the position of the **centromere**, which appears as a constriction along the chromosome. In some cases the centromere is near the middle, while in others it is close to one end.
- patterns of light and dark bands that result from special staining techniques.

In humans, autosomes are identified by the numbers 1 to 22 in order of decreasing size; the number-1 chromosomes are the longest, and the number-21 and number-22 chromosomes are the smallest (figure 6.18). The larger the chromosome, the more DNA it contains and usually the greater the number of genes that it carries. The members of each matching pair of chromosomes, such as the two number-5 chromosomes, are homologous. Non-matching chromosomes, such as a number-5 chromosome and a number-14 chromosome, are non-homologous.

sex chromosomes a pair of chromosomes that differ in males and females of a species; allosomes

autosome any one of a pair of homologous chromosomes that are identical in appearance in both males and females

centromere the position where the chromatids are held together in a chromosome

FIGURE 6.18 An ideogram of human chromosomes. Only one chromosome from each homologous pair is shown. The banding patterns are produced from staining the chromosomes and help in their identification.

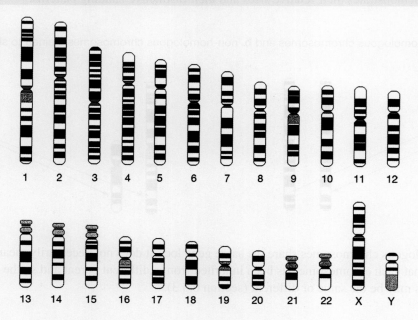

Sex chromosomes

The remaining 'odd' pair of chromosomes are the sex chromosomes. In a human male with a normal set of chromosomes, the 'odd' pair is made up of one larger X chromosome and a smaller Y chromosome. A shorthand way of denoting this is: 46, XY. In a human female with a normal set of chromosomes, a similar arrangement is seen, except that there are two X chromosomes and no Y chromosome. A shorthand way of denoting this is: 46, XX.

Note that in these shorthand abbreviations, the number indicates the total number of chromosomes including the sex chromosomes, and the letters denote the sex chromosomes. A similar pattern is seen in other mammals, where the female typically has two X chromosomes and the male has one X and one Y chromosome. This is not the case in other animal groups.

6.4.3 Sex determination

Mammals: XX/XY system

The X chromosome in humans contains approximately 800 genes (see figure 6.7a) whereas the Y chromosome only has 50. The Y chromosome plays a crucial role in determining the sex of the developing embryo. A gene found at the top of the Y chromosome, *SRY*, codes for a protein that controls the development of male characteristics. Therefore, if an embryo contains a Y chromosome, it will develop into a male (XY) and if not, it will develop into a female (XX). A similar XX/XY situation applies, with a few rare exceptions, to other mammals. During **meiosis** in males, the X and Y chromosomes pair up during metaphase before they are separated into daughter cells. This means there is a 50 per cent chance that a sperm cell produced will have either an X chromosome or a Y chromosome (figure 6.19). This is discussed in further detail in subtopic 6.7.

FIGURE 6.19 Diagram representing sex cell formation from 44 + XX (female) and 44 + XY (male) cells. Half of the sperm cells contain X chromosomes and the other half contain Y chromosomes.

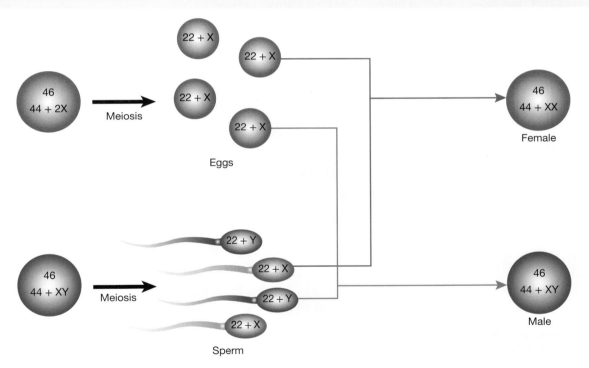

Since all egg cells contain an X chromosome, there is a 50 per cent chance the resulting fertilised cell will be 44 + XX (female sex) and a 50 per cent chance it will be 44 + XY (male sex). The 44 chromosomes are the autosomes. When these autosomes are added to the sex chromosomes, it adds up to the 46 chromosomes present in all human somatic (non-sex) cells.

meiosis a type of cell division that produces haploid gametes

Birds and reptiles: WZ/ZZ system

Sex chromosome differences also occur in birds, but the arrangement is different from mammals. Male birds have two similar sex chromosomes that are known as Z chromosomes. In contrast, the pair of sex chromosomes in female birds comprises one W and one Z chromosome, and so it is the female parent in these groups that determines the sex of the offspring. This WZ/ZZ genetic system of sex determination is also seen in some reptiles, such as snakes and monitor lizards (e.g. goannas), and in amphibians, such as some frog species. Tiger snakes (*Notechis* spp.), for example, have a total of 34 chromosomes, with males having two Z chromosomes and females having one Z and one W chromosome.

FIGURE 6.20 A mating pair of Cardinal birds

Reptiles: environmental sex determination

It was discovered that in some reptiles the sex of offspring depends on the incubation temperature of the eggs. This is an example of **environmental sex determination (ESD)** and is seen in green turtles (*Chelonia midas*). Female turtles lay an average of 110 eggs and bury them in the sandy beaches along Australia's tropical coastline. After laying, the female turtles return to the sea. Male turtle hatchlings result from eggs that are incubated at high temperatures (above 31 °C), female hatchlings are produced when the incubation temperature is lower (27 °C and below), while at intermediate temperatures (around 29 °C) about equal numbers of both sexes are produced.

environmental sex determination (ESD) the sex of the offspring is established by environmental conditions rather than genetic factors

FIGURE 6.21 Environmental sex determination is seen in the endangered green turtle.

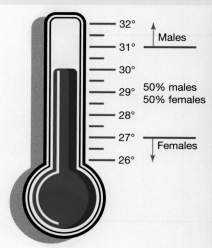

SAMPLE PROBLEM 3 Explaining male sex chromosomes

tlvd-1755

In humans, if the sex chromosomes possessed by an individual are XY, the person is biologically male. Explain why. **(3 marks)**

THINK	WRITE
1. Identify what the question is asking you to do. When being asked to *explain* you need to account for the reasoning why something occurs.	The Y chromosome (1 mark) has the gene *SRY* (1 mark).
2. Your answer should include detailed scientific understanding.	This gene codes for a protein that controls the development of male sexual characteristics (1 mark).

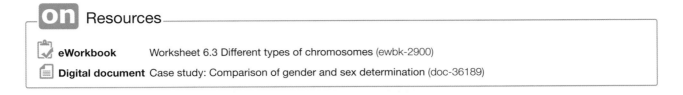

KEY IDEAS

- In humans there are 44 autosomes and 2 sex chromosomes.
- Homologous chromosomes have the same gene loci but not necessarily the same alleles.
- Chromosomes that determine the sex of the organism are called allosomes or sex chromosomes. The other chromosomes are called autosomes.
- In humans, the Y chromosome is much smaller than the X chromosome and its presence controls the development of male sexual characteristics. Therefore, females are XX and males are XY for sex chromosomes.
- The genes that are carried on the autosomes are the same in both males and females.
- There is a 50 per cent chance of a sperm cell carrying an X chromosome and a 50 per cent chance it will carry a Y.

6.4 Activities

learn on

To answer questions online and to receive **immediate feedback** and **sample responses** for every question, go to your learnON title at **www.jacplus.com.au**. A downloadable solutions file is also available in the resources tab.

| 6.4 Quick quiz | 6.4 Exercise | 6.4 Exam questions |

6.4 Exercise

1. Distinguish between the terms *autosome* and *sex chromosome*.
2. The image shows several autosomes that have been stained to show the banding patterns.

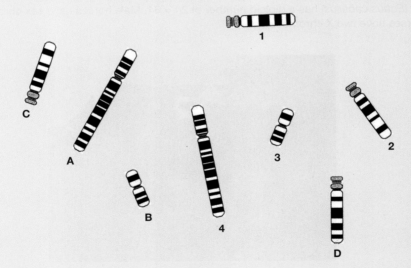

   a. Match each numbered chromosome with its lettered homologous partner.
   b. Identify the three physical features that allowed you to match up the chromosomes in part a.
3. John states that, 'Human females have the same number of homologous chromosomes as human males.' Do you agree with this statement? Explain why.
4. Would the banding patterns on one of the chromosomes from human chromosome pair 1 be identical to the banding pattern on a chromosome from chimpanzee pair 1? Explain why.

5. Klinefelter syndrome is a genetic condition in which individuals have an extra sex chromosome — XXY. It usually results in infertility, weaker muscles, greater height, poor coordination, less body hair, breast growth and less interest in sex. Will Klinefelter syndrome affect both males and females? Explain your answer.

6.4 Exam questions

Question 1 (1 mark)
MC How many pairs of autosomes does a human have?
A. 1　　　　　　　　B. 22　　　　　　　　C. 23　　　　　　　　D. 46

Question 2 (1 mark)
MC The genetic system of sex determination in birds differs from that of humans. Male birds have two similar sex chromosomes, ZZ, while female birds have a non-matching pair, WZ.

Most birds have a diploid number $2n = 80$.

How many pairs of matching chromosomes do most female birds have?
A. 2
B. 40
C. 39
D. 79

Male ruby-throated hummingbird
Sex chromosomes are ZZ

Female ruby-throated hummingbird
Sex chromosomes are WZ

Question 3 (1 mark)
MC Which statement correctly describes homologous chromosomes?
A. They carry identical alleles.
B. They have the same gene loci.
C. They come from the same parent.
D. Their DNA sequences are identical.

Question 4 (2 marks)
MC The horse (*Equus caballus*) has a diploid number of $2n = 64$. Male horses have sex chromosomes X and Y, while female horses have two X chromosomes.

a. How many autosomes does a female horse have?　　　　　　　　　　　　　　　　　　　**1 mark**
　　A. 32　　　　　　　　B. 62　　　　　　　　C. 64　　　　　　　　D. 66
b. How many matching pairs of chromosomes does a male horse have?　　　　　　　　**1 mark**
　　A. 31　　　　　　　　B. 32　　　　　　　　C. 63　　　　　　　　D. 64

> **Question 5** (2 marks)
> How many chromosomes are present in a human
> a. skin cell   1 mark
> b. liver cell?   1 mark
>
> More exam questions are available in your learnON title.

6.5 Variability of chromosomes

KEY KNOWLEDGE

- Variability of chromosomes in terms of size and number in different organisms

Source: VCE Biology Study Design (2022–2026) extracts © VCAA; reproduced by permission.

6.5.1 Do all chromosomes look the same?

In human cells, the DNA is divided up into 46 chromosomes. When examined under an electron microscope, it is clear to see that the chromosomes vary in appearance (figure 6.22).

You may recall from topic 2 that during S-phase of the cell cycle, all the DNA inside the cell must replicate. Once this has occurred the cell can enter prophase, where the chromosomes condense and take on the familiar 'X' shape (figure 6.23). The arms of the 'X'-shaped chromosomes are called **sister chromatids**. The shorter arms are called p arms whereas the longer arms are called q arms.

FIGURE 6.22 SEM image showing physical variation of human chromosomes

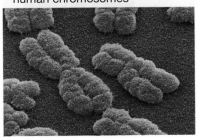

FIGURE 6.23 a. A chromosome before replication consisting of a single molecule of DNA **b.** 'X'-shaped chromosome formed by replication during S-phase

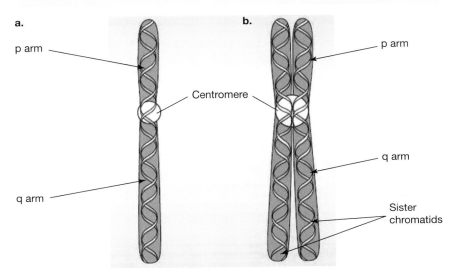

sister chromatids identical copies of DNA formed by the replication of a chromosome

Along the lengths of chromosomes are constriction points called centromeres, which occur in chromosomes both before and after replication. The centromeres are the regions where the spindle fibre proteins attached to move the chromosomes during cell division. Chromosomes can be classified based on the position of their centromere (table 6.4).

TABLE 6.4 Classification of chromosomes based on position of centromere

Appearance	Classification	Description
	Metacentric	• Centromere is positioned in the middle of the chromosome • p and q arms are of equal length
	Submetacentric	• Centromere positioned towards one end • q arms approximately twice the length of the shorter p arms
	Acrocentric	• Centromere is positioned very close to one end • p arms are very short
	Telocentric	• Centromere is positioned at the tip of the chromosome • No p arms

SAMPLE PROBLEM 4 Differences between metacentric and submetacentric chromosomes

Describe the differences between metacentric and submetacentric chromosomes. You may use a diagram to assist in your description. (4 marks)

THINK

1. Identify what the question is asking you to do. In a *describe* question, you need to clearly show factual recall and communicate a concept. If diagrams are allowed, they should be included as they can communicate knowledge very effectively. In your response, clearly contrast the differences that exist between the chromosomes. You must refer to the centromere and p and q arms.

2. Draw clear, simple diagrams with annotations. You must label each chromosome and its centromere. Ensure the diagrams and annotations complement your writing and do not contradict it.

WRITE

Metacentric chromosomes have a centromere in the middle of the chromosome (1 mark), whereas submetacentric chromosomes have a centromere closer to one end, with the q arms being approximately twice the length of the p arms (1 mark).

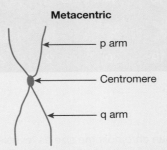

The centromere can be seen in the middle of the chromosome arms (1 mark).

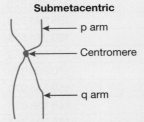

The centromere is positioned towards one end of the chromosome. The p arms can be seen as being only half as long as the larger q arms (1 mark).

6.5.2 Size differences of chromosomes

Human chromosomes, like the chromosomes of many different organisms, show considerable size differences across the 23 pairs found in the nucleus (figure 6.24).

There are approximately 25 000 genes across all the different chromosomes in the human cell, coding for specific proteins. The longest chromosome, chromosome 1, contains around 2000 genes, whereas chromosome 22 only has around 500 genes (figure 6.24). Across many eukaryotic cells the chromosome sizes can differ dramatically, indicating the difference in the number of genes that they contain (table 6.5).

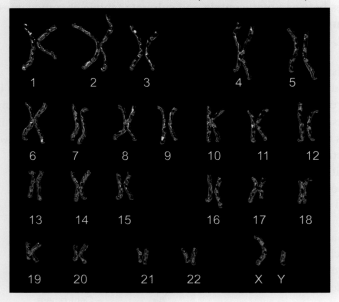

FIGURE 6.24 False-colour image showing differences in sizes of human chromosomes (normal human male)

TABLE 6.5 Human chromosome size as revealed by DNA sequencing

Chromosome	Length (bp)	Number of genes (bp)
1	248 956 422	2000
2	242 193 529	1300
3	198 295 559	1000
4	190 214 555	1000
5	181 538 259	900
6	170 805 979	1000
7	159 345 973	900
8	145 138 636	700
9	138 394 717	800
10	133 797 422	700
11	135 086 622	1300
12	133 275 309	1100
13	114 364 328	300
14	107 043 718	800
15	101 991 189	600
16	90 338 345	800
17	83 257 441	1200
18	80 373 285	200
19	58 617 616	1500
20	64 444 167	500
21	46 709 983	200
22	50 818 468	500
X (sex chromosome)	156 040 895	800
Y (sex chromosome)	57 227 415	50

The specific location where each gene is found is called a locus. Between genes are sections of DNA that do not appear to carry the instructions to produce proteins (figure 6.25). These are called non-coding regions and scientists are still working to fully understand their purpose.

FIGURE 6.25 DNA indicating positions of genes (loci) with non-coding regions between

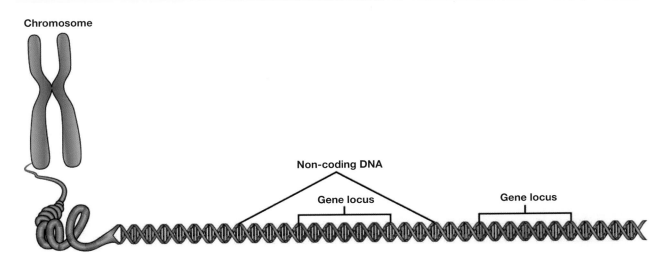

6.5.3 Differences in chromosome number

So far we have shown that the size of chromosomes and position of their centromeres vary. These factors help scientists to identify the type of organism that the chromosomes belong to. For example, human chromosomes (figure 6.24) differ in size and centromere positions to those of a fruit fly (figure 6.26). In the common fruit fly (*Drosophila melonogaster*), the sex chromosomes are chromosome 1.

It is evident that humans and the common fruit fly also have different *numbers* of chromosomes. Humans have 46 whereas the fruit fly only has 8. This difference in number is very important in allowing scientists to identify the species to which chromosomes belong. These differences can be seen when comparing chromosome numbers from different species (table 6.6).

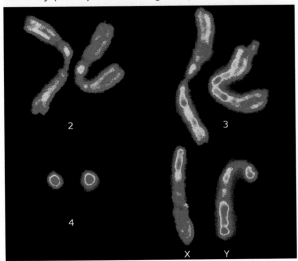

FIGURE 6.26 Chromosomes from the common fruit fly (*Drosophila melanogaster*)

TABLE 6.6 Comparison of number of chromosomes across a selection of different animal and plant species

Species	Number of chromosomes
Animal	
Chicken (*Gallus gallus*)	78
Butterfly (*Lysandra nivescens*)	190
Cat (*Felis catus*)	38

(continued)

TABLE 6.6 Comparison of number of chromosomes across a selection of different animal and plant species *(continued)*

Species	Number of chromosomes
Dog (*Canis familiaris*)	78
Bilby (*Macrotis lagotis*)	19 (male) 18 (female)
Garden snail (*Helix aspersa*)	54
Honey bee (*Apis mellifera*)	32
Housefly (*Musca domestica*)	12
Leopard seal (*Hydrurga leptonyx*)	34
Platypus (*Ornithorhynchus anatinus*)	52
Rabbit (*Oryctolagus cuniculus*)	44
Eastern tiger snake (*Notechis scutatus*)	34
Fruit fly (*Drosophila melanogaster*)	8
Ant (*Myrmecia pilosula*)	2
Plant	
Crimson bottlebrush (*Callistemon citrinus*)	22
Drooping she-oak (*Allocasuarina verticillata*)	26
Edible pea (*Pisum sativum*)	14
Ginkgo (*Ginkgo biloba*)	24
Iceland poppy (*Papaver nudicaule*)	14
Kangaroo paw (*Anigozanthos flavidus*)	12
Ovens wattle (*Acacia pravissima*)	26
Pineapple (*Ananas comosus*)	50
River red gum (*Eucalyptus camaldulensis*)	22
Silky oak (*Grevillea robusta*)	20
Silver wattle (*Acacia dealbata*)	26
Sydney blue gum (*Eucalyptus saligna*)	22
Tomato (*Lycopersicon esculentum*)	24
Corn (*Zea mays*)	20

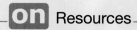 Resources

 eWorkbook   Worksheet 6.4 Chromosome variability (ewbk-2901)

▶ Video eLesson Chromosome structure (eles-4373)

KEY IDEAS

- Once chromosomes have passed through S phase of the cell cycle they replicate, taking on an X-shaped appearance consisting of sister chromatids joined at the centromere.
- There are four major types of chromosomes based on centromere position: metacentric, submetacentric, acrocentric and telocentric.
- Chromosome sizes vary in a eukaryotic cell and between cells of different species.
- Number and sizes of chromosomes can help to identify the species to which a cell belongs.

6.5 Activities

To answer questions online and to receive **immediate feedback** and **sample responses** for every question, go to your learnON title at **www.jacplus.com.au**. A **downloadable solutions** file is also available in the resources tab.

| 6.5 Quick quiz | 6.5 Exercise | 6.5 Exam questions |

6.5 Exercise

1. Draw a submetacentric chromosome and label the centromere, sister chromatids, p and q arms.
2. The image shows all the chromosomes from the cell of an organism.
   a. Using the data from table 6.6, identify which species the chromosomes belong to. Justify your answer.
   b. The geneticist examining the chromosomes classified them as being telocentric. Explain whether you agree or disagree with the geneticist.
3. Explain the function of the centromere.
4. Compare and contrast the appearance of acrocentric and telocentric chromosomes. You may use a diagram to assist in your answer.
5. Human chromosomes contain approximately 20 000 genes spread across the 46 chromosomes. Bananas contain 36 000 genes spread across 33 chromosomes. What does this suggest about the size of the non-coding regions in human chromosomes compared with the sizes of non-coding regions in bananas, assuming all genes were the same size?
6. The non-coding regions that exist between genes are found in all chromosomes. Scientists have some theories about their possible functions. Suggest a possible reason why non-coding regions exist.

6.5 Exam questions

Question 1 (1 mark)

MC If two chromosomes have the same length, centromere position and banding pattern when stained, they are called

A. homozygous.
B. homologous.
C. heterozygous.
D. heterogeneous.

Question 2 (1 mark)

MC The diploid number of the domestic cat is 38. It is reasonable to conclude that
A. somatic cells of the cat would each contain 38 chromosomes.
B. mature gametes of the cat would each contain 38 chromosomes.
C. the chromosomes of the cat in a non-dividing cell would contain sister chromatids.
D. 38 chromosomes would be visible in each somatic cell of a cat.

Question 3 (1 mark)

Source: VCAA 2012 Biology Exam 2, Section A, Q5

MC The following information shows the chromosome number in root tip cells from a range of plants.

Species	Common name	Chromosome number
Arabis holboellii	rockcress	14 or 21 or 28
Nasturtium spp.	flowery peppery goodness	32 or 64
Vitis vinifera	common grape vine	38 or 57 or 76
Viola spp.	violets	12 or 24 or 36 or 48

It is reasonable to conclude that
A. common grape vine plants all show triploidy.
B. only two of the three kinds of rockcress could reproduce by seed.
C. those plants with odd numbers of chromosomes must be haploids.
D. many of the flowery peppery goodness plants would be unable to carry out meiosis.

▶ **Question 4 (1 mark)**
Describe one structural difference between the X and Y chromosomes in a human.

▶ **Question 5 (2 marks)**
Identify two features that can be used to distinguish different non-homologous chromosomes from each other.

More exam questions are available in your learnON title.

6.6 Karyotypes

KEY KNOWLEDGE
- Karyotypes as a visual representation that can be used to identify chromosome abnormalities

Source: VCE Biology Study Design (2022–2026) extracts © VCAA; reproduced by permission.

6.6.1 Organising chromosomes

When viewed down a microscope, chromosomes can often be seen randomly arranged (figure 6.27a). This can make it difficult for their number, size and shape to be analysed properly. By looking carefully at their size, centromere position and staining pattern they can be positioned into homologous pairs. This regular arrangement is known as a **karyotype** (figure 6.27b).

karyotype an image of chromosomes from a cell arranged in an organised manner

FIGURE 6.27 a. The random arrangement of chromosomes in a cell **b.** Coloured light micrograph of the human female karyotype (XX), the complete set of chromosomes. Humans have 46 chromosomes: 23 inherited from the mother and 23 from the father. The sex chromosomes (bottom right) show two XX chromosomes.

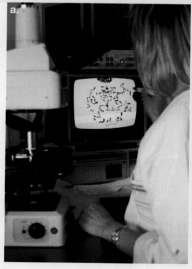
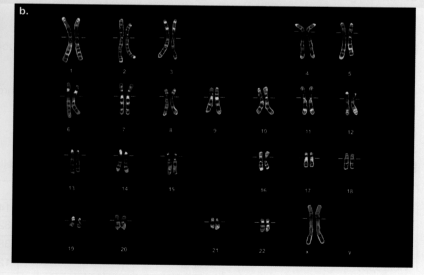

Karyotypes are used to assist in the analysis of the chromosomes that are present in cells. In a karyotype, the chromosome images are organised in a pattern according to an international convention. Such an arrangement enables any abnormality in either number or structure of the chromosomes to be quickly identified.

As well as using conventional stains, a range of probes with fluorescent labels that bind to specific segments of DNA on the different human chromosomes can also be used. Figure 6.28 shows a karyotype with these fluorescent labels. Each chromosome has a distinctive colour.

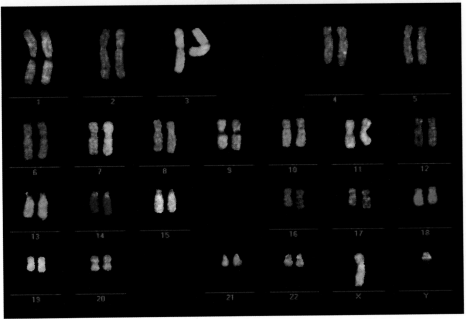

FIGURE 6.28 A spectral karyotype showing the distinctive colours of each human chromosome. Each homologous pair of chromosomes fluoresces with a distinctive colour, as determined by the colour of specific fluorescent probes that bind to specific sequences in the DNA of each particular chromosome.

6.6.2 Analysing karyotypes

Mistakes in chromosome numbers or abnormalities of single chromosomes can produce congenital disorders. In addition, specific chromosome abnormalities are associated with various cancers, and these chromosome changes can indicate the likelihood of remission. Scientists who specialise in the study of human karyotypes are known as **cytogeneticists**.

Today, in hospital cytogenetic laboratories, images of chromosome sets from cells are captured by a camera attached to a microscope (see figures 6.27a and 6.29). The images are then transferred to a computer, where a scientist uses special software — such as CytoVision® — that analyses the chromosomes from cells and automatically generates a karyotype (see figure 6.29). The chromosomes in a minimum of 15 cells must be examined before a karyotype can be decided. This computer-based automation has increased the capacity of hospital laboratories to prepare the karyotypes that are important in diagnosis of conditions such as **Down syndrome (DS)**, where an extra number-21 chromosome is present, or Prader–Willi syndrome, in which a small **deletion** of the number-15 chromosome occurs.

cytogeneticist a scientist who specialises in the study of human karyotypes

Down syndrome (DS) a chromosomal disorder due to the presence of an additional number-21 chromosome, either as a separate chromosome (trisomy or triplo-DS) or attached to another chromosome (translocation DS)

deletion type of chromosome change in which part of a chromosome is lost

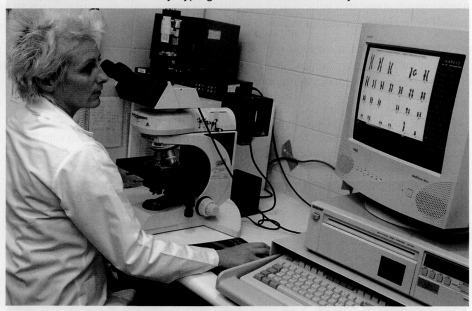

FIGURE 6.29 Automatic karyotype generation in the laboratory

6.6.3 Chromosome abnormalities

Various changes can occur involving chromosomes, including:
- changes in the total number of chromosomes
- changes involving part of one chromosome
- changed arrangements of chromosomes.

Changes in the total number of chromosomes

Some newborn babies have an abnormal number of chromosomes in their cells. A baby may have an additional chromosome, giving a total of 47 instead of the normal 46. One additional chromosome or one missing chromosome typically has deleterious effects on development and, for most chromosomes, death occurs during early development and the pregnancy never proceeds to term.

A pregnancy may still be carried to term if the chromosomal changes involve a few particular chromosomes (see table 6.7).

Two conditions that involve changes in the total number of chromosomes are as follows:
- **Trisomy** refers to when three copies of a chromosome occur, instead of the typical pair of chromosomes. The most common chromosomal anomaly seen in human populations is Down syndrome (DS), in which there is an additional copy of the number-21 chromosome. For this reason it is also known as trisomy 21 (see Case study box).
- **Monosomy** refers to when one member of the typical pair of chromosomes is missing. Monosomy causes embryonic death, except for a monosomy involving the sex chromosomes.

Using the karyotype shorthand mentioned in section 6.4.2 — for example, 46, XX — a missing sex chromosome is usually indicated with the symbol 'O'. If an extra entire autosome is present, this is shown by the chromosome number with a plus sign in front of it — for example, +21. A plus or a minus sign after the chromosome number indicates that only part of a chromosome is either present (+) or missing (−), and either a 'p' (short) or a 'q' (long) symbol is used to denote which arm of the chromosome is involved. Table 6.7 gives some examples of shorthand notations of a karyotype.

trisomy a condition in which a cell or organism has three copies of a particular chromosome that is normally present as a homologous pair

monosomy a condition in which a cell or organism has only one copy of a particular chromosome that is normally present as a homologous pair

TABLE 6.7 Chromosome abnormalities with respective incidence rates

Chromosome change	Resulting syndrome	Approximate incidence rate
Addition: whole chromosome		
Extra number-21 (47, +21)	Down syndrome	1/700 live births
Extra number-18 (47, +18)	Edwards syndrome	1/3000 live births
Extra number-13 (47, +13)	Patau syndrome	1/5000 live births
Extra sex chromosome (47, XXY)	Klinefelter syndrome	1/1000 male births
Extra Y chromosome (47, XYY)	N/a	1/1000 male births
Deletion: whole chromosome		
Missing sex chromosome (46, XO)	Turner syndrome	1/5000 female births
Deletion: part chromosome		
Missing part of short arm of number-4 (46, 4p–)	Wolf–Hirschhorn syndrome	1/50 000 live births
Missing part of short arm of number-5 (46, 5p–)	Cri-du-chat syndrome	1/25 000 live births

CASE STUDY: Down syndrome

People with 47 chromosomes in their body cells instead of the normal 46 due to the presence of an extra number-21 chromosome have the condition Down syndrome (DS). A karyotype of the individual in figure 6.30a is denoted as 47, XY, +21 (where '47' denotes the total number of chromosomes, 'XY' denotes the sex chromosomes present and '+21' denotes the identity of the extra chromosome).

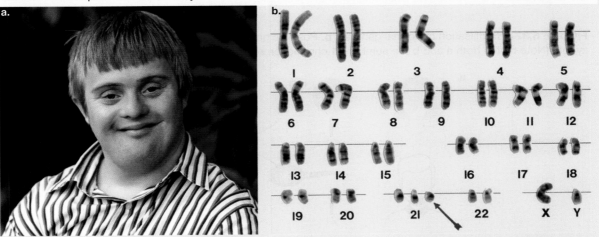

FIGURE 6.30 a. A young man with Down syndrome **b.** A typical karyotype from a DS male. Which chromosome is present as a trisomy?

About one in 700 babies born in Australia has an extra number-21 chromosome, but the rate differs according to the age of the mother (see figure 6.31). The risk of having a DS baby increases with maternal age; the risk for mothers aged 20 is about 1/2300, while the risk for mothers aged 40 is about 1/100. Increased father's age also increases the risk, but to a lesser extent.

The most common form of DS is the trisomy condition, in which three separate copies of the number-21 chromosome are present in the karyotype. In most cases, the extra number-21 chromosome is transmitted via an abnormal egg with 24 chromosomes, including two number-21 chromosomes. If this egg is fertilised by a normal sperm (with 23 chromosomes, including one number-21 chromosome), a DS embryo will result (figure 6.32).

An abnormal egg with 24 chromosomes results when, during the process of egg formation by meiosis in the ovary, the normal separation of the two copies of chromosome 21 to opposite poles of the spindle does not occur. This type of error is known as a **nondisjunction** and is unpredictable. It may also occur during sperm production in the father. The chance of a sibling with DS is low.

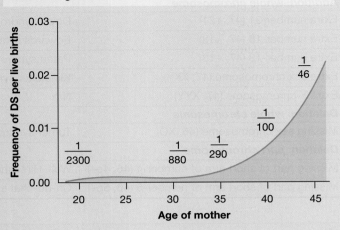

FIGURE 6.31 Incidence of DS births for mothers of different ages

The presence of the extra chromosome in people with DS produces various symptoms including: a fold on the inner margin of the upper eyelid; a smaller than normal mouth cavity; distinctive creases on the palms of the hands and the soles of the feet; poor muscle tone and loose joints; upward slanting, almond-shaped eyes; and a short stature. When a child with DS is born, new challenges arise for the family.

A much rarer form of DS can also occur, called **translocation**. This is discussed in the 'Rearrangements of chromosomes' section later in this subtopic.

nondisjunction failure of normal chromosome separation during cell division

translocation type of chromosome change in which a chromosome breaks and a portion of it reattaches to a different chromosome

FIGURE 6.32 a. Fertilisation of normal gametes **b.** Possible gametes involved in fertilisation to produce a DS zygote (Note that in both a and b the number-21 chromosomes are shown separately.)

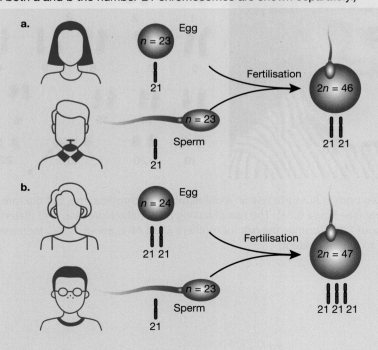

Errors in sex chromosomes

The chromosomes that determine sex are also found in the gametes, or sex cells, of the organism. All females produce egg cells that contain an X chromosome, whereas in males the sperm may contain either an X or a Y chromosome. These cells are formed by meiosis (see subtopic 6.7).

During the formation of gametes, critical events include the orderly disjunction of homologous chromosomes (figure 6.33a) to opposite poles of the spindle. However, if nondisjunction of homologous chromosomes occurs, or if homologous chromosomes fail to separate to opposite poles of their spindle at anaphase, errors in sex chromosomes can occur. For example:
- a gamete may have two copies of one chromosome instead of the normal single copy
- a gamete may be lacking a copy of one chromosome (figure 6.33b).

The fertilisation of an abnormal gamete by a normal gamete will produce a zygote with an imbalance in its sex chromosomes. Table 6.8 shows some possible abnormal outcomes, as well as the normal XX female and the normal XY male outcomes. Remember that the 'O' does not denote a chromosome, it denotes that a chromosome is missing. Gametes that are the result of a nondisjunction at anaphase II of meiosis are shown in red.

FIGURE 6.33 a. The disjunction of a chromosome pair during an error-free meiosis. The duplicated pair of homologous chromosomes in the cell in the top of the diagram undergoes two anaphase separations to produce four gametes, each with a single copy of the chromosome. **b.** The result of a nondisjunction of homologous chromosomes in anaphase II of meiosis

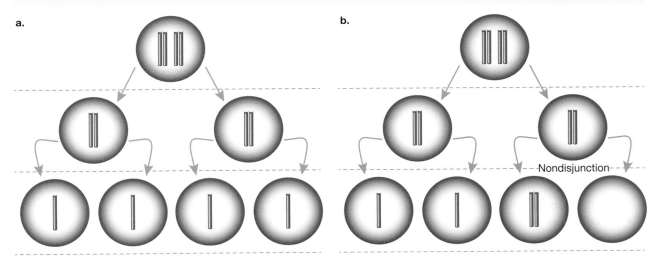

TABLE 6.8 Possible outcomes, in terms of sex chromosomes, of fertilisation of an egg by a sperm. In some cases one gamete is abnormal because of a nondisjunction of the sex chromosomes at anaphase II of meiosis (as shown in red).

Egg of female parent	Sperm of male parent	Resulting zygote
X	X	XX, normal female
X	Y	XY, normal male
XX	X	XXX, triple X female
XX	Y	XXY, Klinefelter syndrome
X	XY	XXY, Klinefelter syndrome
O	X	XO, Turner syndrome

(continued)

TABLE 6.8 Possible outcomes, in terms of sex chromosomes, of fertilisation of an egg by a sperm. In some cases one gamete is abnormal because of a nondisjunction of the sex chromosomes at anaphase II of meiosis (as shown in red). *(continued)*

Egg of female parent	Sperm of male parent	Resulting zygote
X	O	XO, Turner syndrome
O	Y	OY, non-viable
X	YY	XYY, double Y male

Of the possible outcomes, two result in a person with clinical abnormalities:
- Turner syndrome (45, XO): Persons with Turner syndrome are female and display clinical signs that include sterility because of the absence of a uterus.
- Klinefelter syndrome (47, XXY): Persons with Klinefelter syndrome are male and display clinical signs that include sterility and often female-type breast development.

In contrast, females who are 47, XXX and males who are 47, XYY show no clinical signs, are fertile and typically would be unaware of their less usual chromosomal status.

Comparing errors in autosomes and sex chromosomes

If we compare the situation between the autosomes and the sex chromosomes, it becomes apparent that changes to the numbers of autosomes are far more drastic in their effects than changes to the numbers of sex chromosomes.

All cases of monosomy of an autosome are non-viable, resulting in embryonic death, but the monosomy involving a sex chromosome, XO (Turner syndrome), is viable. This indicates that two copies of each autosome are essential for prenatal development. In contrast, even with only one X chromosome, prenatal development proceeds and the affected female survives into adulthood. However, the absence of both X chromosomes creates a non-viable situation. Only a few cases of trisomy of an autosome are viable and of these, only trisomy 21 (Down syndrome) normally survives into adulthood. In contrast, a person with an XXX trisomy shows no clinical signs.

Changes to parts of chromosomes

Changes can occur that involve part of a chromosome, such as:
- **duplication**, in which part of a chromosome is duplicated (see figure 6.34a)
- **deletion**, in which part of a chromosome is missing (figure 6.34b) — as in cri-du-chat syndrome, for example, so named because affected babies have a cat-like cry (see table 6.7).

Rearrangements of chromosomes

Structural changes may occur in which the location of a chromosome segment is altered so that it becomes relocated to a new region within the karyotype. Such a change is known as a translocation. One example is related to a special case of Down syndrome, when part of the number-21 chromosome becomes physically attached to a number-14 chromosome (figure 6.34c). The parental origin of the chromosomes is also important. Normally a child inherits one member of each chromosome pair from each of its parents. If both copies of a particular chromosome are inherited from one parent, instead of the usual one from each parent, abnormalities of development result. For example, **Angelman syndrome**, characterised by poor motor coordination and developmental delay, can result if an embryo inherits a faulty gene on the number-15 chromosome from its mother (as only the maternal copy of this gene is active in certain parts of the brain).

duplication type of chromosome change in which part of a chromosome is repeated

deletion type of chromosome change in which part of a chromosomes is lost

Angelman syndrome a genetic disorder resulting from the loss of function of a gene on chromosome-15 inherited from the mother; it primarily affects the central nervous system

FIGURE 6.34 a. Normal chromosome and the same chromosome showing a duplication b. Normal chromosome and the same chromosome showing a deletion c. An example of a 14/21 translocation

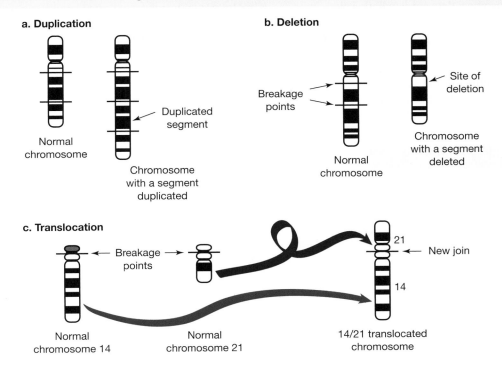

SAMPLE PROBLEM 5 Identifying sex and abnormalities from human karyotypes

tlvd-1757

Identify the following human karyotype, state the sex of the individual and which chromosomal abnormality, if any, is present. Use tables 6.7 and 6.8 to help in your answer. (2 marks)

THINK

1. Check that all the homologous chromosomes are in pairs.

WRITE

Not all the chromosomes are in homologous pairs; therefore a chromosomal abnormality is present.

2. If there are missing or additional chromosomes, determine whether they are autosomes or sex chromosomes. Use the reference tables to determine the name of the abnormality.	There is an additional X chromosome. XXY is a Klinefelter syndrome karyotype (1 mark).
3. Look carefully at the sex chromosomes and determine whether a Y chromosome is present. If so, it is a male.	A Y chromosome is present; therefore it is a male Klinefelter syndrome karyotype (1 mark).

INVESTIGATION 6.2 — online only

Modelling karyotypes

Aim

To model and analyse different karyotypes

Resources

- **eWorkbook** Worksheet 6.5 Analysing karyotypes (ewbk-2902)
- **Video eLesson** Changes in chromosomes (eles-4201)
- **Weblink** Online karyotyping

KEY IDEAS

- Chromosome sets can be ordered into karyotypes.
- Chromosomal changes can occur, including:
  - changes in the total number of chromosomes
  - changes involving part of one chromosome
  - changed arrangements of chromosomes.
- Additional or missing chromosomes can be readily identified by analysis of karyotypes.
- The sex of an individual can be identified from karyotype analysis.

6.6 Activities

To answer questions online and to receive **immediate feedback** and **sample responses** for every question, go to your learnON title at **www.jacplus.com.au**. A **downloadable solutions** file is also available in the resources tab.

6.6 Quick quiz	6.6 Exercise	6.6 Exam questions

6.6 Exercise

1. Describe how the karyotype of a human male would differ from a human female.
2. Distinguish between the terms *monosomy* and *trisomy*.
3. Explain why all those affected with Turner syndrome are female.
4. Explain how nondisjunction may result in a female giving birth to a Down syndrome child.
5. A karyotype of a developing human embryo is examined. The individual is diagnosed with Edwards syndrome. Describe how this karyotype would differ from a normal human karyotype and explain whether Edwards syndrome can affect boys and girls.

6. A newborn baby showed facial abnormalities and other signs, including deformed kidneys and nails, and an unusual way of clenching its fist. Its karyotype was prepared as shown.

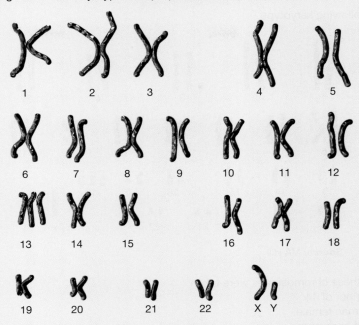

a. What is the total number of chromosomes in the karyotype?
b. What is the sex of the baby?
c. What abnormality is visible in the karyotype?
d. What name is given to this condition?

7. Suggest why the incidence rates of Patau syndrome (1/5000 live births) is lower than Edwards syndrome (1/3000 live births), which is in turn lower than Down syndrome (1/700 live births). Use table 6.7 to help.

6.6 Exam questions

Question 1 (1 mark)

MC What is a karyotype?
A. The data output from DNA sequencing
B. A map of a chromosome showing all of its gene loci
C. The combination of alleles that an individual has for a particular gene
D. The ordered arrangement of images of a diploid set of chromosomes

Question 2 (1 mark)

MC Which cell could provide the data to construct a karyotype?
A. An ovum
B. A sperm cell
C. A skin cell in metaphase
D. A liver cell during interphase

Question 3 (1 mark)

Source: VCAA 2011 Biology Exam 2, Section B, Q5a

The quarter horse, as a breed, originated by selective breeding. The first Australian quarter horses were imported from North America in the 1950s. A genetic condition called Hereditary Equine Regional Dermal Asthemia (HERDA) affects certain individuals. HERDA horses have a reduced life expectancy. Affected horses have a pedigree that is linked to an American stallion, Polo Bueno, which lived in the 1940s.

Which cells in the body of Polo Bueno were affected by the mutation allowing HERDA to be inherited?

Question 4 (1 mark)
Source: VCAA 2016 Biology Exam, Section A, Q31

MC Consider the following karyotype.

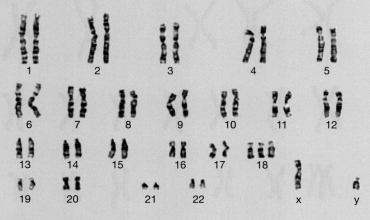

Source: MA Hill

The cell from which these chromosomes were taken
A. has a diploid number of 44.
B. comes from a human female.
C. has two copies of each of the genes found on chromosome 18.
D. has inherited one chromosome number 4 from the mother and inherited one chromosome number 4 from the father.

Question 5 (4 marks)
Consider a karyotype with an additional copy of an entire chromosome, specifically chromosome X.
a. How would this chromosomal change appear in a human karyotype? **2 marks**
b. Explain the differences that may appear between a male and female with an additional X chromosome. **2 marks**

More exam questions are available in your learnON title.

6.7 Production of haploid gametes

KEY KNOWLEDGE

- The production of haploid gametes from diploid cells by meiosis, including the significance of crossing over of chromatids and independent assortment for genetic diversity

Source: VCE Biology Study Design (2022–2026) extracts © VCAA; reproduced by permission.

6.7.1 How many chromosomes?

In sexual reproduction, each parent makes an essentially equal genetic contribution to each of its offspring in the form of a gamete; that is, an egg and a sperm. These gametes, and the cells used to produce them, are often referred to as the germline. If, in the case of humans, these gametes each contained 46 chromosomes, an offspring would have a total of 46 + 46 = 92 chromosomes. Over successive generations, this chromosome number would increase further. However, this doubling of the number of chromosomes does not occur across generations. Each generation of human beings has a constant 46 chromosomes in their somatic cells. In consequence, this means that each normal human gamete must have just 23 chromosomes, so that an offspring receives 23 + 23 = 46 chromosomes in total from its parents (see figure 6.35).

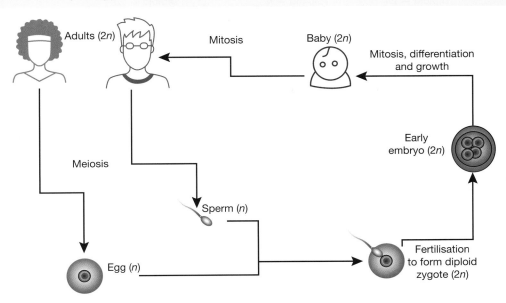

FIGURE 6.35 The life cycle of the human species. In sexual reproduction, offspring result from the fusion of two parental contributions (one egg and one sperm).

In humans, the normal somatic cells have 46 chromosomes, while the gametes (either eggs or sperm) have just 23 chromosomes. Alternatively, dogs have 78 chromosomes, while their gametes have 39 chromosomes. The number of chromosomes in the gametes is the haploid number and is denoted by the symbol n. Hence, in humans $n = 23$ and in dogs $n = 39$. It is then reasonable to conclude that the process of gamete formation in a person involves a **reduction division**. In humans, a starting cell with 46 chromosomes gives rise to gametes, either egg or sperm, that have only 23 chromosomes. This reduction division is a process called meiosis. The fertilisation of an egg by a sperm restores the diploid number (see figure 6.36). We will see in the following section that meiosis is a *non-conservative division* in which the chromosome number is halved and, as we will see later, the genetic information on the chromosomes is shuffled.

A key feature of sexual reproduction is that offspring produced through this mode of reproduction differ genetically from each other and also from their parents. In contrast, asexual reproduction involves a cellular process called mitosis that is conservative, so it faithfully reproduces an exact copy of the genetic information of the single parent cell in the two daughter cells.

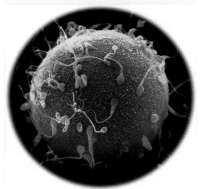

FIGURE 6.36 An egg cell with sperm cells adhering. Only one sperm will succeed in fertilising the egg.

reduction division production of gametes through cellular division in meiosis I that reduces the chromosome number from diploid to haploid

- Sexual reproduction depends upon meiosis, which halves the chromosome number to produce gametes. The offspring produced differ genetically from each other and also from their parents.
- Asexual reproduction involves a cellular process called mitosis that reproduces an exact copy of the genetic information of the single parent cell in the two daughter cells.

6.7.2 Meiosis: the stages

Meiosis is the process that produces gametes with the haploid number of chromosomes; that is, half the number present in somatic cells. After fertilisation, when the nucleus of a sperm fuses with that of an egg, the diploid number of chromosomes is restored (see figure 6.37).

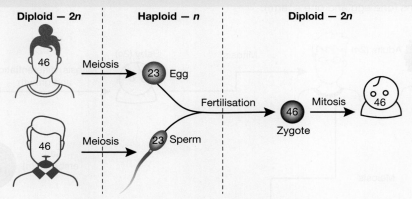

FIGURE 6.37 Diploid animals produce haploid gametes that fuse at fertilisation to produce a diploid zygote.

In order to create haploid daughter cells that are genetically different from each other, a series of carefully coordinated stages of meiosis are necessary. Figure 6.38 shows how a diploid cell with a total of four chromosomes ($2n = 4$) undergoes meiosis to produce four haploid gametes.

6.7.3 Introducing variation

The biological significance of meiosis is that it produces genetic variability among the offspring produced by sexual reproduction involving gametes from two parents. No two offspring are the same!

A litter of pups may consist of an odd mixture of colours and patterns (see figure 6.39). Siblings (brothers and sisters) in a human family can show differences in hair colour, eye colour, blood types and other traits. Inherited differences between offspring can be traced back to the process of meiosis and to the genetic variation that it produces in gametes.

Crossing over

Meiosis produces variation because of two mechanisms:
1. **Crossing over**
2. **Independent assortment**.

Let's look at these two mechanisms in more detail. Firstly, as mentioned briefly in the previous section, crossing over occurs during prophase I. The homologous chromosomes become connected and are known as a **bivalent**. While connected non-sister chromatids become entangled, the DNA breaks and recombines, causing new combinations of alleles to form along the length of chromatids (figure 6.40).

Figure 6.40 shows three genes represented as different letters, each with two alleles signified by upper- or lowercase text. The haploid gametes produced have different combinations of alleles of the three genes. The parental diploid cell had alleles P, Q and R on one chromosome and p, q and r on the other homologous chromosome, whereas the haploid daughter cells have the combinations shown in figure 6.41.

It is possible to appreciate how this process alone could produce a huge genetic variety in the daughter cells. Remember that humans have 22 pairs of autosomes that can all engage in crossing over, shuffling thousands of alleles. Therefore, the offspring that result from fertilisation of two daughter cells do not look identical to their parents since they do not have exactly the same alleles that either parent had; instead they have a combination of alleles from both parents. Similarly, siblings (unless they are identical twins) do not look identical because the gametes from which they arose were not identical. It is amazing to think that every sperm a male produces and every egg a female produces in their lifetime are likely to be unique.

crossing over an event that occurs during meiosis, involving the exchange of corresponding segments of non-sister chromatids of homologous chromosomes
independent assortment the formation of random chromosome combinations during meiosis that contributes towards producing variation
bivalent a pair of homologous chromosomes that are held together by at least one crossover

FIGURE 6.38 Stages in the process of meiosis, from the starting diploid cell (2n = 4) to the final haploid gametes (n = 2)

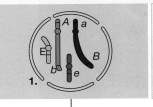

1. Before meiosis, a diploid cell with four chromosomes (2n = 4)

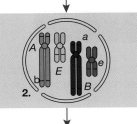

2. **Interphase:** The DNA is replicated, producing the X-shaped chromosomes, each consisting of sister chromatids joined at the centromere **(section 6.5.1)**.

3. **Prophase I** of meiosis: The chromosomes condense and the nuclear envelope degrades.
Crossing over (section 6.7.3) occurs between corresponding sections of homologous chromosomes, causing DNA to move between the homologous chromosomes and therefore alleles are transferred.
This creates new combinations of alleles along the chromosomes.

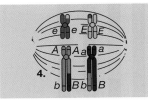

4. **Metaphase I:** The homologous chromosomes line up alongside each other down the equator of the cell.
Anaphase I: They are then separated from each other when their centromeres are pulled to opposite poles of the cell by the spindle fibres.

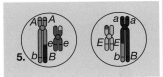

5. **Telophase I:** A nuclear envelope reforms around each set of chromosomes.
The spindle fibres disappear and **cytokinesis** occurs. Cytokinesis involves the division of the cytoplasm, forming two new cells, and the nuclear envelope reforms within each.

Prophase II: During meiosis, another round of division occurs. The chromosomes may re-condense if necessary and the nuclear envelope again breaks down.

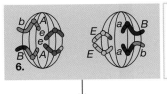

6. **Metaphase II:** The chromosomes again line up along the equator of each cell.
Anaphase II: The replicated chromosomes are separated into single-stranded ones by the spindle fibres pulling the sister chromatids.

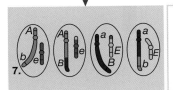

7. **Telophase II:** A new nuclear envelope reforms around each set of chromosomes. Cytokinesis occurs, forming four haploid daughter cells (n = 2).

FIGURE 6.39 Pups from two parents show variation in colour and pattern.

FIGURE 6.40 a. One pair of homologous replicated chromosomes (left) at interphase of meiosis. Note the genetic information that they carry. b. The exchange of segments when crossing over occurs during prophase I of meiosis c. The resulting recombination of genetic material

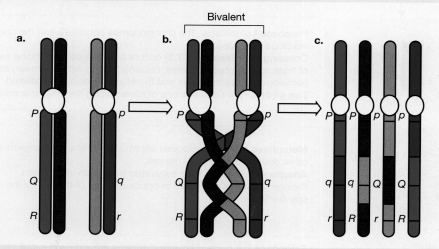

FIGURE 6.41 The combinations present in the haploid daughter cells resulting from the process shown in figure 6.40

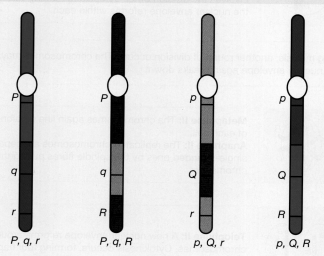

Independent assortment

Another source of considerable variation is independent assortment. This describes how pairs of alleles separate independently from each other during meiosis. Let's look at the example shown in figure 6.42, which has a cell with just two pairs of homologous chromosomes ($2n = 4$).

During metaphase I, the homologous chromosomes line up along the middle of the cell. There are two different possibilities of how the chromosomes could be orientated in relation to each other. If the purple *T*-containing chromosome is positioned above the orange *t*-containing chromosome, then gametes containing *RT* are formed along with *rt* gametes. However, there is an equal chance that the purple and orange chromosomes line up on different sides of the equator of the cell. In this case, different combinations of chromosomes are present in the gametes — *Rt* and *rT*.

FIGURE 6.42 Meiosis showing different outcomes from independent assortment of chromosomes in a cell containing two homologous pairs of chromosomes

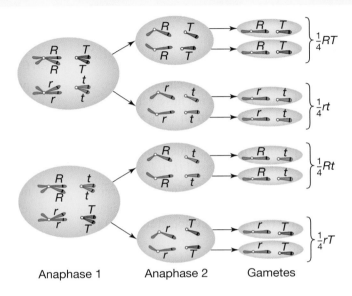

It is possible to imagine that with more chromosomes in a cell, a greater variety of chromosome combinations would occur. In fact, for a human cell containing 23 pairs of chromosomes, there are over 8 million possible chromosome combinations. This is given by the formula 2^n (where n = number of chromosome pairs). We can check this formula using the example in figure 6.42. There are two pairs of chromosomes, therefore $2^2 = 4$. A check of the chromosome combinations reveals there are indeed four possible outcomes.

The advantage of sexual reproduction comes from the genetic diversity that it creates in offspring. In contrast to a population generated by asexual reproduction that is composed of genetically identical organisms, a population of organisms produced by sexual reproduction contains a remarkable level of genetic diversity. The presence of this variation within its gene pool means that such a population is better equipped to survive in changing, unstable environmental conditions, in order to cope with an outbreak of a new viral or bacterial disease, or to survive a disaster.

Genetic diversity

- Meiosis produces sex cells with almost infinite variation due to crossing over and independent assortment.
- Crossing over is the exchange of corresponding segments of non-sister chromatids of homologous chromosomes.
- Independent assortment is the formation of random chromosome combinations.
- The genetic diversity of the resulting offspring is the main advantage of sexual reproduction over asexual reproduction, since this variation allows species to adapt to changing environments.

SAMPLE PROBLEM 6 Identifying the stages of meiosis

tlvd-1758

Four diagrams demonstrating various stages of meiosis are shown. State the correct order and identify the name of each stage. (2 marks)

  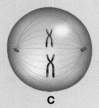 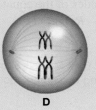

A   B   C   D

THINK

1. Look closely at the number and appearance of the chromosomes. Cells with four chromosomes (A and D) must not have yet gone through one round of division and so must be in an earlier stage of meiosis. Chromosomes still within a nuclear membrane and undergoing crossing over (A) are in prophase I, which is before they line up along the middle of the cell at metaphase I (D).

2. Cells with half the number of chromosomes have already undergone one round of division. During metaphase II the chromosomes again line up in the middle of the cell. The nuclear envelope reforms at the end of the process (telophase II).

WRITE

A — prophase I
D — metaphase I (1 mark)

C — metaphase II
B — telophase II (1 mark)

INVESTIGATION 6.3   online only

elog-0065

Modelling and observing meiosis

Aim

To model the formation of haploid cells through meiosis and observe meiosis under the microscope

---

INVESTIGATION 6.4   online only

elog-0066

Investigating inheritance of chromosomes from grandparents

Aim

To observe how chromosomes are inherited across generations

---

on Resources

- **eWorkbook**   Worksheet 6.6 The process of meiosis (ewbk-2903)
- **Video eLesson**   Stages of meiosis (eles-4216)
- **Interactivity**   Labelling the stages of meiosis (int-8129)
- **Weblink**   What is meiosis?

CASE STUDY: Identical twins

Do you know any identical twins? Can you tell them apart? It can be very difficult to distinguish between identical twins since their physical characteristics are very similar. This is because identical twins are formed after a single sperm fertilises a single egg cell. Rather than dividing by mitosis to produce a single embryo, instead, the resulting zygote divides to produce *two* embryos shortly after fertilisation. Since the embryos are from the union of the same single sperm and egg, their genetic information is also the same. The twins will, therefore, also have the same alleles. This can make it difficult to tell the twins apart.

Identical twins are termed monozygotic — mono meaning 'one' and zygotic meaning 'the organism that develops from the fertilised egg'.

FIGURE 6.43 Identical twins have the same genetic information as they result from the division of a single zygote.

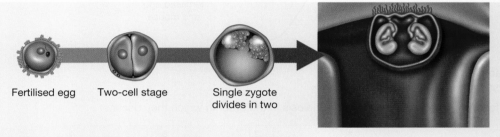

Fertilised egg   Two-cell stage   Single zygote divides in two

KEY IDEAS

- Apart from identical twins, every human is genetically unique.
- Meiosis is the process that produces haploid gametes from a diploid cell.
- Genetic variation in gametes results from crossing over and from independent assortment.
- Crossing over and independent assortment contribute to the recombination of genetic information in gametes.
- Increased genetic diversity from sexual reproduction aids species survival.

6.7 Activities

To answer questions online and to receive **immediate feedback** and **sample responses** for every question, go to your learnON title at **www.jacplus.com.au**. A **downloadable solutions** file is also available in the resources tab.

| 6.7 Quick quiz | 6.7 Exercise | 6.7 Exam questions |

6.7 Exercise

1. What name is given to the haploid gametes produced by human females?
2. An organism has a diploid number of 32.
   a. How many chromosomes are present in a sex cell of this organism?
   b. How many chromosome combinations are possible due to independent assortment? Show your working.
3. What behaviour of chromosomes in meiosis gives rise to the segregation of alleles of one gene into different gametes?
4. Using diagrams, draw the arrangements of chromosomes in a cell ($2n = 4$) undergoing anaphase I and then anaphase II of meiosis.
5. Describe the process of crossing over and explain how it contributes towards genetic variation in a species.
6. Suggest why increased genetic diversity is beneficial to the survival of a species.
7. Horses have a chromosome number of 64 ($2n = 64$). Donkeys have a chromosome number of 62 ($2n = 62$). Horses can breed with donkeys to produce mules that are infertile. Using the information of the number of chromosomes of horses and donkeys, and your knowledge of meiosis, explain why mules are infertile.

6.7 Exam questions

Question 1 (1 mark)
Source: VCAA 2013 Biology Section B, Q8a

Mice have a diploid number of 40.

How many chromosomes are there in each of the following cells?

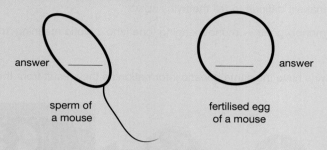

Question 2 (1 mark)
Source: VCAA 2015 Biology Exam, Section A, Q22

MC The following images show stages in meiosis in the order in which they occur.

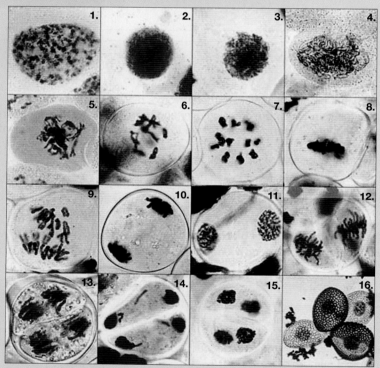

Source: Radboud University Nijmegen, Faculty of Science, www.vcbio.science.ru.nl (Virtual Classroom Biology)

Which one of the following statements is correct?
A. During the stage shown in image 9, chromatids separate.
B. Cells after the stage shown in image 10 are haploid.
C. During the stage shown in image 11, DNA will be replicated.
D. Homologous chromosomes pair up in the stage shown in image 12.

▶ **Question 3** (4 marks)

The cause of abnormal chromosome number is nondisjunction. Nondisjunction occurs during production of a gamete that creates the abnormal baby. It may occur during anaphase of meiosis I or meiosis II.
a. What is nondisjunction? **2 marks**
b. Explain how nondisjunction can lead to an abnormal number of chromosomes in a baby. **2 marks**

▶ **Question 4** (6 marks)

Examine the diagram of the life cycle of a fern. Different stages of the fern's life are indicated by the letters A to D.

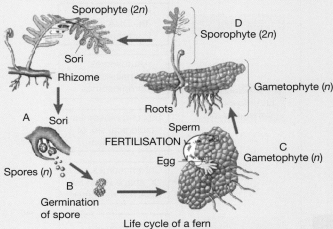

Life cycle of a fern

a. Which letter indicates the stage in the fern's life cycle that performs meiosis? Explain your choice. **2 marks**
b. The diploid number of this fern is $2n = 148$. The sperm fertilises the egg to produce the first cell of the sporophyte. The sporophyte grows by mitosis. How many chromosomes would be present in the fern's:
  i. sporophyte cells **1 mark**
  ii. spores **1 mark**
  iii. gametophyte cells **1 mark**
  iv. sperm and egg? **1 mark**

▶ **Question 5** (5 marks)

Source: VCAA 2011 Biology Exam 2, Section B, Q4

Species of the fruit fly *Drosophila* generally have four pairs of homologous chromosomes.
a. What is meant by the term homologous chromosomes? **1 mark**

The number of chromosomes sometimes varies from the usual four pairs. Karyotypes of two different *Drosophila* are shown in the following diagram. Note that one is a subspecies of the other.

b. Describe an event that could have caused the chromosome differences between *D. nasuta* and *D. nasuta* subspecies *albomicans*. **1 mark**

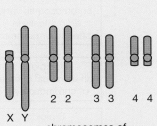

chromosomes of *Drosophila nasuta*

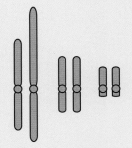

chromosomes of *Drosophila nasuta* subspecies *albomicans*

A cross between a female *D. nasuta* and a male *D. nasuta* subspecies *albomicans* results in offspring.
c. What would be the diploid number of these hybrid flies? **1 mark**
d. Explain how chromosome differences between *Drosophila nasuta* and *Drosophila nasuta* subspecies *albomicans* could result in their reproductive isolation and speciation. **2 marks**

More exam questions are available in your learnON title.

6.8 Review

6.8.1 Topic summary

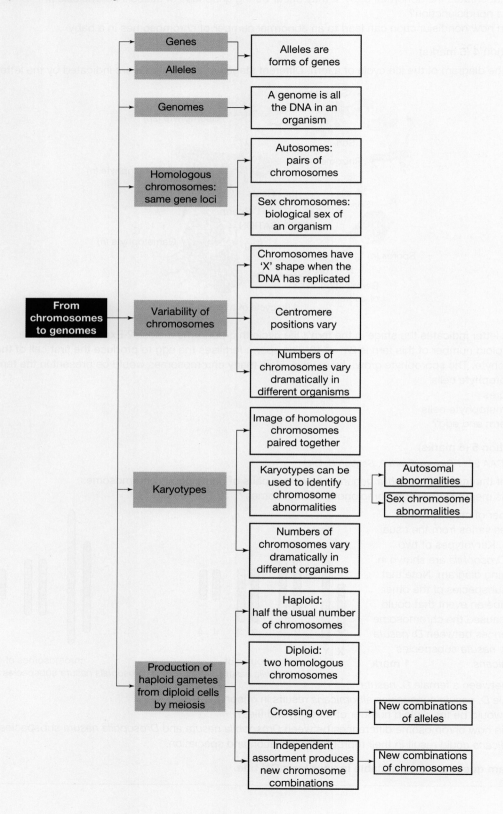

Resources

- **eWorkbook** — Worksheet 6.7 Reflection — Topic 6 (ewbk-2904)
- **Practical investigation eLogbook** — Practical investigation eLogbook — Topic 6 (elog-0166)
- **Digital documents** — Key terms glossary — Topic 6 (doc-34654)
  Key ideas summary — Topic 6 (doc-34665)

6.8 Exercises

To answer questions online and to receive **immediate feedback** and **sample responses** for every question, go to your learnON title at **www.jacplus.com.au**. A **downloadable solutions** file is also available in the resources tab.

6.8 Exercise 1: Review questions

1. If monosomy of an autosome occurs, the zygote is non-viable. However, there is one case of monosomy of a sex chromosome that is viable.
   a. Identify the name of this syndrome and how it is represented.
   b. Explain why all suffers of the condition are the same sex.

2. Where would you locate each of the following?
   a. Human cells undergoing meiosis
   b. Haploid cells in a cat
   c. Cells containing 20 unpaired chromosomes in a mouse
   d. The genetic instruction for your ABO blood type

3. Use words and/or diagrams to identify the differences, if any, between the items in each of the following pairs:
   a. Haploid and diploid
   b. Autosome and sex chromosome
   c. Somatic cell and gamete
   d. Gene and allele.

4. Match the processes in the table to the outcome.

Process	Outcome
a. Meiosis	A. Exchange of segments occurs between homologous chromosomes
b. Crossing over	B. Haploid gametes produced from a diploid cell
c. Disjunction of homologous chromosomes	C. Diploid number of chromosomes restored — one set from each parent
d. Independent disjoining of non-homologous chromosomes	D. Alleles of one gene segregate from each other
e. Fusion of gametes at fertilisation	E. Random combinations of different genes produced in gametes

5. Using a diagram, show how crossing over can result in the production of chromosomes with new allele combinations.

6. A sample of DNA from a human cell is found to contain 28% guanine. What percentage of thymine is present? Show your working

7. a. Referring to the diagram provided, place the stages of meiosis in the correct order.

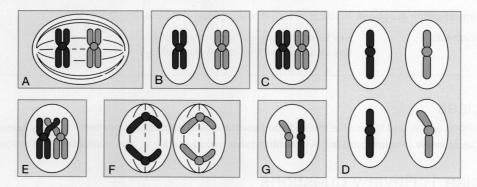

b. What is the diploid number of this organism?
c. What is the haploid number of the gamete produced?

8. Explain why chromosomes have an 'X'-shaped appearance at particular parts of the cell cycle.

9. Explain how the appearance of acrocentric and submetacentric chromosomes differ. You may use an annotated diagram to assist in your answer.

10. In humans, only the sex chromosomes determine the biological sex of the individual. However, in some reptiles, the environment plays a role too. Describe how one environmental factor can determine the sex.

6.8 Exercise 2: Exam questions

on Resources

▶ **Teacher-led videos** Teacher-led videos for every exam question

---

Section A — Multiple choice questions

All correct answers are worth 1 mark each; an incorrect answer is worth 0.

▶ **Question 1**
Source: VCAA 2012 Biology Exam 2, Section A, Q1

Consider the following diagram of a cellular structure.

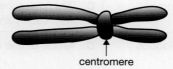
centromere

This structure is

A. visible only using an electron microscope.
B. found only in eukaryotic organisms.
C. found in all living organisms.
D. made up entirely of DNA.

▶ **Question 2**

Crossing over occurs during which phase of meiosis?

A. Telophase II
B. Cytokinesis
C. Prophase I
D. Anaphase I

▶ **Question 3**

Telocentric chromosomes have a centromere position

A. at the centre of the chromosome.
B. at the very end of the chromosome.
C. close to the end of the chromosome.
D. halfway along the p arm.

▶ **Question 4**

Source: VCAA 2016 Biology Exam, Section A, Q19

A biologist was working with a cell culture. He viewed a cell just before it entered mitosis and he counted 18 chromosomes. Later, the nucleus of one of the daughter cells was found to contain 19 molecules of DNA and the nucleus of the other daughter cell contained 17 molecules of DNA.

The most likely explanation for this observation is

A. the microtubules of the spindle apparatus did not connect to the centromeres of two of the chromosomes.
B. during anaphase, sister chromatids of one chromosome failed to separate.
C. during prophase, two of the chromosomes failed to line up.
D. at the end of telophase, cytokinesis failed to occur.

▶ **Question 5**

A cell contains 11% cytosine.

What percentage of guanine does it contain?

A. 11
B. 27
C. 39
D. Impossible to determine

▶ **Question 6**

Source: VCAA 2006 Biology Exam 2, Section A, Q5

In bees, females are diploid and males are haploid.

This means that male bees

A. produce gametes with half the haploid number of chromosomes.
B. produce gametes by meiosis.
C. produce gametes by mitosis.
D. do not produce gametes.

▶ **Question 7**

Source: VCAA 2012 Biology Exam 2, Section A, Q3

The genome of the woodland strawberry *Fragaria vesca* has been recently sequenced to show a relatively small genome of just 206 million base pairs. *F. vesca* is an ancestor of the garden strawberry and is a relative of apples and peaches.

It is expected that an offspring produced from sexual reproduction of *F. vesca*

A. translates all of its 206 million base pairs.
B. has equal numbers of adenine and cytosine nucleotides.
C. receives half of its chromosomes from the female parent.
D. possesses the same combination of alleles as other strawberry plants from the same parents.

Use the following information to answer Questions 8 and 9.

The red kangaroo (*Macropus rufus*) has 20 chromosomes in each of its somatic cells.

▶ **Question 8**

What will be the haploid number?

A. 5
B. 10
C. 20
D. 40

▶ **Question 9**

Which figure is closest to the number of possible chromosome combinations that could be produced during meiosis from a cell of the red kangaroo?

A. 10 000
B. 100 000
C. 1 000 000
D. 10 000 000

▶ **Question 10**

Source: VCAA 2011 Biology Exam 2, Section A, Q17

In leaf-cutting ants, a male develops from an unfertilised egg and a female from a fertilised egg.

It is reasonable to assume that

A. sperm produced by a particular male are genetically identical.
B. males can be either homozygous or heterozygous at any gene locus.
C. unfertilised eggs from a particular female develop into identical males.
D. homologous pairs of chromosomes are found in both male and female ants.

Section B — Short answer questions

Question 11 (5 marks)

Source: VCAA 2015 Biology Exam, Section B, Q6a, Q6bi, Q6bii, Q6c

The diagrams below show a pair of homologous chromosomes during cell division.

Figure 1 shows the whole chromosomes and Figure 2 is an enlarged view of the section circled in Figure 1.

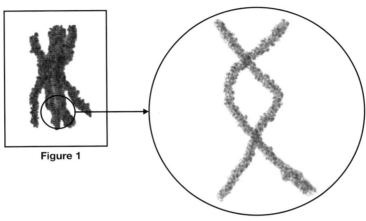

a. Name the type of cell division that would be occurring for this arrangement of chromosomes to be observed. **1 mark**

b. i. What name is given to the process occurring in the circled area in Figure 1? **1 mark**

ii. What is the outcome of this process and what advantage does the result of this process give a species? **2 marks**

c. Sometimes mistakes occur in this process. One such mistake is shown in the diagrams below.

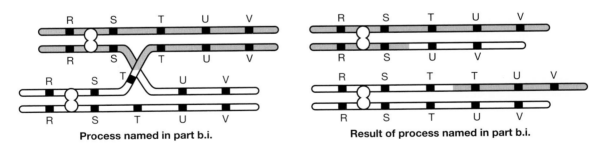

What will be the result of this mistake for the genetic makeup of the daughter cells? **1 mark**

Question 12 (8 marks)

Some chromosomes, when present as a trisomy in a human, produce obvious signs of clinical conditions in the affected person and are typically diagnosed soon after birth.

a. Give three examples of these chromosomes and state the name of each condition. **3 marks**

b. How would these conditions be diagnosed? **1 mark**

c. Briefly explain how a trisomy can be produced. **2 marks**

d. A student stated: 'Strange how the number-1 and the number-2 chromosomes are not examples of clinically recognised trisomies. It must be that they are never involved in a nondisjunction. It looks like only a few chromosomes are subject to that kind of error.'

Carefully consider this statement and indicate whether you agree with this student, giving a reason for your decision. **2 marks**

Question 13 (6 marks)

The DNA from a patient is examined in a hospital laboratory. A karyotype is produced and analysed by geneticists as shown.

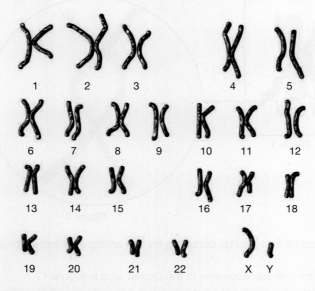

a. The chromosomes have been arranged into homologous pairs. Explain what is meant by the term *homologous chromosomes*. **1 mark**
b. This karyotype is from a male. Explain how the geneticists were able to determine this from only the karyotype. **1 mark**
c. The DNA in a karyotype is in the form of highly condensed chromosomes. Describe how the DNA is packaged to form chromosomes. **2 marks**
d. Explain how the karyotype from one of the male's sperm cells would differ from the one presented. **2 marks**

Question 14 (7 marks)

The sketch shown was made by a student who was observing a cell undergoing meiosis.

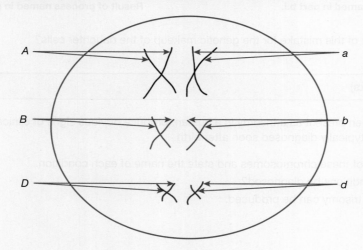

a. What phase of meiosis is the student's sketch representing? **1 mark**
b. What is the diploid number of the cell? **1 mark**
c. Using the labels the student has added, and assuming no crossing over occurs, write the possible chromosome combinations that could be found in the gametes formed. **2 marks**
d. Explain why the chromosome number must be halved during gamete formation. **3 marks**

Question 15 (9 marks)

Scientist A carried out an investigation to determine the mass of DNA in a human somatic cell. They examined a cell from a human kidney and found it contained 6.69 picograms of DNA. Scientist B also carried out the same investigation but chose to examine 1000 human cells from different organs. They concluded that 6.53 picograms of DNA is present in each cell.

a. Explain why scientist B had chosen to examine 1000 cells from different organs rather than just one cell like scientist A. **2 marks**
b. Explain why neither scientist sampled cells from the sex organs. **1 mark**

A third scientist, C, decided to repeat scientist B's experiment. Seven cells were selected at random from their experiments and their masses are shown in the table.

Mass of cells measured by scientists B and C

Scientist	Mass of cells (picograms)						
B	6.51	9.33	6.52	6.44	6.84	6.21	6.41
C	6.32	6.55	6.53	6.52	6.51	6.50	6.52

c. From the sample of results shown, which scientist's results show the most precise data? Explain your answer. **2 marks**
d. The true mass of DNA in a human somatic cell is 6.57 picograms. State whether scientist B or C has obtained more accurate results and explain why. **4 marks**

6.8 Exercise 3: Biochallenge online only

Resources

 eWorkbook Biochallenge — Topic 6 (ewbk-8076)

 Solutions Solutions — Topic 6 (sol-0651)

teach on

Test maker
Create unique tests and exams from our extensive range of questions, including practice exam questions.
Access the assignments section in learnON to begin creating and assigning assessments to students.

Online Resources

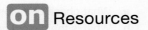

Below is a full list of **rich resources** available online for this this topic. These resources are designed to bring ideas to life, to promote deep and lasting learning and to support the different learning needs of each individual.

eWorkbook

- 6.1 eWorkbook — Topic 6 (ewbk-3164)
- 6.3 Worksheet 6.1 Exploring genes and alleles (ewbk-2898)
  Worksheet 6.2 Genomes and the Human Genome Project (ewbk-2899)
- 6.4 Worksheet 6.3 Different types of chromosomes (ewbk-2900)
- 6.5 Worksheet 6.4 Chromosome variability (ewbk-2901)
- 6.6 Worksheet 6.5 Analysing karyotypes (ewbk-2902)
- 6.7 Worksheet 6.6 The process of meiosis (ewbk-2903)
- 6.8 Worksheet 6.7 Reflection — Topic 6 (ewbk-2904)
  Biochallenge — Topic 6 (ewbk-8076)

Solutions

- 6.8 Solutions — Topic 6 (sol-0651)

Practical investigation eLogbook

- 6.1 Practical investigation eLogbook — Topic 6 (elog-0166)
- 6.2 Investigation 6.1 Extraction of DNA from kiwi fruit (elog-0063)
- 6.6 Investigation 6.2 Modelling karyotypes (elog-0064)
- 6.7 Investigation 6.3 Modelling and observing meiosis (elog-0065)
  Investigation 6.4 Investigating inheritance of chromosomes from grandparents (elog-0066)

Digital documents

- 6.1 Key science skills — VCE Biology Units 1–4 (doc-34648)
  Key terms glossary — Topic 6 (doc-34654)
  Key ideas summary — Topic 6 (doc-34665)
- 6.3 Case study: DNA and its bases (doc-36101)
- 6.4 Case study: Comparison of gender and sex determination (doc-36189)

Teacher-led videos

Exam questions — Topic 6
- 6.2 Sample problem 1 Comparing and contrasting eukaryotic and prokaryotic chromosomes (tlvd-1753)
- 6.3 Sample problem 2 Calculating percentages in genomes (tlvd-1754)
- 6.4 Sample problem 3 Explaining male sex chromosomes (tlvd-1755)
- 6.5 Sample problem 4 Differences between metacentric and submetacentric chromosomes (tlvd-1756)
- 6.6 Sample problem 5 Identifying sex and abnormalities from human karyotypes (tlvd-1757)
- 6.7 Sample problem 6 Identifying the stages of meiosis (tlvd-1758)

Video eLessons

- 6.2 Coiling and supercoiling of DNA (eles-4140)
- 6.5 Chromosome structure (eles-4373)
- 6.6 Changes in chromosomes (eles-4201)
- 6.7 Stages of meiosis (eles-4216)

Interactivity

- 6.7 Labelling the stages of meiosis (int-8129)

Weblinks

- 6.3 The Human Genome Project
- 6.6 Online karyotyping
- 6.7 What is meiosis?

Teacher resources

There are many resources available exclusively for teachers online

To access these online resources, log on to **www.jacplus.com.au**

AREA OF STUDY 1 HOW IS INHERITANCE EXPLAINED?

7 Genotypes and phenotypes

KEY KNOWLEDGE

In this topic you will investigate:

Genotypes and phenotypes
- the use of symbols in the writing of genotypes for the alleles present at a particular gene locus
- the expression of dominant and recessive phenotypes, including codominance and incomplete dominance
- proportionate influences of genetic material, and environmental and epigenetic factors, on phenotypes

Source: VCE Biology Study Design (2022–2026) extracts © VCAA; reproduced by permission.

PRACTICAL WORK AND INVESTIGATIONS

Practical work is a central component of learning and assessment. Experiments and investigations, supported by a **practical investigation eLogbook** and **teacher-led videos**, are included in this topic to provide opportunities to undertake investigations and communicate findings.

7.1 Overview

Numerous **videos** and **interactivities** are available just where you need them, at the point of learning, in your digital formats, learnON and eBookPLUS at **www.jacplus.com.au**.

7.1.1 Introduction

One of the most important areas of research in biology focuses on how an organism's genes and alleles affect its physical characteristics. This is still an area of considerable research and there is much debate over the relative contribution of genetic and environmental factors towards an organism's physical make-up.

FIGURE 7.1 By studying the different characteristics of pea plants, Gregor Mendel's work founded the basis for our current understanding of genes and alleles.

The pioneering work of Gregor Mendel with pea plants in the mid-nineteenth century laid the foundations for our current understanding of how traits are inherited. We can now make predictions about the characteristics of the offspring based on the characteristics of the parents and recognise different patterns of how certain characteristics are expressed. Modern genetics can now explain why some characteristics may be very common in a family and found in every generation, whereas others may suddenly appear. In addition, fascinating research has shown that events in someone's lifetime could change the way their DNA is expressed, and how this expression pattern is passed on to the next generation.

In this topic you will learn how geneticists use symbols to represent alleles of genes and how phenotypes are expressed in a variety of different inheritance patterns. Finally, the contributions that genetic and environmental factors have on the physical characteristics of an organism will be examined alongside epigenetic factors.

LEARNING SEQUENCE

- 7.1 Overview 412
- 7.2 Writing genotypes using symbols 413
- 7.3 Expression of dominant and recessive phenotypes 419
- 7.4 Influences on phenotypes of genetic material, and environmental and epigenetic factors 429
- 7.5 Review 438

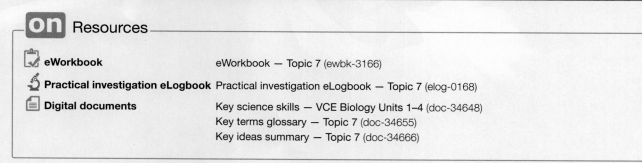

7.2 Writing genotypes using symbols

KEY KNOWLEDGE

- The use of symbols in the writing of genotypes for the alleles present at a particular gene locus

Source: VCE Biology Study Design (2022–2026) extracts © VCAA; reproduced by permission.

7.2.1 Mendel's contribution to genetics

BACKGROUND KNOWLEDGE: Mendel — the father of genetics

Gregor Mendel's work on pea plants transformed our understanding of how characteristics are inherited across generations. His work was based on careful observation and meticulous record keeping. Carried out over eight years, his breeding experiments included starting from known parental traits. Mendel's initial experiment tested only one trait at a time (**monohybrid cross**). Once he recognised the pattern of inheritance for that trait, he crossed plants that differed in two traits (**dihybrid cross**). This allowed him to recognise that each trait was controlled by a pair of factors (such as yellow or green), and that each of these factors were retained across generations, with one factor per gamete produced. Furthermore, these pairs of factors behaved independently of other pairs of factors — this is the principle of independent assortment.

For further information on Mendel's experiments and results, and their significance to the scientific community, please download the digital document.

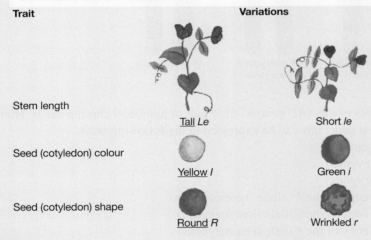

FIGURE 7.2 Some of the variations in pea plants used by Mendel. Dominant traits are underlined.

monohybrid cross a cross in which alleles of only one gene are involved

dihybrid cross a cross in which alleles of two different genes are involved

genotype both the double set of genetic instructions present in a diploid organism and the genetic make-up of an organism at one particular gene locus

phenotype an observable or measurable characteristic of an organism that is the product of genetic and environmental factors

 Resources

Digital document Background knowledge: Mendel — the father of genetics (doc-36114)

7.2.2 What is a genotype?

A **genotype** is the underlying genetic make-up that determines an organism's characteristics (or **phenotype**). A genotype is:
- not visible — only its phenotypic effects can be seen or measured
- the combination of the particular alleles of a gene, or genes, that are present and active in a cell or in an organism and determine a specific aspect of its structure or functioning.

Autosomal genes are those located on an autosome, found in both males and females. While each type of organism shares identical genes, they may differ in the specific alleles they possess. Each individual has two alleles for an autosomal gene — one on each chromosome. As we learned in section 6.3.3, in diploid organisms the genotype is usually written as a pair of alleles.

Figure 7.3 shows three different pairs of homologous chromosomes, each from a different organism and displaying different combinations of alleles (or the genotype). If the two alleles of a gene are identical — for example, *CC* — the genotype (and the organism concerned) is said to be **homozygous**. If the two alleles are different — for example, *Cc* — the genotype is described as **heterozygous**.

> **autosomal genes** have two copies of each gene located on an autosome
>
> **homozygous** a genotype at a particular gene comprising of two identical alleles; for example, *AA* or *aa*
>
> **heterozygous** a genotype at a particular gene comprising of two different alleles; for example, *Aa*

FIGURE 7.3 Different combinations of alleles (*A* and *a*) of a gene: the homozygous *AA* and *aa*, and the heterozygous *Aa*

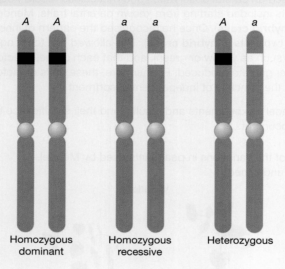

For example, each person has two copies of the *ABO* gene on their pair of number-9 chromosomes. However, people may have different alleles of that gene; this can be expressed in the following ways:
- two copies of the *i* allele (homozygous)
- two copies of the I^A allele (homozygous)
- two copies of the I^B allele (homozygous)
- one copy of the I^A allele and one copy of the I^B allele (heterozygous)
- one copy of the I^A allele and one copy of the *i* allele (heterozygous)
- one copy of the I^B allele and one copy of the *i* allele (heterozygous).

The particular combination of alleles of a gene is that person's genotype in terms of that gene.

The gametes produced by a homozygous person, such as genotype $I^A I^A$, will be identical, with all having the same allele. The gametes produced by a heterozygous person, such as genotype $I^A I^B$, will be of two kinds, with half having the I^B allele and half having the I^A allele.

- A genotype is the combination of the particular alleles of a gene or genes that are present and active in a cell or in an organism.
- The genotype determines a specific aspect of the gene's structure or functioning.

Genotypes for genes on the sex chromosomes

The sex chromosomes (or allosomes) in mammals are the X and the Y chromosomes. The DNA of each chromosome contains genes, and because of their size differences, the X chromosome has many more gene loci than the Y chromosome (see figure 7.4).

For genes on the X chromosome, females may be either homozygous or heterozygous. **Hemizygous** genotypes occur only in males for X-linked and Y-linked genes. For these genes only, the genotype of a normal male consists of just one allele (hemi = half).

People have been aware for a long time that some conditions, such as certain colour vision defects and a blood-clotting disorder (haemophilia) that occur in particular families, appear more often in males than in females. This is because the genes controlling colour vision and blood clotting are located on the X chromosome and are not represented on the Y chromosome. To be affected, females must inherit two copies of the particular allele; however, males are affected if they have just one allele, so they more commonly show the trait. This is discussed in more detail in topic 8.

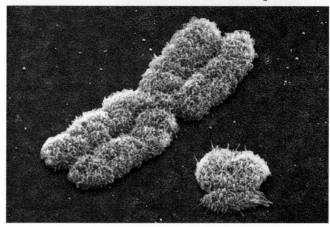

FIGURE 7.4 The human sex chromosomes magnified 10 000 times. The X chromosome (left) is much larger than the Y chromosome and carries about 800 genes; in contrast, the Y chromosome carries about 50 genes.

Resources

- Video eLesson Genotypes (eles-4222)
- Interactivity Genotypes (int-0668)

hemizygous the genotype with respect to any gene carried on either the X or the Y chromosome, which comprises just a single allele for each gene

7.2.3 Writing a genotype

To write the genotype at a gene, we list the combination of alleles. Shorthand notation that uses variants of the same letter(s) of the alphabet is used as follows:
- Where a gene has two phenotypic expressions or alleles — such as 'trait present' and 'trait absent', or 'red' and 'white' flower colour — symbols such as R and r or D and $D´$ might be used, depending on the dominance relationship between the two alleles. Usually, the letter chosen relates to one of the phenotypic expressions of the gene, such as R for red flower colour.
- Where a gene has multiple alleles, each having a different phenotypic expression, a common letter is still used, but with the addition of superscripts — for example, I^A, I^B and i, or C, c^b, c^s and c^a.
- Where a gene is on the X chromosome, the chromosome must be specified, with an allele in superscript. For example, in a gene on the X chromosome related to colour blindness, X^B may be the allele for normal vision and X^b the allele for colour blindness. Females can be homozygous ($X^B X^B$ or $X^b X^b$) or heterozygous ($X^B X^b$). Males only have one allele, so can only have a hemizygous genotype ($X^B Y$ or $X^b Y$). No allele is assigned to the Y chromosome as that particular gene is not present.

To better understand this notation system, let's look at some specific examples.

When using alleles, a **dominant** trait is represented by a capital letter and the lowercase letter represents the allele from the **recessive** trait. Any letter can be used to represent an allele. If using a letter that looks similar in upper and lowercase, a line can be drawn above the lowercase letter when handwritten to avoid confusion — for example, $\overline{c}$.

> **dominant** a trait that is expressed in the heterozygous condition; also a trait that requires only a single copy of the responsible allele for its phenotypic expression
>
> **recessive** a trait that is not expressed but remains hidden or masked in a heterozygous organism

CASE STUDY: Cystic fibrosis

Cystic fibrosis (CF) is a genetic disease that affects humans. The gene responsible is the cystic fibrosis transmembrane conductance regulator (*CFTR*) gene and is located on chromosome number 7. The *CFTR* gene encodes a transmembrane protein that is responsible for allowing chloride ions to move out of cells. This transport of chloride ions helps control the movement of water in tissues that is necessary to produce thin mucus. Mucus is a slippery substance that lubricates and protects the lining of the airways, digestive system, reproductive system, and other organs and tissues. Sufferers of CF have a mutation in the *CFTR* gene that produces a non-functioning transmembrane protein. This means their cells cannot transport chloride ions properly, causing a thick, sticky mucus to build up in their lungs and the organs of the digestive system. This can cause difficulty in breathing and enable infections to more easily develop.

CF is the most common inherited single-gene disorder seen in Caucasians of northern European descent, and in their derived populations in Australia, Canada and New Zealand. It occurs equally in females and in males. The incidence of CF in Caucasians is generally stated to be about 1 in every 2500 live births, but in other populations — such as Asian and Pacific Islander populations — the incidence is much lower.

FIGURE 7.5 Cystic fibrosis produces a non-functioning transmembrane protein, which inhibits the transport of chloride ions and causes a thick, sticky mucus to build up in the lungs and other organs.

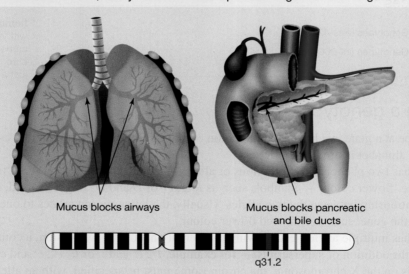

CF cannot be cured but the development of affected babies is carefully monitored; they are closely watched for any bacterial infections of the lungs, which are rapidly treated. Other treatments include pancreatic enzyme replacement and physiotherapy sessions to assist in removing the thick, sticky mucus from the lungs.

In the case of the *CFTR* gene, the allele that produces the normal transmembrane protein can be represented by the letter *C*. If you have a mutation of the *CFTR* gene, it causes the disease, and so is represented by the letter *c*. Remember that the *CFTR* gene is located on chromosome 7 and so each person, male or female, will have two copies of the gene. Sufferers of the condition have the genotype *cc* because two defective alleles, one on each chromosome 7, are needed to develop the disease. The letter *C/c* is chosen as it is the first letter of the disease.

CASE STUDY: Fur colour in rabbits and multiple alleles

The colour of fur in rabbits is controlled by a gene that has multiple different alleles. This produces several different colours (figure 7.6). Because the gene controls fur colour, we will assign the letter F for fur. The addition of superscript letters to the common letter is used when multiple alleles of the gene are present. There are four common alleles:

F — black or brown colour

F^{ch} — chinchilla colouration (greyish colour)

F^h — Himalayan colouration (white body and dark ears, face, feet and tail)

f — albino (white colour)

Therefore, a black rabbit may have the genotype FF. Since the gene is on an autosome, all rabbits of either sex will have two alleles for fur colour.

FIGURE 7.6 Colour variation in the fur of rabbits. The variation is a result of multiple alleles for the same gene. **a.** Black colouration, F **b.** Chinchilla colouration, F^{ch} **c.** Himalayan colouration, F^h **d.** Albino colouration, f

tlvd-1800

SAMPLE PROBLEM 1 Choosing symbols for alleles

Eye colour in some species of terrestrial animals is controlled by one gene with multiple alleles. There are four possible colours: red, green, black and grey. Choose suitable symbols for each eye colour allele. **(1 mark)**

THINK	WRITE
1. Look carefully at what the question is asking. Is the characteristic controlled by one gene with multiple alleles or is it a case of trait present/trait absent?	In this example, eye colour is controlled by one gene with multiple alleles.
2. Choose a capital letter based on the characteristic.	The characteristic is 'eye colour', therefore the capital letter E is chosen.
3. Use superscript letters to distinguish between the different types of characteristic.	E^R is used for red eye, E^G is used for green eye, E^B is used for black eye and E^{GR} is used for grey eye (1 mark).

 Resources

 eWorkbook Worksheet 7.1 Writing genotypes (ewbk-7229)

KEY IDEAS

- The combination of alleles for a particular gene makes up an organism's genotype.
- Where a trait is either absent or present, a capital letter and corresponding lowercase letter are used to represent the genotype.
- If multiple alleles exist for a gene, superscript letters are added to capital or lowercase letters.

7.2 Activities

learn on

To answer questions online and to receive **immediate feedback** and **sample responses** for every question go to your learnON title at **www.jacplus.com.au**.

| 7.2 Quick quiz | 7.2 Exercise | 7.2 Exam questions |

7.2 Exercise

1. Define the term *genotype*.
2. Explain why diploid organisms will have a genotype that is composed of two alleles for a particular gene.
3. What is the advantage of using symbols to represent genotypes over words or terms?
4. In fruit flies *(Drosophila melanogaster)*, body colour is controlled by a single gene with two alleles. Brown body colour is dominant to black body colour. Choose suitable symbols for each body colour allele.
5. Height in tomato plants is controlled by one gene. All tomato plants are either 'tall' or 'dwarf' type. Tall is dominant to dwarf. Choose suitable symbols for the 'tall' allele and 'dwarf' allele.
6. Galactosemia is a rare genetic condition that affects an individual's ability to metabolise the sugar galactose properly. If untreated it can lead to loss of coordination and reduction in bone density.
   In order to suffer from the condition, an individual must have two copies of the disease allele.
   a. Using suitable allele symbols, give the genotype(s) of a sufferer of galactosemia.
   b. Using suitable allele symbols, give the genotype(s) of a non-sufferer of galactosemia.

7.2 Exam questions

Question 1 (1 mark)

MC Consider the gene for height in the tomato plant, where the *D* allele codes for a tall plant, while *d* codes for a dwarf plant.

With respect to the height gene, it is correct to state that in a tomato plant
A. the dwarf trait is dominant to the tall trait.
B. there is one copy of *D* or *d* in each of its somatic cells.
C. the *D* allele has the same DNA nucleotide sequence as the *d* allele.
D. the only possible allele combinations are *DD*, *Dd* or *dd*.

Question 2 (1 mark)

MC In cats, one gene with alternative alleles influences coat colour, with black coat colour being dominant to grey.

Appropriate symbols to denote these alleles would be

A. black (*B*) and grey (*g*).
B. black (*b*) and grey (*g*).
C. black (*B*) and grey (*b*).
D. black (B^1) and grey (B^2).

Question 3 (1 mark)

MC The gene for seed colour in the common edible pea *(Pisum sativum)* is autosomal. It has two alleles: *G* and *g*.

What is the genotype of a pea plant that is heterozygous for seed colour?
A. *Gg*
B. $X^G X^g$
C. *GG* or *gg*
D. $X^G Y$ or $X^g Y$

▶ **Question 4 (1 mark)**
MC For the two sexes of a mammal, all possible genotypes for a gene with two alleles (*A* and *a*) can appropriately be identified as
A. homozygous *AA* or *aa* in males for an autosomal gene.
B. *AA*, *Aa* and *aa* in males for an X-linked trait.
C. homozygous X^aX^a or X^AX^A in females for an X-linked trait.
D. homozygous *AA* or *aa*, or heterozygous *Aa*, in females for an autosomal gene.

▶ **Question 5 (1 mark)**
Source: *VCAA 2008 Biology Exam 2, Section B, Q3b*
Another gene associated with blood groups has the alleles **H** (H substance) and **h** (lack of H substance). The pathway involved can be summarised as follows.

$$\text{mucopolysaccharide precursor} \xrightarrow{\text{in the presence of at least one } \mathbf{H} \text{ allele}} \text{H substance}$$

Write the genotype of an individual who fails to produce H substance.

More exam questions are available in your learnON title.

7.3 Expression of dominant and recessive phenotypes

KEY KNOWLEDGE

- The expression of dominant and recessive phenotypes, including codominance and incomplete dominance

Source: VCE Biology Study Design (2022–2026) extracts © VCAA; reproduced by permission.

7.3.1 What is a phenotype?

A phenotype is the observable or measurable characteristics of an organism. It is a product of the genotype and the **environment**.

environment the external conditions (both biotic and abiotic factors) that surround and affect an organism

For example, in section 7.2.3 we learned about cystic fibrosis being caused by a mutation to the *CFTR* gene, which codes for a non-functioning transmembrane protein. If a person has the genotype *cc* — that is, two alleles for the non-functioning transmembrane protein — their phenotype will be a sufferer of cystic fibrosis because this is an observable characteristic. If they have at least one copy of the functioning *CFTR* gene, their phenotype will be non-sufferer.

Figure 7.6 shows several different phenotypes of rabbit fur. In this case, a genotype of $F^{ch}F^{ch}$ will produce the phenotype of chinchilla colouration (greyish colour). Again, this colouration is an example of a phenotype as it is an observable or measurable characteristic.

- A phenotype is an observable or measurable characteristic in an organism that is the result of its genotype and its environment.
- It is an observable expression of the organism's genotype.

Other examples of phenotypes include:
- short leg length and normal leg length in sheep (*Ovis aries*) (figure 7.7a)
- blood types A, B, AB and O in humans
- purple, white and yellow kernel colour in corn (*Zea mays*) (section 6.3.3; figure 6.14b)
- red, green, orange and yellow fruit colours in capsicums (section 6.3.3; figure 6.15)
- beaked, round, flattened and elongated fruit shapes in tomatoes (*Solanum lycopersicum*) (figure 7.7b)
- presence and absence of fur and eye pigment in kangaroos (*Macropus* spp.)
- ability and inability to differentiate between the colours red and green in humans
- requirement or non-requirement for the amino acid arginine for growth in yeast (*Saccharomyces cerevisiae*).

FIGURE 7.7 Examples of phenotypes include **a.** leg length in sheep, with the abnormally short legs of an Ancon sheep seen here in comparison with the normal leg length, and **b.** shape and colours of tomato fruits.

CASE STUDY: Phenotype differences in prokaryotes

Phenotypic differences can be seen not only in eukaryotic species, but also in microbial species. One significant phenotypic difference that affects human health is the emergence of resistance to antibiotics in many bacterial species. For example, *Staphylococcus aureus* (also known as golden staph) includes some strains that are sensitive to one or more antibiotic drugs and some that are resistant to virtually all the current useful antibiotics. These antibiotics include methicillin, which interferes with the synthesis of a component of the bacterial cell wall, and erythromycin and streptomycin, which attack the bacterial ribosomes.

Different strains of one bacterial species, as well as different bacterial species, can be distinguished as being either drug-sensitive or drug-resistant using antibiotic sensitivity testing (see figure 7.8). In this test, a specific bacterial species is allowed to grow across the surface of a gel in a Petri dish. The gel contains all the nutrients required for growth and the bacterial growth appears as a white film. Discs impregnated with different antibiotics (or with different concentrations of the same antibiotic) are placed on the plate and the antibiotic will diffuse from the disc into the gel. If the bacteria are sensitive to the antibiotic on a disc, the bacteria in the area will die, and this appears as a clear area around the disc.

FIGURE 7.8 Results of antibiotic sensitivity testing of various antibiotics against two strains of bacteria. Areas of no growth around a disk show that a bacteria is sensitive to the antibiotic.

> **Resources**
>
> ▶ **Video eLesson** Phenotypes (eles-4223)

7.3.2 Relationship between expression of alleles

If you are simultaneously given two instructions that are mutually exclusive, such as 'Turn left at the next corner!' and 'Turn right at the next corner!', you can carry out only one of these instructions. However, for two other instructions, such as 'Paint the door yellow' and 'Paint the door red', you might carry out both instructions by producing a bicoloured yellow and red door. Or you might mix the colours and produce an orange door. Similar situations can be recognised for the phenotypes produced by genes.

Complete dominance

As we learned in section 7.2.3 in the Case study of cystic fibrosis (CF), there are two alleles for the *CFTR* gene — one that codes for a working transmembrane protein, represented by the notation *C*, and another that codes for a faulty protein, represented by the notation *c*.

A person with the heterozygous *Cc* genotype has two mutually exclusive genetic instructions, but such a person produces an active transporter protein and can produce normal mucus secretions, and so is free of cystic fibrosis. From this, we can conclude that the disease-free condition with normal mucus secretions is dominant to cystic fibrosis with abnormal thick mucus. The relationship between the alleles of the *CFTR* gene are shown in table 7.1. Note that a person with the *cc* genotype can produce only an inactive abnormal transporter protein and so has cystic fibrosis.

TABLE 7.1 Relationship between alleles of the *CFTR* gene

Genotype	Phenotype
Homozygous *CC*	Normal mucus (unaffected)
Heterozygous *Cc*	Normal mucus (unaffected)
Homozygous *cc*	Thick, sticky mucus (cystic fibrosis)

In this case, the production of normal mucus (being unaffected by CF) is referred to as dominant. In dominant traits, only one allele is required to express the trait (as seen in table 7.1). Recessive traits need two copies of a specific allele in order to be expressed. If an allele for a dominant trait is present, it masks the recessive trait. In this example, we say that unaffected is dominant to cystic fibrosis.

> To decide whether a trait is dominant or recessive, the phenotype of a heterozygous organism is identified. The trait that is expressed in this phenotype is the dominant trait.

Alleles that control dominant traits are usually symbolised by a capital letter; for example, the allele *S* controls the dominant trait of short fur length in cats. Alleles that control recessive traits are symbolised by the same letter in lowercase; for example, the allele *s* controls the recessive trait of long fur length in cats.

Some additional examples of dominant and recessive traits are shown in figure 7.9 and table 7.2.

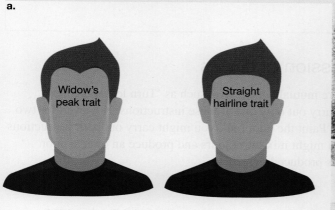

FIGURE 7.9 a. Widow's peak hairline and straight hairline. A widow's peak is dominant to a straight hairline. **b.** Two parents with their seven sons and one daughter. Most show a widow's peak hairline, which is an inherited trait.

TABLE 7.2 Some dominant and recessive human traits. Alleles for dominant traits are shown with a capital letter and those for recessive traits with a lowercase letter.

Dominant trait	Recessive trait
Peaked hairline (widow's peak) (*W*) (see figure 7.9)	Straight hairline (*w*)
Free ear lobes (*F*)	Attached ear lobes (*f*)
Mid-digital hair present (*G*)	Mid-digital hair absent (*g*)
Shortened fingers (brachydactyly) (*S*)	Normal length fingers (*s*)
Normal pigmentation (*A*)	Pigmentation lacking (albinism) (*a*)
Non-red hair (*R*)	Red hair (*r*)
Normal secretions (*C*)	Cystic fibrosis (*c*)
Dwarf stature (achondroplasia) (*N*)	Average stature (*n*)
Rhesus positive (Rh +ve) blood (*D*)	Rhesus negative (Rh −ve) blood (*d*)

Being a carrier

In genetics, the term **carrier** refers to a heterozygote that has the allele for a recessive trait but does not show the trait in their phenotype. In people, alleles may be carried for hidden recessive traits that do not affect normal functioning, such as straight hairline (figure 7.9) and blood type O. However, some alleles that are carried by heterozygotes are for recessive disorders, such as cystic fibrosis or albinism (table 7.2).

> **carrier** an organism that has inherited an allele for a recessive genetic trait but usually does not display that trait or show symptoms of the disease

Heterozygotes are most often *not* aware of their carrier status for an allele controlling a recessive trait. Parents may realise they are carriers only when they have a baby with a recessive disease. Parents of children with cystic fibrosis may have been unaware that they were both genotype *Cc*, and so were carriers of an allele that resulted in the disease, until their child was born.

Heterozygotes carry alleles for recessive traits, but their effects are not expressed.

Codominance

Codominance is when a heterozygote expresses both the dominant and the recessive trait of a gene in its phenotype. This is a type of partial dominance.

Codominance examples can be found throughout the plant and animal kingdoms. For example, the gene that controls flower colour in rhododendron plants has multiple alleles. The allele for red flowers (C^R) is codominant to white (C^W). When a homozygous red-flowered plant ($C^R C^R$) is crossed with a homozygous white-flowered plant ($C^W C^W$), a plant with both red and white flowers will be produced ($C^R C^W$) (figure 7.10).

> **codominance** the relationship between two alleles of a gene such that a heterozygous organism shows the expression of both alleles in its phenotype

FIGURE 7.10 Codominance in rhododendron plants **a.** A red-flowered plant $C^R C^R$ crossed with **b.** a white-flowered plant $C^W C^W$ produces **c.** a red-and-white-flowered plant $C^R C^W$.

$C^R C^R$

$C^W C^W$

$C^R C^W$

CASE STUDY: Codominance in blood

The *ABO* gene, located on the number-9 chromosome, has three alleles that determine antigen production. Antigen A and antigen B occur on the surface of the red blood cells of some people. Depending on which antigens are present, blood is typed as group A, B, AB or O (figure 7.11). The presence (or absence) of a particular antigen is inferred by adding specific antibodies and observing the result (see figure 7.12). Antibodies used to type blood are anti-A antibodies and anti-B antibodies. Anti-A antibodies cause clumping or agglutination of red blood cells with antigen A. Anti-B antibodies cause clumping of red blood cells with antigen B.

FIGURE 7.11 Blood is typed depending upon which antigens occur on the surface of red blood cells.

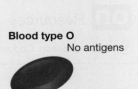

Look at the reaction of the blood sample in column 3 of figure 7.12.
- When anti-A antibodies were added the blood clumped, so the red blood cells have antigen A and the allele I^A.
- When anti-B antibodies were added the same sample also clumped, so the red blood cells have antigen B and must also have the allele I^B.

This genotype is heterozygous $I^A I^B$. Because both traits are expressed in the heterozygote, these two alleles show codominance. Alleles showing codominance are denoted by a capital letter with a superscript added to distinguish between them. Table 7.3 shows that the phenotypic actions of both the I^A and I^B alleles are dominant to the action of the *i* allele, and so blood group O is recessive to the other blood types.

FIGURE 7.12 The addition of specific antibodies to blood samples causes cells to 'clump' or agglutinate when the corresponding antigen is present on the surface of the red blood cells.

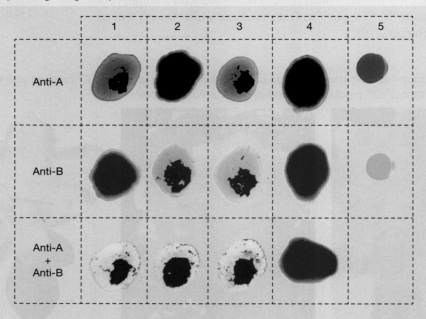

TABLE 7.3 Relationship between genotypes and phenotypes for the *ABO* gene. What relationship exists between the I^A and the I^B alleles? What relationship exists between the I^B and the *i* alleles?

Genotype	Instructions carried by alleles	Phenotype
Homozygous $I^A I^A$	'Produce antigen A'	Blood type A
Homozygous $I^B I^B$	'Produce antigen B'	Blood type B
Homozygous *ii*	'Produce neither antigen A nor B'	Blood type O
Heterozygous $I^A I^B$	'Produce antigen A' and 'produce antigen B'	Blood type AB
Heterozygous $I^A i$	'Produce antigen A' and 'produce neither antigen'	Blood type A
Heterozygous $I^B i$	'Produce antigen B' and 'produce neither antigen'	Blood type B

 Resources

▶ **Video eLesson** Codominance (eles-4224)

Incomplete dominance

Prior to Mendel's work, the common belief was that all characteristics were blended. Mendel showed that this was not the case and that some traits were dominant over other recessive ones. However, blending of characteristics does occur for the heterozygotes of both animals and plants for some characteristics, and this is known as **incomplete dominance**.

incomplete dominance the appearance in a heterozygote of a trait that is intermediate between either of the trait's homozygous phenotypes

In humans, incomplete dominance is seen in hair type — straight, curly or wavy hair.
- $H^S H^S$ genotype are individuals with straight hair.
- $H^C H^C$ genotype are individuals with curly hair.
- $H^S H^C$ genotype are individuals who inherit one straight and one curly allele, and have an intermediate hair shape between straight and curly, or a blend of the two shapes, producing wavy hair (figure 7.13).

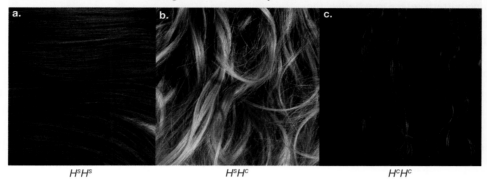

FIGURE 7.13 Human hair shape with corresponding genotypes. Wavy hair $H^S H^C$ is an intermediate form of straight $H^S H^S$ and curly $H^C H^C$.

Incomplete dominance can also be observed in plants. A good example is exhibited when looking at snapdragons *(Antirrhinum majus)*, where multiple alleles control flower colour.
- $C^R C^R$ genotype is a red-flowered snapdragon plant.
- $C^W C^W$ genotype is a white-flowered snapdragon plant.
- $C^R C^W$ genotype is the genotype produced by the cross of a red-flowered snapdragon plant with a white-flowered snapdragon plant. It produces an intermediate form with a blend of the two colours — that is, a pink-flowered snapdragon (figure 7.14).

By understanding whether traits are inherited in a codominant or incomplete dominant manner, plant breeders and horticulturalists can intentionally breed certain plants together to create a huge amount of variety not only in flower colour, but for other traits too.

FIGURE 7.14 Snapdragon plants. What is the genotype for colour for each plant shown?

elog-0792

INVESTIGATION 7.1

online only

Comparing phenotypes

Aim

To compare phenotypes with other members of the class and determine the genotype of traits based on these phenotypes

TOPIC 7 Genotypes and phenotypes **425**

SAMPLE PROBLEM 2 Determining dominance patterns

Gregor Mendel carried out many experiments breeding tall pea plants with small, or dwarf, pea plants. He found that the offspring were always either tall or small. Explain what this shows about the dominance pattern for size in the pea plants. **(2 marks)**

THINK	WRITE
1. Look carefully at what the question is asking. Dominance pattern refers to whether the alleles involved are inherited in a complete dominant, codominant or incomplete dominant manner. Now, examine the characteristics of the offspring.	All the offspring are either tall or small.
2. Are any of the offspring of an intermediate form?	The offspring are not an intermediate form of their parents — they are only tall or small. They do not display incomplete dominance.
3. Do they display both characteristics of the parents?	They cannot display both characteristics of tall and small. They do not display codominance (1 mark).
4. Determine the dominance pattern.	Size in pea plants must be of a complete dominance pattern (1 mark).

 Resources

 eWorkbook Worksheet 7.2 Dominant and recessive phenotypes (ewbk-7231)
Worksheet 7.3 Codominance and incomplete dominance (ewbk-7233)

 Interactivities Generating the phenotypes (int-0178)
Making families (int-0681)

KEY IDEAS

- A phenotype is an observable or measurable characteristic of an organism determined by the genotype and the environment.
- Alleles are expressed through various dominance patterns.
- Completely dominant traits are expressed over recessive traits in an organism's phenotype. This can be seen in heterozygous individuals.
- Codominant traits are expressed together, and both alleles contribute towards the phenotype.
- Incomplete dominance refers to an intermediate or blending of parental characteristics in the offspring's phenotype.

7.3 Activities

To answer questions online and to receive **immediate feedback** and **sample responses** for every question go to your learnON title at **www.jacplus.com.au**. A **downloadable solutions** file is also available in the resources tab.

| 7.3 Quick quiz | 7.3 Exercise | 7.3 Exam questions |

7.3 Exercise

1. Using table 7.2, state the genotype of an individual with red hair.
2. Explain why a carrier for cystic fibrosis is not a sufferer of the disease.
3. Cattle with white coats can be bred with cattle with red coats. This produces offspring with the genotype $C^R C^W$, which means they have a red-and-white coat. State what type of inheritance this displays. Justify your answer.
4. If a horse with a cream coat breeds with a horse with a red coat, offspring with a palomino or buckskin coat, which is golden/yellowish in colour, are produced. Explain what this shows about the dominance pattern for coat colour in horses.
5. Fur colour in rabbits is controlled by multiple alleles. F — black or brown colour is dominant over F^{ch} — chinchilla colouration (greyish colour), which in turn is dominant over F^h — Himalayan colouration (white body and dark ears, face, feet and tail), which is again dominant over f — albino (white colour).
    a. State the genotype of an albino rabbit.
    b. Give all possible genotypes of a rabbit with chinchilla colouration.
6. Suggest why it is important that doctors know the phenotype of their patient's blood.
7. a. Using the table, identify — as fully as possible and including the sexes — the phenotypes that correspond to the following genotypes:

    i. Dd ii. $I^A i$ iii. aa iv. $X^h Y$.

TABLE Some human genes and their common alleles. Note that for alleles for X-linked genes, the X chromosome notation has been omitted for simplicity.

Gene and its function		Chromosomal location	Common alleles	
ABO	Encodes A B antigens	9	I^A I^B i	Produces antigen A Produces antigen B Produces neither antigen
CBD	Encodes green-sensitive pigment	X	V v	Produces green-sensitive pigment Lacks pigment (colour vision defect)
DMD	Encodes muscle protein (dystrophin)	X	M m	Produces normal muscle protein Produces abnormal muscle protein
F8	Encodes factor VIII blood-clotting protein	X	H h	Produces factor VIII No factor VIII (haemophilia)
HBB	Encodes beta chains of haemoglobin	11	T t	Produces beta chains Beta chains missing (thalassaemia)
RHD	Encodes Rhesus D antigen	1	D d	Rhesus positive Rhesus negative
TYR	Encodes tyrosinase enzyme	11	A a	Produces tyrosinase enzyme No enzyme (albinism)

   b. Using the allelic notation in the table, write possible genotypes for the following persons:
      i. A male hemizygous for red–green colour blindness
      ii. A male with normal beta chains but whose child has thalassaemia
      iii. A female carrier of Duchenne muscular dystrophy
      iv. A female homozygous for Rhesus positive blood type
      v. A male with group O blood type

7.3 Exam questions

Question 1 (1 mark)
MC In humans, the *CBD* gene controls colour vision. The gene locus is on the X chromosome. There are two alleles, X^B and X^b. Consider two individuals with the following phenotypes:
- Individual 1 is a male with normal colour vision.
- Individual 2 is a male with defective colour vision.

From this information it can be stated that
A. normal colour vision is recessive to defective colour vision.
B. there are three possible genotypes in males with respect to the *CBD* gene.
C. individuals 1 and 2 are both homozygous at the *CBD* locus.
D. the phenotype of a heterozygous female will confirm the dominant phenotype.

Question 2 (1 mark)
MC In humans, the *TYR* gene produces the enzyme tyrosinase. Tyrosinase controls pigment production. People who lack tyrosinase cannot produce pigment. They have albinism and have no pigmentation in their hair, skin or eyes. There are two alleles for the *TYR* gene: *A* and *a*. Heterozygotes for *TYR* can produce tyrosinase and have normally pigmented skin, hair and eyes.

From this information it can be concluded that
A. albinism is dominant over normal pigmentation.
B. albinism is recessive to normal pigmentation.
C. production of tyrosinase is needed to show albinism.
D. the genotype of an albino is *AA*.

Question 3 (1 mark)
Source: VCAA 2006 Biology Exam 2, Section B, Q1a

In humans the presence of a dimple in the chin is dominant. The first child of a couple, each with a dimple, does not have a dimple but their second child does have a dimple.

Use alleles **D** and **d** to show the genotype corresponding to the phenotype of the parents and the child without a dimple.

genotype of parents:

genotype of child without a dimple:

Question 4 (1 mark)
MC Stem texture in tomatoes is controlled by one gene with two alleles, smooth and rough, with smooth being dominant to rough.

Based on this information, which of the following is the most reasonable conclusion?
A. Rough-stemmed plants would have one of two possible genotypes.
B. Smooth-stemmed plants would have one of two possible genotypes.
C. All rough-stemmed plants are heterozygous.
D. All smooth-stemmed plants are heterozygous.

Question 5 (3 marks)
In the tomato (*Lycopersicon esculentum*), red fruit colour is dominant to yellow fruit colour. The gene for fruit colour is autosomal.
a. Assign allele symbols of your choice to represent the two alleles for the fruit colour gene and state clearly which symbol represents each allele. **1 mark**
b. With respect to fruit colour in the tomato, what is the:
   i. genotype of a heterozygote **1 mark**
   ii. phenotype of a heterozygote? **1 mark**

More exam questions are available in your learnON title.

7.4 Influences on phenotypes of genetic material, and environmental and epigenetic factors

> **KEY KNOWLEDGE**
>
> - Proportionate influences of genetic material, and environmental and epigenetic factors, on phenotypes
>
> **Source:** VCE Biology Study Design (2022–2026) extracts © VCAA; reproduced by permission.

7.4.1 The environment and phenotypes

Do you know any identical twins? If so, do they look absolutely identical? Most people who do know identical twins have usually developed some way to tell them apart. This is because the phenotype of an individual is not always produced by the genotype alone. In some cases, the phenotype is a result of the *interaction* between the genotype and environment.

$$\text{Genotype} + \text{environment} \rightarrow \text{phenotype}$$

Many environmental factors, both external and internal, act on an organism so that its final phenotype is due to varying contributions of genotype and environment.

The phenotype due to a particular genotype may appear only in one specific environment. In a different environment, another phenotype may appear. Let's examine some examples of how environmental factors can affect the phenotype.

CASE STUDY: PKU and dietary-controlled phenotype

The inherited disorder phenylketonuria (PKU) results from the action of the gene that controls production of an enzyme known as phenylalanine hydroxylase. Babies that inherit the homozygous recessive genotype *pp* from their parents cannot produce this enzyme. If these babies are fed diets including proteins that contain normal quantities of the amino acid phenylalanine (Phe), the babies will suffer brain damage and be severely developmentally delayed. Therefore, early diagnosis is critical. Brain damage can be prevented in these babies — and they will show a normal phenotype — if they are fed a special diet that includes proteins with very low levels of Phe.

In this case, the phenotype of a child (PKU or normal) with genotype *pp* depends on the internal environment that is controlled by the diet:

$$\text{Genotype } pp + \text{HIGH Phe diet} \rightarrow \text{phenotype: PKU}$$

$$\text{Genotype } pp + \text{LOW Phe diet} \rightarrow \text{phenotype: normal}$$

CASE STUDY: Acid-sensitive colouration in plants

Hydrangea plants (figure 7.15) produce blooms with colours that depend on the acidity or alkalinity (pH) of the soil in which they are growing. The colour is due to pigments known as anthocyanins, which are located in membrane-bound sacs within the petal cells. In soil with an acidic pH these pigments are a bright blue, while at alkaline pH they are a pink or red.

FIGURE 7.15 The colour of a hydrangea varies with soil pH.

CASE STUDY: Temperature-sensitive colouration in mammals

At birth, the Siamese kittens shown in figure 7.16a were all white. A few weeks later the kittens began to develop pigmentation along the edges of their ears. Gradually the pigment spread until the kittens showed the characteristic Siamese cat colouring on their faces, ears, feet and tails. This pattern of colour change is due to an interaction between the cats' genotypes and their environment.

Siamese cats have a particular form of a gene that codes for the production of tyrosinase. This enzyme catalyses one step in the production of pigment:

$$\text{precursor} \xrightarrow{\text{tyrosinase}} \text{pigment}$$

In Siamese cats, the particular form (allele) of this gene produces a tyrosinase enzyme that is heat-sensitive. This enzyme can catalyse the step in the production of pigment when the temperature is lower than the core body temperature only. Siamese kittens undergo embryonic development in a warm uterine environment and so are born unpigmented (white). Pigment appears first on the coolest parts of their bodies — the ear margins — and then on other extremities (figure 7.16b).

FIGURE 7.16 a. Red point and seal point Siamese kittens. What colour were they at birth? **b.** Temperature-sensitive colouration develops after birth.

7.4.2 Epigenetics: above genetics

The term **epigenetics** literally means 'above genetics', 'in addition to genetics' or 'on top of genetics' Epigenetics is the study of how cells with identical genotypes can show different phenotypes.

These differences:
- are not due to differences in the base sequences of the DNA of their genes
- are stable within an organism
- in some cases can be transmitted across generations.

Hence, in addition to traditional Mendelian inheritance, another kind of inheritance exists — namely, **epigenetic inheritance**.

epigenetics the study of changes in organisms caused by modifications of gene expression rather than alteration of the genetic code itself

epigenetic inheritance the inheritance of epigenetic tags across generations

Epigenetic factors

Epigenetics refers to all changes to genes — apart from changes to their base sequences — that bring about phenotypic changes. These changes can be brought about by **epigenetic factors**. Such factors are external to DNA, but act on DNA and turn genes permanently 'on' or 'off'. Epigenetic factors may underlie some of the differences seen in the phenotypes of identical twins, since these differences cannot be explained by differences in their genotypes.

FIGURE 7.17 As a result of his time on the International Space Station, Scott Kelly's epigenetic factors are now different to that of his identical twin, Mark.

Resources

Weblink NASA Twins Study

Epigenetic factors can change how DNA in cells is packaged or how it is labelled.
- Packaging of DNA in cells may be tight or may be open (see subtopic 6.2). Genes in segments of DNA that are tightly packaged are silenced, while genes in segments of DNA with open packaging are active and translated into protein.
- Labelling DNA is like adding a 'tag' that does not alter the base sequences of genes, but can either silence genes or make them active. **Methyl groups** ($-CH_3$) are one example of an epigenetic tag. The addition of methyl groups is called **methylation**, and they can be added to any cytosine (C) base alongside a guanine (G) base in DNA (see figure 7.18). Active genes are found to have fewer methyl groups than inactive genes, so it appears that tagging genes by the addition of methyl groups to their C–G bases can change gene expression and permanently switch those genes 'off'. This epigenetic mechanism is shown in figure 7.19.

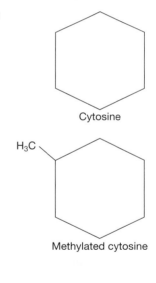

FIGURE 7.18 The addition of a methyl group to a cytosine base is an epigenetic tag.

FIGURE 7.19 DNA methylation tagging

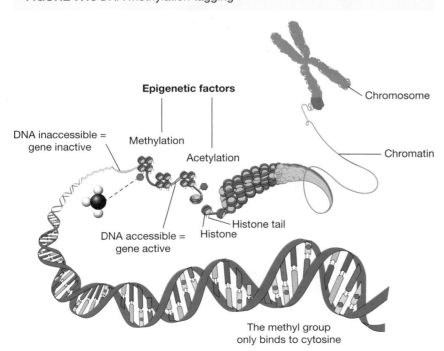

The methyl group only binds to cytosine

epigenetic factors external factors which change genes, but not the base DNA sequence

methyl group a group of one carbon atom bonded to three hydrogen atoms ($-CH_3$)

methylation the addition of a methyl group ($-CH_3$) to a cytosine base of DNA, usually to repress gene transcription

TOPIC 7 Genotypes and phenotypes

Once established, epigenetic tags remain for the life of a cell and are transmitted to all daughter cells derived from that cell. Usually they are not passed on to the next generation as typically the DNA of a fertilised egg is cleared of the epigenetic tags. In some cases, however, the epigenetic tags on the DNA are not erased but instead are conserved and passed to the next generation(s).

Examples of epigenetic inheritance

Cell differentiation

The cells of a human embryo, and later a foetus, are all derived from a single fertilised egg by a series of mitotic cell divisions. During embryonic development, cells will develop along different pathways; for example, some cells will differentiate as brain cells (neurons), some will develop into smooth muscle cells and some into liver cells. All the various cell types — more than 200 cell types in total — have the same genotypes, but different sets of genes are active in each cell type. Epigenetic factors produce the changes that start various stem cells down different developmental paths.

FIGURE 7.20 All cells have the same genotype but epigenetic factors initiate the development of different cell types.

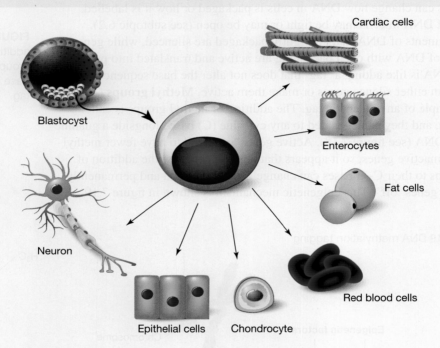

X-inactivation

The somatic cells of normal female mammals have two copies of the X chromosome. Early in embryonic development, one of the X chromosomes in each somatic cell is inactivated, switching off all its genes. The epigenetic tags that cause the inactivation of a particular X chromosome are transmitted to all the daughter cells produced by subsequent mitotic cell divisions, so that the same X chromosome remains inactive.

Imprinted genes

Imprinted genes refers to those whose expression is affected by their parental origin. Children born with a small deletion of one of their number-15 chromosomes will show one of two different phenotypes. The phenotype displayed depends on whether the chromosome with the deletion came from the mother or from the father. If from the father, the clinical phenotype displayed is that of Prader–Willi syndrome; if from the mother, the clinical phenotype displayed is Angelman syndrome (see figure 7.21).

FIGURE 7.21 a. People with Prader–Willi syndrome become constantly hungry from about the age of two years old, which often leads to obesity and type 2 diabetes. b. People with Angelman syndrome have jerky movements, smile frequently and laugh often.

Chemical action

Vinclozolin is a commercial fungicide. If ingested by mammals, it interferes with sperm formation. In an experiment, pregnant rats were injected with vinclozolin and, as expected, the male offspring produced reduced numbers of sperm with lower than normal mobility. However, an interesting finding was that if these male rats managed to mate, their sons showed the same defect, and this was then passed to males in the next generation. The chemical did not cause a mutation of the DNA of the original male. What was observed in the three generations of male rats was a change in the level of methylation of their DNA.

If chemicals such as vinclozolin have been shown to affect future generations of rats, could the same be observed in humans? Research of this kind is much more difficult to conduct, but some studies have tried to determine whether the diet and lifestyle of one generation could affect the health of the next. The Överkalix study looked at three generations of people in Sweden from various families. Some interesting results were found:

- A father's poor food supply and a mother's good food supply were associated with *lower* risks of cardiovascular disease in their children.
- The body mass index of sons at 9 years old was *higher* if their fathers had started smoking early in life (by age 11). In fact, their sons had an extra 5 to 10 kilograms of body fat by the age of 13 compared to sons whose fathers started smoking later in life.
- When paternal grandparents ate a large quantity of food during their childhood, their grandsons had a *shorter* lifespan by an average of six years, dying more often of cancer.

The exact reasons to explain each of these observations are still under investigation, and other lifestyle factors including stress, alcohol consumption and physical activity are also thought to have effects on future generations through the way the DNA is expressed. The environment of our immediate ancestors may impact our phenotype, including traits relating to our health.

Resources

eWorkbook Worksheet 7.4 The effects of environment on phenotype (ewbk-7235)
Worksheet 7.5 The effects of epigenetic factors on phenotype (ewbk-7237)

Video eLesson Epigenetics — switching genes on and off (eles-4210)

Weblink Epigenetics

SAMPLE PROBLEM 3 Designing an experiment

tlvd-1760

There are four different species of flamingos, each with a distinctive appearance. One area of significant variation is seen in their feather colour, which can have a variety of shades from white to pink. Scientists hypothesised that concentration of carotenoids in the diet rather than the genotype of the flamingos caused this variation in feather colour. Design an experiment to test this hypothesis. **(5 marks)**

THINK

1. Determine the independent and dependent variable in the investigation.

2. Determine how you will measure your dependent variable.

3. Determine variables that need to be controlled.

4. Write a clear experimental method that is reproducible, with identifiable quantities and an obvious control of variables.

WRITE

Independent variable: carotenoid concentration
Dependent variable: colour of flamingo feathers
(1 mark for identifying IV and DV)

Dependant variable measurement: scoring colour of feathers against a colour chart

Variables to control: temperature, mass of food, time of day to be fed, level of physical activity, flamingos have same age range

1. Select 50 individuals of a single flamingo species and divide into five groups at random.
2. Score the feather colour of each flamingo against a colour chart.
3. Feed each group of flamingos a specific concentration of carotenoid for a particular period of time; for example, six months.
4. After the designated period of time, score the feather colour of each flamingo against a colour chart.
5. Take a mean of each colour score from each of the carotenoid concentration groups.
6. Compare the mean results to determine whether significant colour change was observed in feather colour at the start and end of the period.
7. Repeat the experiment with other flamingo species (1 mark for reproducible method, 1 mark for multiple trials/large sample size, 1 mark for control of variables and 1 mark for measuring the DV).

INVESTIGATION 7.2

elog-0794

online only

Environmental influences on phenotypes

Aim

To investigate the influence of environmental conditions on the phenotypes of different organisms

KEY IDEAS

- The phenotype of an organism can be the result of an interaction between its genotype and the environment.
- Epigenetics is the study of how cells with identical genotypes can show different phenotypes.
- Epigenetic factors act on DNA but do not change the base sequence.
- Methylation in DNA is one kind of epigenetic modification.
- Packaging of DNA is another means by which epigenetic factors can act.
- Epigenetic modifications may be inherited.

7.4 Activities

learn**on**

To answer questions online and to receive **immediate feedback** and **sample responses** for every question go to your learnON title at **www.jacplus.com.au**. A **downloadable solutions** file is also available in the resources tab.

| 7.4 Quick quiz on | 7.4 Exercise | 7.4 Exam questions |

7.4 Exercise

1. Distinguish between the terms *genotype* and *phenotype*.
2. Describe how soil pH can produce different-coloured flowers in genetically identical hydrangea plants.
3. Give two examples of an epigenetic modification that can be transmitted across generations.
4. Explain how the addition of methyl groups could alter the expression of genes.
5. Bisphenol A (BPA) is an industrial chemical widely used in the production of common food and drink containers. It has been suggested by some scientists that if animals, including humans, are exposed to BPA in their diet, it may induce epigenetic changes of DNA due to methylation. These changes are hypothesised to contribute to a range of conditions such as obesity, infertility and various cancers in offspring. Design an experiment to determine whether BPA is linked to obesity.
6. The leaves of white oak trees (*Quercus alba*) can show two different phenotypes, as shown in the figures provided.

Figure A

Figure B

Further examination revealed that the leaves with the shape shown in figure a grow on areas of trees that are exposed to full sunlight, while the leaves shaped as in figure b are located in areas shaded from the Sun.
   a. Do the leaf cells on areas of the same tree exposed to sunlight have the same genotype as those growing in shaded areas? Briefly explain.
   b. Which of the following is the best biological explanation for the variation in leaf shape on the same tree?
      i. The tree wants to increase the surface area of the shaded leaves in order to maximise photosynthesis.
      ii. The phenotype of the leaves is due not only to the genotype, but is also influenced by an environmental factor — namely, light intensity.
      iii. The genotypes of the leaves change in response to the different environments in which the leaves are growing.

7. You are given the following information about a particular species of flowering plant:
   - Flower colour in this plant species is under genetic control.
   - Two flower colours, red and white, are seen in flowers of this plant.

   a. Based on this information, identify a probable explanation for the inheritance of flower colour in this plant.
   b. You are later told that, as well as white and red flower colours, other colours are seen in this plant species, including pale pink, medium pink and intense pink. Given this additional information, how, if at all, would you change your explanation for the inheritance of flower colour in this plant species?

8. *'Epigenetic markers are always removed from the fertilised egg.'* Does evidence from the Överkalix study support this hypothesis? Explain your answer.

7.4 Exam questions

Question 1 (1 mark)

MC DNA methylation and histone modification can produce
A. changes to the DNA sequences of the entire parental organism.
B. changes in how genes are switched 'on' and 'off', that can be passed to offspring.
C. changes to the DNA sequences of a grandparent.
D. new combinations of alleles.

Question 2 (1 mark)

Source: VCAA 2003 Biology Exam 2, Section A, Q22

MC In a group of organisms, individuals genetically identical at a particular single gene locus show a variety of phenotypes for the trait.

It is reasonable to conclude that the variation in phenotypes for this trait is the result of
A. codominance.
B. polygenic inheritance.
C. environmental influences.
D. multiple alleles at the locus.

Question 3 (1 mark)

Source: VCAA 2014 Biology Exam, Section A, Q32

MC Flamingos are birds that live by lakes. The feather colour of flamingos may vary from white to pink to red. To investigate the inheritance of feather colour, a scientist performed the following crosses and recorded the feather colour of all the offspring when one year old. The diet of the offspring was also recorded.

Cross	Feather colour of parents	Feather colour of all one-year-old offspring	Diet of offspring
1	white × white	white	aquatic plants
2	red × white	white	aquatic plants
3	white × white	pink	algae and crustaceans
4	red × white	pink	algae and crustaceans

Based on this information, a correct conclusion would be that
A. both the parents in cross 1 must be homozygous for white feather colour.
B. white feather colour is recessive to red feather colour.
C. the feather colour of flamingos is influenced by their environment.
D. two parents, both with pink feather colour, would produce one-year-old offspring with only pink feather colour.

▶ Question 4 (1 mark)

MC During the first four days of human embryonic development, totipotent cells in the zygote and morula divide. These cells have the potential to become any type of cell in the human body.

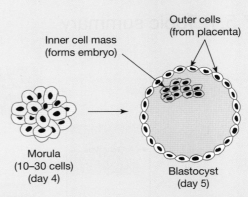

After day 4, the cells continue to divide and now they specialise into the pluripotent cells of the mesoderm, endoderm and ectoderm in the blastocyst.

From week 3, pluripotent cells continue dividing and now they differentiate into multipotent cells of the nervous system, blood and heart, muscle and other organ systems.

By week 9, the multipotent cells have differentiated further into all the highly specialised cells that are needed to construct a human baby.

Embryonic cell differentiation is a result of genes switching 'on' and 'off' selectively in different cells, in response to chemicals, nearby cell types and other factors in each cell's environment.

Cell differentiation is an example of
A. epigenetic change.
B. genotypic change.
C. an inherited mutation.
D. a somatic cell mutation.

▶ Question 5 (1 mark)

MC A gardener obtained six identical pots and filled them with soil. To each pot the gardener added a different quantity of lime (calcium oxide) to alter the pH of the soil. In each pot he planted a stem-cutting, taken from the same parent hydrangea plant.

The pots were kept in a well-lit area and watered regularly. The cuttings grew leaves, roots and branches. Several months later, flowers appeared on all of the cuttings. The flowers in each pot had a different colour as shown in the table.

pH of soil in pot	Colour of hydrangea flowers
4.5	Deep, bright blue
5.0	Medium blue
5.5	Purple
6.0	Purplish-pink
6.8	Medium pink
7.0	Deep, bright pink

A logical conclusion from this information is that
A. acidic soils promote pink colour in hydrangea flowers.
B. the plant in each pot had a different genotype.
C. the difference in flower colour must be a result of genes switching 'on' or 'off'.
D. if a hydrangea cutting from the same plant is grown in soil of pH 5.2 it would produce bluish-purple flowers.

More exam questions are available in your learnON title.

7.5 Review

7.5.1 Topic summary

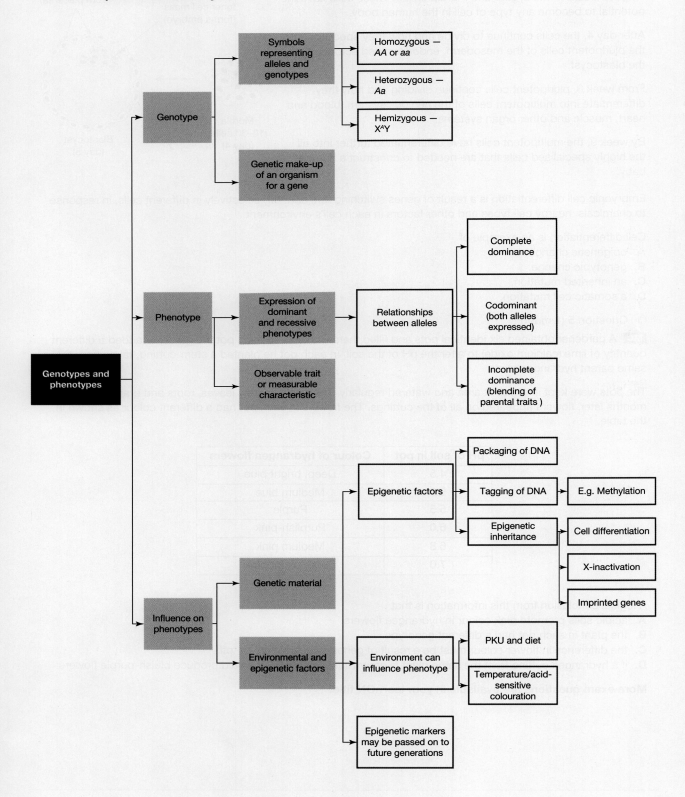

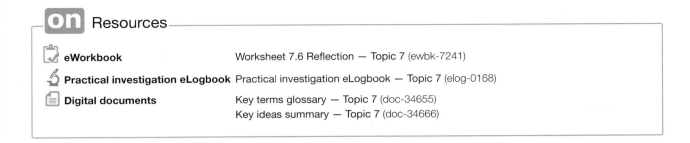

7.5 Exercises

To answer questions online and to receive **immediate feedback** and **sample responses** for every question go to your learnON title at **www.jacplus.com.au**. A **downloadable solutions** file is also available in the resources tab.

7.5 Exercise 1: Review questions

1. What does the term *epigenetics* mean?
2. Distinguish between the terms *homozygous* and *heterozygous*.
3. Haemophilia is an inherited condition resulting from a deficiency of a blood-clotting protein. Sufferers of haemophilia experience abnormal bleeding episodes because their blood fails to clot. Haemophilia is an X-linked recessive trait. The two alleles for the haemophilia gene are represented by X^H and X^h.
   a. Write the genotypes possible for a female with respect to haemophilia.
   b. For each genotype that you listed in part (a), identify whether it is homozygous or heterozygous.
   c. Explain what is meant by the term hemizygous and use a genotype for the haemophilia trait to support your answer.
4. In humans, red hair (r) is recessive to non-red hair (R). Explain why it is easier to determine the genotype of those with red hair compared to those without.
5. 'The genotype alone is responsible for the phenotype.' Explain whether you agree or disagree with this statement.
6. Using examples, explain the difference between codominant and incomplete dominant expression of alleles.
7. Explain why the genotype $I^A I^A$ produces the same blood group as the genotype $I^A i$.
8. Explain how the expression of genes may be affected by epigenetic markers inherited from the previous generation.
9. Briefly design an experiment to determine whether the relationship in the expression of blue and yellow alleles for flower colour in snapdragon plants is complete dominant, codominant or incomplete dominant.

7.5 Exercise 2: Exam questions

Resources

Teacher-led videos Teacher-led videos for every exam question

Section A — Multiple choice questions

All correct answers are worth 1 mark each; an incorrect answer is worth 0.

Question 1

Cystic fibrosis is a recessive autosomal condition.

A sufferer will have the genotype

A. *cc.* B. *CC.* C. *Cc.* D. *cC.*

Use the following information to answer Questions 2 and 3.

Human ear lobes can either be 'free' or 'attached', as represented by the diagram shown.

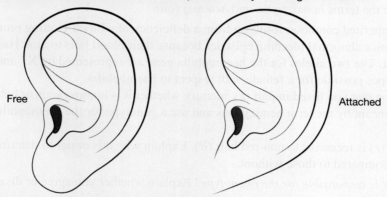

This characteristic is controlled by a gene with two alleles. The allele for 'free' earlobes (*F*) is dominant over 'attached'.

Question 2

A suitable letter to represent the 'attached' allele would be

A. *a.* B. *A.* C. *FF.* D. *f.*

Question 3

A heterozygous individual would show which of the following phenotype(s)?

A. Free
B. Attached
C. Intermediate between free and attached
D. Both free and attached

Question 4

In blood groups, blood groups A and B are codominant. Both are dominant to blood type O.

If an individual is blood group B, their genotype could be

A. *BB* or *BO.* B. $I^B I^B$ or $I^B i.$ C. $I^B i$ only. D. *BO* only

Use the following information to answer Questions 5 and 6.

All humans have two copies of the huntingtin gene, which is present on chromosome 4. The gene codes for the Huntingtin protein, which is needed for a normally functioning nervous system. Having a single mutated form of the gene will lead to Huntington's disease because the disease allele is completely dominant over the healthy allele. The disease results in various symptoms such as lack of coordination and jerky movements.

▶ Question 5

A non-sufferer of the condition will have the genotype

A. *HH*. B. *Hh*. C. *hh*. D. $H^S H^s$.

▶ Question 6

Which of the following statements about Huntington's disease is correct?

A. Sufferers of the disease could either be homozygous for the disease allele or heterozygous.
B. The disease is controlled by one gene with multiple alleles.
C. The alleles for the huntingtin protein have a codominant relationship.
D. Homozygous individuals for the disease will not demonstrate a lack of coordination and jerky movements.

▶ Question 7

An example of an epigenetic marker is

A. stress.
B. a reduction in the number of histones present within a chromosome.
C. a mutation in the base sequence of a gene.
D. methylation of cytosine bases.

▶ Question 8

Which of the following statements regarding epigenetics is correct?

A. Epigenetic markers are always cleared from the fertilised egg.
B. The effects of epigenetic markers have been observed in several generations.
C. Epigenetic markers cause cytosine bases to mutate other nitrogenous bases.
D. How tightly the DNA is packaged around histones is not an epigenetic marker.

▶ Question 9

An experiment is conducted to investigate how the alleles for flower colour in a plant species are expressed in relation to each other. Homozygous plants with blue and green flowers are bred together.

Which of the following statements is correct?

A. If all the offspring are purple, the traits have a codominant relationship.
B. If all the offspring are spotted blue and green, the traits have an incomplete dominant relationship.
C. If all of the offspring are green, green is completely dominant over blue.
D. The offspring with blue flowers must have the phenotype $C^B C^G$.

▶ Question 10

Why is studying identical twins useful for scientists to determine the relative contributions that genetics and the environment have on phenotype?

A. They will have different methylation patterns on their DNA.
B. Their genetics are identical but the environments in which they were raised may be different.
C. Their genetics and the environments in which they were raised will be the same.
D. The environments in which they were raised are the same, but their genetics will be different.

Section B – Short answer questions

▶ Question 11 (3 marks)

Phenotypes seen in bacterial species include sensitivity to particular antibiotic drugs. The antibiotic streptomycin interferes with bacterial ribosomes.

a. Identify a reason that streptomycin is effective in killing some bacteria.   **1 mark**
b. Would you predict that streptomycin would be effective for treating just one infection caused by one kind of bacteria only, or would you predict that it would be generally effective against several bacterial infections? Briefly explain your decision.   **2 marks**

▶ Question 12 (7 marks)

The colour of seed coats in a species of plant was investigated to determine the relationship between the expression of different alleles. Five different phenotypes exist: grey, black, green, yellow and white. Thousands of plants with these different seed colours were bred together and the offspring were analysed. The grey allele was found to be dominant over the black allele, which in turn was dominant over green. The green allele was dominant over yellow, which in turn was dominant over white.

a. Define the term *phenotype*.   **1 mark**
b. State the genotype of a white-seeded plant. Chose suitable allele symbols.   **1 mark**
c. Write down the possible genotypes that a grey-seeded plant may possess.   **2 marks**
d. Another scientist carrying out the same experiment obtained the same results, except they found that the yellow and green alleles were expressed in a codominant pattern.
    i. Describe the evidence the scientist must have collected for them to reach this conclusion. Explain your answer.   **2 marks**
    ii. Using suitable allele symbols, give the genotype of a plant produced for seed colour when a yellow-seeded plant is crossed with a green-seeded plant.   **1 mark**

▶ Question 13 (7 marks)

Atherosclerosis is a condition in which the arteries become blocked with deposits known as plaques. These plaques are composed of fat, cholesterol and other chemicals. Blocking of the arteries can lead to severe medical problems such heart attacks and strokes. An investigation was carried out to investigate the possible epigenetic links.

Two groups of mice that were susceptible to high blood cholesterol and atherosclerosis were selected. One group was fed a normal diet whilst the other was fed a diet high in fat and cholesterol. After several weeks of feeding, the bone marrow of the mice from each group was extracted and transplanted into mice that were closely related. These mice were then fed a normal diet for several months and the progression of atherosclerosis was measured around the heart and neck.

DNA methylation markers were measured and shown to be significantly lower around several genes in mice that had bone marrow from the high-fat diet mice compared to mice who received bone marrow from the normal diet mice. As a result of this, those who received high-fat diet bone marrow were at much greater risk of developing atherosclerosis compared to those who received normal diet bone marrow, despite both being fed a normal diet.

a. Other than DNA methylation, give one other example of an epigenetic change.   **1 mark**
b. State whether DNA methylation was higher or lower in mice that were fed a normal diet.   **1 mark**
c. Suggest why the mice that received the transplanted bone marrow needed to be closely related.   **2 marks**
d. It has been suggested that the diet of our parents may have an influence on our health. Do the results of this study support this theory?   **3 marks**

Question 14 (7 marks)

Fur colour in Himalayan rabbits is controlled by temperature. At normal body temperature the enzyme tyrosinase is produced but does not function. At low temperatures the enzyme is active, producing melanin and darker fur.

a. In this example, which environmental factor is having an effect on the phenotype of the rabbits? **1 mark**
b. Design an experiment to demonstrate a possible relationship between temperature and fur colour. **6 marks**

Question 15 (8 marks)

In humans, the *TYR* gene produces the enzyme tyrosinase. Tyrosinase controls pigment production. People who lack tyrosinase cannot produce pigment and have the albinism trait. These individuals lack pigmentation in their hair, skin and eyes. There are two alleles for the gene, *A* and *a*.

The heterozygote for this condition can produce tyrosinase and has normally pigmented skin, hair and eyes.

a. Is the albinism trait dominant or recessive? Explain. **2 marks**
b. Write the genotypes of the following individuals:
   i. an albino woman **1 mark**
   ii. a homozygous normally pigmented man. **1 mark**
c. Individuals with normal pigmentation often observe that areas of their skin change colour from winter to summer. Explain the cause of the colour change. **1 mark**
d. A biology student asked her teacher whether seasonal changes in skin colour were an example of an epigenetic change. What additional information would the teacher need in order to decide whether or not the change in skin colour is an epigenetic change? **2 marks**
e. In a human population there is a great range of skin colours between individuals. What type of genes are responsible for the enormous range of skin colour among humans? **1 mark**

7.5 Exercise 3: Biochallenge online only

Resources

eWorkbook Biochallenge — Topic 7 (ewbk-8077)

Solutions Solutions — Topic 7 (sol-0652)

teach on

Test maker
Create unique tests and exams from our extensive range of questions, including practice exam questions.
Access the Assignments section in learnON to begin creating and assigning assessments to students.

Online Resources

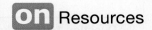

Below is a full list of **rich resources** available online for this topic. These resources are designed to bring ideas to life, to promote deep and lasting learning and to support the different learning needs of each individual.

eWorkbook

- 7.1 eWorkbook — Topic 7 (ewbk-3166)
- 7.2 Worksheet 7.1 Writing genotypes (ewbk-7229)
- 7.3 Worksheet 7.2 Dominant and recessive phenotypes (ewbk-7231)
- Worksheet 7.3 Codominance and incomplete dominance (ewbk-7233)
- 7.4 Worksheet 7.4 The effects of environment on phenotype (ewbk-7235)
- Worksheet 7.5 The effects of epigenetic factors on phenotype (ewbk-7237)
- 7.5 Worksheet 7.6 Reflection — Topic 7 (ewbk-7241)
- Biochallenge — Topic 7 (ewbk-8077)

Solutions

- 7.5 Solutions — Topic 7 (sol-0652)

Practical investigation eLogbook

- 7.1 Practical investigation eLogbook — Topic 7 (elog-0168)
- 7.3 Investigation 7.1 Comparing phenotypes (elog-0792)
- 7.4 Investigation 7.2 Environmental influences on phenotypes (elog-0794)

Digital documents

- 7.1 Key science skills — VCE Biology Units 1–4 (doc-34648)
- Key terms glossary — Topic 7 (doc-34655)
- Key ideas summary — Topic 7 (doc-34666)
- 7.2 Background knowledge: Mendel — the father of genetics (doc-36114)

Teacher-led videos

- Exam questions — Topic 7
- 7.2 Sample problem 1 Choosing symbols for alleles (tlvd-1800)
- 7.3 Sample problem 2 Determining dominance patterns (tlvd-1759)
- 7.4 Sample problem 3 Designing an experiment (tlvd-1760)

Video eLessons

- 7.2 Genotypes (eles-4222)
- 7.3 Phenotypes (eles-4223)
- Codominance (eles-4224)
- 7.4 Epigenetics — switching genes on and off (eles-4210)

Interactivities

- 7.2 Genotypes (int-0668)
- 7.3 Generating the phenotypes (int-0178)
- Making families (int-0681)

Weblinks

- 7.4 NASA Twins Study
- Epigenetics

Teacher resources

There are many resources available exclusively for teachers online

To access these online resources, log on to **www.jacplus.com.au**

AREA OF STUDY 1 HOW IS INHERITANCE EXPLAINED?

8 Patterns of inheritance

KEY KNOWLEDGE

In this topic you will investigate:

Patterns of inheritance

- pedigree charts and patterns of inheritance, including autosomal and sex-linked inheritance
- predicted genetic outcomes for a monohybrid cross and a monohybrid test cross
- predicted genetic outcomes for two genes that are either linked or assort independently.

Source: VCE Biology Study Design (2022–2026) extracts © VCAA; reproduced by permission.

PRACTICAL WORK AND INVESTIGATIONS

Practical work is a central component of learning and assessment. Experiments and investigations, supported by a **practical investigation eLogbook** and **teacher-led videos**, are included in this topic to provide opportunities to undertake investigations and communicate findings.

8.1 Overview

Numerous **videos** and **interactivities** are available just where you need them, at the point of learning, in your digital formats, learnON and eBookPLUS at **www.jacplus.com.au**.

8.1.1 Introduction

Most people can agree that Iceland is a country of outstanding natural beauty. However, for geneticists around the world — and perhaps for the people of Iceland itself — the most remarkable feature of the country is the incredibly detailed genealogical records stretching back over 1000 years. The Íslendingabók website allows Icelanders to easily construct detailed family trees, and an app even allows people to 'bump' their mobile phones together to determine how closely related they are. What really excites scientists is the ability to couple the genealogical records with health data to allow them to understand the inheritance pattern of particular diseases.

FIGURE 8.1 Iceland's combination of detailed written genealogical records stretching back many centuries and its relative isolation have helped to make tracing Icelandic ancestry much easier than almost all other countries.

How genetic diseases are inherited is no different to the way in which other non-disease traits are inherited. It was the work of Mendel in the 1860s, through thousands of experiments and careful recording of data, that showed that characteristics are inherited in specific patterns and that certain traits are dominant over others. When we combine our understanding of inheritance with genealogical records, we can begin to make predications about the likelihood of family members having or developing particular traits.

In this topic, you will study pedigree charts and look for patterns in the way certain traits are inherited. You will also study genetic crosses involving alleles of one gene and understand how to predict the genotypes and phenotypes of the offspring. These predictions will extend to two genes and how they change depending on the proximity of the genes to each other.

LEARNING SEQUENCE

- 8.1 Overview 446
- 8.2 Pedigree charts and patterns of inheritance 447
- 8.3 Predicting genetic outcomes: a monohybrid cross and a monohybrid test cross ... 460
- 8.4 Predicting genetic outcomes: linked genes and independently assorted genes ... 474
- 8.5 Review 488

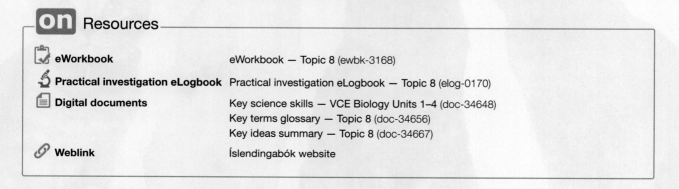

8.2 Pedigree charts and patterns of inheritance

KEY KNOWLEDGE

- Pedigree charts and patterns of inheritance, including autosomal and sex-linked inheritance

Source: VCE Biology Study Design (2022–2026) extracts © VCAA; reproduced by permission.

8.2.1 What is a pedigree chart?

A family portrait is an important record of events in the life of a particular family, but a different kind of family picture can be made to show the inheritance of a particular trait. This kind of picture is called a **pedigree**. In drawing a human pedigree, specific symbols are used (see figure 8.2). Each individual in a pedigree can be identified using the generation number (I, II) and the position in the generation (1, 2). For example, the first individual in the second generation is referred to as II-1.

To provide evidence in support of a given mode of inheritance, a pedigree should show:
1. an adequate number of persons, both affected and unaffected
2. persons of each sex
3. preferably, a minimum of three generations.

FIGURE 8.2 A pedigree and explanation of symbols used

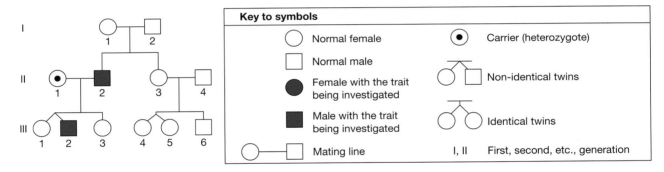

The pattern of inheritance of a trait in a pedigree may provide information about the trait. The pattern may indicate whether:
- the trait concerned is dominant or recessive
- the controlling gene is located on an autosome or on a sex chromosome.

However, a single pedigree may not always provide conclusive evidence. As well as this, carriers are not always represented on a pedigree, as many individuals do not know if they are a carrier for a certain trait.

8.2.2 Inheritance of autosomal dominant traits

Figure 8.3 shows the pattern of appearance of familial hypercholesterolaemia in a family. This is an inherited condition in which affected individuals have abnormally high levels of cholesterol in their blood. Those affected are at risk of suffering a heart attack in early adulthood. The gene concerned is the *LDLR* gene located on the short arm of chromosome 19. Abnormally high blood cholesterol level (*B*) is dominant to normal levels (*b*). So, hypercholesterolaemia is expressed in a heterozygote.

pedigree a graphic representation, using standard symbols, showing the pattern of occurrence of an inherited trait in a family

FIGURE 8.3 Pedigree for highly elevated blood cholesterol levels, an autosomal dominant trait. Does every affected person have at least one affected parent?

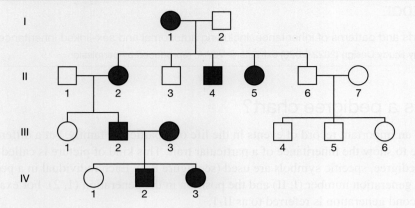

An important question to answer is how could you tell this condition has an autosomal dominant pattern of inheritance just by looking at the pedigree?

Features of an autosomal dominant trait in a pedigree

- Both males and females are affected in broadly equal numbers across a large pedigree.
- All affected individuals have at least one affected parent.
- Once the condition disappears from a branch of the family tree, it never reappears.
- Fathers can pass the condition on to both sons and daughters, as can mothers.

FIGURE 8.4 A section of the pedigree from figure 8.3, showing the disappearance of disease

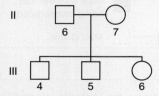

Let's look closely at figure 8.3 to determine if the points above are satisfied from the pedigree.
- Three males and four females are affected — these are roughly equal.
- All members do have an affected parent.
- Individuals II-6 and II-7 have no affected children — the condition has disappeared from this branch of the family tree and did not appear in any offspring (figure 8.4).
- There is evidence that both fathers and mothers are able to pass the condition on to sons and daughters.

As all four indicators are satisfied, this indicates it is an autosomal dominant pattern of inheritance.

8.2.3 Inheritance of autosomal recessive traits

Figure 8.5 is a pedigree showing the pattern of inheritance of oculocutaneous **albinism**, a condition in which pigmentation is absent from skin, eyes and hair. Albinism (*a*) is an autosomal recessive trait.

FIGURE 8.5 Pedigree for albinism, an autosomal recessive condition. Can a person show an albino phenotype even though neither parent has the condition?

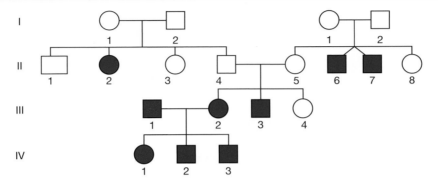

Again, being able to recognise this pattern of inheritance from a pedigree alone is an important skill.

A different set of indicators are needed to recognise autosomal recessive traits.

Features of an autosomal recessive trait in a pedigree

- Both males and females are affected in broadly equal numbers across a large pedigree.
- Two unaffected parents can have an affected child.
- All the children of parents who both have the condition will be affected.
- The trait can 'skip' a generation. In other words, it may disappear from a branch of the pedigree and reappear in later generations.

Let's look at figure 8.5 and see if the above points are satisfied from the pedigree.
- Three females and six males are affected. Whilst not broadly equal, the pedigree size is small.
- There are multiple instances of unaffected parents having affected children.
- III-1 and III-2 are affected parents and all their children are affected.
- There is insufficient evidence clearly showing it skipping generations.

Although not all four indicators are satisfied, this may be the result of insufficient evidence (small sample size). However, the results available — particularly multiple instances of unaffected parents having affected children, and all of the children of affected parents are also affected — does suggest it is an autosomal recessive trait.

TIP: Not all dot points need to be satisfied to determine the pattern of inheritance, although the more that are satisfied, the more confident you can be. It may be helpful to rule out other modes of inheritance when analysing a pedigree to help increase your certainty.

8.2.4 Inheritance of X-linked dominant traits

Figure 8.6 shows a pedigree for one form of vitamin D-resistant rickets in a family example of an **X-linked gene**. This form of rickets is an X-linked dominant trait, controlled by the *XLHR9* gene located on the X chromosome, and is expressed even if a person has just one copy of the allele responsible.

albinism inherited condition in which pigment production does not occur normally

X-linked gene a gene located on the X chromosome; also refers to a trait that is controlled by such a gene

FIGURE 8.6 Pedigree for vitamin D-resistant rickets, an X-linked dominant condition. Look at the daughters and sons of affected males. What pattern is apparent? Could daughters of person III-3 show the trait?

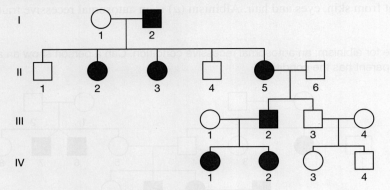

In this instance a new set of points need to be examined to confirm this pattern of inheritance.

Features of an X-linked dominant trait in a pedigree

- A male with the trait passes it on to all his daughters and none of his sons.
- Every affected individual has at least one parent with the trait.
- Once the trait disappears from a branch of the family tree, it never reappears.
- In a large pedigree there are more affected females than males.

Check to see if the pedigree in figure 8.6 satisfies each of these points. Remember, the more points satisfied, the more confident you can be that you have identified the correct pattern.

8.2.5 Inheritance of X-linked recessive traits

Figure 8.7 shows the pattern of appearance of favism in a family. Favism is a disorder in which the red blood cells are rapidly destroyed if the person with this condition comes into contact with certain agents. These agents include substances in broad beans and mothballs.

Favism results from a missing enzyme and is controlled by the *G6PD* gene on the long arm of the X chromosome (Xq28). Favism (f), which occurs when the G6PD enzyme is absent from red blood cells, is recessive to the unaffected condition (F), which occurs when the enzyme is present. The condition of favism is inherited as X-linked recessive and is expressed only in females with the homozygous X^fX^f genotype or in males with the hemizygous X^fY genotype.

FIGURE 8.7 Pedigree for favism — an X-linked recessive disorder. Can this disorder 'skip a generation'?

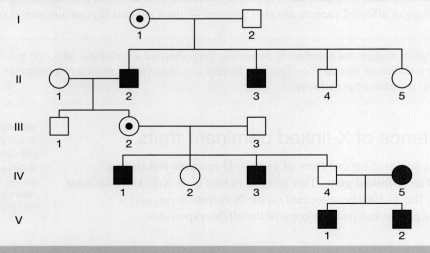

There are many factors that can help to determine this type of inheritance.

Features of an X-linked recessive trait in a pedigree

- Affected females always produce affected sons.
- The daughters of affected males are always carriers.
- All the children with two affected parents will show the trait.
- Across a large pedigree, many more males than females will be affected.

Again, check to see if the pedigree in figure 8.7 satisfies each of these points.

EXTENSION: mtDNA — maternal pattern of inheritance

Most of the DNA present in diploid human cells is present in the nuclei of these cells. The nuclear DNA contains more than three thousand million base pairs and, during cell division, this DNA is organised into chromosomes. In all, this DNA carries about 21 000 genes. A tiny package of DNA, termed **mitochondrial DNA (mtDNA)**, is present in the mitochondria of the cell cytosol and contains just 16 569 base pairs that carry 37 genes.

Inherited disorders due to mitochondrial genes include:
- Leber hereditary optic neuropathy (LHON), which causes loss of central vision; onset is typically seen in young adults
- Kearns–Sayre syndrome, which is expressed as short stature and degeneration of the retina, as well as other associated conditions, including heart, central nervous system and skeletal system conditions, deafness, endocrine disorders and renal failure
- myoclonic epilepsy with ragged-red fibers (MERRF) syndrome, which involves deficiencies in the enzymes concerned with energy transfers.

The transmission of mtDNA, along with the genes it carries, follows a distinctive pattern as shown in figure 8.8. Traits on mtDNA mostly show a pattern of **maternal inheritance**, with mitochondrial genes being transmitted from a mother via her egg to all her offspring. Only her daughters can pass this trait on to their children.

FIGURE 8.8 mtDNA pattern, where females pass on the trait to all of their children

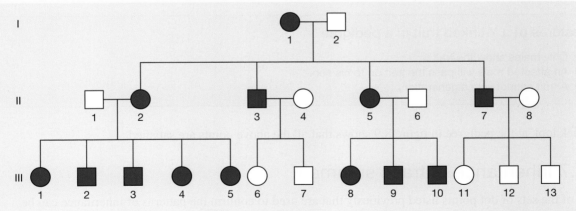

An idealised pattern of inheritance of traits controlled by genes on the mtDNA includes the following features:
- Each mtDNA-controlled trait passes from a mother to all offspring, both female and male.
- While males can receive the trait from their mothers, they cannot pass it on to their children.
- Only females can transmit mtDNA traits to their children.

mitochondrial DNA (mtDNA) the DNA within mitochondria

maternal inheritance inherited traits that only occur through the mother's egg to her offspring; only her daughters can pass the trait on

8.2.6 Inheritance of Y-linked traits

Y-linked genes are inherited in a pattern that differs from that seen in autosomal and X-linked genes. All Y-linked genes show a pattern of **paternal inheritance** in which the DNA of Y-linked genes is transmitted exclusively from males to their sons only. The *AMELY* gene that controls the organisation of enamel during tooth development is located on the Y chromosome, and so is Y-linked. Figure 8.9 shows the pedigree for inheritance of the defective tooth enamel trait. Other genes that are Y-linked include the *SRY* gene that encodes a testis-determining factor and some other genes that affect sperm formation.

Y-linked gene a gene located on the Y chromosome; also refers to a trait that is controlled by such a gene

paternal inheritance inherited traits that only occur through the transmission of Y-linked genes from the father's sperm to his sons

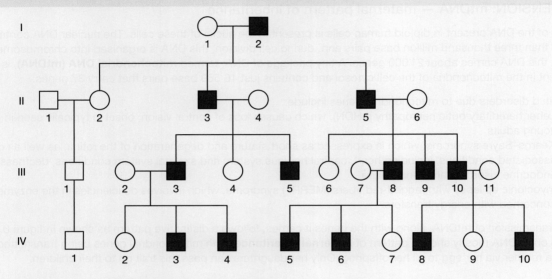

FIGURE 8.9 Pedigree for Y-linked inheritance of defective tooth enamel trait

Identifying Y-linked inheritance is somewhat easier than the other patterns of inheritance.

Features of a Y-linked trait in a pedigree

- Only males show the trait.
- An affected male will pass the trait on to his son.
- A trait cannot skip a generation.

A quick look at the pedigree in figure 8.9 shows that all the above points are satisfied.

8.2.7 Inheritance of traits summary

Each of the sets of dot points listed previously that are used to confirm the patterns of inheritance can be summarised and presented as a flow chart to help make the process somewhat simpler.

TIP: When you are analysing pedigrees, you should support your statements by identifying the individuals by generation and number (such as I-3, II-2) to justify findings.

FIGURE 8.10 Flow chart to determine inheritance of traits

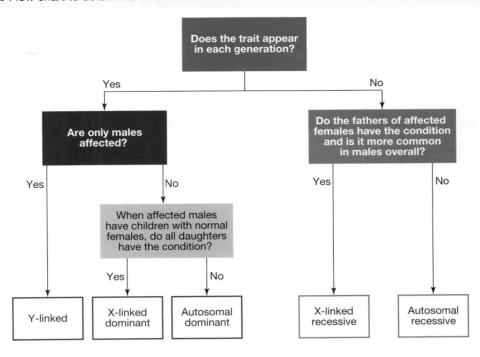

SAMPLE PROBLEM 1 Determining patterns of inheritance

tlvd-1761

Examine the following pedigree for a genetic disorder and determine the most likely pattern of inheritance. Explain your choice. **(3 marks)**

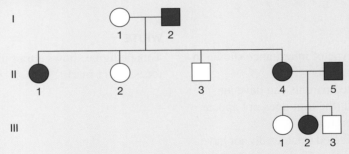

THINK

1. Look carefully at the pedigree and follow the steps in the flow chart shown in figure 8.10.

2. Rule out other types of inheritance based on your findings to further support your assumptions.

3. Check the previous dot points for the suspected inheritance pattern.

WRITE

Males and females are affected, and the father of the affected female has the trait.
The trait is present in each generation.
Not all the daughters of affected males and normal females have the condition.
As two affected parents (II-4 and II-5) have an unaffected child, the trait cannot be recessive (1 mark).
A male also has a trait (I-2) but does not pass it on to all of his daughters (II-2), so the trait cannot be X-linked dominant (1 mark).
It is likely to be autosomal dominant (1 mark).

TOPIC 8 Patterns of inheritance 453

8.2.8 Using pedigrees to determine genotypes

Upon determining the type of inheritance, you can also use this to determine the genotypes of the individuals, remembering that each parent passes one of their alleles on to their children.

It can be helpful to consider the different possible genotypes before you commence.

TABLE 8.1 Genotypes for different types of inheritance

	Female with trait (shaded)	Male with trait (shaded)	Female without trait (unshaded)	Male without trait (unshaded)
Autosomal dominant	BB or Bb	BB or Bb	bb	bb
Autosomal recessive	bb	bb	BB or Bb	BB or Bb
X-linked dominant	$X^B X^B$ or $X^B X^b$	$X^B Y$	$X^b X^b$	$X^b Y$
X-linked recessive	$X^b X^b$	$X^b Y$	$X^B X^B$ or $X^B X^b$	$X^B Y$

In some cases, an individual may have an unknown genotype if they have the dominant trait (they may be *BB* or *Bb*). If a pedigree does not provide the information to determine this, the genotype may be written as *B–*, with the '–' representing an unknown allele.

SAMPLE PROBLEM 2 Determining genotypes from pedigrees

tlvd-1762

For the following pedigrees, determine the genotype of each individual shown. (6 marks)

a.
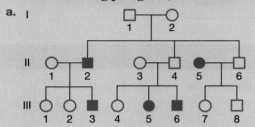

b.
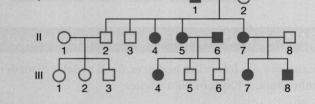

THINK

a. 1. Determine the type of inheritance shown in the pedigree.
   - Two unaffected parents can have an affected child (the trait does not appear in every generation).
   - Fathers of affected females do not have the condition (thus it cannot be X-linked recessive).
   - Both sexes affected equally.
2. Identify the possible genotypes for each individual (the letters used are arbitrary in this case).
3. Fill in the genotypes of individuals that can only be one genotype (shaded individuals as *bb*).

WRITE

This pedigree shows a trait that is autosomal recessive (1 mark).

Shaded individuals are *bb*. Non-shaded individuals are *BB* or *Bb*.

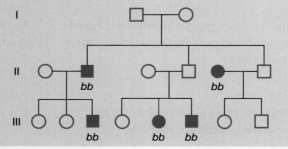

4. Determine the genotype of other individuals by looking at their parents and/or offspring. In this case, if a non-shaded individual has a shaded child or parent, they must be *Bb* (as they have inherited or passed on a *b* allele).
   TIP: If you find that a certain combination is not working, you may have chosen the wrong type of inheritance.

5. If there is not enough information for an individual, you can note the genotype as *B–*.

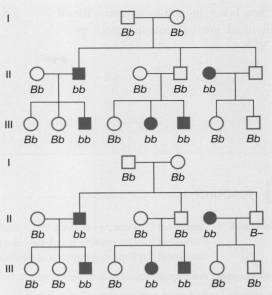

(1 mark for shaded, 1 mark for non-shaded)

b. 1. Determine the type of inheritance shown in the pedigree.
   - It appears in every generation (an affected individual has an affected parent).
   - Two affected parents have an unaffected child.
   - An affected male has all daughters affected, but no sons affected.
   - There are more females affected than males.

   This pedigree most likely shows a trait that is X-linked dominant (1 mark).
   (Not that it could also be autosomal dominant, but X-linked dominant is most likely due to the skew in biological sex where more females are affected than males.)

2. Identify the possible genotypes for each individual (the letters used are arbitrary in this case). Ensure you consider biological sex in this case.

   Shaded females are $X^B X^B$ or $X^B X^b$.
   Shaded males are $X^B Y$.
   Unshaded females are $X^b X^b$.
   Unshaded females are $X^b Y$.

3. Fill in the genotypes of individuals that can only be one genotype (shaded males, unshaded females and unshaded males).

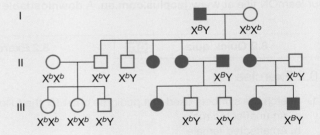

4. Determine the genotypes of other individuals by looking at their parents and/or offspring. In this case, if a non-shaded individual has a shaded child or parent they must be *Bb* (as they have inherited or passed on a *b* allele).

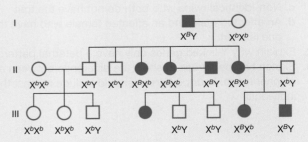

5. If there is not enough information for an individual, you can note the genotype as X^BX^-.

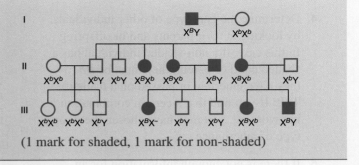

(1 mark for shaded, 1 mark for non-shaded)

Resources

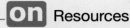

eWorkbook Worksheet 8.1 Pedigree analysis (ewbk-7539)
Worksheet 8.2 Determining genotypes and phenotypes from pedigrees (ewbk-7541)
Worksheet 8.3 Exploring patterns of inheritance — family portraits (ewbk-7543)

Video eLesson Autosomal recessive disorders (eles-4221)

Interactivity Pedigrees and genotypes (int-8122)

KEY IDEAS

- A pedigree uses symbols to show the appearance of an inherited trait across generations.
- Features of the inheritance may allow conclusions to be drawn regarding its pattern of inheritance.
- Common patterns of inheritance include autosomal dominant, autosomal recessive, X-linked dominant and X-linked recessive.
- Y-linked genes show a paternal pattern of inheritance.

8.2 Activities

learn

To answer questions online and to receive **immediate feedback** and **sample responses** for every question, go to your learnON title at **www.jacplus.com.au**. A **downloadable solutions** file is also available in the resources tab.

| 8.2 Quick quiz | 8.2 Exercise | 8.2 Exam questions |

8.2 Exercise

1. Sketch the symbols used in a pedigree chart for the following:
   a. An unaffected male
   b. An affected female
   c. Non-identical twins who both do not have the trait
   d. An affected male and an affected female who have three affected children. Two of the children are boys and one is a girl.
2. Explain why Y-linked genes only have a paternal pattern of inheritance.
3. Sketch two pedigrees to show the difference between autosomal dominant and autosomal recessive patterns of inheritance. Your pedigrees should include at least three generations, including the genotype of each individual.

4. Examine the pedigree shown.

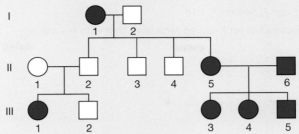

   a. Determine the most likely pattern of inheritance shown in this pedigree. Explain your choice.
   b. Determine the genotypes of each individual shown in this pedigree.
5. Genetic counsellors are medical professionals that are particularly interested in studying pedigrees. Suggest why a couple who are thinking of starting a family might want to seek the advice of a genetic counsellor.
6. Examine the pedigree shown.

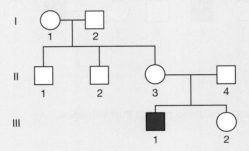

The pedigree shows the inheritance for a disease. Scientist A states the pedigree shows an autosomal recessive pattern of inheritance, whereas scientist B states it shows an X-linked recessive pattern. Do you agree with either of the scientists? Explain your answer.
7. a. Copy the pedigree from figure 8.6 showing an X-linked dominant trait. Note down the genotype of each individual.
   b. Copy the pedigree from figure 8.7 showing an X-linked recessive trait. Note down the genotype of each individual.

8.2 Exam questions

Question 1 (1 mark)
Source: VCAA 2010 Biology Exam 2, Section A, Q9

MC The following pedigree shows the inheritance of an X-linked dominant trait in a family.

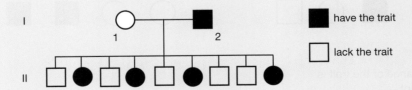

It is reasonable to assume that the
A. mother of I 1 had the trait.
B. father of I 1 had the trait.
C. mother of I 2 had the trait.
D. father of I 2 had the trait.

▶ **Question 2** (1 mark)

Source: VCAA 2011 Biology Exam 2, Section A, Q16

MC Red-green colour blindness is an X-linked recessive trait with the alleles

X^R: normal colour vision

X^r: red-green colour blind

Examine the following pedigree.

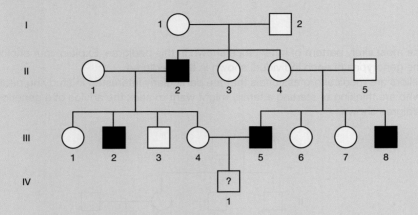

With respect to this gene, it is reasonable to predict that individual
A. II3 must be $X^R X^R$.
B. III4 must be $X^R X^r$.
C. II4 has a two in three chance of being $X^R X^R$.
D. IV1 has a one in four chance of being colour blind.

▶ **Question 3** (1 mark)

Source: VCAA 2012 Biology Exam 2, Section A, Q7

MC In the following pedigree, shaded individuals have a particular genetic trait.

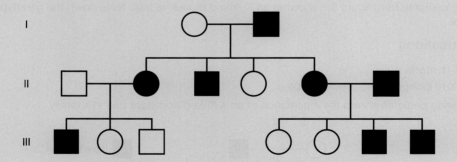

The mode of inheritance of the trait is
A. X-linked dominant.
B. X-linked recessive.
C. autosomal recessive.
D. autosomal dominant.

▶ **Question 4** (2 marks)
Source: Adapted from VCAA 2015 Biology Exam, Section B, Q8b

Hereditary retinoblastoma is a rare autosomal dominant trait. The pedigree below shows the trait appearing in a family with no prior history of the condition before generation III.

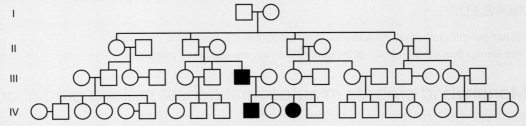

Explain the appearance of this trait in generations III and IV.

▶ **Question 5** (8 marks)
Source: VCAA 2007 Biology Exam 2, Section B, Q4

The pedigree below represents a family of rabbits. The shaded rabbits have an inherited disease. The phenotypes of rabbits I-1, I-2, I-3 and I-4 are not known.

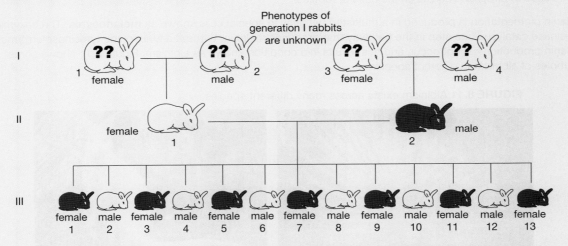

On the basis of the offspring produced by II-1 and II-2 it has been suggested that the disease is inherited as an X-linked dominant characteristic.

a. What evidence from generations II and III supports the suggestion made about the mode of inheritance of the disease? **1 mark**

b. If the mode of inheritance suggested is correct, complete the table below to show the possible phenotypes and genotypes of rabbits I-1, I-2, I-3 and I-4. For each rabbit's phenotype, select from
 • has the disease
 • does not have the disease
 • impossible to tell from the information given.
Use the following symbols to represent the alleles involved: X^D, X^d, Y. **4 marks**

	i. Phenotype	ii. Possible genotype(s)
rabbit I-1		
rabbit I-2		
rabbit I-3		
rabbit I-4		

Assume each female in generation III was mated to a male from the same litter.

c. What genotypic and phenotypic ratios would you expect in the offspring? **3 marks**

More exam questions are available in your learnON title.

8.3 Predicting genetic outcomes: a monohybrid cross and a monohybrid test cross

KEY KNOWLEDGE

- Predicted genetic outcomes for a monohybrid cross and a monohybrid test cross

Source: VCE Biology Study Design (2022–2026) extracts © VCAA; reproduced by permission.

8.3.1 Monohybrid crosses in autosomal genes: making predictions

CASE STUDY: Making melanin pigment

The *TYR* gene is just one of many genes on the human number-11 chromosome. This gene encodes a protein that functions as the enzyme tyrosinase. This enzyme catalyses a step in the pathway that produces the pigment **melanin**. Melanin pigment is seen in the hair, skin and irises of a person's eyes. Melanin pigment is present not only in people, but also in other vertebrate species in their skin, eyes and fur (in the case of mammals), feathers (in the case of birds) and scales (in the case of reptiles).

Melanin pigmentation is produced in a multistep pathway in special cells known as **melanocytes**. The enzyme tyrosinase catalyses one step in the pathway that produces melanin. Without a functioning tyrosinase enzyme, melanin production cannot occur, and this results in a condition known as albinism. Figure 8.11 shows some examples of albinism in different species.

FIGURE 8.11 Albinism exists across many different species.

Alleles of the *TYR* gene

The *TYR* gene in humans has two common alleles:
- Allele *A*, which produces normal tyrosinase enzyme, resulting in normal pigmentation
- Allele *a*, which produces a faulty protein that cannot act as the tyrosinase enzyme, resulting in a lack of pigment (albinism).

melanin the pigment produced by melanocytes; in humans it gives skin, hair and eyes their colour

melanocytes the pigment-producing cells in the basal layer of the epidermis

Persons with the genotypes *AA* or *Aa* have normal pigmentation. In contrast, persons with the *aa* genotype lack a functioning tyrosinase enzyme and so have albinism. From this, we can conclude that the normal pigmentation phenotype is dominant to albinism.

Making predictions: a monohybrid cross of the *TYR* gene in humans

A cross involving the alleles of the *TYR* gene is an example of a **monohybrid cross** because it involves the alleles of just one gene at a time. When the alleles of two genes are involved, the cross is termed a **dihybrid cross** (see section 8.4.1). A monogenic cross involves the segregation of alleles of the same gene into separate gametes. This segregation, or separation, occurs when homologous chromosomes disjoin at metaphase I of meiosis (refer to figure 6.42).

monohybrid cross a cross in which alleles of only one gene are involved

dihybrid cross a cross in which alleles of two different genes are involved

Punnett square a diagram used to determine the probability of an offspring having a particular genotype

Tracey and John are planning their next pregnancy. They have non-identical twins, Fiona and Tim. Fiona has the condition of albinism and the parents want to know about the chance of this condition appearing in their next child. Figure 8.12 shows the chromosomal portraits of Tracey, John and their twins. Both parents are carriers of albinism and have the genotype *Aa*.

FIGURE 8.12 The genotypes and phenotypes of Tracey, John, Fiona and Tim for the *TYR* gene controlling pigment production

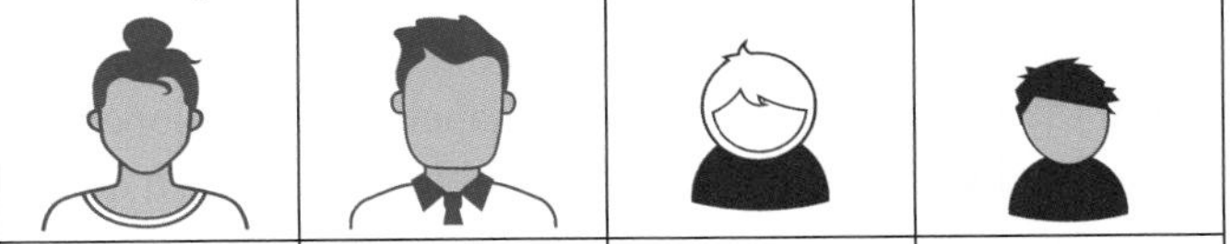

	Tracey	John	Fiona	Tim
Chromosomes	A a 11 11	A a 11 11	a a 11 11	A A 11 11
Genotype	Aa	Aa	aa	AA
Phenotype	Normal pigment	Normal pigment	Albino	Normal pigment

For the *TYR* gene on the number-11 chromosome, which controls pigment production, the cross can be seen in figure 8.13.

During meiosis, the pair of number-11 chromosomes disjoin, carrying the alleles to different gametes. Tracey's eggs have *either* the *A* allele *or* the *a* allele. This also applies to the sperm cells produced by John. This separation of the alleles of one gene into different gametes that occurs during meiosis is known as the *segregation* of alleles. For each parent, the chance of a gamete with *A* is 1 in 2 and the chance of a gamete with *a* is also 1 in 2.

FIGURE 8.13 Cross of Tracey and John for the *TYR* gene

These probabilities can be incorporated into a **Punnett square** (figure 8.14). The *AA* and the *Aa* genotypes both result in normal pigmentation. The *aa* genotype causes albinism.

A Punnett square shows the *chance* of each possible outcome, not what will happen. This chance can be expressed as a ratio, a fraction or a percentage.

The Punnett square shows that the chance that Tracey and John's next child will have:
- albinism, *aa*, is 1 in 4, or $\frac{1}{4}$
- normal pigmentation, *AA* or *Aa*, is $\frac{3}{4}$.

What is the chance that the next child will have normal pigmentation but will be a heterozygous carrier of albinism? Looking at the Punnett square, there are three ways a child with normal pigmentation can result: *AA*, *Aa* and *Aa*. So, the chance that this child will be a heterozygous *Aa* carrier of albinism is $\frac{2}{3}$.

The fractions, such as $\frac{1}{2}$ or $\frac{1}{4}$, that appear in a Punnett square identify the *chance or probability of various outcomes*.

FIGURE 8.14 Punnett square for a monohybrid cross: *Aa* × *Aa*

Gametes	Tracey's eggs	
	$\frac{1}{2}$ A	$\frac{1}{2}$ a
$\frac{1}{2}$ A (John's sperm)	$\frac{1}{4}$ AA Normal	$\frac{1}{4}$ Aa Normal
$\frac{1}{2}$ a	$\frac{1}{4}$ Aa Normal	$\frac{1}{4}$ aa Albino

Genotypes: $\frac{1}{4}$ AA : $\frac{2}{4}$ Aa : $\frac{1}{4}$ aa

Phenotypes: 3 normal : 1 albino

- Tracey is heterozygous *Aa* with two different alleles at the *TYR* gene **loci**. The chance that her *A* allele will go to a particular egg cell is $\frac{1}{2}$; likewise, the chance that her *a* allele will go to that egg cell is also $\frac{1}{2}$. (This situation is like tossing a coin. There are two possible outcomes, namely heads (H) and tails (T), so when a coin is tossed, the chance of H is $\frac{1}{2}$ and the chance of T is $\frac{1}{2}$.)
- Since John is also heterozygous *Aa*, a similar situation exists in regard to his sperm cells: each of his sperm cells must carry either the *A* allele or the *a* allele. This accounts for the $\frac{1}{2}$ *A* and the $\frac{1}{2}$ *a* entries along the top and down the left-hand side of the Punnett square.
- The $\frac{1}{4}$ entries within the Punnett square are the chance of two independent events occurring, such as Tracey's egg with an *A* allele being fertilised by John's sperm with an *A* allele. The chance, or probability, of two independent events occurring is the product of the chance of each separate event; that is, $\frac{1}{2} \times \frac{1}{2} = \frac{1}{4}$.

CASE STUDY: ABO blood type

The four blood groups in the ABO blood system are A, B, AB and O. These different blood group phenotypes are controlled by the *ABO* gene on the number-9 chromosome. Each phenotype is determined by the presence or absence of specific proteins, known as **antigens**, on the plasma membrane of the red blood cells (see figure 8.15).

In the Australian population, type O is the most common blood group (about 49 per cent) and type AB is the rarest (about 3 per cent). Table 8.2 shows the antigens that determine the ABO blood types.

TABLE 8.2 ABO blood types and corresponding antigens present on red blood cells. What antigens are present in a person with O-type blood?

Antigens present on red blood cells	ABO blood group	Frequency in Australian population*
Antigen A	A	38%
Antigen B	B	10%
Antigens A and B	AB	3%
Neither antigen	O	49%

* based on Australian Red Cross data

FIGURE 8.15 Blood cells consist mainly of red blood cells. Also present here are white blood cells (yellow) and platelets.

locus the position of a gene on a chromosome; plural = loci

antigens proteins on the plasma membrane of the red blood cells, the presence or absence of which determines phenotype

Making predictions: a monohybrid cross of the *ABO* gene in humans

Tracey and John are both blood group B and their twins are both blood group O. This means that both parents have the heterozygous genotype $I^B i$, while the twins have the genotype ii (see figure 8.16). The cross between Tracey and John for the *ABO* gene is shown in figure 8.17.

FIGURE 8.16 Genotypes and phenotypes of Tracey, John, Fiona and Tim for the *ABO* gene controlling ABO blood types

	Tracey	John	Fiona	Tim
Chromosomes	I^B i 9 9	I^B i 9 9	i i 9 9	i i 9 9
Genotype	$I^B i$	$I^B i$	ii	ii
Phenotype	Group B	Group B	Group O	Group O

FIGURE 8.17 Cross of Tracey and John for the *ABO* gene

Tracey $I^B i$ × John $I^B i$

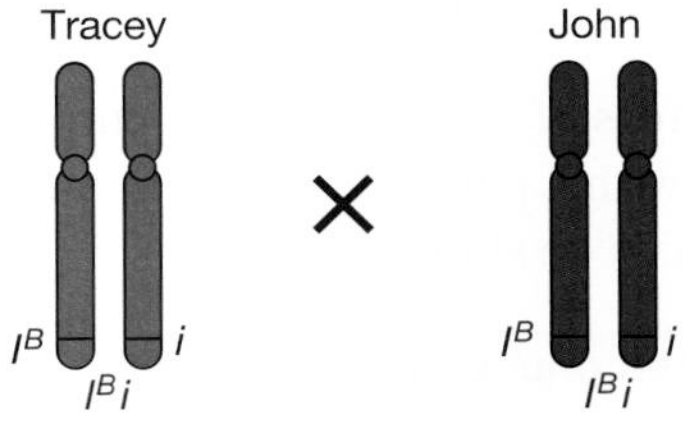

During the formation of the gamete during meiosis, disjunction of the number-9 chromosome means that both Tracey's eggs and John's sperm have *either* one I^B allele *or* one i allele, and the chance of each type is $\frac{1}{2}$. Again, we can show the chances of the various outcomes in a Punnett square (see figure 8.18). In this and the following examples, the outcomes are shown as a ratio.

FIGURE 8.18 Punnett square showing outcome of the cross: $I^B i \times I^B i$

	Gametes	Tracey's eggs I^B	i
John's sperm	I^B	$I^B I^B$ Blood group B	$I^B i$ Blood group B
	i	$I^B i$ Blood group B	ii Blood group O

Genotypes: 1 $I^B I^B$: 2 $I^B i$: 1 ii

Phenotypes: 3 blood group B : 1 blood group O

The Punnett square shows that the chance that Tracey and John's next child will be:
- blood group B, $I^B I^B$ or $I^B i$, is $\frac{3}{4}$
- blood group O, ii, is $\frac{1}{4}$.

Making predictions: monohybrid crosses in plants

In sweet peas (*Lathyrus odoratus*), purple flower colour (*P*) is dominant to white (*p*) (figure 8.19). What happens if a pure-breeding purple-flowering plant is crossed with a pure-breeding white-flowering plant? Pure breeding means that the plant is homozygous for the allele in question, and such a plant can produce only a single kind of gamete.

FIGURE 8.19 One gene in sweet peas controls flower colour and has the alleles *P* (purple) and *p* (white), with purple being the dominant phenotype.

The cross of a pure-breeding purple-flowering plant with a pure-breeding white-flowering plant can be shown as:

Parental phenotypes	Purple	×	White
Parental genotypes	PP		pp
Possible gametes	P		p
Possible offspring	All are purple, Pp		

We can see that the offspring from this cross are heterozygous purple (*Pp*).

Consider now, the outcome if these heterozygous purple plants are crossed:

Parental phenotypes	Purple	×	Purple
Parental genotypes	Pp		Pp
Possible gametes	P and p		P and p

We can now use a Punnett square to show the possible outcome of this cross (see figure 8.20).

Thus, from the cross of two heterozygous purple plants, the chance that the offspring will show the:
- dominant purple colour, *PP* or *Pp*, is $\frac{3}{4}$
- recessive white phenotype, *pp*, is $\frac{1}{4}$.

What is the chance that a purple-flowered offspring will be homozygous *PP*?

8.3.2 Monohybrid crosses: X-linked genes

So far, we have looked at monohybrid crosses involving autosomal genes. What happens in a monohybrid cross when the gene involved is located on the X chromosome?

FIGURE 8.20 Punnett square showing the possible outcome of crossing two heterozygous purple plants

Gametes	P	p
P	PP Purple	Pp Purple
p	Pp Purple	pp White

Genotypes: 1 *PP* : 2 *Pp* : 1 *pp*
Phenotypes: 3 purple : 1 white

CASE STUDY: Colour blindness in humans

Humans normally have three colour receptors (red, green and blue) in the retina of their eyes. These receptors allow us to differentiate colours, such as red from green. Inherited defects in colour receptors cause various kinds of colour blindness, which can be identified by specific screening tests. One such test, administered by a professional under controlled conditions, involves the use of coloured images known as Ishihara colour plates (see figure 8.21).

The *CBD* gene on the human X chromosome controls one form of red–green colour blindness. Normal colour vision (*V*) is dominant to red–green colour blindness (*v*). Table 8.3 shows the genotypes and phenotypes for this X-linked *CBD* gene. Because the Y chromosome pairs with and then disjoins from the X chromosome during meiosis, it is included here.

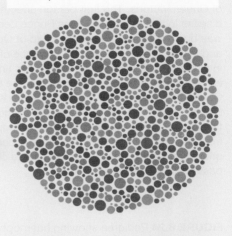

FIGURE 8.21 A person with normal vision sees the number 74, a person with red–green colour blindness may see this as 21. (Note: This reproduction of an Ishihara colour plate is not a valid test for colour vision.)

TABLE 8.3 Genotypes and phenotypes for the X-linked *CBD* colour-vision gene.

Sex	Genotype	Phenotype
Female	$X^V X^V$	Normal colour vision
	$X^V X^v$	Normal colour vision
	$X^v X^v$	Red–green colour blindness
Male	$X^V Y$	Normal colour vision
	$X^v Y$	Red–green colour blindness

Note that:
- females have two copies of the X chromosome and so must have two copies of any X-linked gene. This means that females will be either homozygous or heterozygous for X-linked genes.
- males have only one X chromosome and so must have just one copy of any X-linked gene. This means that males are hemizygous for X-linked genes.

Making predictions: a monohybrid cross of the X-linked *CBD* colour-vision gene

Could two people with normal colour vision produce a child with red–green colour blindness? If so, what is the chance of this outcome?

Consider the cross of a heterozygous, $X^V X^v$, female and a male who has normal colour vision. Could a child from this cross be colourblind?

The details of the parents are as follows:

Parental phenotypes	Female: normal vision	×	Male: normal vision
Parental genotypes	$X^V X^v$		$X^V Y$
Possible gametes	X^V and X^v		X^V and Y

The gametes of the male parent will have *either* an X chromosome with the *V* allele *or* a Y chromosome.

We can now use a Punnett square to show the expected outcomes of this cross (figure 8.22).

FIGURE 8.22 Punnett square showing the possible outcomes of crossing a heterozygous female with a male who has normal colour vision

Gametes	X^V	X^v
X^V	$X^V X^V$ Female Normal vision	$X^V X^v$ Female Normal vision
(Y)	$X^V Y$ Male Normal vision	$X^v Y$ Male Colourblind

Genotypes: 1 $X^V X^V$: 1 $X^V X^v$: 1 $X^V Y$: 1 $X^v Y$

Phenotypes: 2 females with normal colour vision : 1 male with normal colour vision : 1 colourblind male

Thus, from the cross of a heterozygous female and a male with normal vision:
- the offspring could be a colourblind son, but not a colourblind daughter. The chance of a colourblind son is $\frac{1}{4}$.
- the chance of a colourblind daughter is 0.

CASE STUDY: The royal X-linked recessive disorder

Haemophilia is sometimes called the royal disease due to it being found in several of the descendants of Queen Victoria. The first person affected with the disease in this royal family was Prince Leopold, who was born in 1853 — the eighth child of Queen Victoria and Prince Albert. There is no evidence of the disease in either of their family trees before this and so it is assumed that a mutation occurred in the sperm of Queen Victoria's father, Edward Duke of Kent. This meant Queen Victoria inherited the haemophilia allele from her father and a normal, non-haemophilia allele from her mother. She was, therefore, a carrier — phenotypically normal but possessing a chance to pass the disease on to her sons.

FIGURE 8.23 Queen Victoria with her children

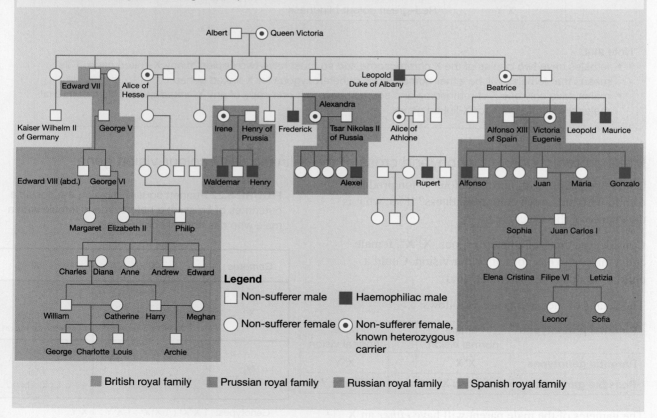

FIGURE 8.24 Pedigree showing haemophilia incidence in Queen Victoria and her descendants

What is haemophilia?

Haemophilia is an X-linked recessive disorder. Like the *CBD* gene, which is responsible for colour blindness, the haemophilia gene is found on the X chromosome. This means that if males have a faulty allele, they will develop the disease. Females, of course, require two disease alleles to develop haemophilia since they have two X chromosomes. If they have only one disease allele, they are carriers and phenotypically normal. Queen Victoria and her descendants likely had haemophilia B, resulting in a deficiency of the protein factor IX, which is needed for blood to clot properly. The condition can lead to excessive blood loss from relatively minor wounds, which can result in death. There was no treatment for the condition until the 1960s.

Prince Leopold was diagnosed with haemophilia as a child. He went on to marry a German princess, Helena of Waldeck-Pyrmont, but died at the age of 31 from a fall that caused a cerebral haemorrhage (a bleed on the brain). Leopold had two children, a boy and a girl. His son Rupert was also affected, and his daughter Alice was a carrier.

FIGURE 8.25 Prince Leopold shortly before his death

Queen Victoria had a desire to foster peace across Europe, and so she arranged for many of her children to marry other European royals. This means that many of the European royal families — including the Prussian, the Russian and the Spanish — were descended from Queen Victoria and carried the haemophilia allele (figure 8.24). However, as her son Edward VII did not possess the haemophilia allele, and none of the people who subsequently married into the family were affected, the current British royal family are free from haemophilia.

What were the chances of Queen Victoria's children being haemophiliacs?

Non-disease allele: X^H
Haemophilia allele: X^h
Genotype of Queen Victoria: $X^H X^h$
Genotype of Prince Albert: $X^H Y$

FIGURE 8.26 Possible genotypes of Queen Victoria and Prince Albert

Gametes	X^H	X^h
X^H	$X^H X^H$	$X^H X^h$
Y	$X^H Y$	$X^h Y$

Thus, from the cross of Queen Victoria (heterozygous female) and Prince Albert (male with a normal genotype):
- there was a 50 per cent chance any sons born to Queen Victoria would have haemophilia
- there was a 50 per cent chance any daughters born to Queen Victoria would be carriers
- the chance of a daughter with haemophilia was 0.

Prince Leopold would have had the genotype $X^h Y$.

8.3.3 Variations to monohybrid crosses

We have seen in the previous sections that the expected result from the cross of two parents heterozygous at an autosomal gene locus is a 3 : 1 ratio of offspring showing the dominant phenotype to those showing the recessive trait.

Variations from the expected 3-dominant : 1-recessive phenotypic ratio from a cross of two heterozygotes can occur. This can happen, for example, when:
- the relationship between the alleles of the gene is one of codominance; or
- one of the alleles is lethal in the homozygous condition.

Codominant alleles

Codominance refers to a situation in which both alleles of a heterozygous organism are expressed in its phenotype (see section 7.3.2). For example, in cattle (*Bos taurus*), coat colour is controlled by an autosomal gene with the alleles C^R and C^W. The relationship between these two alleles is one of codominance. The coat colour of heterozygous cattle is called roan and it consists of a mixture of red hairs and white hairs. Figure 8.27 shows the phenotypes of the three possible genotypes.

FIGURE 8.27 Genotypes and phenotypes in cattle for alleles of a coat-colour gene that have a codominant relationship. Note that both alleles are expressed in the heterozygous roan cattle.

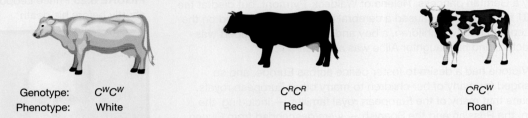

Genotype:	$C^W C^W$	$C^R C^R$	$C^R C^W$
Phenotype:	White	Red	Roan

Consider the cross of two heterozygous roan cattle. Will the outcome be the typical 3 : 1 ratio?

We can show the parental details as follows:

Parental phenotypes	Roan	×	Roan
Parental genotypes	$C^R C^W$		$C^R C^W$
Possible gametes	C^R and C^W		C^R and C^W

We can now use a Punnett square (figure 8.28) to show the outcome of the cross of two heterozygous roan cattle.

FIGURE 8.28 Punnett square showing the possible outcomes of crossing two heterozygous roan cattle

Gametes	C^R	C^W
C^R	$C^R C^R$ Red	$C^R C^W$ Roan
C^W	$C^R C^W$ Roan	$C^W C^W$ White

Genotypes: 1 $C^R C^R$: 2 $C^R C^W$: 1 $C^W C^W$
Phenotypes: 1 red : 2 roan : 1 white

Where the allelic relationship is one of codominance, the expected result from the cross of two heterozygotes is a ratio of 1 : 2 : 1 of red : roan : white (rather than 3 : 1).

Lethal genes

Some genotypes can result in death early in embryonic development. Offspring with such lethal genotypes do not develop and so do not appear among the offspring of a cross. How does this affect the expected ratio of offspring? Let's look at an example.

One gene in mice has the alleles A^y (yellow coat colour) and a (agouti coat colour). Yellow coat colour is dominant to the agouti coat colour (see figure 8.29). However, the A^y allele in a double dose (homozygous A^yA^y) is lethal, causing death in early embryonic development. In terms of lethality, the A^y allele is a recessive lethal.

What would be expected from the cross of two mice, each with a heterozygous A^ya yellow genotype?

We can show the parental details as follows:

FIGURE 8.29 An A^ya yellow mouse (right) with the common agouti aa mouse (left)

Parental phenotypes	Yellow	×	Yellow
Parental genotypes	A^ya		A^ya
Possible gametes	A^y and a		A^y and a

We can now use a Punnett square to show the outcome of the cross of these heterozygous yellow mice (figure 8.30).

FIGURE 8.30 Punnett square showing the possible outcomes of crossing two heterozygous yellow mice. Which genotype does not appear?

Gametes	A^y	a
A^y	A^yA^y	A^ya Yellow
a	A^ya Yellow	Aa Agouti

Genotypes: 2 A^ya : 1 aa

Phenotypes: 2 yellow : 1 agouti

The potential A^yA^y offspring die in very early embryonic development, even before they are implanted in the wall of the uterus. As a result, these potential offspring are never detected. Evidence of the recessive lethal nature of the A^y allele comes from the fact that litter sizes from these crosses are smaller than average and, in addition, yellow mice are always heterozygous, never homozygous.

8.3.4 Monohybrid test crosses

An organism that shows the dominant phenotype may have one or two copies of the allele that determines that phenotype; that is, it may be homozygous TT or heterozygous Tt. This situation can be denoted by the genotype $T-$, where the dash symbol denotes either T or t. In this case, a monohybrid **test cross** can be used to identify the organism's genotype.

test cross a cross used to determine the genotype of an individual with an unknown genotype by crossing it to an individual with a homozygous recessive genotype

A monohybrid test cross is a special kind of cross in which an organism of uncertain genotype is crossed with a homozygous recessive organism. As a monohybrid cross, it involves only one gene and determines whether an organism with a dominant phenotype is heterozygous or homozygous for that gene.

In reality, it is possible today to identify genotypes more directly by examining the relevant part of an organism's genome, without the need to carry out a test cross and produce numbers of offspring.

However, let's consider a male black cat whose parentage is unknown. Such a cat has dense pigmentation and so could have the genotype either *DD* or *Dd*. What are the possible outcomes of a test cross of this male black cat with a homozygous recessive female grey cat with dilute colouring? The result will reveal whether the black cat is homozygous dense or heterozygous dense.

- If the black cat is homozygous *DD*, he cannot sire any kittens with dilute-coloured (grey) fur as he can pass on only a *D* allele to all his offspring and, in consequence, all kittens must have black fur.
- If the black cat is heterozygous *Dd*, he can produce two types of gamete: *D*-carrying sperm and *d*-carrying sperm. When combined with the mother's gametes, which must all be *d*-carrying eggs, both black kittens and grey kittens can result (see figure 8.31).

Note that the two possible phenotypes are expected to appear in equal proportions, so the chance of a:
- grey kitten is $\frac{1}{2}$
- black kitten is also $\frac{1}{2}$.

These probabilities do *not* mean that in a litter of four kittens, two will be black and two will be grey. The appearance of just one grey kitten is *sufficient* evidence to conclude that the black-furred parent is heterozygous *Dd*. It would not be valid to conclude that the black-furred parent was homozygous *DD* on the evidence of a litter of five black kittens because this outcome can occur by chance (see figure 8.32).

FIGURE 8.31 Possible outcomes of a monohybrid test cross of a heterozygous black male cat with a grey female cat

Gametes	*D* (from male)	*d* (from male)
d (from female)	*Dd* Black	*dd* Grey

Genotypes: 1 *Dd* : 1 *dd*
Phenotypes: 1 black : 1 grey

FIGURE 8.32 The outcome of this test cross depends on the genotype of the parent showing the dominant trait — in this case, black fur colour. If homozygous *DD*, this cat cannot sire any grey kittens; if heterozygous *Dd*, the cat could sire grey kittens.

SAMPLE PROBLEM 3 Predicting outcomes using Punnett squares

Cystic fibrosis is an autosomal recessive disorder.
a. Use a Punnett square to list the genotypic and phenotypic outcomes of a couple who are both carriers of cystic fibrosis. **(4 marks)**
b. Determine the percentage chance of the couple having an affected child. **(2 marks)**

THINK

a. 1. Determine the genotype of the parents, choosing suitable allele symbols.

2. Construct a Punnett square using parental genotypes and fill in potential offspring genotypes.

3. Analyse the offspring genotypes and identify the affected genotype.

b. To determine the chance for a couple who are both carriers to have an affected child, divide the number of affected phenotypes (1; cc) by the total number of offspring genotypes (4; CC, Cc, Cc, cc) and multiply by 100.

WRITE

Both parents are carriers and therefore must be heterozygotes — Cc (1 mark).
c is used as the disease allele because it is the first letter of the disease.

Gametes	C	c
C	CC	Cc
c	Cc	cc

(1 mark)

Affected genotype = cc
Genotypes: 1 CC : 2 Cc : 1 cc (1 mark)
Phenotypes: 3 unaffected : 1 affected (1 mark)

$\frac{1}{4} \times 100 = 25\%$ (2 marks)

INVESTIGATION 8.1 `online only`

Genetics in *Drosophila*

Aim
To investigate inheritance in *Drosophila melanogaster*

INVESTIGATION 8.2 `online only`

What's the chance of being Rhesus positive?

Aim
To collect and analyse simulated data relating to genetic crosses on one gene

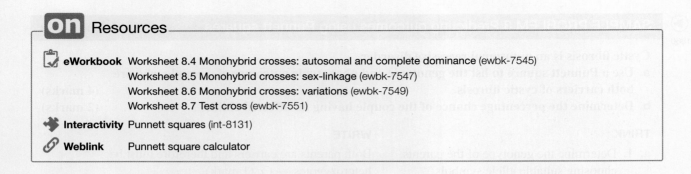

KEY IDEAS

- Monohybrid crosses involve the alleles of one gene.
- Punnett squares can be used to help in determining the outcomes of genetic crosses.
- A test cross is the cross of an organism with a known homozygous recessive genotype and another that shows the dominant phenotype, but whose genotype is unknown.
- Patterns of inheritance of X-linked genes are characterised by an unequal occurrence of phenotype by sex.

8.3 Activities

To answer questions online and to receive **immediate feedback** and **sample responses** for every question, go to your learnON title at **www.jacplus.com.au**. A **downloadable solutions** file is also available in the resources tab.

| 8.3 Quick quiz | 8.3 Exercise | 8.3 Exam questions |

8.3 Exercise

1. Describe why Punnett squares are used by geneticists.
2. Achondroplasia is a genetic condition in humans that results in dwarfism. It is caused by inheriting a single disease allele on chromosome 4. Therefore, those affected have the genotype (*Dd*). Children who are homozygous dominant (*DD*) do not survive beyond 12 months old.

   An affected person and an unaffected person have decided to try for a child. Use a Punnett square to show the genotypic and phenotypic outcomes.

3. Fur colouration in rabbits is controlled by multiple alleles of a single gene. Black (*C*) is dominant over chinchilla colouration (C^{ch}), which in turn is dominant over Himalayan colouration (C^h), which in turn is dominant over albino (*c*). A purebred black rabbit is crossed with a purebred Himalayan-coloured rabbit. Use a Punnett square to show the genotypic and phenotypic outcomes.

4. A couple wishing to start a family have a family history of Duchenne muscular dystrophy. This is an X-linked recessive genetic disorder in which the muscles severely weaken over time. The male and female are both unaffected, but the female knows she is a carrier. As such, they worry any children they have may develop the disease. They have made an appointment to see you, a genetic counsellor, to determine the percentage chance of them having an unaffected baby. Identify the genotypic and phenotypic outcomes and calculate the percentage chance of an unaffected child.

Athletes with achondroplasia taking part in the Paralympic Games in 2016

5. In guinea pigs, fur colour is controlled by a single gene with two alleles. Black fur (*B*) is dominant over white fur (*b*). John has a black guinea pig but is unsure of the genotype. Explain how he could determine the genotype of his guinea pig without using any laboratory equipment.
6. A woman who is blood group A has a child who is blood group O. Two men are claiming to be the biological father of the child. One of the men is blood group A and the other is blood group B. Who is more likely to be the father of the child? Explain your answer.

8.3 Exam questions

▶ Question 1 (1 mark)

Source: *VCAA 2012 Biology Exam 2, Section A, Q6*

MC A couple, each phenotypically normal, have a child with phenylketonuria, an autosomal recessive trait.

The chance that their second child will have the trait is

A. one in four.
B. two in three.
C. one in three.
D. three in four.

▶ Question 2 (1 mark)

Source: *VCAA 2012 Biology Exam 2, Section A, Q10*

MC In humans, the ABO blood group has a single autosomal gene locus with three possible alleles. There are four different blood group types. The different blood group types and their genetic make-up are shown in the following table.

Blood group type	Possible alleles
Group O	ii
Group A	$I^A I^A$ or $I^A i$
Group B	$I^B I^B$ or $I^B i$
Group AB	$I^A I^B$

A woman of blood group A, whose genotype is unknown, and a man of blood group O have a child.

Genetically, this is an example of a

A. self cross.
B. test cross.
C. dihybrid cross.
D. sex-linked cross.

▶ Question 3 (4 marks)

Source: *VCAA 2011 Biology Exam 2, Section B, Q2a, b*

In a particular plant species, an autosomal gene controls flower colour, with yellow flower colour being dominant to blue.

a. i. Assign relevant symbols to the alternative traits. **1 mark**
Allele symbols: Yellow flower _____
Blue flower _____
A test cross was carried out using a heterozygous yellow flower and a blue flower.
ii. Indicate the genotypes of the parents in the test cross. **1 mark**
iii. Show the expected genotype(s) and phenotype(s) of the offspring. **1 mark**
b. Outline how a test cross might be used to identify whether a yellow flower plant is homozygous or heterozygous. **1 mark**

▶ Question 4 (7 marks)

Source: *VCAA 2005 Biology Exam 2, Section B, Q2*

Bay scallops (*Argopecten irradians*) have three shell colours: orange, yellow and black. It is known that the colour is under the control of one gene locus with three alleles.

a. What is a gene locus? **1 mark**
b. If you compared the alleles of this locus at the molecular level, what would be different? **1 mark**
c. What is the name given to the process which gives rise to new alleles? **1 mark**

Several crosses of scallops were carried out and the number and phenotype of the offspring were recorded. These crosses and results are shown in the table below.

Cross	Phenotypes of scallop parents	Offspring
1	yellow X yellow	27 yellow : 9 black
2	black X black	all black
3	orange X orange	30 orange : 10 black

d. Using the allelic symbols p^Y for yellow, p^O for orange and p^b for black, show the genotypes of the parents and offspring in cross 1. **3 marks**

Parent's phenotype	yellow	X	yellow
Parent's genotype(s)		X	
Offspring phenotype	yellow		black
Offspring genotype(s)			

e. What term is used to describe the genotype of the black parents and offspring in cross 2? **1 mark**

▶ Question 5 (3 marks)

Source: VCAA 2006 Biology Exam 2 Section B, Q1d, e

Coat colour in mice is under the control of a single gene with two alleles. Many crosses between yellow-coated mice and mice with grey coats gave the following results. The mice with grey coats were known to be homozygous.

Parental cross   yellow x grey
First generation   50% yellow : 50% grey

Many crosses were carried out between the first generation yellow mice.

a. What genotypic ratio and phenotypic ratio would we expect to see in the offspring of the cross between the first generation yellow mice? Make sure you indicate the allelic symbols you are using for this gene locus. **2 marks**

Scientists performed this cross many times and the result they observed was always a ratio of 2 yellow to 1 grey mouse.

b. How can this result be explained? **1 mark**

More exam questions are available in your learnON title.

8.4 Predicting genetic outcomes: linked genes and independently assorted genes

KEY KNOWLEDGE

- Predicted genetic outcomes for two genes that are either linked or assort independently

Source: VCE Biology Study Design (2022–2026) extracts © VCAA; reproduced by permission.

8.4.1 Dihybrid crosses: two genes in action

A dihybrid cross involves the alleles of two different genes. Let us first consider two genes that are located on different (non-homologous) chromosomes, such as the *ABO* gene on the number-9 chromosome and the *TYR* gene on the number-11 chromosome.

Genes on non-homologous chromosomes are said to be unlinked. When two genes are unlinked, the alleles at one gene locus behave independently of those at the second locus in their movements into gametes. Unlinked genes can be shown by using a semicolon to separate the alleles of one gene from those of the second gene — for example, *Aa*; *Bb*.

A dihybrid cross between *ABO* and *TYR* genes in humans

Let's now return to Tracey and John and look at their genotypes for the *ABO* and the *TYR* genes. Both Tracey and John are heterozygous at each gene locus (see figure 8.33).

FIGURE 8.33 All the possible normal gametes produced by meiosis in a person with the heterozygous genotype $I^B i$; *Aa*

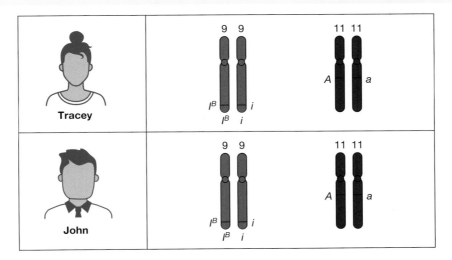

What are the possible outcomes of a cross between Tracey and John?

Because the genes are unlinked, each parent can produce four kinds of gametes in equal numbers, as shown in figure 8.34. The various combinations of alleles of different genes are the result of allele segregation and of the independent assortment of genes in meiosis.

We can summarise the genetic information about Tracey and John as follows:

FIGURE 8.34 All the possible normal gametes produced by meiosis in a person with the heterozygous genotype $I^B i$; *Aa*

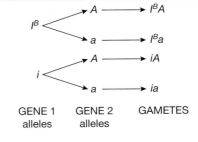

GENE 1   GENE 2   GAMETES
alleles   alleles

Parental phenotypes	Blood group B; normal pigment	×	Blood group B; normal pigment
Parental genotypes	$I^B i$; *Aa*		$I^B i$; *Aa*
Possible gametes	$I^B A$, $I^B a$, *IA* and *IaI*$^B A$, $I^B a$, *IA* and *Ia*		

The dihybrid cross between Tracey and John can be shown in two different ways:
1. A dihybrid cross can be shown as a combination of two monohybrid crosses (figure 8.35). The chance that Tracey and John's next child will have albinism is $\frac{1}{4}$ and the chance that the next child will be blood group O is also $\frac{1}{4}$. So the combined probability that the next child will have albinism and be blood group O is $\frac{1}{4} \times \frac{1}{4} = \frac{1}{16}$.

FIGURE 8.35 Looking at a dihybrid cross. What is the chance of an $aaI^B I^B$ offspring?

Locus 1	Locus 2	Offspring and chance
$Aa \times Aa$	$I^B i \times I^B i$	$AAI^B I^B : \frac{1}{4} \times \frac{1}{4} = \frac{1}{16}$
↓	↓	$Aaii : \frac{1}{2} \times \frac{1}{4} = \frac{1}{8}$
$\frac{1}{4} AA : \frac{1}{2} Aa : \frac{1}{4} aa$	$\frac{1}{4} I^B I^B : \frac{1}{2} I^B i : \frac{1}{4} ii$	$AaI^B i : \frac{1}{2} \times \frac{1}{2} = \frac{1}{4}$
		$aaii : \frac{1}{4} \times \frac{1}{4} = \frac{1}{16}$
		etc.

2. A dihybrid cross can be shown as a Punnett square, with all the possible parental gametes shown across the top and down the side of the square (figure 8.36). Combining these various gametes in a Punnett square gives all the possible genotypes that can occur in offspring of these parents. These 16 different genotypes can be grouped into 4 different phenotypes as shown at the bottom of the Punnett square.

Note that in the overall summary statement below the Punnett square, the 'dash' symbol (–) means that the second allele in the genotype can be either of the two possible alleles. So, for example, $A-$ denotes both AA and Aa.

FIGURE 8.36 Punnett square showing the dihybrid cross of Tracey and John

$I^B i\, Aa$ × $I^B i\, Aa$
Tracey   John

Tracey's eggs

Gametes	$\frac{1}{4} I^B A$	$\frac{1}{4} I^B a$	$\frac{1}{4} iA$	$\frac{1}{4} ia$
$\frac{1}{4} I^B A$	$\frac{1}{16} I^B I^B AA$	$\frac{1}{2} I^B I^B Aa$	$\frac{1}{16} I^B i AA$	$\frac{1}{16} I^B i Aa$
$\frac{1}{4} I^B a$	$\frac{1}{16} I^B I^B Aa$	$\frac{1}{16} I^B I^B aa$	$\frac{1}{16} I^B i Aa$	$\frac{1}{16} I^B i aa$
$\frac{1}{4} iA$	$\frac{1}{16} I^B i AA$	$\frac{1}{16} I^B i Aa$	$\frac{1}{16} ii AA$	$\frac{1}{16} ii Aa$
$\frac{1}{4} ia$	$\frac{1}{16} I^B i Aa$	$\frac{1}{16} I^B i aa$	$\frac{1}{16} ii Aa$	$\frac{1}{16} ii aa$

(John's sperm on the left axis)

Overall summary: $\frac{9}{16} I^B - A-$: $\frac{3}{16} I^B - aa$: $\frac{3}{16} ii A-$: $\frac{1}{16} ii aa$

↓ ↓ ↓ ↓

| 9 Group B Normal pigment | 3 Group B Albino | 3 Group O Normal pigment | 1 Group O Albino |

In summary, the expected phenotypic result from a dihybrid cross of two heterozygotes is 9 : 3 : 3 : 1, where:
- $\frac{9}{16}$ refers to offspring showing both dominant traits
- the two $\frac{3}{16}$ refer to offspring showing one of the dominant traits
- the $\frac{1}{16}$ refers to offspring showing both recessive traits.

A dihybrid cross between fur colour and length of fur in cats

Let's review another cross, this one in cats, where:
- one gene controls the presence of white spotting and has the alleles *W* for white spotting and *w* for absence of white spots. White spotting is dominant to absence of white spots.
- a second unlinked gene controls fur length and has the alleles *S* for short fur and *s* for long fur. The short fur phenotype is dominant to long fur.

What is the expected outcome of the cross of two cats that are heterozygous at each gene locus; that is, with genotypes *Ww*; *Ss*?

Based on the summary above, we can predict that the expected outcome is in the ratio 9 : 3 : 3 : 1. That is:

9 white-spotted with short fur : 3 white-spotted with long fur : 3 no white spots with short fur : 1 no white spots with long fur.

These ratios can be expressed as probabilities, such as a $\frac{9}{16}$ chance of showing both dominant traits, a $\frac{3}{16}$ chance of showing one only of the dominant traits, a $\frac{3}{16}$ chance of showing the other dominant trait and a $\frac{1}{16}$ chance of showing both recessive traits.

Let's test that theory (figure 8.37).

Two cats, heterozygous at each gene locus: *WwSs* × *WwSs*

FIGURE 8.37 Possible outcomes from dihybrid cross of fur colour and length in cats

Gametes	$\frac{1}{4}$ WS	$\frac{1}{4}$ Ws	$\frac{1}{4}$ wS	$\frac{1}{4}$ ws
$\frac{1}{4}$ WS	$\frac{1}{16}$ WWSS	$\frac{1}{16}$ WWSs	$\frac{1}{16}$ WwSS	$\frac{1}{16}$ WwSs
$\frac{1}{4}$ Ws	$\frac{1}{16}$ WWSs	$\frac{1}{16}$ WWss	$\frac{1}{16}$ WwSs	$\frac{1}{16}$ Wwss
$\frac{1}{4}$ wS	$\frac{1}{16}$ WwSS	$\frac{1}{16}$ WwSs	$\frac{1}{16}$ wwSS	$\frac{1}{16}$ wwSs
$\frac{1}{4}$ ws	$\frac{1}{16}$ WwSs	$\frac{1}{16}$ Wwss	$\frac{1}{16}$ wwSs	$\frac{1}{16}$ wwss

From the possible outcomes, we can predict that:
- $\frac{9}{16}$ will have white spots with short fur
- $\frac{3}{16}$ will have white spots with long fur
- $\frac{3}{16}$ will have an absence of white spots with short fur
- $\frac{1}{16}$ will have an absence of white spots with long fur.

This again gives the predicted 9 : 3 : 3 : 1 ratio of phenotypes.

Resources

Interactivity Dihybrid cross (int-0180)

8.4.2 Genes can be linked

Genes do not float around the nucleus like peas in soup. Each gene has a chromosomal location or locus.

The genes that are located on one chromosome form a **linkage group**. Figure 8.38 shows three of the linkage groups in the tomato (*Lycopersicon esculentum*). The total number of linkage groups in an organism corresponds to the haploid number of chromosomes. Thus, the tomato with a diploid number of 24 has 12 linkage groups.

linkage group genes that are physically close to each other on a chromosome and are likely to be inherited together as a single unit

In humans, there are 22 autosomal linkage groups plus the X and Y-linkage groups, making a total of 24. Figure 8.39 shows just a few of the genes in the linkage groups on five human chromosomes.

FIGURE 8.38 Three of the twelve linkage groups in the tomato

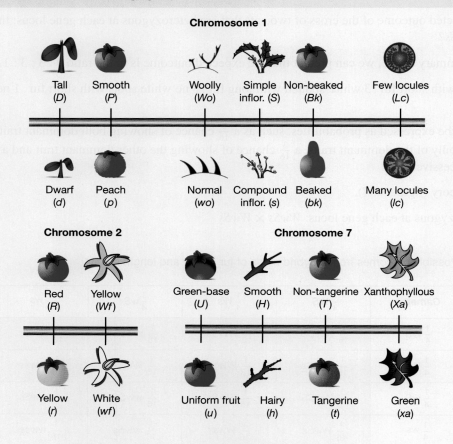

FIGURE 8.39 A chromosome map showing a very small sample of the genes that form part of five linkage groups

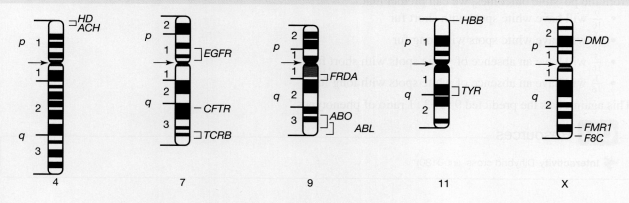

Behaviour of linked genes

Earlier in this subtopic (section 8.4.1) we explored the inheritance of unlinked genes. Such genes are typically located on non-homologous chromosomes and their alleles assort independently into gametes.

Linked genes are located close together on a chromosome. The combinations of their alleles on homologous chromosomes tend to stay together (figure 8.40), but they can on occasions be separated by *crossing over* during meiosis. The closer together on a chromosome that the allelic forms of two different genes are, the more tightly they are linked, and the less likely they are to be separated by crossing over during gamete formation by meiosis. The consequence is that the more widely separated on a chromosome, the more likely that the alleles of two different genes will be separated by crossing over.

Making predictions: linked genes in blood

The *RHD* gene, which controls Rhesus blood type, and the *EPB41* gene, which controls the shape of red blood cells, are close together on the number-1 chromosome and so are said to be linked.

FIGURE 8.40 Alleles of closely linked genes are more likely to be inherited together than the alleles of widely separated genes.

Sarah has elliptical red blood cells and is Rhesus positive; her genotype is *DdEe*. Dave has normal red blood cells and he is Rhesus negative; his genotype is *ddee*. Figure 8.41 shows the arrangements of the alleles of the two genes on the chromosomes in Sarah and David. Their genotypes can be written as *DE/de* and *de/de* respectively. Because the *RHD* and the *EPB41* genes are physically close together on the number-1 chromosome, alleles of these two genes do *not* behave independently, as is the case for the *TYR* and the *ABO* genes.

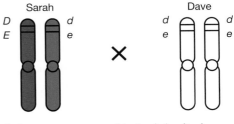

FIGURE 8.41 Independent assortment of alleles of two linked genes

Because the loci of two linked genes are physically close, the particular combination of alleles of the genes that are present on parental chromosomes tend to be inherited together more often than alternative combinations. These combinations of alleles can, however, be broken by crossing over during meiosis so that new combinations of alleles are generated. The chance that this occurs depends on the distance between the two linked genes.

Sarah can produce both *DE* eggs and *de* eggs and the chance of each type is equal. These eggs are called **parental (noncrossover) gametes** because they are identical to the original allele combinations present in Sarah. Crossing over and an exchange of segments can occur anywhere along the paired number-1 chromosomes. When an exchange occurs between the *RHD* locus and the *EPB41* locus, **recombinant (crossover) gametes** result. Sarah's recombinant eggs are *De* and *dE*.

Crossing over occurs between all paired chromosomes during meiosis. A crossover point is more likely to occur between two genes that are widely separated on a chromosome than between two gene loci that are closer together. *The closer the genes, the smaller the chance of a crossover.* Because the loci of the *RHD* gene and the *EPB41* gene are very close on the number-1 chromosome, the chance of a crossover occurring between them is small. As a result, Sarah's eggs are more likely to transmit the parental than the recombinant types.

linked genes describes genes whose loci are located on a given chromosome

parental (noncrossover) gametes sex cells that contain the same combination of alleles along their chromosomes as were present in the parental cells

recombinant (crossover) gametes sex cells that contain new combinations of alleles along their chromosomes than were present in the parental cells

We will assume that for these two genes the chance of each kind of parental (noncrossover) gamete is 0.49 and the chance of each type of recombinant gamete is 0.01.

Since Dave is homozygous, he produces only *de* sperm cells — a 1.0 chance.

A Punnett square can be drawn up to show the chance of production of each kind of gamete and the chance of each possible offspring (figure 8.42).

For Sarah and Dave, what is the chance of a child with Rhesus negative blood and elliptical red blood cells? This phenotype corresponds to the genotype *dE/de*. The chance of this offspring is 0.01 or $\frac{1}{100}$.

FIGURE 8.42 Punnett square showing the chance of production of each kind of gamete and the chance of each possible offspring for Sarah and Dave

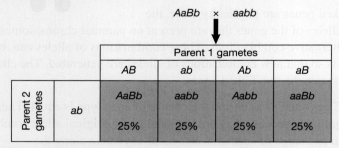

Detecting linkage

Are two gene loci linked? This can be explored by a particular test cross of a known double heterozygote (*AaBb*) with a double homozygous recessive (*aabb*).

- If the two gene loci are not linked, the genes will assort independently, and the outcome of the test cross will be four classes of offspring in equal proportions (figure 8.43).
- If the two gene loci are linked, the outcome of the test cross can reveal that linkage. There will be four classes of offspring but the proportions of these will not be equal. Instead, there will be an excess of offspring from parental gametes and a deficiency of offspring from recombinant gametes.

FIGURE 8.43 Test cross to determine gene linkage of a known double heterozygote (*AaBb*) with a double homozygous recessive (*aabb*)

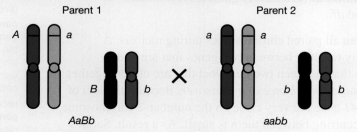

The ratio of offspring is 25 per cent for each phenotype. Since the chances of offspring being produced from parental and recombinant gametes is equal, the genes assort independently and so are not linked.

Estimating distance between linked genes

From the results of a test cross with linked genes, it is possible to estimate the distance between the gene loci. This estimate is based on the percentage of recombinant offspring.

Estimating the distance between gene loci

$$\text{Distance between loci} = \frac{100 \times \text{number of recombinant offspring}}{\text{total number of offspring}}$$

The percentage of recombinant offspring corresponds to the number of map units separating the two genes. So, if there are 12 per cent total recombinant offspring, then the loci of the two genes are separated by about 12 map units (see figure 8.44).

FIGURE 8.44 Determining the gene linkage and distance between the gene loci

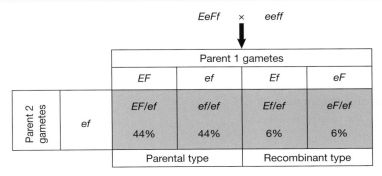

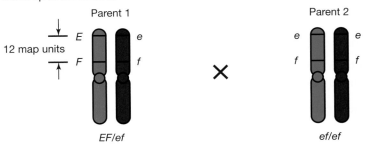

In this example cross, the percentage of offspring with genotypes that are produced from recombinant gametes (*Ef/ef* and *eF/ef*) are less than those produced from parental gametes. This means the genes are linked. The distance of 12 map units is calculated using the formula:

$$\text{Distance between loci} = \frac{100 \times \text{number of recombinant offspring}}{\text{total number of offspring}}$$

$$= \frac{100 \times (6+6)}{100}$$

$$= 12 \text{ map units}$$

SAMPLE PROBLEM 4 Conducting a test cross and calculating the distance between two genes

Two genes in a species of plant are thought to be linked on the same chromosome but scientists do not know how far apart they are. Red flower colour (*R*) is dominant over white (*r*), and green seeds (*G*) are dominant over brown (*g*).

a. Conduct a test cross to show the genotypes of the offspring, indicating which are the parental and recombinant gametes. **(3 marks)**

b. Explain how you would calculate the distance between these two genes in map units if the recombinant genotypes make up 20 per cent of the offspring. **(2 marks)**

THINK

a. 1. Conduct a cross between a heterozygote individual (*RG/rg*) and one that is homozygous recessive for each gene (*rg/rg*).

WRITE

Gametes	RG	Rg	rG	rg
rg	RG/rg	Rg/rg	rG/rg	rg/rg

(1 mark)

2. Identify which are the parental and recombinant gametes.

Parental gametes = *RG* and *rg* (1 mark)
Recombinant gametes = *Rg* and *rG*
(1 mark)

b. Calculate the distance between the genes, which will give you the map unit distance, using the formula.

If the recombinant genotypes make up 20 per cent of the offspring, we can assume that for a total of 100 offspring, there will be 20 recombinant genotypes.

$$\text{Distance} = \frac{100 \times \text{number of recombinant offspring}}{\text{total number of offspring}}$$

$$= \frac{100 \times 20}{100} \text{ (1 mark)}$$

$$= 20 \text{ map units (1 mark)}$$

Making predictions: outcomes of crosses for linked genes

When a test cross is carried out with two genes that are known to be linked and are separated by a known number of map units (but fewer than 40), the outcome of the cross can be predicted.

For example, if two linked genes are separated by 8 map units, then a test cross involving these genes will produce about 8 per cent of the recombinant type offspring and about 92 per cent of the parental type offspring. The actual genotypes and phenotypes of the recombinant offspring depend on which alleles of the two genes were together originally on the one chromosome in the heterozygous parent, before any crossing over occurred during gamete formation. For example, the test cross *RT/rt* × *rt/rt* gives 8 per cent recombinant offspring. The cross *Rt/rT* × *rt/rt* also gives 8 per cent recombinant offspring. However, the genotypes of the recombinant offspring differ in each case, as shown in figure 8.45.

Note that the outcome of a test cross involving two linked genes allows you to deduce how the alleles of the genes are arranged on the chromosomes of the heterozygous parent.

FIGURE 8.45 Variations in recombinant offspring genotype depend upon the alleles of the heterozygous parent.

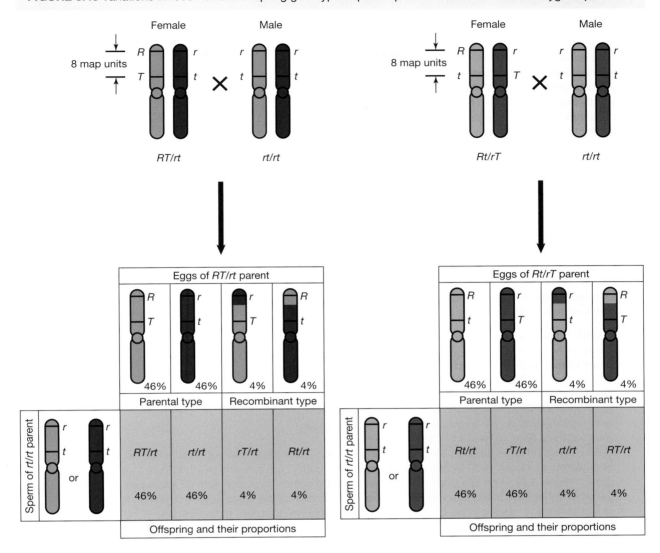

The recombinant offspring each contribute 4 per cent in each example cross. Therefore, in each case:

$$\text{Distance between loci} = \frac{100 \times \text{number of recombinant offspring}}{\text{total number of offspring}}$$

$$= \frac{100 \times (4 + 4)}{100}$$

$$= 8 \text{ map units}$$

elog-0836

INVESTIGATION 8.3 online only

Two genes at a time

Aim

To collect and analyse simulated data relating to genetic crosses on two genes

EXTENSION: Genetic screening and testing

Genetic screening is a test on segments of a population, as part of an organised program, for the purpose of detecting inherited disorders.

genetic screening testing of persons to detect those with the allele responsible for a particular genetic disorder

It is carried out when:
1. members of the population being screened can benefit from early detection of an inherited disorder
2. a reliable test exists that, in particular, does not produce false negative results (a false negative occurs when an affected person fails to be detected by the test)
3. the benefit is balanced against costs (including financial cost)
4. appropriate systems are in place to provide treatment and other follow-up services.

Newborn screening

In Australia, newborn babies are screened shortly after birth for a number of inherited disorders. These conditions are rare, do not show symptoms at birth and most commonly occur in babies where there is no family history of the disorder. If not identified early, these disorders can have serious negative consequences on a baby's mental and physical development. Parents must give their informed consent for the genetic screening of their baby.

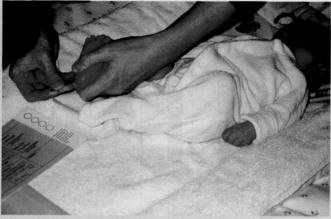

FIGURE 8.47 Newborn genetic screening uses a blood sample taken from the baby's heel.

The genetic disorders for which screening is performed in Australia are:
- *phenylketonuria (PKU)* — occurs in about one in every 10 000 babies. An affected person fails to produce a liver enzyme (phenylalanine hydroxylase) and so is unable to metabolise the amino acid phenylalanine (Phe) to tyrosine. The build-up of Phe results in brain damage. With early detection of the condition and supply of a diet with low levels of phenylalanine, a baby with PKU will develop normally, both physically and mentally (see topic 7).
- *cystic fibrosis (CF)* — occurs in about one in every 2500 babies. A person with CF produces abnormal secretions that have a serious adverse effect on the function of the lungs and on digestion (see topic 7).
- *galactosaemia* — occurs in about one in every 40 000 babies. Lactose, a disaccharide in milk, is digested into galactose and glucose, which then enter the bloodstream. A baby with galactosaemia lacks the enzyme that metabolises galactose and dies if untreated because of the build-up of galactose in the blood. Prompt treatment with special milk that does not contain lactose completely prevents the development of this condition. (This is another example of an interaction between genotype and environment determining a person's phenotype.)
- *congenital hypothyroidism* — a disorder caused by a small, improperly functioning or absent thyroid gland, which occurs in about one in every 3500 babies. Untreated, a baby with hypothyroidism lacks the thyroid hormone and has impaired growth and brain development. Early treatment with a daily tablet of thyroid hormone means the baby develops normally, both physically and mentally.

Genetic testing

Genetic testing is the scientific testing of an individual's genotype.

It may be carried out on people who are at high risk of inheriting a faulty gene based on their family history of occurrence of an inherited disorder. For example, development of breast or ovarian cancer in women with a strong family history of the disease can be associated with two faulty genes, *BRCA1* and *BRCA2*. If a woman has inherited one of these genes, she would be at higher risk of developing breast and/or ovarian cancer. Knowing her genotype enables a woman to make an informed decision about any strategies to reduce her risk. Huntington's disease (HD) is an inherited disease with a dominant phenotype. Onset of the disease occurs in adulthood and causes progressive dementia and involuntary movements. Persons in their late teens or twenties who are at risk of HD may wish to have presymptomatic genetic testing to find out if they have inherited the HD allele from their affected parent.

To find out more about genetic screening and testing, including procedures and ethical implications, please download the digital document.

on Resources

📄 **Digital document** Extension: Genetic screening and testing (doc-36115)

KEY IDEAS

- Dihybrid crosses involve alleles of two different genes.
- The expected outcome of a dihybrid cross of two heterozygotes is four phenotypes in the ratio $9:3:3:1$.
- A dihybrid test cross may be used to determine whether two genes are linked or not.
- The closer two genes are on a chromosome, the less chance there is that crossing over will separate the alleles.
- When genes are linked, parental gametes are formed in addition to smaller numbers of recombinant gametes.
- The total percentage of recombinant gametes formed in a dihybrid test cross is equal to the map units distance the genes are apart from each other.

on Resources

eWorkbook Worksheet 8.8 Dihybrid cross (ewbk-7553)
Worksheet 8.9 Linked or unlinked? (ewbk-7555)

Weblink Victorian Clinical Genetics Services

8.4 Activities

learn on

To answer questions online and to receive **immediate feedback** and **sample responses** for every question, go to your learnON title at **www.jacplus.com.au**. A **downloadable solutions** file is also available in the resources tab.

| 8.4 Quick quiz on | 8.4 Exercise | 8.4 Exam questions |

8.4 Exercise

1. Distinguish between the terms *monohybrid cross* and *dihybrid cross*.
2. In rabbits, the gene for black fur (F) is dominant over white fur (f). The gene that controls fur length is on a different chromosome and is therefore unlinked to fur colour. Long fur (L) is dominant over short fur (l). Using a Punnett square, calculate the ratio of phenotypic outcomes if two heterozygous individuals are crossed.
3. Explain why genes that are closer together on a chromosome are more likely to be inherited together.
4. A heterozygote individual (*WwYy*) is crossed with a homozygous recessive individual (*wwyy*). The genes are not linked. Determine the percentage of each genotype that will be produced.
5. Two genes in a species of fungi are being investigated for possible linkage. The ability to produce the enzyme xylanase (Y) is dominant over the inability to produce it (y), and a protein channel in the membrane (P) is dominant over the lack of the protein (p). Explain how you could determine whether these genes were linked or unlinked.
6. Crab apple plants (*Malus sylvestris*) have two genes that are suspected as being linked. One gene controls height, and tall plants (T) are dominant over dwarf plants (t). The second gene is fruit colour, with green fruit (G) being dominant over yellow fruit (g). A series of test crosses are carried out between a heterozygous plant and a homozygous plant. Out of 526 offspring produced, just 13 had a recombinant genotype. Of these recombinant individuals, 7 were tall plants with yellow fruit and 6 were short plants with green fruit.
   a. Use a Punnett square to determine the genotypes from the cross.
   b. Calculate how far apart the genes are in map units.

8.4 Exam questions

Question 1 (1 mark)
Source: VCAA 2012 Biology Exam 2, Section A, Q13

MC Two genes for coat colour in dogs have the following alleles.

	Gene 1	Gene 2
	B : black	**S** : solid colour
	b : brown	**s** : white spotting

It is reasonable to conclude that a dog with the genotype
A. **BB Ss** would be black with white spotting.
B. **Bb Ss** would be brown with white spotting.
C. **bb SS** would be a solid brown colour.
D. **bb ss** would be a solid black colour.

Question 2 (1 mark)
Source: VCAA 2009 Biology Exam 2, Section A, Q18

MC In humans, smooth chin and straight hairline are each inherited as autosomal recessive traits. The alleles for each of the genes involved are

chin line – **S** : cleft chin   hairline – **W** : widow's peak
          s : smooth chin   **w** : straight hairline

A mother and son each have a smooth chin and a straight hairline. The father of the boy has a cleft chin and a widow's peak.

The father's genotype must be

A. *Ss Ww*.   B. *SS WW*.   C. *ss WW*.   D. *ss ww*.

Question 3 (1 mark)

MC Pea plants *(Pisum sativum)* can have either tall stems (*T*) or short stems (*t*). The flower position can be either axial (*F*) or terminal (*f*), as shown in the diagram.

Axial *F*

Terminal *f*

Many crosses were performed between a heterozygous, tall-stemmed, axial-flowered plant (*TtFf*) and a short-stemmed, terminal-flowered plant (*ttff*). The seeds were collected, planted, watered and monitored until they grew and flowered.

The total number of each phenotype observed in the offspring was recorded. The results are shown in the table.

Phenotype of pea plant offspring	Number of pea plants
tall stems, axial flowers	115
tall stems, terminal flowers	121
short stems, axial flowers	119
short stems, terminal flowers	116

What can be concluded from the results?
A. The genes controlling stem length and flower position are linked.
B. Alleles for stem length and flower position tend to be inherited together.
C. The gene loci for stem length and flower position are on different chromosomes.
D. Tall-stemmed, terminal-flowered plants are a result of crossing over during meiosis in the parents.

Question 4 (4 marks)
Source: VCAA 2013 Biology Exam, Section B, Q8c

Mice have a diploid number of 40.

In mice, black hair colour is dominant to white hair colour. Another gene controls the length of the tail; short tails are dominant to long tails. The genes are not linked.

a. What is meant by linked genes? **1 mark**

The symbols for the alleles for the two genes are shown below.

      B : black hair   **S** : short tail
      b : white hair   **s** : long tail

b. A mouse with black hair and a short tail is crossed with a mouse with white hair and a long tail. The black-haired, short-tailed mouse is heterozygous at both gene loci.

Predict the genotypes and phenotypes of the offspring of these two mice. Show all working. **2 marks**

c. Explain whether you would expect the same genotypes and phenotypes in the offspring if the two genes had been linked. **1 mark**

Question 5 (3 marks)
Source: VCAA 2011 Biology Exam 2, Section B, Q2c

In dogs, two gene loci on different autosomes have the following alleles.

  Gene 1   **B** : black coat colour   Gene 2   **T** : no white on coat
        b : grey coat colour   **t** : white areas on coat

Two dogs, dog **F** and dog **G**, were mated. The litter of four pups that resulted had the following phenotypes.

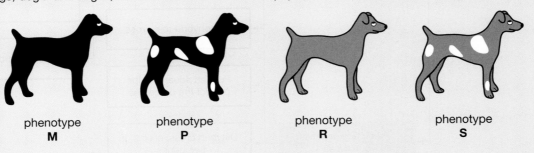

   phenotype   phenotype   phenotype   phenotype
     M   P   R   S

a. What is the genotype of pup **S**? **1 mark**
b. Explain whether all pups with phenotype **P** would have the same genotype. **1 mark**
c. What are the genotypes of the parent dogs? **1 mark**
 Dog **F** _____
 Dog **G** _____

More exam questions are available in your learnON title.

8.5 Review
8.5.1 Topic summary

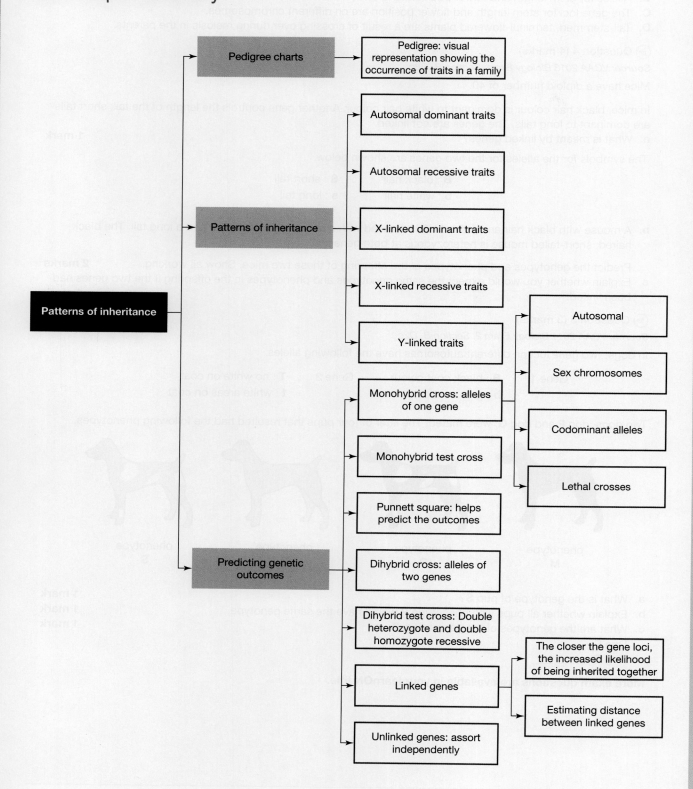

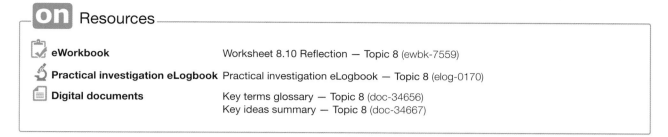

8.5 Exercises

To answer questions online and to receive **immediate feedback** and **sample responses** for every question, go to your learnON title at **www.jacplus.com.au**. A **downloadable solutions** file is also available in the resources tab.

8.5 Exercise 1: Review questions

1. Explain why X-linked recessive traits are more common in males than females in Australia.
2. Explain what is meant by the term *linkage group*.
3. In cats, one autosomal gene controls fur length, with short fur (*S*) being dominant to long fur (*s*). A second autosomal gene unlinked to the first controls the density of colour and has the alleles black (*D*) and grey (*d*).
   a. Are either of these genes located on one of the sex chromosomes?
   b. A grey male cat with long fur (cat 1) is crossed with a black female cat with short fur (cat 2).
      i. Write the genotype of cat 1 and draw a simple chromosomal picture showing this genotype.
      ii. How many kinds of gametes can cat 1 produce?
      iii. What is the phenotype of cat 2?
      iv. List all possible genotypes for cat 2.
      v. If cat 2 produced many kittens and, regardless of the phenotype of the other parent, the kittens were always black and short-furred, what does this suggest about the genotype of cat 2?
4. A pedigree is shown for a genetic disease.

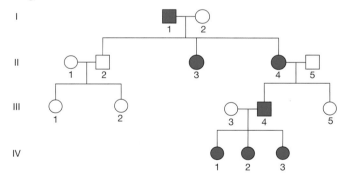

   a. What is the relationship between II-3 and III-5?
   b. Identify the pattern of inheritance of this disease.
   c. Write the genotypes of the following individuals:
      i. I-1
      ii. II-4
      iii. IV-3.
5. In a certain breed of dog, the coat colouring is either black or brown (also known as liver). A number of observations were made about this breed as follows:
   - Brown by brown matings always produced brown pups.
   - Black by black matings sometimes produced all black pups, but in other similar matings produced both black and brown pups.

   Suggest a possible genetic basis for this colouring in dogs.

6. In sunflower plants, brown seed colour (G) is dominant over white (g). Use a Punnett square(s) to show how you could identify the unknown genotype of a brown-seeded plant.
7. Two genes are 6 map units apart. What would be the percentage of recombinant and parental gametes produced?
8. A person has the genotype AB/ab. Give the genotypes of the gametes that can be produced by this person, stating which are the parental and which are the recombinant.
9. The allele for curly leaf (C) shape in a plant species is dominant over straight (c), and green leaf colour (G) is dominant over red (g). A cross is carried out between a homozygous dominant plant for both genes and a heterozygous plant. The genes are unlinked.
    a. Give the genotype of both parents.
    b. Use a Punnett square to show the genotypes produced from the cross.
    c. State the resulting phenotypic ratio.
10. A male with normal colour vision and a female with normal colour vision, but who is a carrier of a colour vision defect, have children.
    a. Make simple drawings to show the relevant chromosome pair(s) and appropriate alleles for both the male and the female.
    b. Identify the types of gametes that they would be expected to produce.
    c. What different kinds of children, in terms of colour vision capability, could they produce?
    d. Could this couple produce a colourblind daughter? Explain.
11. Two genes located close together on chromosome 9 are the *ABO* gene, with the alleles I^A, I^B and i, and the *NPS* gene, with the alleles N (deformed nails) and n (normal nails). Fred is blood type O and is homozygous for normal nails. Gina is blood type AB and is heterozygous for deformed nails.
    a. What is Fred's genotype?
    b. What kind(s) of sperm can Fred produce?
    c. Can crossing over affect the kinds of sperm that Fred produces? Explain.
    d. What is Gina's genotype?
    e. Draw Gina's pair of number-9 chromosomes, showing one possible arrangement of the alleles of the two genes.
    f. What kinds of parental (noncrossover) gametess can Gina produce?
    g. What kinds of recombinant (crossover) gametes can Gina produce?
12. Consider the typical features in the pedigree for inheritance of an X-linked dominant trait. Explain how this pattern would change if all affected males died in infancy.
13. Scientists are trying to find the relative position of three genes — A, B and C — on a chromosome. A series of dihybrid test crosses was carried out. The first cross involved an individual who was *AB/ab* with another who was *ab/ab*. Of the 150 offspring produced, 12 had recombinant genotypes. A second cross was carried out between an individual who was *AC/ac* with another who was *ac/ac*. On this occasion, of the 180 offspring, 20 had recombinant genotypes. Drawing a line to represent a chromosome, show the position of the three genes and the distance in map units between each of the genes.

8.5 Exercise 2: Exam questions

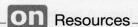

Teacher-led videos Teacher-led videos for every exam question

Section A — Multiple choice questions

All correct answers are worth 1 mark each; an incorrect answer is worth 0.

Question 1

On a pedigree chart, what does the symbol shown represent?

A. Affected male
B. Unaffected female
C. Affected father
D. Unaffected identical twin

Question 2

Which diagram is used to determine the probability of an offspring having a particular genotype?

A. Dihybrid test cross
B. Monohybrid test cross
C. Pedigree chart
D. Punnett square

Question 3

John has Rhesus positive blood (*Dd*) and he has a child with Ann, who is Rhesus negative (*dd*). What is the percentage chance of their child being Rhesus positive?

A. 0%
B. 25%
C. 50%
D. Impossible to determine

Question 4

Two genes are suspected as being linked. When a dihybrid test cross is performed, 8 recombinant genotypes are produced out of 200. Which one of the following is closest to the gene loci distance in map units?

A. 4
B. 8
C. 16
D. 20

Question 5

If genes are unlinked, what percentage of the offspring would be recombinant?

A. 0%
B. 25%
C. 50%
D. 100%

Question 6

Two genes are unlinked. What are the expected phenotypic ratios of offspring produced from a cross of two heterozygous parents?

A. 3 : 1
B. 3 : 3 : 1
C. 9 : 3 : 3 : 1
D. 9 : 3 : 1

Question 7
Which process, occurring in meiosis, is responsible for producing recombinant gametes?

A. Independent assortment
B. Crossing over
C. Cytokinesis
D. Prophase I

Question 8
A monohybrid test cross is conducted between two plants. All 1000 of the offspring have the same phenotype as one of the parents for the trait. What useful information does this provide?

A. Both parents are homozygous recessive for the trait.
B. One parent is homozygous recessive and the other is homozygous dominant for the trait.
C. Both parents are heterozygous for the trait.
D. One parent is homozygous recessive and the other is heterozygous for the trait.

Question 9
Source: VCAA 2012 Biology Exam 2, Section A, Q11

In humans, the ABO blood group has a single autosomal gene locus with three possible alleles. There are four different blood group types. The different blood group types and their genetic make-up are shown in the following table.

Blood group type	Possible alleles
Group O	ii
Group A	$I^A I^A$ or $I^A i$
Group B	$I^B I^B$ or $I^B i$
Group AB	$I^A I^B$

Examine the following pedigree, which shows the phenotype with respect to the ABO gene locus of each individual.

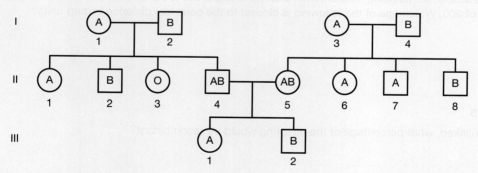

Individuals that would be homozygous at the ABO gene locus include

A. I3.
B. II2.
C. II6.
D. III2.

Question 10

Source: VCAA 2010 Biology Exam 2, Section A, Q15

Duchenne Muscular Dystrophy (DMD) is inherited as an X-linked recessive condition in which cells fail to produce normal dystrophin protein.

The following pedigree shows a family in which some members have DMD.

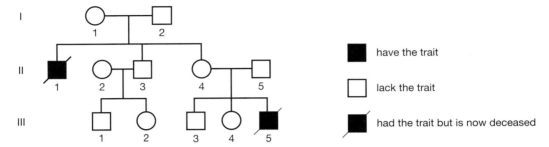

From the pedigree it is reasonable to conclude that

A. II 4 is homozygous normal at the DMD locus.
B. I 1 is heterozygous with respect to the DMD allele.
C. I 1 and I 2 have a one-in-four chance of producing a daughter with DMD.
D. II 4 and II 5 have a one-in-three chance of producing a daughter who is a carrier of DMD.

Section B — Short answer questions

Question 11 (8 marks)

The pedigree shows the incidence of an autosomal disease within a family. Those affected with the disease have the genotype ee.

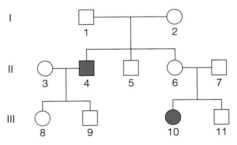

a. State the genotype of individual 7. **1 mark**

b. Individual 10 has a child with a man who is a carrier for the condition.
   i. Use a Punnett square to show genotypes of the offspring. **2 marks**
   ii. State the genotypic and phenotypic ratios of the offspring. **2 marks**

c. Explain why it is impossible to be certain of the genotype of individual 11. **3 marks**

Question 12 (5 marks)

Source: VCAA 2011 Biology Exam 2, Section B, Q5b–d

The quarter horse, as a breed, originated by selective breeding. The first Australian quarter horses were imported from North America in the 1950s. A genetic condition called Hereditary Equine Regional Dermal Asthemia (HERDA) affects certain individuals. HERDA horses have a reduced life expectancy. Affected horses have a pedigree that is linked to an American stallion, Polo Bueno, which lived in the 1940s.

Examine the following pedigree.

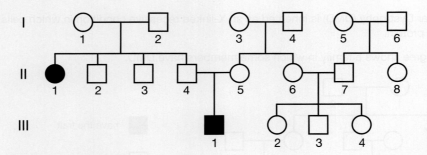

a. What is the mode of inheritance of HERDA? **1 mark**

b. After the horses arrived in Australia, mare II5 is mated to a normal stallion from a different family with no family history of HERDA. Using appropriate symbols for genotypes, show all possible outcomes of this mating. Include the phenotypes and genotypes of parents and foals. **2 marks**

A young woman, Emma, purchases a quarter horse, called Penny, for breeding. Testing shows that Penny is heterozygous for HERDA as well as another genetic condition, Overo Lethal White Syndrome (OLWS). Any horse homozygous for OLWS dies at birth.

c. Penny is mated to a stallion which is also heterozygous for both conditions.
  i. What is the chance that Penny has a foal that dies at birth? **1 mark**
  ii. What is the chance that Penny has a foal that is phenotypically normal? **1 mark**

▶ Question 13 (7 marks)

A black rat and a white rat were bred together, giving all black offspring in the first generation. Two of these black rats were then selected at random and interbred. This second generation produced eight black rats and two white rats.

a. Using a series of Punnett squares, account for the results of the first and second generations. Show clearly the parental genotypes. **5 marks**

b. If the two white rats were bred together, what would be the colour of the offspring? Explain your answer using a Punnett square. **2 marks**

▶ Question 14 (12 marks)

Examine the pedigree below that shows the inheritance of favism in a family. This trait is known to be X-linked recessive.

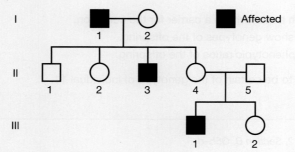

■ Affected

a. Only one child in generation II shows the trait. Explain whether it is reasonable to conclude that the son (II-3) inherited this trait from his affected father. **3 marks**

b. Child III-1 was found to have favism before his sister was born. His parents could not understand how this could have occurred since they were both unaffected. Explain how this occurred. **2 marks**

c. Use the allele symbols X^F and X^f to denote all possible genotypes of the following individuals:
   i. I-1 **1 mark**
   ii. II-2 **1 mark**
   iii. II-5 **1 mark**
   iv. a female with favism. **1 mark**

d. i. Construct a Punnett square to predict all possible genotypes of children of the two parents in generation I (I-1 and I-2). **2 marks**
   ii. What is the chance of this couple producing a daughter with favism? **1 mark**

Question 15 (6 marks)

Two genes, each with two alleles (*Mm* and *Nn*), are being investigated for linkage. A cross between a heterozygous organism (*MN/mn*) for both alleles and a homozygous recessive organism for both alleles is carried out.

a. Define the term *allele*. **1 mark**

The genotypes of the offspring and their percentages are shown.

MmNn	Mmnn	mmNn	mmnn
44%	6%	6%	44%

b. The results show that the genes are linked. Calculate how far apart the genes are. **2 marks**

c. A different scientist conducts a similar experiment looking at whether another gene with two alleles (*Pp*) is linked to gene *M*. The genes are determined to be unlinked. Complete the table with the results that would support this conclusion from the test cross giving 100 offspring. Explain your answer. **3 marks**

MmPp	Mmpp	mmPp	mmpp

8.5 Exercise 3: Biochallenge online only

Resources

- **eWorkbook** Biochallenge — Topic 8 (ewbk-8078)
- **Solutions** Solutions — Topic 8 (sol-0653)

teach on

Test maker
Create unique tests and exams from our extensive range of questions, including practice exam questions.
Access the assignments section in learnON to begin creating and assigning assessments to students.

Online Resources

Below is a full list of **rich resources** available online for this topic. These resources are designed to bring ideas to life, to promote deep and lasting learning and to support the different learning needs of each individual.

eWorkbook

- 8.1 eWorkbook — Topic 8 (ewbk-3168)
- 8.2 Worksheet 8.1 Pedigree analysis (ewbk-7539)
  Worksheet 8.2 Determining genotypes and phenotypes from pedigrees (ewbk-7541)
  Worksheet 8.3 Exploring patterns of inheritance — family portraits (ewbk-7543)
- 8.3 Worksheet 8.4 Monohybrid crosses: autosomal and complete dominance (ewbk-7545)
  Worksheet 8.5 Monohybrid crosses: sex-linkage (ewbk-7547)
  Worksheet 8.6 Monohybrid crosses: variations (ewbk-7549)
  Worksheet 8.7 Test cross (ewbk-7551)
- 8.4 Worksheet 8.8 Dihybrid cross (ewbk-7553)
  Worksheet 8.9 Linked or unlinked? (ewbk-7555)
- 8.5 Worksheet 8.10 Reflection — Topic 8 (ewbk-7559)
  Biochallenge — Topic 8 (ewbk-8078)

Solutions

- 8.5 Solutions — Topic 8 (sol-0653)

Practical investigation eLogbook

- 8.1 Practical investigation eLogbook — Topic 8 (elog-0170)
- 8.3 Investigation 8.1 Genetics in *Drosophila* (elog-0832)
  Investigation 8.2 What's the change of being Rhesus positive? (elog-0834)
- 8.4 Investigation 8.3 Two genes at a time (elog-0836)

Digital documents

- 8.1 Key science skills — VCE Biology Units 1–4 (doc-34648)
  Key terms glossary — Topic 8 (doc-34656)
  Key ideas summary — Topic 8 (doc-34667)
- 8.2 Extension: Genetic screening and testing (doc-36115)

Teacher-led videos

Exam questions — Topic 8
- 8.2 Sample problem 1 Determining the patterns of inheritance (tlvd-1761)
  Sample problem 2 Determining genotypes from pedigrees (tlvd-1762)
- 8.3 Sample problem 3 Predicting outcomes using Punnett squares (tlvd-1860)
- 8.4 Sample problem 4 Conducting a test cross and calculating the distance between two genes (tlvd-1763)

Video eLessons

- 8.2 Autosomal recessive disorders (eles-4221)

Interactivities

- 8.2 Pedigrees and genotypes (int-8122)
- 8.3 Punnett squares (int-8131)
- 8.4 Dihybrid cross (int-0180)

Weblinks

- 8.1 Íslendingabók website
- 8.2 Punnett square calculator
- 8.3 Victorian Clinical Genetics Services

Teacher resources

There are many resources available exclusively for teachers online

To access these online resources, log on to **www.jacplus.com.au**

UNIT 2 | AREA OF STUDY 1 REVIEW

AREA OF STUDY 1 How is inheritance explained?

OUTCOME 1
Explain and compare chromosomes, genomes, genotypes and phenotypes, and analyse and predict patterns of inheritance.

PRACTICE EXAMINATION

STRUCTURE OF EXAMINATION		
Section	Number of questions	Number of marks
A	20	20
B	5	30
	Total	50

Duration: 50 minutes

Information:
- This practice examination consists of two parts. You must answer all question sections.
- Pens, pencils, highlighters, erasers, rulers and a scientific calculator are permitted.

SECTION A — Multiple choice questions

All correct answers are worth 1 mark each; an incorrect answer is worth 0.

1. The total genetic information belonging to an organism is referred to as its
    A. chromosomes.
    B. genome.
    C. alleles.
    D. genes.

2. Alleles for a gene
    A. may have different nucleotide sequences.
    B. are found on the same chromosome.
    C. must contain the same number of nucleotides.
    D. code for the same information.

3. Rye (*Secale cereal*), pea (*Pisum sativum*), barley (*Hordeum vulgare*) and *Aloe vera* are four types of plants that each have seven pairs of homologous chromosomes. Which of the following is a correct statement that can be made when comparing the four plants?
    A. Chromosomes in each of the plants would be the same length.
    B. The four plants would each have the same number of genes.
    C. Each of the four plants has a diploid number of 14.
    D. The four plants would have similar phenotypes.

4. Consider the following diagram of chromosomes, labelled 1–8. The alleles for one gene on each of the chromosomes have been labelled.

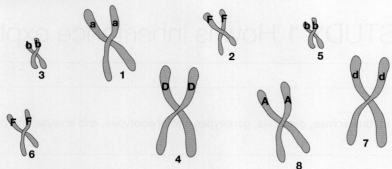

Chromosome 4 is homologous to
A. chromosome 1.
B. chromosome 2.
C. chromosome 6.
D. chromosome 7.

5. The failure of homologous chromosomes to separate during anaphase 1 is referred to as
A. non-disjunction.
B. crossing over.
C. cytokinesis.
D. mitosis.

6. Consider the following karyotype.

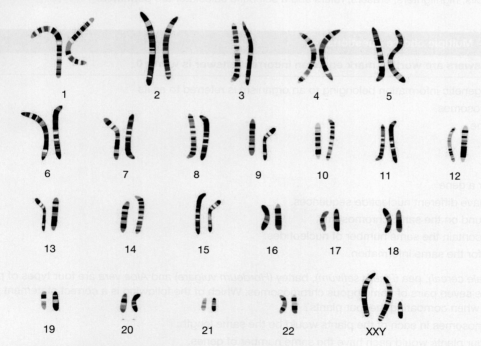

How many autosomes are seen in the karyotype?
A. 2
B. 3
C. 22
D. 44

7. Which one of the following statements is true about the following human karyotype?

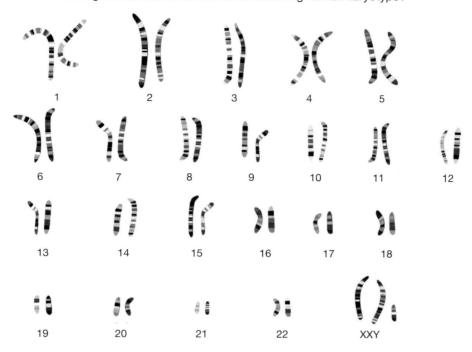

   A. The person is a female.
   B. The person has 48 chromosomes.
   C. Each non-dividing cell has two copies of each gene found on chromosome 1.
   D. The person has less genetic information than a person with Turner syndrome.

8. A chicken with white feathers is crossed with a chicken with black feathers. All the offspring produced have some white feathers and other feathers that are black.
   The genotype for the offspring would most correctly be written as
   A. *Wb*.
   B. *Bw*.
   C. *WwBb*.
   D. $F^W F^B$.

9. In pea plants the following crosses were performed and the offspring examined.

Cross	Phenotypes of parents	Phenotypes of offspring
1	Plants with green pods × plants with green pods	Plants with green pods and some plants with yellow pods
2	Plants with yellow pods × plants with yellow pods	All plants with yellow pods
3	Plants with green pods × plants with green pods	All plants with green pods

   Plants that must be heterozygous for the colour of the pod would be found
   A. in parents of cross 1.
   B. in parents of cross 2.
   C. in offspring of cross 2.
   D. in offspring of cross 3.

10. A human gene is known to have four different alleles. How many different genotypes would be possible?
    A. Three
    B. Four
    C. Six
    D. Ten

11. Consider two characteristics that follow the dominant/recessive pattern of inheritance. The symbols for the alleles for the genes of these two characteristics are shown.
    Characteristic 1: dominant trait T, recessive trait t
    Characteristic 2: dominant trait R, recessive trait r
    An organism with the genotype TtRR would have the same phenotype as an organism with the genotype
    A. ttrr.
    B. ttRr.
    C. TtRr.
    D. TTrr.

12. Which of the following is a human trait that could be influenced by the environment?
    A. Rhesus blood type
    B. The ability to roll tongue
    C. Colour-blindness
    D. Weight

13. Identify an example of a change bought about by an epigenetic factor.
    A. The deletion of a section of a chromosome
    B. The removal of introns from pre-mRNA
    C. DNA molecules winding more tightly around histone molecules
    D. A change in the nucleotide sequence of a gene

14. A gardener took a cutting from their favourite plant and grew it successfully in a pot. The plant in the pot did not produce flowers of exactly the same colour as the gardener's favourite plant.
    This difference in colour may be due to a difference in
    A. the way the gardener thought about the plant.
    B. soil the plant was grown in.
    C. size of chromosomes of each plant.
    D. number of genes in the nucleus of each plant.

15. The seed shape in plants is genetically inherited. The gene for the trait has two alleles.
    A test cross was carried out using two plants. Which of the following is a correct statement about the cross?
    A. One plant will be homozygous for the recessive trait.
    B. The phenotype of one of the plants in the cross will be determined.
    C. Both plants in the cross will be heterozygous.
    D. One plant must be homozygous for the dominant trait.

16. Consider the following pedigree.

    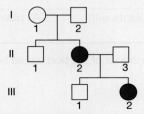

    The most likely pattern of inheritance of the trait shown by the two individuals is
    A. autosomal dominant.
    B. autosomal recessive.
    C. sex-linked dominant.
    D. sex-linked recessive.

17. Consider the following pedigree.

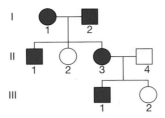

The trait inherited by the individuals who are shaded is a dominant trait. The trait must be autosomal because

A. I-1 has the trait.
B. II-2 lacks the trait.
C. II-4 lacks the trait.
D. III-1 has the trait.

18. Consider the following pedigree.

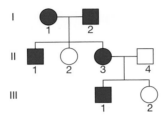

The trait is inherited as an autosomal dominant trait. Males heterozygous for the trait would include

A. I-2.
B. II-1.
C. II-3.
D. II-4.

19. Consider the following pedigree.

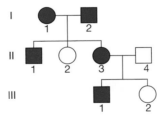

The trait is inherited as an autosomal dominant trait. The chance that a third child of parents II-3 and II-4 will have the trait is

A. zero.
B. one in two.
C. one in three.
D. one in four.

20. Consider the following pedigree.

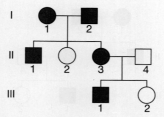

The trait is inherited as an autosomal dominant trait.
The chance that a fourth child of parents I-1 and I-2 will have the trait is
A. zero.
B. one in two.
C. one in four.
D. three in four.

SECTION B — Short answer questions

Question 21 (4 marks)

Independent assortment of pairs of homologous chromosomes occurs in metaphase I of meiosis.
a. Draw a diagram that clearly illustrates the process of independent assortment of two pairs of homologous chromosomes. Add labels to your diagram. 2 marks
b. Explain the significance of independent assortment to the organism. 2 marks

Question 22 (5 marks)

The hair colour of some horses is cream. Other horses have red hair. The colour of their hair is under genetic control.

Appropriate symbols for the alleles for the gene controlling hair colour in these horses would be:

Cream: H^C

Red: H^R

The following crosses between different horses produced many offspring.

Parents	Phenotypes of offspring
Cream x cream	All cream
Red x red	All red
Red x cream	All golden

a. Explain how it would be possible to get golden haired horses that are the offspring of a red-haired horse and a cream-haired horse. 2 marks
b. If a horse with golden hair is mated with a horse with red hair, what are the possible genotypes and phenotypes of their offspring? Show your working. 3 marks

Question 23 (5 marks)

a. Scientists examined three different cells in the reproductive tissue of a sheep. The number of DNA molecules found within the nucleus of each of the three cells was compared. The results of the comparison are shown in the table below.

Cell	Number of DNA molecules
Cell 1	Least number of molecules (X molecules)
Cell 2	Twice as many as in cell 1 (2X molecules)
Cell 3	Four times as many as in cell 1 (4X molecules)

For each cell identify the stage(s) of meiosis that the cell could be going through. Justify your answers. 3 marks

b. Consider the following photograph of a chromosome.

If this photograph represented a chromosome at the end of prophase I of meiosis would you expect the sister chromatids to be identical to each other? Justify your response. 2 marks

Question 24 (9 marks)

Individuals II-2 and II-3 in the pedigree below have a trait that follows a sex-linked recessive pattern of inheritance.

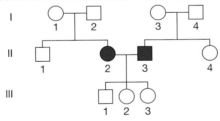

a. Shade all other individuals within the pedigree who must also have the trait. 2 marks

b. Choose a suitable notation for alleles in a sex-linked pattern of inheritance to write the genotypes of the following individuals. 3 marks
   i. I-1
   ii. I-4
   iii. III-3

c. The female III-3 has a male partner who does not have the trait.
   Complete the Punnett square to show the genotypes possible in their offspring. 2 marks

Female gametes	Male gametes	

d. How many different phenotypes are possible in the offspring? Name the possible phenotypes. 2 marks

Question 25 (7 marks)

In goats the colour of hair is controlled by a gene with two alleles. The shape of horns on the goat is controlled by a different gene with two alleles.

The symbols for the alleles for each of the genes are shown below:

Hair colour: *B* brown, *b* black

Horn shape: *H* straight, *h* curved.

A goat heterozygous for both genes is crossed with a goat that has black hair and curved horns.

a. What is the genotype of a goat with black hair and curved horns? **1 mark**

b. A goat heterozygous for both genes and a goat that has black hair and curved horns reproduce. Determine the genotypes and the phenotypes for their offspring showing your working in a Punnett square. State the ratio of the phenotypes of the offspring. **5 marks**

c. When the cross was carried out and a large number of offspring produced, the ratio of the phenotypes was found to be 4:4:1:1. This should be different to the ratio you obtained in part b.
Explain how this ratio may have arisen in this cross. **1 mark**

END OF EXAMINATION

UNIT 2 | AREA OF STUDY 1

PRACTICE SCHOOL-ASSESSED COURSEWORK

ASSESSMENT TASK — A data analysis of collated secondary data to investigate patterns of inheritance

In this task you will analyse collated secondary data to identify chromosomal abnormalities, interpret patterns of inheritance and predict outcomes of genetic crosses.
- This practice SAC comprises three questions; you must complete all question parts. Pens, pencils, highlighters, erasers, rulers and a scientific calculator are permitted.
- Mobile phones and/or any other unauthorised electronic devices including wrist devices are NOT permitted.

Total time: 50 minutes (5 minutes reading time, 45 minutes writing time)
Total marks: 40 marks

ANALYSING AND PREDICTING PATTERNS OF INHERITANCE

Question 1 (8 marks)

Renal phosphate transport disorder is a disease caused by abnormal vitamin D metabolism. This can result in bone and teeth abnormalities. The disease is caused by a mutation of the *PHEX* gene. A study was undertaken to try to understand the pattern of inheritance of the condition. A pedigree was produced from the study which is shown below.

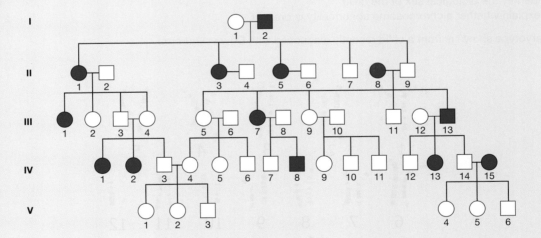

a. How many people affected with the condition are present in generation II? **1 mark**
b. Using the data from the pedigree, determine the most likely pattern of inheritance of the disease and explain your answer. **3 marks**
c. An affected female with the condition and a non-affected male want to have a child. Using a Punnett square, determine the percentage chance of them having a child who is affected by the disorder. Give examples of all possible parental genotypes for the disorder. **4 marks**

Question 2 (15 marks)

Edwards syndrome is a serious genetic condition diagnosed in childhood. Babies with the disease have a range of physical abnormalities including small head size, club feet, curvature of the spine and organ malfunctions.

A newborn baby suspected of having Edwards syndrome was tested. A karyotype was produced and is shown below.

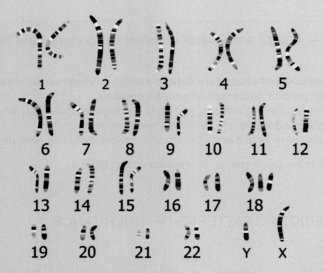

a. A karyotype arranges chromosomes into homologous pairs. Describe what is meant by the term *homologous chromosomes*. **1 mark**

b. Using information from the karyotype
   i. identify the biological sex of the child **1 mark**
   ii. explain whether a chromosome abnormality is present. **2 marks**

The karyotype shown is from a child recently diagnosed with Down syndrome.

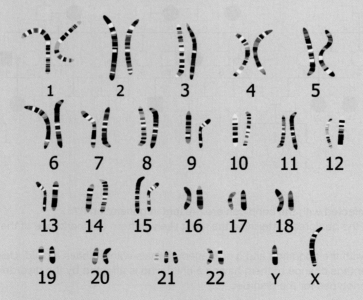

c. Chromosome 1 is described as 'metacentric' whereas chromosome 15 is 'acrocentric'. Explain what these terms tell us about the position of the centromeres and the relative lengths of the p and q arms. **3 marks**

Down syndrome, along with other trisomies, can be caused by non-disjunction. It is hypothesised that the age of the mother may be related to the condition. A study was conducted to test the hypothesis and the data collected is shown below in a graph.

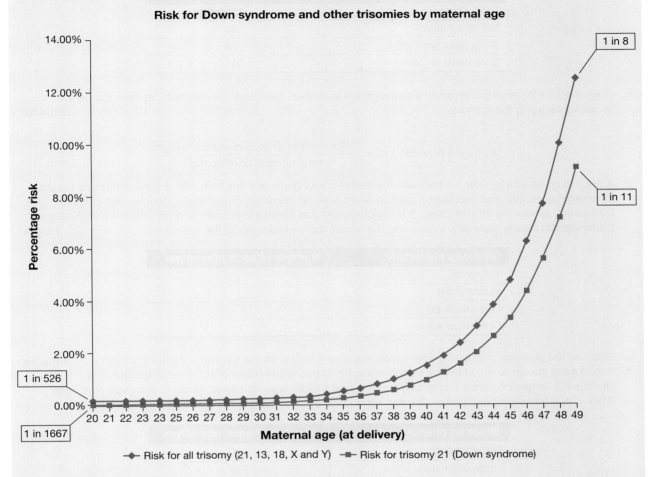

Risk for Down syndrome and other trisomies by maternal age

d. Using the data from the graph, explain if the hypothesis is supported or not. **4 marks**

e. Explain how non-disjunction may lead to trisomy conditions such as Down syndrome. You may use an annotated diagram in your answer. **4 marks**

Question 3 (17 marks)

For many years scientists have studied the genome of maize (corn). One study has tried to identify the relative positions of two genes that control two different characteristics of the kernels. They crossed maize plants and examined the characteristics of the offspring to produce linkage maps. The scientists took a maize plant that was homozygous for yellow kernels and homozygous smooth kernels. This was bred with another maize plant that was homozygous for colourless kernels and homozygous wrinkled kernels.

All the offspring in the F_1 generation were yellow and smooth.

a. Choosing suitable allele symbols, give the genotype of both the parental plants
  i. homozygous for yellow kernels and homozygous for smooth kernels **1 mark**
  ii. homozygous for colourless kernels and homozygous for wrinkled kernels. **1 mark**

b. Using a Punnett square, demonstrate how the parental cross can produce an F_1 generation consisting entirely of maize plants with yellow and smooth kernels. **3 marks**

To determine whether or not the gene for kernel colour (yellow or colourless) is on the same chromosome as the gene for kernel shape (smooth or wrinkled), test crosses are carried out. Homozygous recessive individuals for both traits (colourless, wrinkled) and heterozygous individuals (yellow, smooth) for both traits were crossed.

The characteristics of the offspring for these traits are examined. The results of the test crosses are shown below.

Offspring phenotype	Number of each phenotype
Yellow smooth	237
Yellow wrinkled	84
Colourless smooth	78
Colourless wrinkled	201

c. From the data it can be determined that the genes are linked. Calculate the distance between the genes in map units using the formula: **2 marks**

$$\text{Distance between loci} = \frac{100 \times \text{number of recombinant offspring}}{\text{total number of offspring}}$$

d. A second cross was carried out between the kernel colour gene, but this time with a gene controlling height. The height gene also has two alleles: *tall* and *short*, with tall dominating over short. Another test cross is carried out, producing 420 offspring. It is hypothesised that these genes are found on different chromosomes. Complete the results table below, showing the results that would support the hypothesis. **1 mark**

Offspring phenotype	Number of each phenotype
Yellow tall	
Yellow short	
Colourless tall	
Colourless short	

e. Describe the process of crossing over, explaining how it increases the genetic variation in offspring. **3 marks**

f. A third gene known to be on the same chromosome as the kernel colour and kernel shape genes is identified. This gene controls kernel size and has two alleles: *large* and *small.* Small is dominant over large. When a test cross involving kernel size and kernel shape is conducted, the following results are obtained.

Offspring phenotype	Number of each phenotype
Large smooth	10
Large wrinkled	205
Small smooth	202
Small wrinkled	15

The scientists used all the data to construct a map of the three genes along the chromosome. Complete the map below indicating the relative position of the gene, giving the distance in map units between **each of the 3 genes**. The locations do not need to be to scale. Show all of your working. **6 marks**

Kernel shape

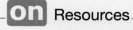 Resources

Digital document Unit 2 Area of Study 1 School-assessed coursework (doc-35093)

AREA OF STUDY 2 HOW DO INHERITED ADAPTATIONS IMPACT ON DIVERSITY?

9 Reproductive strategies

KEY KNOWLEDGE

In this topic you will investigate:

Reproductive strategies
- biological advantages and disadvantages of asexual reproduction
- biological advantages of sexual reproduction in terms of genetic diversity of offspring
- the process and application of reproductive cloning technologies

Source: VCE Biology Study Design (2022–2026) extracts © VCAA; reproduced by permission.

PRACTICAL WORK AND INVESTIGATIONS

Practical work is a central component of learning and assessment. Experiments and investigations, supported by a **practical investigation eLogbook** and **teacher-led videos**, are included in this topic to provide opportunities to undertake investigations and communicate findings.

9.1 Overview

Numerous **videos** and **interactivities** are available just where you need them, at the point of learning, in your digital formats, learnON and eBookPLUS at **www.jacplus.com.au**.

9.1.1 Introduction

Figure 9.1 shows a scanning electron microscope image of *Saccharomyces cerevisiae*, a species of fungi that is a yeast, best known as brewer's or baker's yeast. Its name is derived from the Greek and means 'sugar fungus'. It is used extensively in fermentation during winemaking and baking, and you may recognise it as a thin, white film visible on dark-skinned grapes and plums. It is one of the most studied eukaryotic organisms. In biological research, it is used as a homolog to study human proteins, including cell cycle proteins, signalling proteins and protein-processing enzymes. *Saccharomyces cerevisiae* exists in two forms — haploid and diploid. The haploid cells have a life cycle of mitosis and growth, reproducing asexually through budding, and do not tolerate conditions of high stress. However, in stressed conditions the diploid forms will enter meiosis to produce haploid gametes in order to undergo sexual reproduction.

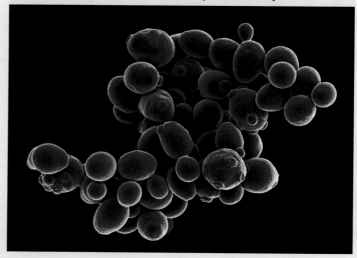

FIGURE 9.1 *Saccharomyces cerevisiae* is a yeast (kingdom: Fungi) that plays an important role in biological research and industry. It can reproduce asexually and sexually.

Both asexual and sexual reproduction have advantages and disadvantages, which you will learn about in this topic. You will discover the variety of strategies used by organisms to contribute to future generations and ensure the longevity of the species, and the way that humans interfere with nature to create clones and selectively bred organisms.

LEARNING SEQUENCE

9.1 Overview	510
9.2 Asexual reproduction	511
9.3 Advantages of sexual reproduction	529
9.4 Reproductive cloning technologies	542
9.5 Review	553

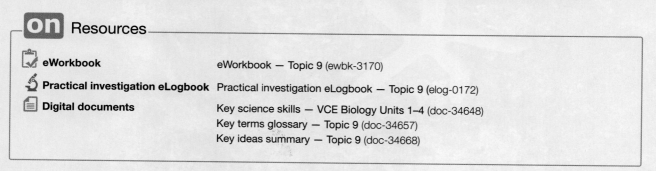

9.2 Asexual reproduction

> **KEY KNOWLEDGE**
>
> - Biological advantages and disadvantages of asexual reproduction
>
> **Source:** VCE Biology Study Design (2022–2026) extracts © VCAA; reproduced by permission.

9.2.1 Asexual reproduction

Asexual reproduction is the ability to produce new offspring without having to find a mate. As there is only one organism contributing genetic material, and the offspring are genetically identical to the parent, they are **clones**. Recall from topic 2 that **binary fission** and **mitosis** create genetically identical daughter cells — therefore their offspring are clones of the parent cell.

Asexual reproduction can occur in a variety of ways, across a range of organisms — from Komodo dragons to the *E. coli* bacterium. It occurs in prokaryotes (bacteria and archaea) and in eukaryotes (animals, plants and fungi).

There are various methods of asexual reproduction used by different organisms, allowing for genes to be passed on to successive generations. Some of these methods are summarised in table 9.1 and will be discussed in detail in this subtopic.

asexual reproduction type of reproduction that does not require the fusion of gametes, where offspring arise from a single parent and are genetically identical to that parent

clones groups of cells, organisms or genes with identical genetic make-up

binary fission process of cell multiplication in bacteria and other unicellular organisms in which there is no formation of spindle fibres and no chromosomal condensation

mitosis process involved in the production of new cells genetically identical with the original cell; an essential process in asexual reproduction

TABLE 9.1 Summary of the different methods of asexual reproduction

Reproduction method	Key characteristics	Example
Binary fission — prokaryotes	Division in half. Daughter cells are the same size as the parent cells.	Most common form of replication on prokaryotic organisms, including red halophilic bacteria, *Salinibacter ruber* (see Case study box in section 9.2.2)
Binary fission — eukaryotes	Division in half of single-celled eukaryotic organisms	Unicellular eukaryotes, such as *Paramecium*, replicate via mitosis
Mitosis	The duplication of cells (see topic 2)	Spore formation in algae and fungi; replacement of body cells in eukaryotes
Vegetative propagation	The plant releases runners from the parent plant that establish their own root system to obtain water and nutrients from the soil. The daughter cells are initially smaller than the parent cells.	Strawberry plant
Budding	A daughter organism is created from a growth on the parent (or bud). The daughter cells are smaller than the parent cells.	*Hydra* (small, freshwater organism) and yeast
Fragmentation	An organism breaks into smaller parts, which develop into new daughter organisms. The daughter organisms are smaller than the parent organisms.	Starfish
Parthenogenesis	Also referred to as the 'virgin birth'. The embryo can develop without fertilisation.	Komodo dragons (see Case study box in section 9.2.4)

9.2.2 Binary fission in prokaryotes

In topic 2 you learned about how bacteria replicate through the process of binary fission (*binary* = two; *fission* = splitting), an example of asexual reproduction. The binary fission of a bacterial cell involves:
- replication of the circular molecule of DNA of the cell
- attachment of the two DNA molecules to the plasma membrane
- lengthening of the cell
- division of the cell into two via a constriction across the middle of the cell and the formation of a septum, so that each new cell contains one circular molecule of DNA (figure 9.2).

Your body has symbiotic relationships with many bacteria. Your skin is covered in 'good' bacteria that are constantly reproducing via binary fission to outcompete 'bad' bacteria from entering your body. Bacteria can be your friend! Your digestive system is reliant on bacteria to assist in the breakdown of food so that you can obtain the nutrients and energy required to survive.

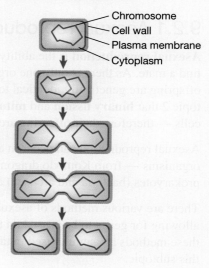

FIGURE 9.2 Binary fission in a bacterial cell. Replication of the circular DNA molecule is followed by cell lengthening and then its division into two.

- Chromosome
- Cell wall
- Plasma membrane
- Cytoplasm

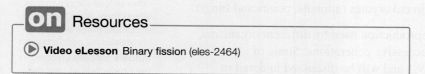

Resources

▶ **Video eLesson** Binary fission (eles-2464)

tlvd-1732

SAMPLE PROBLEM 1 The symbiotic relationship between bacteria and humans

Bacteria can have a detrimental effect on the human body. *E. coli* is a regular cause of food poisoning and *Neisseria meningitidis* (meningococcus) can cause life-threatening sepsis. With so many bacteria causing illness, how can bacteria have a symbiotic relationship with humans? (4 marks)

THINK

1. In *how* questions you need to address what it does — in this case, what role do bacteria play in the body?

2. Recall that symbiotic means beneficial to both organisms. How do both humans and bacteria benefit from the relationship?

TIP: As this question is worth four marks, you need to ensure that you include four clear points in your response. You may use either a paragraph (as shown) or bullet points to show your response.

WRITE

Bacteria are essential in breaking down food in the gut that humans are unable to break down (1 mark).
They also aid in the immune response by outcompeting with bad bacteria, protecting us from infection (1 mark).

Bad bacteria are unable to enter the body as they are outcompeted for resources by good bacteria (1 mark). Humans obtain nutrients from food that would otherwise be unobtainable. Bacteria in return receive nutrients and space to survive and reproduce (1 mark).

CASE STUDY: Bacteria turn a lake pink

Lake Hillier (figure 9.3), renowned for its bright-pink colour, is on Middle Island off the southern coast of Western Australia. The traditional owners of this land, the Noongar people, were present when European explorer Matthew Flinders arrived in the region in 1802. The island was named after William Hillier, a member of the HMS *Investigator* crew who died on the second journey to the island.

This inland body of water — separated from the Southern Ocean by paperbark trees, eucalypt trees and a narrow beach — has a high salt content, limiting the organisms that can survive in this extreme environment. Red halophilic bacteria found in the salt crusts at the edge of the lake and

FIGURE 9.3 Lake Hillier, the pink lake of Western Australia

Dunaliella salina, a species of green algae, are thought to be what creates the year-round pink colour of the water. DNA analysis of lake-water samples indicate up to ten species of bacteria, with the species *Salinibacter ruber* making up 33 per cent of the total bacteria in the lake. All these organisms are known as **halophiles** — organisms that thrive in a saline environment. Lac Rose in Senegal, Africa is another pink lake. Unlike Lake Hillier, fish are found in Lac Rose that have mechanisms to pump the extra salt from their bodies.

Both the red halophilic bacteria and the green algae can reproduce asexually (green algae can also reproduce sexually) and, with little competition for resources due to the extreme environment, they are able to maintain their population numbers so that Lake Hillier is pink year-round.

Despite the colour, the water is safe to swim in; however, permission is required from the Department of Lakes and Wildlife should you be planning a visit. Whilst the colour of the water is akin to a strawberry milkshake, it is not recommended that you drink the water due to the high salt content. Ingesting the fluid would cause crenation (or shrivelling) of somatic cells, as water would exit the cells via osmosis. This is because salt, being a polar molecule, is unable to cross the plasma membrane unaided. The kidneys, our internal filtration system, are only able to produce urine that has a lower saline concentration than salt water. This means you would excrete more water than you ingested, leading to dehydration.

9.2.3 Binary fission in eukaryotes

Unicellular eukaryotes

Let's split into two! Some eukaryotic unicellular organisms — such as *Amoeba* (figure 9.4), *Euglena* and *Paramecium* (figure 9.5) — live in freshwater ponds and are less than the size of a pinhead. These unicellular organisms can reproduce asexually by splitting into two (figures 9.4b and 9.5). Although this is occurring in eukaryotic cells, the process is also known as binary fission. However, the process of binary fission in these unicellular eukaryotes is different from that which occurs in bacteria. In eukaryotes, the formation of new cells by binary division involves the process of mitosis.

halophile an organism that grows in or can tolerate saline conditions

FIGURE 9.4 a. Phase contrast microscope image of an *Amoeba* sp. **b.** Binary fission process in *Amoeba* spp.

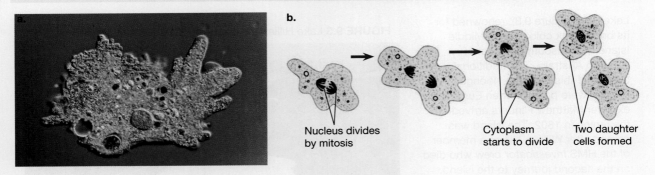

FIGURE 9.5 Longitudinal binary fission occurs in *Paramecium* spp.

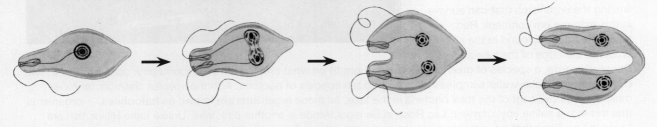

Amoebae (singular = amoeba) can also undergo **multiple fission**. Mitosis occurs repeatedly and many nuclei form within a single cell. Each nucleus becomes enclosed within a small amount of cytoplasm and forms a **spore**. Spores can later develop into new amoebae.

> **multiple fission** process of division in which multiple cells are produced from a single starting cell
>
> **spore** in bacteria, a reproductive structure that is resistant to heat and desiccation; also formed by fungi and some plants

Multicellular eukaryotes

Simple multicellular animals, such as flatworms, anemones and coral polyps, can also reproduce asexually by splitting into two (figure 9.6). Each of the parts then grows into a complete animal. This kind of splitting does not occur in other multicellular organisms because their structure is more complex, being built of many different tissues and organs.

FIGURE 9.6 A sea anemone in the process of reproducing by splitting in two. Note the narrowing (centre of image) that marks the point where it will split.

> Binary fission in eukaryotes involves mitosis.

514 Jacaranda Nature of Biology 1 VCE Units 1 & 2 Sixth Edition

SAMPLE PROBLEM 2 Binary fission in prokaryotes and eukaryotes

Compare and contrast binary fission in prokaryotes and eukaryotes. **(2 marks)**

THINK

1. *Compare* means to find the similarities. *Contrast* means to find the differences.
2. Compare the two modes of reproduction.
3. Contrast the two modes of reproduction.

WRITE

Binary fission in prokaryotes and eukaryotes involves the replication of DNA, division of the cytoplasm and the production of two identical daughter cells (1 mark).

In prokaryotes, the circular DNA is replicated and the two strands bind to separate parts of the plasma membrane. In eukaryotes, the linear DNA is replicated and the cell goes through the stages of mitosis, ending in cytokinesis (1 mark).

9.2.4 Mitosis in eukaryotes

Replacement of body cells

Recall from topic 2 that mitosis is the process by which cells are replaced. At the point of fertilisation, all multicellular organisms (including you) were one cell in size. Through the process of asexual reproduction of your cells, otherwise known as mitosis, you have been able to grow to your current size. Even while you are reading this, your body is in the process of asexual reproduction as it replaces old or damaged cells through the process of mitosis. For example, red blood cells are replaced every 120 days, whilst epithelial cells of the intestines have a lifespan of just 5 days!

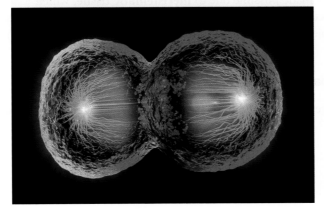

FIGURE 9.7 Cytokinesis — one of the stages of mitosis that cells in your body are currently going through. In this stage the cytoplasm divides into two, producing two new daughter cells.

Spore formation in fungi

Whilst the most common form of asexual reproduction in eukaryotes is the replacement of body cells, it is also asexual reproduction that forms spores in algae and some fungi. These spores are true asexual spores produced by mitosis. After dispersal, the spores develop into new organisms that are genetically identical to each other and to the parent.

The bread mould fungus *Rhizopus stolonifer* provides an example of asexual spore formation in a fungal species (figure 9.8). Spores are formed by mitosis in aerial structures called **sporangia**. When released from a sporangium, the spores are carried away by air currents. If a spore lands on a moist location, such as a slice of bread, the spores germinate and form a branching structure or fungal network (**mycelium**). Soon after, new sporangia containing spores develop. And so the cycle is completed.

FIGURE 9.8 The bread mould fungus *Rhizopus stolonifer* is a common fungus. Note the 'stalks' that raise the sporangia above the substrate surface. Why is this important?

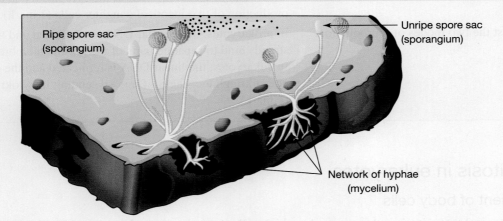

Plants such as mosses and ferns also produce spores as part of their life cycles. However, because the spores are produced by **meiosis**, not mitosis, the spores produced by one moss or one fern are not genetically identical to the plant that produced them, or to each other. Each spore then develops into a new plant, called a gametophyte, by the process of mitosis.

Vegetative propagation

Asexual reproduction is common in plants. This is also known as **vegetative propagation**. This process is made possible due to **meristematic tissue**, which consists of undifferentiated cells and is found in stems, leaves and tips of roots. These meristematic tissue cells rapidly divide by mitosis to allow fast (and often extensive) plant growth.

Types of vegetative propagation

There are many types of vegetative propagation, each with varying benefits and uses. The main types include:
- **runners**
- **cuttings**
- **rhizomes**
- **tubers**
- **bulbs**
- **corms**
- **plantlets**.

Each of these are outlined in table 9.2.

sporangium a plane aerial structure where spores are formed by mitosis

mycelium the vegetative part of a fungus, consisting of a network of fine, white filaments

meiosis a type of cell division that produces haploid gametes

vegetative propagation asexual reproduction in plants

meristematic tissue plant tissue found in tips of roots and shoots that is made of unspecialised cells that can reproduce by mitosis

runners stem-like growths extending from a mother plant's growing point

cuttings a type of vegetative propagation that involves taking pieces of shoots, roots or leaves and planting them

rhizomes horizontal underground stems

tuber a thickened underground part of a stem or rhizome

bulb an underground storage organ with short stems, a central bud and many closely packed, fleshy leaves

corm a rounded underground storage organ present in plants, consisting of a swollen stem base covered with scaly leaves

plantlets tiny young plants that develop from the meristem tissue along plant margins

TABLE 9.2 Types of vegetative propagation

Type of vegetative propagation	Description	Examples
Runner	<ul><li>Stem-like growths from parent plant that run along the ground</li><li>New buds develop into roots, leaves, flowers and fruit</li></ul>	<ul><li>Strawberries</li><li>Water hyacinth</li></ul>
Cuttings	<ul><li>Some plants can be cloned by taking cuttings of shoots, roots or leaves and planting them.</li></ul>	<ul><li>Lavender</li><li>Geraniums</li><li>Hydrangea</li><li>Sage</li><li>Oregano</li></ul>
Rhizomes	<ul><li>Underground stems that grow horizontally (figure 9.9a)</li><li>Buds and roots sprout from nodes along a rhizome and produce new daughter plants.</li><li>Can be distinguished from plant roots by the presence of buds, nodes and often tiny, scale-like leaves</li><li>Typically thick in structure because they have a food reserve, mainly in the form of starch</li></ul>	<ul><li>Irises</li><li>Grasses, such as kikuyu grass (*Pennisetum clandestinum*) and couch grass (*Cynodon dactylon*)</li><li>Austral bracken, an Australian fern (*Pteridium esculentum*) (figure 9.10a)</li><li>Many types of reeds</li></ul>
Tubers	<ul><li>Swollen underground stems from which buds sprout</li><li>If cut, a piece of tuber with a bud can grow into a new plant (figure 9.9b).</li><li>A type of asexual reproduction known as fragmentation</li></ul>	<ul><li>Potatoes</li><li>Yams</li></ul>
Bulbs	<ul><li>Underground structures with short stems, a central bud and many closely packed, fleshy leaves (figure 9.9c)</li><li>The leaves are the food source for the plant.</li></ul>	<ul><li>Onions</li><li>Garlic</li><li>Daffodils</li><li>Tulips</li><li>Hyacinths</li></ul>
Corms	<ul><li>Enlarged, bulb-like underground stems, with a solid stem tissue typically surrounded by papery leaves</li></ul>	<ul><li>Taro (figure 9.10b)</li><li>Gladioli</li></ul>
Plantlets	<ul><li>Tiny young plants that develop from the meristem tissue along plant margins</li><li>When they reach a particular size, the plantlets drop from the parent plant and take root.</li></ul>	<ul><li>*Asplenium bulbiferum*, a fern native to Australia and New Zealand (figure 9.11a)</li><li>*Bryophyllum* spp. (figure 9.11b).</li></ul>

FIGURE 9.9 a. Rhizomes, as in bracken and some grasses; **b.** tubers, as in potatoes; and **c.** bulbs, as in onions. In each case, the new plants are genetically identical to the parent plant.

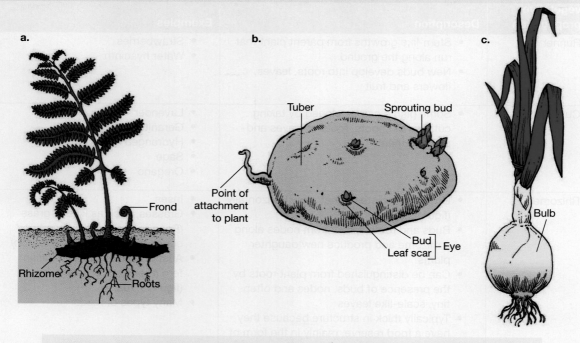

FIGURE 9.10 a. Austral bracken reproduces asexually from underground stems. After a bushfire, it quickly becomes re-established in a burnt area. **b.** Taro root, showing visible corms

FIGURE 9.11 a. Asexual reproduction in the fern *Asplenium bulbiferum*. Note the new plantlets on the fern fronds. **b.** New plants form on the leaf margin of *Bryophyllum* sp.

CASE STUDY: Vegetative propagation — runners

Over ten years, one strawberry plant (*Fragaria ananassa*) grew into the strawberry patch shown in figure 9.12a. How did one small plant grow into such a large patch? Strawberry plants have runners, which are special stems that grow over the ground. The runner grows away from the parent plant and, at alternate nodes on the runner, new buds give rise to roots, leaves, flowers and fruit (see figure 9.12b).

FIGURE 9.12 a. In ten years, one strawberry plant grew by asexual reproduction into this strawberry patch. **b.** The strawberry plant (*Fragaria ananassa*) has runners, special stems that grow over the ground.

Another example of a plant that spreads by runners — this time in water — is the water hyacinth (*Eichhornia crassipes*) (figure 9.13). This is a declared noxious weed that infests wetlands, lakes and rivers in Australia. A variation of 'runners' occurs in blackberry plants (*Rubus* spp.) that propagate when their long stems (canes) bend over and make contact with the ground. Shoots and roots grow from the point where the tips of the stems make contact with the ground.

FIGURE 9.13 The water hyacinth

INVESTIGATION 9.1

online only

elog-0838

Vegetative propagation — reproduction without sex

Aim

To investigate different ways plants can undergo vegetative propagation

Fragmentation

Fragmentation is exactly as the name suggest. It is a form of asexual reproduction where an organism is split into fragments and each fragment can develop into a mature organism. As this is asexual reproduction, offpring are identical to the parent. It is common in filamentous cyanobacteria, moulds and many plants, as well as in animals such as sponges, acoel flatworms, some annelid worms and sea stars.

FIGURE 9.14 Fragmentation in a starfish. The absent arms have broken off and will become a new daughter organism, and the starfish will regenerate the missing parts.

Some types of vegetative propagation occur through fragmentation, such as tubers (figure 9.9b), as the daughter plant grows from a fragment of the parent plant. This is especially useful in horticulture. A plant with desirable characteristics can produce numerous offspring that will also show the same desirable characteristics, such as large or drought-resistant potatoes.

Budding

Budding is a form of asexual reproduction where a new organism develops, through cell division, from an outgrowth on the parent. Budding can be seen in sponges, which are common in many marine habitats. Each sponge is made of thousands of cells but has no specialised organs or nervous system. Sponges are able to reproduce asexually from small groups of cells formed by mitosis that bud or break away from the main organism, and are carried by currents to other locations where they settle and develop into new sponges. The small group of cells settles on some substrate, reproduce by mitosis and develop into a new sponge. Other simple animals, such as *Hydra*, also undergo budding (figure 9.15).

budding asexual reproduction where a new organism develops from cell division at an outgrowth from the parent

FIGURE 9.15 Asexual reproduction involving budding from one parent occurs in *Hydra*.

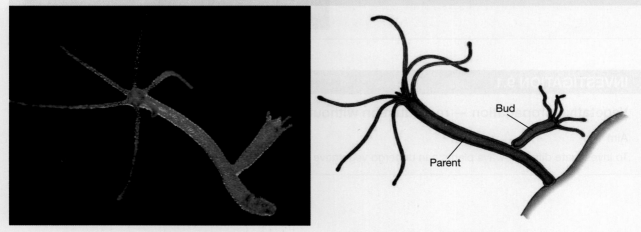

INVESTIGATION 9.2

Asexual reproduction in yeast and amoeba

Aim

To observe budding in yeast and binary fission in amoeba under the microscope

Parthenogenesis

An unusual form of asexual reproduction in animals is **parthenogenesis** (from the Greek, *parthenos* = virgin; *genesis* = birth), which is also referred to as 'virgin birth'. Parthenogenesis is defined as reproduction without fertilisation and almost always involves the development of an unfertilised egg. Offspring are produced from unfertilised eggs — no sperm is necessary. These eggs are produced by mitosis and develop into offspring identical to the female parent. This type of reproduction is seen in aphids when conditions are favourable (figure 9.16a). In contrast to other insects that typically lay eggs, aphids give birth to live young.

Parthenogenesis is seen in many invertebrate animals. It is rare in vertebrate species, but has been reported in several reptile species, such as Komodo dragons and whiptail lizards (figure 9.16b), and in some shark species.

FIGURE 9.16 a. Adult female aphid giving birth. The numerous offspring have developed from unfertilised eggs. **b.** *Aspidoscelis tesselata* is one of the many species of whiptail lizard that reproduce by parthenogenesis.

Populations that reproduce using parthenogenesis are typically all-female. This can be **obligate parthenogenesis**, meaning that this is the only way in which a species can reproduce. This is the case for about one third of the 50-plus species of whiptail lizard of the genus *Aspidoscelis*. These obligate, unisexual lizard species consist only of females. Other reptile species that show a complete absence of male contribution to reproduction include some snakes, rock lizards (*Lacerta* spp.) and Australian geckos (*Heteronotia* spp.).

> **parthenogenesis** one form of asexual reproduction in which new individuals are produced from unfertilised eggs
>
> **obligate parthenogenesis** a type of parthenogenesis in which a species can reproduce only through asexual reproduction

CASE STUDY: Parthenogenesis in Komodo dragons

The Komodo dragon (*Varanus komodoensis*) is a large reptile native to the island of Komodo in Indonesia. The female can grow up to 2.3 m in length and weigh as much as 70 kg. They feed on small mammals and birds. They are currently classified as being vulnerable to extinction (figure 9.17). A female Komodo dragon living in a zoo in Chester, England laid eight eggs without any contributing genetic material from a male. With no access to a viable mate, she was able to clone her eggs to continue to repopulate the species. In this process, the DNA of the eggs is duplicated to enable a viable embryo to be created. This process is known as parthenogenesis, also referred to as the 'virgin birth'.

FIGURE 9.17 The Komodo dragon, *Varanus komodoensis*

In Komodo dragons the sex chromosomes are W and Z. The females are **heterogametic** with WZ, whilst the males are **homogametic** with ZZ. When the embryos are created, the genetic material in the female's eggs is duplicated. As a WW egg is not viable, only the eggs with ZZ will produce a live birth. The Komodo dragons born are all genetically identical to the egg they were formed from, but all offspring will be male.

In this case, it is **facultative parthenogenesis**, meaning that it is a reproductive strategy that is only used when required, such as when no males are around. When males reappear, the species return to sexual reproduction.

on Resources

 eWorkbook   Worksheet 9.1 Types of asexual reproduction (ewbk-7561)

 Weblinks   How some animals have 'virgin births'
Birds Do It. Bees Do It. Dragons Don't Need To.

9.2.5 Advantages of asexual reproduction

Organisms that can reproduce asexually are able to quickly colonise an area. The bacteria *E. coli* is able to reproduce via binary fission every 20 minutes (figure 9.18). With exponential growth, they can rapidly cover an available area. For areas that have been impacted by natural disasters, the ability to repopulate quickly is advantageous to ensure the longevity of the species' survival within the area.

heterogametic organisms with two different sex chromosomes; the prefix hetero, from the Greek term *heteros*, means to be different

homogametic organisms with two of the same sex chromosomes; the prefix homo, from the Greek term *homos*, means same

facultative parthenogenesis a type of parthenogenesis in which females can reproduce via both sexual and asexual reproduction

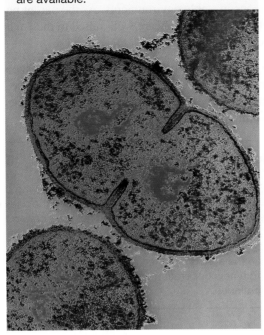

FIGURE 9.18 Bacteria replicating through binary fission. Bacterial growth is exponential while space and resources are available.

The best of both worlds

For organisms that have become isolated from the main population, the ability to reproduce asexually allows them to produce offfspring without having to find a mate. The Komodo dragon maintained in captivity at the Chester Zoo is an example of an organism that reproduced when a mate was unavailable (see Case study box in the previous section). Species that can use both sexual and asexual reproduction include various insects (such as aphids; figure 9.19a), some crustaceans (such as fairy shrimp), algae (such as *Volvox* spp.; figure 9.19b) and almost all fungal species. The switch from asexual to sexual reproduction may occur in advance of a seasonal change to less favourable and more unstable conditions (an example of this is shown in the Case study on *Volvox carteri*).

FIGURE 9.19 a. Aphids are able to reproduce both sexually and asexually, depending on the environmental conditions. b. Volvox sp. daughter colonies produced by asexual reproduction in summer are visible inside the spherical parent colonies. The daughter colonies are released into the water when the parent disintegrates.

Being able to reproduce asexually both conserves energy and time spent looking for a mate, as well as ensuring that the organism's genes are passed on to future generations. As plants are immobile, they have to rely on pollen dispersal from animals or the wind, enabling them to repopulate self-sufficiently.

Asexual reproduction is advantageous in stable environments. As the parent has been able to survive and reproduce, it is likely that due to their genome, they are well-suited to the environment, and therefore their identical offspring will also be well-suited to it.

Asexual reproduction can also be advantageous in unstable environments, should the organism be suited to the changes in the environment, as no time is wasted seeking a mate. For example, drought provides opportunities for organisms that have lower water requirements to colonise the area and outcompete other organisms for resources. These advantages are summarised in table 9.3.

> **CASE STUDY: The reproduction of *Volvox carteri***
>
> *Volvox carteri* (shown in figure 9.19b) is found in shallow pools that form during the spring rains but that dry out during the late summer. When the pools first form, this species uses the asexual mode of reproduction to produce offspring that are clones of the single parent. However, not long before the ponds dry out, *V. carteri* switches to sexual reproduction that involves genetic contributions (egg and sperm) from two parents. Sexual reproduction produces dormant zygotes that can survive through both the hot, dry conditions of the summer and the cold conditions of winter. When the spring rains return and the ponds re-form, these zygotes emerge from their dormant state and give rise to a new generation of *V. carteri* that reproduce asexually until the ponds start to dry out again. And so the cycle continues.

9.2.6 Disadvantages of asexual reproduction

Organisms produced via asexual reproduction are usually genetically identical to their parents. A disadvantage of asexual reproduction is that if conditions were to change, the population lacks **genetic variation**. There are continual changes in environments, whether that be from natural events such as fires, floods or droughts (see figure 9.20b), or human impact such as clearing of land or increased use of chemicals (see figure 9.20a). Should a population produced via asexual reproduction not be suited to the new environmental conditions, all members of the population are susceptible.

FIGURE 9.20 a. Land clearing provides opportunities and challenges for new organisms to populate the area. **b.** Drought provides opportunities for organisms that have lower water requirements but challenges for organisms that require more water.

> **genetic variation** variation exhibited among members of a population owing to the action of genes

The ability of an organism to reproduce asexually and populate quickly can increase competition for resources within the population. Environments have a finite amount of resources available, and exponential growth of organisms can place pressure on such resources. These disadvantages are summarised in table 9.3.

Comparing the advantages and disadvantages

TABLE 9.3 Advantages and disadvantages of asexual reproduction

Advantages of asexual reproduction	Disadvantages of asexual reproduction
Can reproduce quickly	Lack of genetic variation reduces the chance of a population adapting to new environmental conditions
No energy expended finding a mate	Pressure on availability of resources
Do not have to rely on other organisms/means to spread pollen/seeds	If conditions change, entire population can be lost
Well-suited to the environment	
Able to colonise cleared areas rapidly	

 Resources

 eWorkbook Worksheet 9.2 Advantages and disadvantages of asexual reproduction (ewbk-7563)

KEY IDEAS

- Two modes of reproduction exist: sexual and asexual.
- Asexual reproduction involves a single parental organism that produces offspring that are (usually) genetically identical to each other and to the parent.
- Many different modes of asexual reproduction exist.
- Asexual reproduction occurs by binary fission in prokaryotes.
- Asexual reproduction in all eukaryotes occurs through mitosis.
- Unicellular eukaryotes reproduce by binary fission that involves mitosis.
- Simple multicellular organisms reproduce asexually through budding.
- Parthenogenesis is the development of offspring from unfertilised eggs.
- Certain fungi produce spores as part of their cycle of asexual reproduction.
- Various types of asexual reproduction occur in plants, including runners, rhizomes and tubers.
- Cutting and planting a piece of an existing plant can be used to create a new plant.
- Asexual reproduction involves a single parent, and it is faster, requires less energy and can lead to rapid population growth in favourable and stable conditions.
- Asexual reproduction has the disadvantage that it cannot produce any new genotypes to enable adaptation to changing environments.
- Asexual reproduction can place pressure on environments due to the finite resources available to support exponential growth.

9.2 Activities

To answer questions online and to receive **immediate feedback** and **sample responses** for every question, go to your learnON title at **www.jacplus.com.au**. A **downloadable solutions** file is also available in the resources tab.

| 9.2 Quick quiz | 9.2 Exercise | 9.2 Exam questions |

9.2 Exercise

1. How many parents are required for asexual reproduction?
2. How does fragmentation differ from budding?
3. If the eggs of the Komodo dragon replicate by mitosis in parthenogenesis, explain why a female dragon cannot be born using this method of asexual reproduction.
4. Explain why asexual reproduction does not always produce genetically identical offspring.
5. Identify a key difference between binary fission in a microbe and the binary fission that occurs in an amoeba.
6. Starting with one bacterial cell, calculate how many cells would be expected from six cycles of binary fission.
7. What benefit does an organism derive from undergoing fragmentation?
8. Create a Venn diagram comparing binary fission and mitosis.
9. Describe the process in which spores in mould are produced and how they continue to reproduce in a new cycle.
10. Explain why it is easy for plants that reproduce using runners to quickly become a pest species.
11. Provide a reason why the use of cuttings would be beneficial in commercial horticulture.
12. List the conditions under which asexual reproduction is best suited.
13. Describe three advantages of asexual reproduction.
14. The graph shows two different populations.

    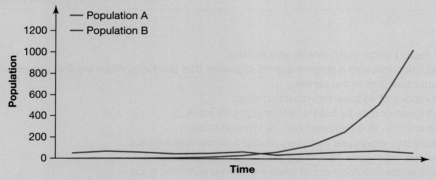

    Which population, A or B, is most likely to reproduce asexually? Explain.
15. Why is it a disadvantage to invest energy in attracting a mate?

9.2 Exam questions

Question 1 (2 marks)

Anemones are simple, marine animals that attach to rocks below the low-tide mark. Anemones can reproduce asexually by splitting into two.

MC a. This form of reproduction would enable anemones to
A. adapt to rapid changes in their environment.
B. quickly populate a rocky marine habitat.
C. grow larger as individuals.
D. avoid competition with other anemones for food.

MC b. What is a possible disadvantage of rapid population growth for the anemone?
A. There will be competition between anemones for attachment sites.
B. It would be difficult for anemones to locate mates.
C. There is a lot of genetic variation between offspring.
D. There will be more anemones for predators to eat.

Question 2 (1 mark)

MC The simple, freshwater organism *Hydra vulgaris* reproduces asexually by budding, as shown in the diagram. In one *Hydra* lake habitat, the water temperature has been unusually high for several years.

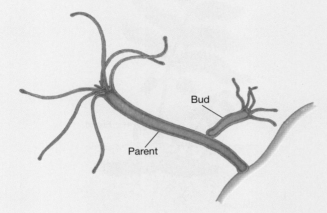

The likely consequence of increasing water temperature on the *Hydra* population of this lake is that the population size will
A. increase because budding occurs faster in warmer temperatures.
B. increase in the absence of competition for resources from other organisms.
C. decrease since all individuals will be negatively affected by rising water temperature.
D. remain stable since all offspring are different, so some will die but others will survive.

Question 3 (1 mark)

MC The unicellular organism amoeba reproduces asexually by a form of binary fission as shown in the diagram. In a jar of pond water there is one single amoeba. This organism undergoes five binary fission events to produce 32 amoeba.

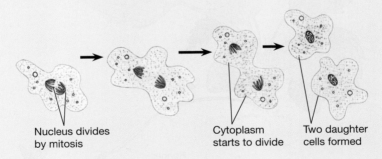

If the amount of DNA present in the first amoeba is represented by X, the amount of DNA present in each of these 32 amoeba is
A. X.
B. $\frac{1}{2}$X.
C. $\frac{1}{32}$X.
D. 32X.

Question 4 (4 marks)

A farmer cleared Austral bracken *(Pteridium esculentum)* from his paddocks by burning it to the ground and bulldozing the soil. A few weeks later the bracken plants re-appeared in the paddock in greater numbers than before.

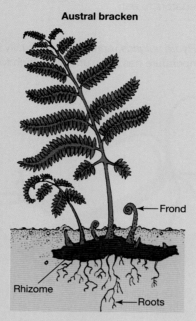

Austral bracken

a. Explain the rapid appearance of bracken plants after burning and bulldozing the paddock.   **2 marks**
b. Suggest the survival advantage of this form of reproduction for bracken in the Australian environment.   **2 marks**

Question 5 (3 marks)

A strawberry plant *(Fragaria ananassa)* in a garden produced offspring plants by runners. All of the plants produced by the runners grew strawberries of a similar size and flavour to the original strawberry plant.

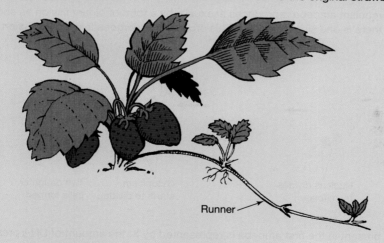

However, when the gardener planted the seeds from the original plant she was disappointed to discover that all the plants resulting from these seeds had small, flavourless fruit.

How would you explain to the gardener the difference in strawberry plants grown by runners compared to the offspring grown from the seeds?

More exam questions are available in your learnON title.

9.3 Advantages of sexual reproduction

KEY KNOWLEDGE

- Biological advantages of sexual reproduction in terms of genetic diversity of offspring

Source: VCE Biology Study Design (2022–2026) extracts © VCAA; reproduced by permission

9.3.1 Sexual reproduction: the fusion of gametes

Thus far, we have looked at reproduction without a mate — asexual reproduction. This is where there is only one parent organism, and the product of this is usually a clone of the parent. When two individuals are contributing genetic material to the offspring, this is known as sexual reproduction.

In mammals, the genetic material from the two contributing parents comes from gametes, which are produced in specialised organs called gonads. In animals, gametes are:
- the eggs produced by the female in the ovaries
- the sperm produced by the male in the testes.

The cells produced in the gonads are known as germline cells. You may remember from topic 6 that these gametes are produced via meiosis.

In sexual reproduction, two parental contributions (egg and sperm/pollen) fuse to produce a zygote that then develops into an animal or a plant. Even in the case of animals that are **hermaphrodites**, the gametes typically come from two separate animals, rather than self-fertilisation occurring.

9.3.2 External fertilisation

External fertilisation occurs when animals release their gametes into the external environment so that fertilisation occurs outside the body of females. Features of external fertilisation include the following:
- Very large numbers of gametes are produced.
- Large numbers of gametes increase the chance of fertilisation but also mean there is much gamete wastage.
- In nature, it is limited to animals that either live in aquatic environments or reproduce in a watery environment, as sperm need a watery environment to swim to an egg. Oysters, for example, each produce about 500 million eggs in a single season.
- It occurs in aquatic invertebrates (such as coral polyps), bony fish and amphibians (such as frogs and toads).
- There are various strategies to increase the chances of fertilisation.

hermaphrodite organism with both male and female reproductive organs

external fertilisation union of sperm and egg occurring outside the body of the female parent

External fertilisation in nature can be a chancy process. Some species that use external fertilisation have developed strategies to increase the chance that eggs and sperm will meet and that fertilisation will occur.

CASE STUDY: Coral spawning

Mass release of gametes by coral polyps (or spawning) on the Great Barrier Reef occurs every year after the full moon in October and in November. For coral polyps on the Ningaloo Reef off the Western Australian coast, this mass spawning occurs after the full moon in March.

Figure 9.21a shows the release of eggs by a colony of coral polyps, the animals that build coral reefs. All the polyps in one region release their eggs and sperm into the sea in a synchronised manner, creating a 'soup' of gametes (figure 9.21b). The high concentration of gametes during spawning increases the likelihood of eggs and sperm colliding and fertilisation occurring.

FIGURE 9.21 a. Close-up of the release of gametes by a group of colonial animals known as coral polyps **b.** Mass spawning of eggs and sperm bundles by coral polyps occurs at the same time (synchronously) each year.

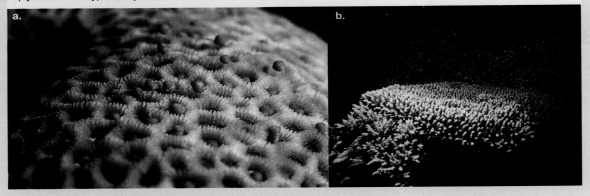

Increasing the chance of fertilisation in fish

Fish depend on external fertilisation. If a female fish releases her eggs independently of the release of sperm by a male fish of the same species, then the chance of fertilisation of her eggs is extremely low or even nil.

Behaviours such as courtships before spawning can lead to an improved chance of fertilisation. Through a courtship display a female can recognise that a male is a member of her species and, if she is ready to release her eggs, this occurs in close proximity to a male. In turn, he will release sperm nearby so that the likelihood of fertilisation is increased. Again, courtship has an energy cost.

Increasing the chance of fertilisation in frogs

Frogs and toads have external fertilisation. The chance of fertilisation is increased by a behavioural adaptation of males. When a female frog is ready to lay eggs she goes into a nearby pond or pool. A male tightly clasps her and remains on her back until she releases her eggs (figure 9.22). As the female releases her eggs, the male is stimulated to release sperm over the eggs, fertilising them. The fertilised eggs then undergo embryonic development and give rise to larvae, known as tadpoles, that later metamorphose to frogs.

FIGURE 9.22 Ornate burrowing frogs (*Limnodynastes ornatus*). The male is clasping the female tightly and when the female releases her eggs into the water, the male releases sperm over the eggs. The eggs are suspended on a raft of bubbles.

In one Australian frog species, fertilised eggs are swallowed by females and develop in their stomachs. They are regurgitated later as tiny froglets. In another Australian frog species, following fertilisation, the eggs are transferred into moist pouches on the hips of the male of the species, where embryonic development occurs.

9.3.3 Internal fertilisation

Internal fertilisation occurs when males deliver sperm directly into the reproductive tract of females so that fertilisation of eggs occurs inside the body of females. Features of internal fertilisation include the following:
- It has an energy cost of finding, attracting and securing a female mate.
- It has the benefit of increasing the chances of the gametes meeting, and therefore increases the chances of fertilisation.
- It occurs in some aquatic organisms as well as in terrestrial organisms.
- All terrestrial animals use internal fertilisation, except for amphibians, such as frogs and toads that mate in the water.

internal fertilisation union of sperm and egg occurring inside the body of the female parent

Marine animals

In sharks, the male of the species has claspers that are appendages of his pectoral fins (figure 9.23). The male shark inserts his claspers into the vagina of a female and releases his sperm inside her, via her cloaca, which is carried by water pressure that he generates.

FIGURE 9.23 Ventral (belly) view of a male shark showing its two claspers

Octopuses use internal fertilisation. In the male octopus, one of his eight arms is shorter than the other arms. This arm is specialised for the direct transfer of a parcel of sperm to a female. Figure 9.24 shows a male octopus that has mounted a female and inserted his specialised arm into her mantle cavity, where he will deposit a package of sperm.

FIGURE 9.24 **a.** Photograph of a smaller male blue-ringed octopus (*Hapalochlaena lunulata*) transferring a sperm package into the mantle cavity of a larger female **b.** Diagram of this process showing the smaller male blue-ring octopus (at left) inserting his specialised arm carrying a sperm package into the mantle cavity of the larger female octopus (at right)

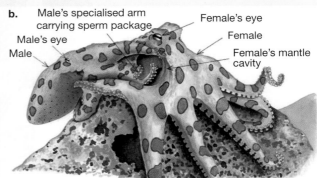

Insects

Internal fertilisation occurs in insects. Complex genital structures at the end of the abdomen of a male insect enable him to couple with a female and transfer sperm packages into her reproductive tract. Some insects, such as dragonflies, mate on the wing, but most insects keep their feet (all 12 in a mating couple) on the ground (see figure 9.25).

FIGURE 9.25 a. Dragonflies can mate in flight or on narrow plant stems or leaves. **b.** Cockroaches mate on the ground.

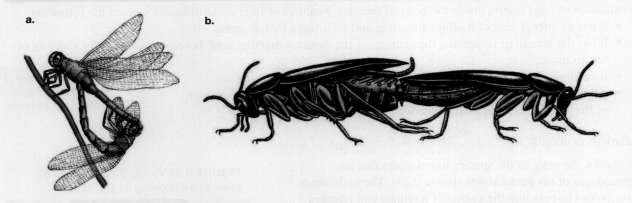

Reptiles

Ancestral reptiles were the first vertebrates to evolve a male copulatory organ, the penis (figure 9.26). The presence of a penis enabled reptiles to transfer sperm directly into the reproductive tract of a female. Apart from sharks, reptiles were the first vertebrates that did not need to release their gametes into water for fertilisation to occur. Internal fertilisation freed the ancestral terrestrial reptiles and their descendants, such as turtles, from the need to return to water for reproduction.

FIGURE 9.26 View of the urogenital organs of a male turtle. Note the presence of the penis.

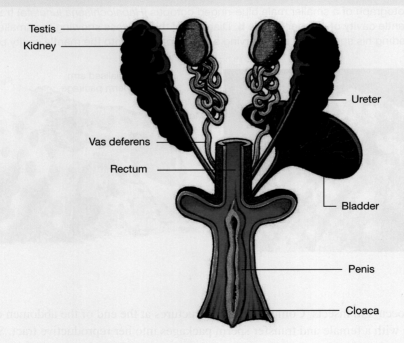

Birds

Birds also have internal fertilisation, but male birds lack a penis. Instead, sperm transfer takes place through close contact of the urogenital openings (cloacae) of male and female birds. Like their reptilian ancestors, birds produce eggs in which their embryos develop (figure 9.28). The production of a shell amniote occurs as the fertilised egg passes through the genital tract of a female bird to the cloaca, from where the egg is laid.

EXTENSION: The development of eggs

The development of eggs in reptiles and birds ensured several reproductive advantages:
- Female animals produce a small number of eggs (or even a single egg) in each reproductive cycle. Less gamete wastage occurs and the chance of fertilisation for a given egg is quite high.
- The eggs have a protective outer shell, a series of internal membranes (the amnion) and a food supply for the developing embryo (the yolk and albumen).
- Eggs that contain the fluid-filled membrane known as the amnion are called amniotic eggs.
- Fluid within the amnion prevents the embryo from drying out, and wastes are stored in a sac called the allantois as insoluble uric acid.
- Amniotic eggs (figure 9.27) enabled reptiles to reproduce on dry land without the dependence on the presence of free water (as exists for amphibians).

FIGURE 9.27 Amniotic egg of a snake

FIGURE 9.28 Bird developing within an amniotic egg

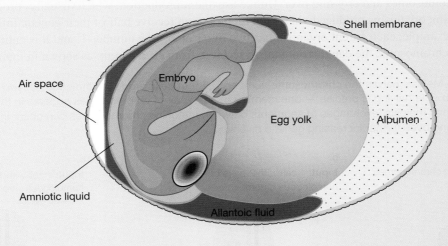

Mammals

Monotremes are mammals that lay eggs. The only surviving monotremes are the platypus and the echinda. In these animals the fertilised egg becomes enclosed within a shell and embryonic development occurs within the shelled egg (figure 9.29), similar to reptiles. Unlike other mammals, they feed their young through pores in the belly that secrete milk, as opposed to teats.

FIGURE 9.29 Echidna egg

monotremes the order of non-placental mammals that lay leathery-shelled eggs and secrete milk through pores in the skin

TOPIC 9 Reproductive strategies 533

In contrast, female **marsupials** and placental mammals retain the fertilised eggs in their bodies and embryonic development proceeds within the uterus. Young marsupials are born at a very undeveloped stage. Marsupials are unique, as their young continue to develop in a pouch. Perhaps the most well-known marsupial is the kangaroo, renowned for carrying their young — called a joey — for 9 months. The joey is still reliant on its mother up until 17 months of age, obtaining high-nutrient milk from her to enable it to develop.

Young placental mammals are born at a much more developed stage. The biological advantage for the young is that by being more developed at birth, they are better able to cope with the environment, increasing their chance of survival. The young survive *in utero* through being connected to the mother via the placenta, receiving food and oxygen and returning waste before being born. Humans are an example of a placental mammal!

> **on Resources**
>
> **eWorkbook** Worksheet 9.3 Comparing fertilisation methods (ewbk-7565)

9.3.4 Genetic diversity

Recall from topic 6 that the production of gametes via meiosis results in genetic variation due to independent assortment of chromosomes and crossing over. Unlike asexual reproduction, which (usually) produces a clone of the parent organism, organisms produced via sexual reproduction receive half of their genetic information from each parent. For example, each parent organism has two of chromosome number-1, and it is entirely random which number-1 chromosome the daughter organism receives from each parent, as shown in figure 9.30.

FIGURE 9.30 a. Meiosis halves the chromosome number. The four cells produced indicate that two cell divisions have occurred. **b.** The halving is precise — one member of each pair of chromosomes appears in the cell products. **c.** During meiosis, exchanges of segments between matching chromosomes can occur in a process known as crossing over, further increasing genetic diversity.

a.
Input: 1 cell with 2n chromosomes

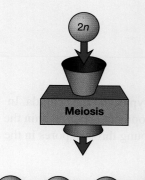

Output: 4 cells, each with n chromosomes

b.
Input: 1 cell with 2 pairs of homologous chromosomes

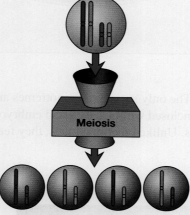

Output: 4 cells with one member only of each pair of homologues

c.
Input: 1 cell with 1 pair of homologous chromosomes with 2 linked genes

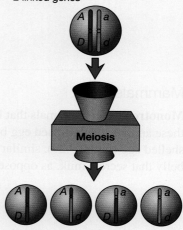

Output: 4 cells, some with 'splicing' of homologous chromosomes to create some new gene combinations

> **marsupials** the order of non-placental mammals that are born at a very early stage of development and then grow inside their mother's pouch

In addition, these chromosomes can also exchange genetic information with each other via crossing over. The point at which this occurs is called the **chiasma**, as shown in figure 9.31 (showing two chiasmata).

FIGURE 9.31 a. One pair of homologous replicated chromosomes at interphase of meiosis. Note the genetic information that they carry. **b.** The exchange of segments when crossing over occurs during prophase I of meiosis **c.** The resulting recombination of genetic material

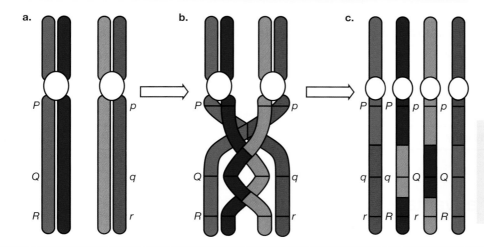

chiasma a point at which paired chromosomes remain in contact during the first metaphase of meiosis, and at which crossing over and exchange of genetic material occur between the strands

Meiosis produces genetic variation

The biological significance of meiosis is that it results in genetic variation among the offspring produced by sexual reproduction involving gametes from two parents.

Gametes carry unique genetic combinations because of:
1. crossing over between homologous (matching) chromosomes
2. independent assortment (separation) of non-matching chromosomes during meiosis.

CASE STUDY: The Habsburgs — the importance of genetic diversity

Charles II of Spain had a pronounced underbite, an oversized tongue that prevented him from closing his mouth properly, and he suffered from regular bouts of vomiting and diarrhoea. In the 1700s, such symptoms were attributed to witchcraft or a curse being placed upon him. With the knowledge that we now have about the importance of variation in a gene pool, it is likely that the ailments from which Charles II suffered were due to inbreeding within the royal family.

Charles II was from the Habsburgs, a family who ruled parts of Western Europe in the 1700s. They believed that for their dynasty to survive, they had to marry within their own royal bloodlines. Those of royal descent were forbidden from marrying 'commoners', or those without a title. With most marriages arranged to ensure maintenance of power and acquisition of land, members of royal families had to look within their family tree for a potential suitor. For Charles II, his mother was also his father's niece. In other words, his mother married her uncle!

FIGURE 9.32 Charles II of Spain, the last Habsburg ruler of the Spanish empire. He is renowned for his deformities caused by inbreeding.

Despite being married twice, Charles was unable to produce an heir to the throne. When he died at the age of 39, the rule of the Habsburg family in Europe ended.

9.3.5 Differences between asexual and sexual reproduction

The key differences between asexual and sexual reproduction are shown in table 9.4.

TABLE 9.4 Differences between asexual and sexual reproduction

Asexual reproduction	Sexual reproduction
One parent contributes genetic material	Two parents contribute genetic material
Daughter cells are usually identical to the parent cell	Daughter cells are different to the parent cells
No exchange of genetic material	Exchange of genetic material through crossing over of chromosomes
Chromosomes are not assorted	Chromosomes are assorted independently
Less energy required	Energy invested in finding a mate
Does not affect genetic diversity	Increases genetic diversity in a population

9.3.6 Biological advantages and disadvantages of sexual reproduction

The advantage of sexual reproduction comes from the genetic diversity that it creates in offspring. In contrast to a population generated by asexual reproduction that is composed of genetically identical organisms, a population of organisms produced by sexual reproduction contains a remarkable level of genetic diversity. The presence of this variation within its gene pool means that such a population is better equipped to survive in changing, unstable environmental conditions; to cope with an outbreak of a new viral or bacterial disease; or to survive a natural disaster.

Disadvantages of sexual reproduction (relative to asexual reproduction) include the commitment of energy required to find, attract and secure a mate — a process that for some species may involve elaborate courtship displays of mating calls and dances, colourful feathers or contests between males for mating rights. All are things that require a significant investment of energy with no guarantee of being able to mate. The bowerbird, *Ptilonorhynchus violaceus* (figure 9.33), is one example of a bird that spends much of its time and energy resources collecting blue items as part of the courting ritual to attract a female.

However, the sexual mode of reproduction dominates the world of eukaryotic organisms, indicating that its advantages outweigh its combined disadvantages.

FIGURE 9.33 The bowerbird (*Ptilonorhynchus violaceus*) is renowned for its efforts to attract a female with blue items obtained from the environment. A considerable amount of energy is invested in the hope of attracting a mate.

The advantages and disadvantages of sexual reproduction are summarised in table 9.5.

TABLE 9.5 Comparison of advantages and disadvantages of sexual reproduction

Advantages of sexual reproduction	Disadvantages of sexual reproduction
Genetic diversity within the species	Energy expended to find a mate
Variation increases survival chances should conditions change	Some organisms can be injured (or killed) in competition for a mate
Variation between members of the same family due to crossing over and independent assortment during meiosis	
More traits to select for when choosing a mate — allows natural selection to occur	

As you can see, the benefits of sexual reproduction outweigh the disadvantages in regard to the longevity of a species in changing environments.

tlvd-1734

SAMPLE PROBLEM 3 Advantages of 'sneaker' fish

Many marine species of animal rely on the ocean currents to enable sperm and eggs released into the water to meet. Prior to the release of gametes, organisms of the same species send signals to one another to ensure that their gametes are not wasted.

In a small number of species, a male fish who is not part of the courtship ritual will 'sneak' in and release his sperm at the same time as the male fish in the courtship ritual. This has been observed in the Minckley's cichlid (*Herichthys minckleyi*), which lives off the coast of Mexico. What benefits does the action of the 'sneaker' have to both himself and the overall genetic diversity of the species? **(2 marks)**

THINK

1. Consider what the benefits of the sneaker's actions are to himself, as he does not take part in the courtship ritual.
2. Consider what the benefits of the sneaker's actions are to the species.
3. How does this contribute to genetic diversity?

WRITE

The sneaker benefits as he has not had to invest energy in the courtship process (1 mark).

Two males fertilising the eggs increases the genetic diversity in the offspring.

Without the sneaker's actions, the offspring will be the product of one female and one male producing gametes via meiosis that are subject to independent assortment and crossing over. The sneaker's actions result in an increase in the variety of gametes available, with offspring produced from one female and one of two male gametes, which increases the genetic diversity in the offspring (1 mark).

EXTENSION: Genetic issues with pedigree pets

Whilst sexual reproduction produces variation within a species, purebred pets have higher levels of genetic diseases due to inbreeding.

A champion male dog of the show ring will often become a stud dog, siring numerous litters and sharing its genes within these populations. The ancestry of Peppa (figure 9.34), a 4-year-old purebred golden retriever, can be traced with her pedigree papers. She is the product of champion show dogs used for selective breeding to maintain the traits of the breed standard for a golden retriever.

FIGURE 9.34 Peppa, the golden retriever. She is the product of selective breeding that has led to the expression of hip dysplasia.

As she is the cumulation of a long line of dogs with similar traits, her gene pool is not as diverse as other dogs with different gene pools contributing to the genome. Upon her yearly check-up with the vet, it was identified that she had poor muscle tone on her hind legs, an indication of hip dysplasia. Following a series of x-rays and a meeting with a specialist, her diagnosis was confirmed. Whilst selective breeding enables the creation of organisms with desired traits for consumers, selective breeding also perpetuates the higher incidence of genetic diseases such as hip dysplasia. Hip dysplasia in canines occurs when the components of the ball-and-socket joint of the hip do not fit together properly, resulting in friction, which causes the joint to deteriorate over time. This is caused by inconsistent growth of the ball and the socket in puppies. Large dogs are at a greater risk of hip dysplasia due to their increased rate of growth in their first 12 months. Like many genetic diseases, the environment also contributes to the phenotype, with factors such as the age of neutering and exercise levels of puppies contributing. In Peppa's case, she will be subjected to weight management and possibly surgery in her later years should her condition worsen. As she has been neutered, there is no risk of her passing the defective gene to a future generation of golden retrievers.

As seen in the next Case study box, increasing the available gene pool of organisms is essential in ensuring the long-term health of the breeds.

DISCUSSION

Should stud dogs be limited in the number of litters they sire to increase genetic diversity in purebred dogs?

CASE STUDY: Increasing the gene pool using artificial insemination

There are two breeds of corgi — the Pembroke Welsh corgi made famous by Queen Elizabeth II, and the lesser known Cardigan Welsh corgi, which was bred to drive cattle across the hills of Wales. Cardigan Welsh corgis are now an endangered breed in Britain, their country of origin, and numbers are also declining in Australia. The small gene pool available locally makes breeding challenging because genetic diversity must be maintained to ensure healthy dogs.

While the gene pool in a country can be increased by importing dogs into the country, the Australian gene pool of this breed was increased using frozen semen imported from suitable stud dogs. Frozen semen has several advantages; it increases the breeding life of males and, as animals do not move between countries, no quarantine is needed.

Jan West from Deakin University carried out artificial insemination of a female Cardigan Welsh corgi, Gem. Gem was born in England and imported to Australia with the long-term goal of increasing the Cardigan Welsh corgi gene pool in Australia. Gem produced a litter of eight puppies, which were conceived using frozen semen. The thawed semen was implanted in the uterus using a technique called transcervical insemination (TCI). The ultrasound in figure 9.35 was taken four weeks after fertilisation and shows sacs of amniotic fluid enclosing the puppies.

After a gestation period of 61 days, the puppies were born. Puppies have a well-developed sense of smell and a strong 'suck' reflex. Their suckling stimulates the mother's pituitary gland to release oxytocin, the hormone that triggers her mammary glands to release milk. Milk is high in calories and nutrients and the pups doubled their birth weight in their first week. Figure 9.36 shows some of the pups aged two months, well on the path of developing into healthy adult Cardigan Welsh corgis.

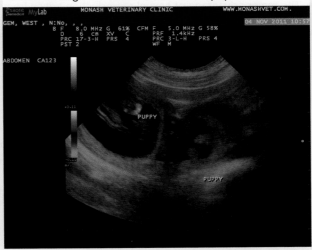

FIGURE 9.35 Ultrasound at week four of Gem's pregnancy. The coloured structure is the movement of blood through the heart of one of her pups.

FIGURE 9.36 Gem and three of her pups, aged two months

Resources

eWorkbook Worksheet 9.4 Advantages and disadvantages of sexual reproduction (ewbk-7567)

KEY IDEAS

- Compared with asexual reproduction, sexual reproduction has a higher energy cost.
- In animals, fertilisation may be internal or external.
- Crossing over and independent assortment in meiosis provides genetic variation.
- The major advantage of sexual reproduction arises from the genetic diversity that it produces in populations of sexually reproducing organisms.
- Genetic variation increases the survival chances of a species should conditions change.
- Variation allows for natural selection to occur, with those most suited to the environment being selected to mate with.

9.3 Activities

learn on

To answer questions online and to receive **immediate feedback** and **sample responses** for every question, go to your learnON title at **www.jacplus.com.au**. A **downloadable solutions** file is also available in the resources tab.

| 9.3 Quick quiz on | 9.3 Exercise | 9.3 Exam questions |

9.3 Exercise

1. Using an example, explain the difference between internal and external fertilisation.
2. Explain the purpose of the albumen in an egg.
3. Describe why external fertilisation is limited to watery environments.
4. Describe three advantages of sexual reproduction.
5. Define the term *chiasma*.
6. Explain how independent assortment and crossing over increase genetic variation in an organism.
7. Explain how sexual reproduction allows for allele frequencies to change in populations.
8. The mating of two animals with a diploid number of 20 produces a litter of 10 offspring. Is it reasonable to assume that at least two of these offspring will be genetically identical? Briefly explain.
9. The bowerbird invests a significant amount of time adorning its nest with blue items and practising intricate dances in the hope of attracting a mate. What benefit does the bowerbird species derive from its investment in courtship behaviours?

9.3 Exam questions

Question 1 (1 mark)

MC A farmer used a pesticide to spray crops that were infested by a crop-eating fly. Immediately after the first spray treatment the fly population was greatly reduced. However, successive treatments were less and less effective at killing the flies. Eventually the pesticide no longer killed the flies. They were just as numerous as before the treatments.

Which explanation accounts for the observed increase in the fly population after the successive insecticide treatments?

A. After each spraying treatment, flies from neighbouring farms flew in to replace those killed by the insecticides.
B. The insecticide became less effective with each spraying treatment as it approached its 'use-by' date.
C. The individual flies became accustomed to the insecticide and eventually were tough enough to withstand the spray.
D. A few flies possessed insecticide-resistant genes. They survived, reproduced and passed on these genes to their offspring.

Question 2 (2 marks)
Source: VCAA 2013 Biology Exam, Section B, Q8b
Briefly state the biological significance of the process of meiosis.

Question 3 (1 mark)
Source: VCAA 2007 Biology Exam 2, Section A, Q14

MC New alleles arise in a sexually reproducing population by
A. mutations in DNA sequences prior to meiosis.
B. random fertilisation of gametes during reproduction.
C. random assortment of homologous chromosomes during meiosis.
D. exchange of chromatin between homologous chromosomes during meiosis.

Question 4 (2 marks)
Source: VCAA 2009 Biology Exam 2, Section B, Q3e

The endangered pygmy possum (*Burramys parvus*) lives in three restricted alpine areas, Mt Buller, Bogong High Plains and Mt Kosciusko.

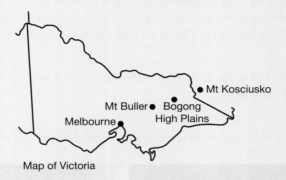
Map of Victoria

About 2000 individuals remain in the wild. Studies show that there is a lot of genetic diversity between the three populations. Due to the isolation of these populations, scientists think that each population has a separate gene pool.

Explain how exchange of genetic material may be beneficial in the survival of endangered species like the pygmy possum.

Question 5 (2 marks)

Several small, young wattle trees (*Acacia* spp.) were observed growing around the base of a large, mature wattle tree. Repeated observations of the young trees were made over a period of a few weeks. The young Acacia trees showed slight differences in their leaf colour and shape, as well as differences in their rates of growth.

A biology student suggested that the trees were suckers, growing by vegetative propagation from the roots of the mature tree. Her classmate disagreed, suggesting that the young trees had grown from seeds produced by sexual reproduction.

Which student was correct? Justify your decision.

More exam questions are available in your learnON title.

9.4 Reproductive cloning technologies

> **KEY KNOWLEDGE**
>
> - The process and application of reproductive cloning technologies
>
> **Source:** VCE Biology Study Design (2022–2026) extracts © VCAA; reproduced by permission.

9.4.1 Cloning in horticultural practice

Using the technology of **plant tissue culture** in the laboratory, many identical copies — or clones — of a plant can be produced starting from a small amount of tissue from one plant. This technique is used with ornamental plants, such as orchids and carnations, and Australian native plants, such as bottlebrush (*Callistemon* spp.), the flannel flower (*Actinotus helianthi*) and various eucalypts (*Eucalyptus* spp.). Tissue culturing can also be used with endangered or very rare plants, such as the Wollemi pine (*Wollemia nobilis*), since only a few specimens exist in the wild.

Figure 9.37a shows flannel flowers in tissue culture in a laboratory and figure 9.37b shows mass plantings of flannel flowers propagated by tissue culture.

FIGURE 9.37 a. Flannel flowers in tissue culture **b.** Large-scale bed plantings of flowers propagated by tissue culture

Tissue culture cloning of plants has several advantages:
- Slow-growing plants can be produced in large numbers.
- Plants can be cultured all year round in controlled conditions of temperature and day length, rather than relying on seasonal growth.

plant tissue culture technique used to clone plants in large numbers

- Virus-free tissue can be used to produce a large number of plants that do not carry the virus. (Viruses are responsible for many plant diseases that can affect commercial crops.)
- Cultured plants can be transported from country to country. The sterile conditions in which they are cultured ensures that the plants are pest-free, so that lengthy quarantine periods are avoided.

How does tissue culture work? It begins with a small piece of a healthy plant, such as a piece of leaf, bud or stem. The plant that supplies the tissue is selected because it has particular desirable characteristics, such as flower colour, disease resistance or timber quality.

Pieces of a plant to be cultured must contain meristematic tissue because this is the only plant tissue that is capable of cell division by mitosis. The tissue culture procedure in plants is summarised in the flow chart in figure 9.38.

FIGURE 9.38 Tissue culture procedure. Here, many *Eucalyptus* plants are produced from a parent plant selected for its genetically determined timber quality. Will the plants produced by tissue culture be expected to have this quality when they mature?

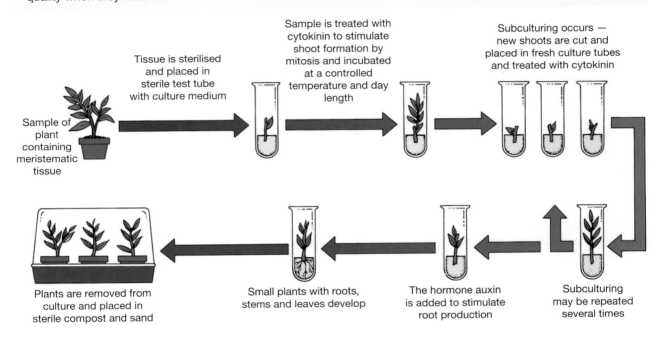

The success of tissue culturing of plants depends on the fact that asexual reproduction produces genetically identical clones of the parent, whether the parent is a whole organism, a tissue sample or even a single cell. This exact copying is a result of the precision of cell division by mitosis.

INVESTIGATION 9.3

elog-0842

Plant tissue cultures

Aim

To use plant tissue cultures to produce clones of a plant

9.4.2 Artificial cloning of mammals

Reproduction in mammals in their natural setting is sexual, involving fertilisation of an egg by a sperm, and one fertilisation event typically produces a single offspring.

Artificial cloning of mammals is a recent development and several techniques have been used:
- cloning using **embryo splitting**
- cloning using **somatic cell nuclear transfer**.

embryo splitting process of separating the totipotent cells of a very early embryo, so that the resultant cells are each able to form a complete embryo

somatic cell nuclear transfer a cloning technique that involves the nucleus of a somatic cell being transferred into the cytoplasm of an enucleated cell that is then stimulated to divide

Embryo splitting to make identical copies

Cloning by embryo splitting occurs when the cells of an early embryo are artificially separated, typically into two cell masses. This process mimics the natural process of embryo splitting that produces identical twins or triplets. Embryo-splitting technology has been used for stockbreeding for many years. It has become a relatively simple technique, but is limited to twinning.

Features of embryo splitting include the following:
- The parents are chosen because of desirable inherited commercial characteristics that they exhibit, such as high milk yield or high milk fat content in dairy cattle, or muscle formation or fat distribution in beef cattle.
- Typically, the embryos to be split are produced through *in-vitro* fertilisation (IVF) — for example, the *in-vitro* fertilisation of a cow's egg by bull sperm.
- Using a very fine glass needle, an embryo at an early stage of development is divided into two smaller embryos.
- The small embryos from the splitting of one embryo are identical, as will be the adults that develop from them. Each small embryo is then implanted into the uterus of a surrogate female parent, where embryonic development continues. Figure 9.39 shows an outline of this process.

FIGURE 9.39 The process of embryo splitting to clone dairy cattle. The cow may first be treated with hormones to cause the release of several eggs (superovulation). Does the surrogate mother make any genetic contribution to the embryo?

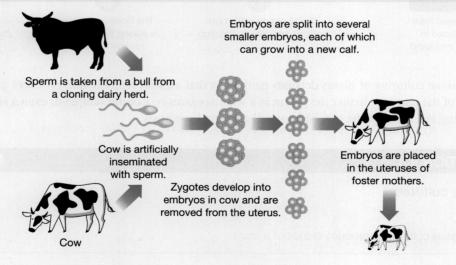

Embryo splitting has been used for some years in the livestock industry. In cattle, for example, embryo splitting enables the genetic output from several matings of a top bull and a prize cow to be doubled. Instead of just one calf from each such mating, two calves can be produced. This process depends on the use of surrogate mothers.

The two offspring from the splitting of one embryo are not genetically identical copies of either the cow that produced the egg or the bull that provided the sperm used to produce the embryo that was split. The two offspring are genetically identical to each other and are identical copies of the *fertilised* egg from which the embryo came.

The eggs produced by one cow are not genetically identical to each other, nor are the sperm from one bull. This is because these gametes are produced by meiosis, not mitosis. As we saw in topic 6, the process of meiosis juggles and reassorts the genes of eggs and sperm. This means that the offspring from the splitting of embryos derived from different fertilised eggs from the same mating will be genetically different.

Somatic cell nuclear transfer (SCNT)

Some possibilities exist to manipulate cells and their nuclei using somatic cell nuclear transfer (SCNT). The processes are summarised in figure 9.40:
a. The nucleus of a cell is removed, creating an **enucleated cell**
b. The nucleus of another cell can be transferred to an enucleated cell to form a redesigned nucleated cell.
c. An enucleated cell can be fused with a somatic cell using a short electrical pulse.

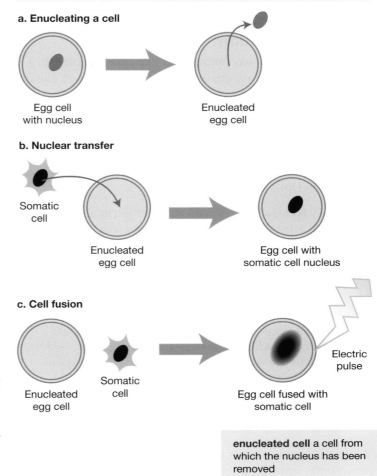

FIGURE 9.40 a. Producing an enucleated cell **b.** Nuclear transfer between two cells **c.** Cell fusion. Fusion of two cells is commonly done using a short electric pulse.

enucleated cell a cell from which the nucleus has been removed

FIGURE 9.41 a. Megan and Morag, two Welsh mountain ewes, born in August 1995 **b.** Megan and Morag were created using nuclear transfer cloning.

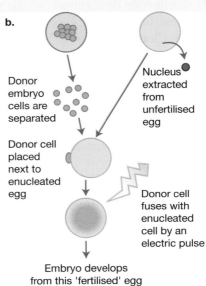

TOPIC 9 Reproductive strategies **545**

The birth of two sheep, Megan and Morag (see figure 9.41), in 1995 marked a significant scientific milestone. These two sheep were the first mammals ever to be cloned using nuclear transfer technology. Each of these sheep developed from an unfertilised enucleated egg cell that was fused with an embryonic cell that contained its nucleus. In each case, the embryonic cell used came from the culture of one embryonic cell line; as a result, Megan and Morag were identical twins.

... and then came Dolly!

The scientific world was stunned after the announcement in February 1997 of the existence of Dolly, a Finn–Dorset female lamb born the previous year in Scotland (figure 9.42). The scientists who created Dolly were Ian Wilmut, Keith Campbell and their colleagues from the Roslin Institute, which is part of the University of Edinburgh in Scotland.

Why was Dolly the lamb famous? Read what Ian Wilmut wrote about Dolly:

> Dolly seems a very ordinary sheep ... yet, as all the world acknowledged ... she might reasonably claim to be the most extraordinary creature ever to be born ...
>
> Dolly has one startling attribute that is forever unassailable: *she was the first animal of any kind to be created from a cultured, differentiated cell taken from an adult.* Thus she confutes once and for all the notion — virtual dogma for 100 years — that once cells are committed to the tasks of adulthood, they cannot again be totipotent.

Source: I Wilmut, K Campbell and C Tudge, *The Second Creation: Dolly and the Age of Biological Control*, Harvard University Press, Cambridge, Mass., 2000

FIGURE 9.42 Dolly and her first lamb, Bonnie. Bonnie was born in April 1998 after a natural mating of Dolly with a Welsh mountain ram.

While cloning via nuclear transfer had occurred successfully in the past, those earlier cases involved embryonic or foetal cells — never **adult somatic cells**. The use of adult somatic cells, such as skin cells, to construct new organisms represents remarkable human intervention in the evolutionary processes. Through this means, cells from sterile animals or from animals past their reproductive period — or even stored cells from dead animals — can provide all the genetic information of new organisms. In nature, the normal evolutionary processes would not allow these events to occur. Cloning has shown that adult somatic cells can be **totipotent**.

adult somatic cells body cells that have differentiated into their cell type

totipotent a cell that is able to give rise to all different cell types

How was Dolly created?

The artificial cloning of mammals involves:
- obtaining the nucleus from a somatic (body) cell of an adult animal — this is the 'donor' nucleus
- removing the nucleus from an unfertilised egg cell, typically of the same species — this is the enucleated egg cell
- transferring the donor nucleus into the enucleated egg cell
- culturing the egg cell with its donor nucleus until it starts embryonic development
- transferring the developing embryo into the uterus of a surrogate animal, where it completes development.

The genetic information in the cloned animal comes from the nucleus of the adult body cell and so the genotype of the cloned animal is determined by the donor nucleus, not by the egg into which the nucleus is transferred.

The procedure in the case of Dolly is shown in figure 9.43. An unfertilised egg from a Scottish Blackface ewe had its nucleus removed. A cell was taken from the culture of mammary cells derived from the udder (mammary gland) of a Finn–Dorset ewe. Using a short electric pulse, the cultured mammary cell was fused with the enucleated egg cell to form a single cell. This reconstructed cell was cultured for a short time and was then implanted into the uterus of a surrogate Scottish Blackface ewe, where the embryo developed. At 5 pm on 5 July 1996, this surrogate Scottish Blackface ewe gave birth to Dolly, a Finn–Dorset lamb, the first mammal to be produced by cloning using an adult somatic cell.

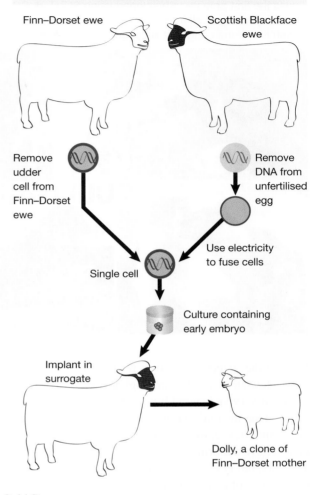

FIGURE 9.43 Technique of animal cloning by somatic cell nuclear transfer

After Dolly — what next?

- Matilda the sheep was the first lamb to be cloned in Australia and was born in April 2000 (see figure 9.44a).
- Mayzi and Suzi were Australia's first calves to be artificially cloned from the skin cells of a cow foetus (see figure 9.44b). Mayzi and Suzi are identical twins but were born two weeks apart in April 2000.
- Zhong Zhong and Hua Hua, two crab-eating macaques, were the first cloned primates (figure 9.44c). The nuclei used to create Zhong Zhong and Hua Hua were obtained from foetal cells rather than embryonic cells.
- CC (short for carbon copy) was the first cat to be artificially cloned, using a cumulus cell from an adult female cat named Rainbow, as announced by a group of US scientists in February 2002 (see figure 9.44d).
- Snuppy was the first dog to be artificially cloned, using an ear cell from a three-year-old Afghan hound, as announced by a group of South Korean scientists in August 2005 (see figure 9.44e). Snuppy is short for Seoul National University puppy.

Cloning: the downside

The success rate in initiating development of the egg cell after transfer of the donor nucleus is low.

For example:
- in the case of an artificially cloned calf known as Second Chance, 189 implantations were made into surrogate cows before a pregnancy was achieved. This case, however, was remarkable because the adult cell that provided the donor nucleus came from a 21-year-old Brahman bull called First Chance. This was an extremely old adult cell to use as the starting point for cloning. Because of testicular disease, First Chance had been castrated so that he was sterile when one of his body cells was successfully cloned.
- the kitten CC, produced by somatic cell cloning, was the only one of 87 embryos implanted into surrogate mothers that survived to term
- to get Snuppy, 123 dog embryos were surgically implanted into surrogate females and, of these, only three survived for a significant period, with one dying before birth, one dying soon after birth, and the sole survivor being Snuppy
- Dolly was the only live birth from a series of 277 cloned embryos.

FIGURE 9.44 Mammalian clones since Dolly: **a.** Matilda was the first sheep to be cloned in Australia. **b.** Suzi, one of two genetically identical Holstein calf clones derived from the skin cell of one cow foetus **c.** A crab-eating macaque was the first primate to be cloned using SCNT. **d.** CC (right) was the world's first cat produced by somatic cell cloning, using a cell from Rainbow (left). **e.** Snuppy, the world's first dog produced by somatic cell cloning, is shown with the Afghan dog that supplied the ear cell (left) and his surrogate mother, a Labrador (right).

Despite the success of Dolly's birth over twenty years ago, somatic cell cloning is still far from routine. This is due to low success rates and high levels of developmental abnormalities. Currently, 0 to 10 per cent of cloned embryos survive beyond birth. Of the clones that do survive, many have abnormalities that can cause death early in life. One institution reported in 2003 that, for every healthy lamb clone born, about five had abnormalities. Abnormalities reported include impaired immune system function and the 'large offspring syndrome', in which clones have abnormally large organs. Premature death is common in cloned animals. Sheep typically live for 12 years, but Dolly was euthanised at six years old because of a deteriorating lung disease and arthritis, unusual conditions for a sheep of Dolly's age and one that was housed indoors. Matilda, the Australian cloned lamb (figure 9.44), died when she was less than three years old.

The low success rate is currently thought to be largely due to failures in reprogramming the donor genome. Reprogramming is the resetting of the DNA and associated proteins in the nucleus of the somatic cell to allow genes to coordinate the early development of the embryo. The adult somatic cell nucleus needs to respond to the egg cytoplasm and act like the nucleus of a zygote — that is, like a pluripotent embryonic cell. Errors in reprogramming lead to abnormal gene expression.

Attitudes to reproductive cloning

SCNT has the potential to be an important process in animal agricultural industries due to the opportunity to select traits that can enhance agricultural yields, such as milk production in dairy cows. It also has the potential to save the genetic information of species facing extinction. However, general public attitudes to reproductive cloning of animals are mixed. While it may bring economic benefits, some people oppose the concept for various reasons, such as their belief that cloning is interfering with nature.

When people are questioned about the reproductive cloning of human beings, there is a very high level of opposition to it. Over 30 countries, including Australia, have banned experiments directed to producing human clones. In Australia this is covered by the *Prohibition of Human Cloning Act 2002*, which was amended in 2006.

The *Research Involving Human Embryos Regulations 2017* was enacted to regulate the use of excess human embryos created during assisted reproductive technologies, and to ensure that it is illegal in Australia to clone a human. However, excess human embryos are used in stem cell and regenerative medicine to generate cells and tissues, as described in topic 2. SCNT is used only in some stem cell research laboratories.

SAMPLE PROBLEM 4 Types of cloning

Scientists took tissue from three cattle that they believed had been created through a cloning process. The cattle all had identical genomes; however, their genome was not identical to the mother who provided her genetic material via her eggs. What type of cloning has been used to create the three cattle? (3 marks)

THINK	WRITE
1. What are the types of cloning?	Embryo splitting involves two parents contributing genetic material. SCNT involves one parent contributing genetic material. Both use a surrogate for the pregnancy.
2. The three cattle offspring had the same genome, so how were they produced?	As the three cattle offspring all have the same genome, they must have come from one fertilisation event with one sperm and one egg (1 mark).
3. However, the three cattle had a different genome to their mother. How could this have occurred?	As they differ from the mother, a second parent must also have contributed genetic material (1 mark).
4. Answer the question by linking to the type of cloning.	The type of cloning used must be embryo splitting if two parents contribute genetic material (1 mark).

Resources

- **eWorkbook** Worksheet 9.5 Reproductive cloning technologies (ewbk-7569)
- **Video eLesson** Somatic cell nuclear transfer (eles-2465)
- **Weblink** 20 Years after Dolly the Sheep Led the Way — Where Is Cloning Now?

KEY IDEAS

- Artificial cloning of plants involves subdividing cultured plant tissue.
- Cloning of plants produces organisms that are genetically identical to each other and to the original cultured plant tissue.
- Artificial cloning of mammals is a new technology of asexual reproduction.
- Two types of artificial cloning techniques are embryo splitting and somatic cell nuclear transfer.
- Embryo splitting involves the artificial separation of embryo cells *in vitro*.
- Nuclear transfer involves the transfer of a nucleus from an adult somatic cell to an egg cell that has had its own nucleus removed. Problems with somatic cell cloning include a low success rate and high levels of abnormalities in offspring.
- Legislation of the Australian Parliament prohibits human cloning.

9.4 Activities

To answer questions online and to receive **immediate feedback** and **sample responses** for every question, go to your learnON title at **www.jacplus.com.au**. A **downloadable solutions** file is also available in the resources tab.

| 9.4 Quick quiz | 9.4 Exercise | 9.4 Exam questions |

9.4 Exercise

1. Every cell of a plant contains a copy of the entire plant genome; however, meristematic tissue is used in tissue culture cloning. Explain why this is the case.
2. List three advantages of tissue culture cloning in plants.
3. Create a flow chart that summarises the process of somatic cell nuclear transfer (SCNT).
4. Contrast embryo splitting and SCNT.
5. Describe the effect that being able to use adult somatic cells to create a clone has on the diversity of the gene pool.
6. Describe why it is important to have government legislation regulating cloning in Australia.

9.4 Exam questions

Question 1 (1 mark)

Source: VCAA 2008 Biology Exam 2, Section A, Q3

MC The following diagram summarises the steps involved in the production of a cloned sheep.

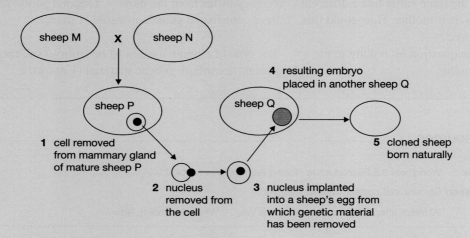

The chromosomes in the cells of the cloned sheep will be identical with those in the cells of
A. sheep M.
B. sheep N.
C. sheep P.
D. sheep Q.

Question 2 (1 mark)

MC Which one of the following is an example of clones?
A. Identical twin calves
B. A human brother and sister who are twins
C. Embryos conceived during in-vitro fertilisation
D. The offspring produced by the seeds of one tomato plant

Question 3 (1 mark)

A flower grower has a spectacular red carnation growing in her nursery. She wishes to produce many carnations identical to this plant to sell to customers.

Name a method that the grower could use to clone many copies of this spectacular red carnation.

Question 4 (6 marks)
Source: VCAA 2013 Biology Exam, Section B, Q12

Early research into animal cloning included an experiment on the frog tadpole *Xenopus laevis*.
a. What is a clone? **1 mark**

The early cloning experiment is summarised in the flow chart below.

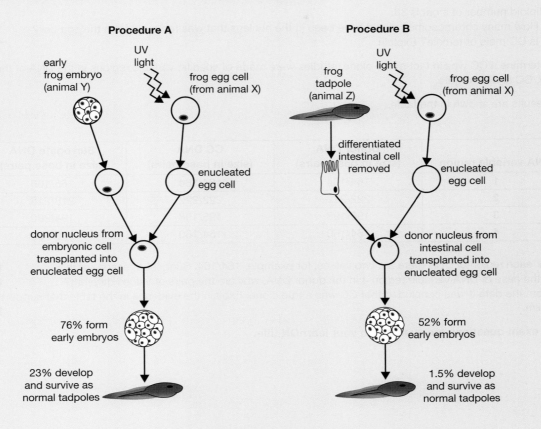

b. i. Why were the frog egg cells from animal X exposed to UV light? **1 mark**
   ii. Describe the genetic make-up of the final tadpole in procedure A. Justify your answer, referring to the information in the flow chart. **2 marks**
c. State the hypothesis being tested in the experiment and explain whether the results support the hypothesis. **2 marks**

Question 5 (6 marks)
Source: VCAA 2006 Biology Exam 2, Section B, Q5

'CC' for Carbon Copy is the name of the first cloned kitten born in 2001. The nucleus of a cat's egg cell was removed. It was replaced by a nucleus from a somatic cell of a donor female cat. Once development commenced the egg cell was transferred into a surrogate female.

a. What is meant by the term cloning? **1 mark**

The diploid number of a cat is 38.

b. i. How many chromosomes would have been in the nucleus that was removed from the egg cell? **1 mark**
ii. Is CC male or female? Explain. **1 mark**

To determine if CC was in fact a true clone, studies were made of specific variable regions in the DNA of the donor, CC and surrogate.

The results are shown in the table.

DNA variable region	Donor DNA (size in base pairs)	CC DNA (size in base pairs)	Surrogate DNA (size in base pairs)
1	164/164	164/164	166/166
2	222/222	222/222	218/218
3	196/198	196/198	194/200
4	154/160	154/160	160/162

c. For each region of DNA there are two values, for example, 164/164. Suggest a reason for this. **1 mark**
d. In the case of DNA variable region 4 in the donor DNA, why are the pairs of values different? **1 mark**
e. From the data it was concluded that CC was a true clone. Explain the evidence in the table that supports this claim. **1 mark**

More exam questions are available in your learnON title.

9.5 Review

9.5.1 Topic summary

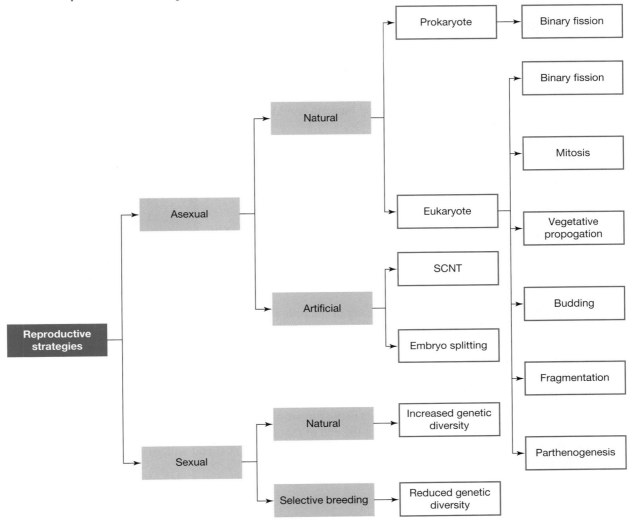

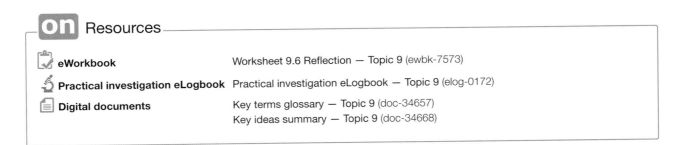

9.5 Exercises

To answer questions online and to receive **immediate feedback** and **sample responses** for every question, go to your learnON title at **www.jacplus.com.au**. A **downloadable solutions** file is also available in the resources tab.

9.5 Exercise 1: Review questions

1. Refer to the figure which shows binary fission in the bacterial species *E. coli*. Assume that the time required for binary fission in these bacteria is an average of 20 minutes at 15 °C.

   a. Starting with 10 bacteria, about how many bacteria would be present at the end of 2 hours of binary fission?
   b. The rate of binary fission doubles (or halves) for each rise (or fall) of 10 degrees in temperature. Assume that, at 25 °C, binary fission becomes twice as fast and is completed in 10 minutes. Also assume that, at 5 °C, binary fission is slowed and requires 40 minutes for completion.
      i. At 25 °C, starting with 10 bacterial cells, how many bacterial cells would be present after 2 hours?
      ii. At 5 °C, starting with 10 bacterial cells, how many bacterial cells would be present after 2 hours?

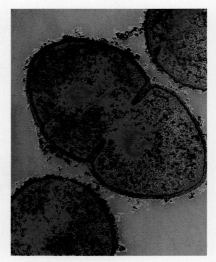

2. Give an explanation in biological terms for each of the following observations.
   a. Cooked meats should be stored in a refrigerator, rather than at room temperature.
   b. After a bushfire, one of the first plants to reappear is the Austral bracken (*Pteridium esculentum*).
   c. An amoeba can be considered to be immortal (unless it is eaten by a predator).
   d. Populations of some species of whiptail lizard are entirely female.
   e. Some species can reproduce without fertilisation.

3. A means of reproduction used by some plants is the formation of underground bulbs. Bulbs are formed by parent plants during a growing season and, at the end of that time, the parent plants typically die back. Examine the figure that shows the typical structure of a bulb. The apical bud will give rise to leaves and a flower; the lateral buds will produce shoots.

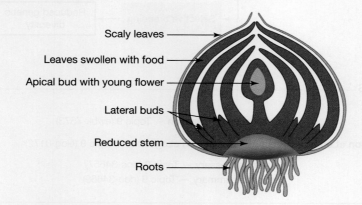

   a. What parts of the new plant are present in a bulb?
   b. What mode of reproduction does bulb formation exemplify?
   c. What role do the leaves within a bulb serve?
   d. What cellular process will transform the embryonic plant within the bulb into a mature plant?
   e. Give the name of a bulb that might be part of a meal — raw or cooked.

4. Cultured plants can be transported from country to country. As they are grown in sterile conditions, there is very little chance of them being infested with pest organisms or being kept in quarantine. With introduced species upsetting the natural balance of ecosystems, should cultured plants be used to repopulate areas affected by natural events? Discuss.

5. Plesiosaurs are extinct reptiles that lived in the sea. What prediction, if any, can be made about their manner of reproduction?

6. External fertilisation in a marine environment frees a sexually reproducing organism from having to find, court and secure a potential mate.
    a. What is the energy downside to this type of external fertilisation?
    b. List three different ways in which gametes can be made to meet.

7. Cloning of mammals using the technique of somatic cell nuclear transfer has a very low success rate. One study has identified the success rate as between 0.1 and 3.0 per cent.
    a. Consider the various steps involved in this technique, and identify at what points the technique could fail.
    b. Use the success rate given above to identify the maximum and minimum number of successful live births expected if 1000 nuclear transfers were carried out.
    c. What were the success rates for the following cloned mammals?
        i. Snuppy the dog
        ii. Dolly the sheep
        iii. Second Chance the bull
        iv. CC the cat
        v. Zhong Zhong and Hua Hua, the macaques

9.5 Exercise 2: Exam questions

**Resources**

▶ **Teacher-led videos** Teacher-led videos for every exam question

Section A — Multiple choice questions

All correct answers are worth 1 mark each; an incorrect answer is worth 0.

▶ Question 1

Which statement correctly describes the offspring produced by sexual reproduction?

A. They are genetically varied.
B. They are clones of the parent.
C. They have a haploid set of chromosomes.
D. They are not able to produce gametes.

▶ Question 2

Source: VCAA 2008 Biology Exam 2, Section A, Q7

An amoeba only reproduces asexually.

Variation in an amoeba most commonly occurs through

A. mutation.
B. random fusion of gametes.
C. chiasmata formation and crossing over.
D. independent assortment of chromosomes.

Question 3
Which of the following is *not* true of asexual reproduction?

A. It gives the ability to repopulate an area quickly.
B. Time is not wasted looking for a mate.
C. Offspring are well-suited to the current environment.
D. Offspring have significant genetic variation.

Question 4
Which of the following organisms is able to reproduce through budding?

A. Starfish
C. *Hydra*
B. Bacteria
D. Komodo dragon

Question 5
What is the genotype of the sex chromosomes of Komodo dragons born via parthenogenesis?

A. ZZ
C. WW
B. ZW
D. XY

Question 6
The point in a chromosome where crossing over occurs is called the

A. chiasma.
C. kinetochore.
B. telomere.
D. sister chromatid.

Question 7
Which of the following is true of internal fertilisation?

A. The chances of fertilisation occurring is lower than for external fertilisation.
B. A high number of gametes are wasted.
C. It occurs more frequently in marine species.
D. A small number of eggs are produced.

Question 8
What conditions do fungal spores require if they are to germinate and produce a mycelium?

A. Cool and dry
B. Warm and dry
C. Warm and moist
D. Any conditions — fungus can grow anywhere

Question 9
Aphids are able to reproduce both asexually and sexually.

What advantage do they derive from sexual reproduction?

A. Genetic variation in their offspring
B. A greater chance of their offspring surviving should conditions change
C. Increasing the number of alleles available in the population for natural selection to act upon
D. All of the above

Question 10
Which of the following organisms are created as a result of gametes fusing?

A. Bacteria
B. Cloned sheep
C. Bracken plants growing from rhizomes
D. Two puppies born in the same litter

Section B — Short answer questions

Question 11 (6 marks)

E. coli is a rod-shaped prokaryotic organism that is normally found in the intestinal tract of humans and animals. There are over 700 known strains of *E. coli*, and it can grow either with or without oxygen. It is also commonly used to produce insulin for human use. When conditions are favourable, *E. coli* is able to undergo rapid cell division.

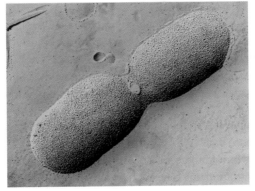

a. Name the process by which *E. coli* replicates. **1 mark**
b. Describe the steps that occur during the replication of *E. coli*. **3 marks**
c. Provide two pieces of evidence that show that the process of cell division in prokaryotes differs from somatic cell division in eukaryotes. **2 marks**

Question 12 (6 marks)

The occurrence of facultative parthenogenesis in some reptile populations was unexpected; for example, in the case of the Komodo dragon.

a. What is facultative parthenogenesis? **2 marks**
b. How does it differ from obligate parthenogenesis? **1 mark**
c. Identify an example of a reptile that is an obligate parthenogen. **1 mark**
d. A female Komodo dragon in a zoo has not had recent contact with any males. She produces some eggs, and a live baby dragon hatches from one of the eggs. One person said that this birth could be explained as being due to sperm that was stored in the female's reproductive tract. Another person said that the birth is a case of parthenogenesis. What data would be needed to decide whether or not the baby dragon was the result of a true 'virgin birth'? **2 marks**

Question 13 (7 marks)

A Year 11 Biology student was interested in growing new plants from cuttings. In his garden he had a geranium plant, and he wanted to see if he could propagate a new plant genetically identical to the one in his garden. He wanted to make sure that he had the correct conditions to grow his plant in, so he decided to test the effect of different soil pH on geranium plants.

a. For the student's experiment, identify:
   i. the independent variable **1 mark**
   ii. the dependent variable **1 mark**
   iii. controlled variables. **1 mark**
b. To obtain his results, the student measured the height of each plant every five days. What type of data had he collected? Explain your reasoning. **2 marks**
c. After his cuttings had been growing for ten days, he realised that he had been watering one plant with twice as much water as the other 19 plants. What type of error is this? Provide a reason. **2 marks**

Question 14 (14 marks)

a. Define the term *cloning* in a reproductive sense.   **1 mark**

b. A somatic cell was taken from a bull (Benny) and its nucleus was successfully fused with the fertilised egg from a cow (Jenny) from which the nucleus had been removed. The egg was then transferred into the uterus of another cow (Minnie). The egg developed successfully and, months later, Minnie gave birth to a calf. Explain whether this newborn calf should be called Lenny (for a bull) or Linda (for a cow)?   **2 marks**

c. What inherited traits, if any, would the calf have inherited from Minnie? Explain.   **2 marks**

d. The diploid number of a cow or a bull is $2n = 60$. How many chromosomes were present in:
   i. the somatic cell taken from Benny   **1 mark**
   ii. the fertilised egg from Jenny, before the nucleus had been removed   **1 mark**
   iii. the fertilised egg from Jenny, after its nucleus had been removed but before the somatic cell nucleus from Benny was fused with it   **1 mark**
   iv. the fertilised egg from Jenny after the somatic cell nucleus from Benny was fused with it   **1 mark**
   v. each somatic cell of the newborn calf?   **1 mark**

e. Suggest two reasons why scientists or farmers would wish to clone farm animals, such as cattle.   **2 marks**

f. Give two arguments against cloning farm animals, such as cattle.   **2 marks**

Question 15 (3 marks)

Scientists are attempting to refine cloning methods to produce animals for medical research. For example, cloned mice could be used to trial new drugs for disease treatment.

One method of cloning is embryo splitting. In this method, a few cells are removed from an embryo at a very early stage of its development. Each removed cell is placed in a separate test tube. Given the correct nutrients, these cells divide and form new embryos. The newly formed embryos are then implanted into the uterus of a surrogate mother mouse. Six weeks later the cloned mice are born.

a. Consider the cloned mice born from one treatment of embryo splitting. How do the cloned mice compare to each other in their genetic make-up?   **1 mark**

b. What is the main advantage of using cloned mice in controlled experiments to trial drug treatments, rather than mice produced by normal sexual reproduction?   **1 mark**

c. Suggest one disadvantage of using cloned mice in experiments to trial drug treatments.   **1 mark**

9.5 Exercise 3: Biochallenge online only

on Resources

 eWorkbook Biochallenge — Topic 9 (ewbk-8079)

 Solutions Solutions — Topic 9 (sol-0654)

teach on

Test maker
Create unique tests and exams from our extensive range of questions, including practice exam questions. Access the assignments section in learnON to begin creating and assigning assessments to students.

Online Resources

Below is a full list of **rich resources** available online for this topic. These resources are designed to bring ideas to life, to promote deep and lasting learning and to support the different learning needs of each individual.

eWorkbook

- 9.1 eWorkbook — Topic 9 (ewbk-3170)
- 9.2 Worksheet 9.1 Types of asexual reproduction (ewbk-7561)
  Worksheet 9.2 Advantages and disadvantages of asexual reproduction (ewbk-7563)
- 9.3 Worksheet 9.3 Comparing fertilisation methods (ewbk-7565)
  Worksheet 9.4 Advantages and disadvantages of sexual reproduction (ewbk-7567)
- 9.4 Worksheet 9.5 Reproductive cloning technologies (ewbk-7569)
- 9.5 Worksheet 9.6 Reflection — Topic 9 (ewbk-7573)
  Biochallenge — Topic 9 (ewbk-8079)

Solutions

- 9.5 Solutions — Topic 9 (sol-0654)

Practical investigation eLogbook

- 9.1 Practical investigation eLogbook — Topic 9 (elog-0172)
- 9.2 Investigation 9.1 Vegetative propagation — reproduction without sex (elog-0838)
  Investigation 9.2 Asexual reproduction in yeast and amoeba (elog-0840)
- 9.4 Investigation 9.3 Plant tissue cultures (elog-0842)

Digital documents

- 9.1 Key science skills — VCE Biology Units 1–4 (doc-34648)
  Key terms glossary — Topic 9 (doc-34657)
  Key ideas summary — Topic 9 (doc-34668)

Teacher-led videos

- Exam questions — Topic 9
- 9.2 Sample problem 1 The symbiotic relationship between bacteria and humans (tlvd-1732)
  Sample problem 2 Binary fission in prokaryotes and eukaryotes (tlvd-1733)
- 9.3 Sample problem 3 Advantages of 'sneaker' fish (tlvd-1734)
- 9.4 Sample problem 4 Types of cloning (tlvd-1735)

Video eLessons

- 9.2 Binary fission (eles-2464)
  Amoeba (eles-2694)
  Euglena (eles-2695)
- 9.4 Somatic cell nuclear transfer (eles-2465)

Weblinks

- 9.2 How some animals have 'virgin births'
  Birds Do It. Bees Do It. Dragons Don't Need To.
- 9.4 20 Years after Dolly the Sheep Led the Way — Where is Cloning Now?

Teacher resources

There are many resources available exclusively for teachers online

To access these online resources, log on to **www.jacplus.com.au**.

AREA OF STUDY 2 HOW DO INHERITED ADAPTATIONS IMPACT ON DIVERSITY?

10 Adaptations and diversity

KEY KNOWLEDGE

In this topic you will investigate:

Adaptations and diversity
- the biological importance of genetic diversity within a species or population
- structural, physiological and behavioural adaptations that enhance an organism's survival and enable life to exist in a wide range of environments
- survival through interdependencies between species, including impact of changes to keystone species and predators and their ecological roles in structuring and maintaining the distribution, density and size of a population in an ecosystem
- the contribution of Aboriginal and Torres Strait Islander peoples' knowledge and perspectives in understanding adaptations of, and interdependencies between, species in Australian ecosystems.

Source: VCE Biology Study Design (2022–2026) extracts © VCAA; reproduced by permission.

PRACTICAL WORK AND INVESTIGATIONS

Practical work is a central component of learning and assessment. Experiments and investigations, supported by a **practical investigation eLogbook** and **teacher-led videos**, are included in this topic to provide opportunities to undertake investigations and communicate findings.

10.1 Overview

Numerous **videos** and **interactivities** are available just where you need them, at the point of learning, in your digital formats, learnON and eBookPLUS at **www.jacplus.com.au**.

10.1.1 Introduction

Since the formation of the Earth, the environment has been constantly changing. From the Neoproterozoic era (about 550 million years ago) until the Carboniferous period (about 335 million years ago) a supercontinent existed called Gondwana, which dominated the southern half of the Earth and included what is now Australia.

FIGURE 10.1 The widespread distribution in the fossil record of *Glossopteris* is significant in the reconstruction of the supercontinent Gondwana.

Evidence for the existence and subsequent break-up of Gondwana comes from fossil evidence of the tree genus *Glossopteris*, which has been found across all of the now detached southern continents of South America, Africa, India, Australia, New Zealand, and Antarctica. *Glossopteris* was a genus of seed-bearing trees or shrubs. Although no whole tree fossils have been discovered, it is thought they grew in very wet conditions, with specially adapted aeration roots to allow them to grow in marshy conditions. Research into species of *Glossopteris* has shown both their adaptability, with diverse pollen assemblages, and their role in supporting diverse forest communities. They existed for nearly 50 million years as the dominant plant of Gondwana, and Gondwana was named after the region of India where *Glossopteris* were found.

Organisms, such as *Glossopteris*, that existed for long periods of time and survived this separation of land — and subsequent movement to different regions of the Earth — demonstrate the adaptability of organisms to survive and flourish in changing ecosystems. In this topic you will explore adaptations of both plants and animals, and the importance of genetic diversity to allow species survival and adaptations to their environment.

LEARNING SEQUENCE

10.1 Overview	562
10.2 Importance of genetic diversity to survive change	563
10.3 Adaptations for survival	567
10.4 Survival through interdependence between species	593
10.5 Australia's First Peoples	627
10.6 Review	634

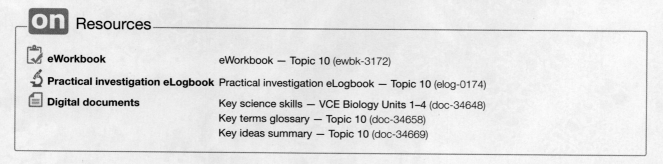

10.2 Importance of genetic diversity to survive change

> **KEY KNOWLEDGE**
>
> - The biological importance of genetic diversity within a species or population
>
> *Source:* VCE Biology Study Design (2022–2026) extracts © VCAA; reproduced by permission.

10.2.1 Biodiversity versus genetic diversity

Biodiversity is the variety of all living organisms on planet Earth. It includes the different animals, plants, fungi, protists and microbes (bacteria and archaea); the genetic information that they contain; and the **ecosystems** of which they form a part. Biodiversity is essentially linked to the physical settings in which organisms live and with which they interact.

Biodiversity may be identified at three levels:
- **Genetic diversity** refers to the variety of genes or the number of different inherited characteristics present in a species. Different **populations** of the same species may have different levels of genetic diversity. Populations with a high level of genetic diversity are more likely to have some individuals that can survive and reproduce when environmental conditions change. Genetic diversity is created through mutations, or changes, in an organism's DNA that are passed on to the next generation.
- **Species diversity** refers to the variety of different kinds of organism living in a particular habitat or region; for example, the diversity of plant species of the Gibson Desert.
- **Ecosystem diversity** refers to the variety of physical environments in which organisms live and with which they interact. They range from biologically rich ecosystems, such as coral reefs and rainforests, to biologically sparse ecosystems, such as the Antarctic landmass.

10.2.2 The importance of genetic diversity

Genetic diversity is created when mutations, or changes, arise in the DNA during meiosis in sexually reproducing organisms or during mitosis in asexually reproducing organisms. The word *mutation* is often perceived as something that will be detrimental to an organism. However, biologically, mutations can be advantageous to both an individual organism and a species. Inherited mutations, which are those that are passed on to future generations, give rise to new alleles that provide variation within populations.

When a mutation occurs in DNA, it is like changing a letter in a word that alters the meaning of that word. In DNA, changing a nucleotide can alter the way that it codes for a specific gene. These different forms of genes are called alleles. Mutations can be either small or large. A substitution mutation is when a single nucleotide is changed — for example, swapping a thymine for a guanine. Large deletions can occur when whole sections of chromosomes change places. The deletion of one nucleotide causes a significant impact on the gene being expressed, as every nucleotide after that is read incorrectly. This is called a frameshift mutation.

For an individual within a species, an advantageous inherited mutation increases their chance of survival. The longer an individual can survive, the greater the likelihood of them reproducing and passing the advantageous allele on to the next generation.

biodiversity total variety of life forms, their genes and the ecosystems of which they are a part; biodiversity can relate to planet Earth or to a given region

ecosystems biological units comprising the community living in a discrete region, the nonliving surroundings and the interactions occurring within the community and between the community and its surroundings

genetic diversity the variety of genes or the number of different inherited characteristics present in a species

population members of one species living in one region at a particular time

species diversity the number of different species — that is, different populations — in a community

ecosystem diversity the variety of physical environments in which organisms live and with which they interact

FIGURE 10.2 Examples of mutations

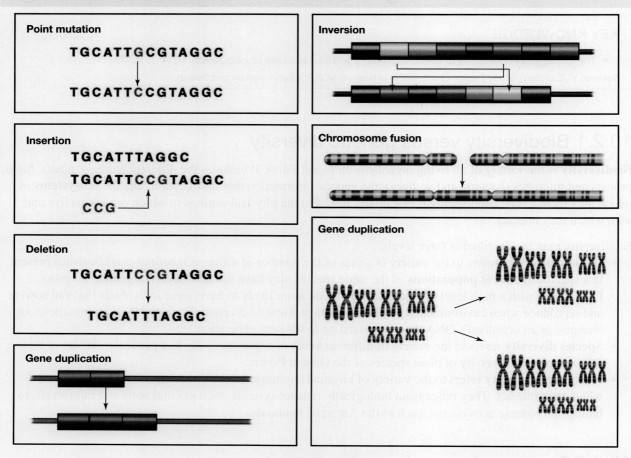

Genetic diversity:
- increases species survival following environmental change
- increases the number of individuals that can survive and reproduce
- increases biodiversity in ecosystems
- means natural selection can act, as there are many alleles to select from.

With Australia prone to natural disasters — such as bushfires, flood and droughts — genetic diversity is essential to ensure that some members of the species are able to survive and repopulate an area.

CASE STUDY: Eucalypt regeneration after bushfire

Many of the 700 species of *Eucalyptus* trees are native to Australia. Just below the soil, they have a swelling called a lignotuber (figure 10.3). The lignotuber holds dormant buds that are activated by fire. Figure 10.3 was taken four years after the 2009 Kinglake fires in Victoria. It shows a new stem emerging from bud development in a eucalypt lignotuber. In addition, the lignotuber also protects the stem of the plant from fire and contains stores of starch. This enables the plant to regenerate after fire when it is unable to undergo **photosynthesis** due to the lack of chlorophyll-containing cells. The biological advantage of this is that the increased survival rate of the trees ensures that the gene pool of the population is retained. However, as fires are increasingly more frequent events, some trees do not have sufficient time to regenerate their dormant seed supply, and they exhaust stored starch reserves. This poses a threat to the long-term genetic diversity of trees such as the eucalypts.

FIGURE 10.3 A new stem has emerged from the lignotuber of this eucalypt.

photosynthesis process by which plants use the radiant energy of sunlight trapped by chlorophyll to build carbohydrates from carbon dioxide and water

---

tlvd-1826

SAMPLE PROBLEM 1 The importance of lignotubers to maintain genetic diversity of a species

Australian eucalypts hold dormant buds in lignotubers that are stimulated to germinate following a fire event. How does the presence of these dormant buds maintain the diversity of the species? **(2 marks)**

THINK	WRITE
1. In *how* questions you need to address what it does; in this case, what is the role of the lignotubers?	The lignotubers hold buds that contain genetic information from the parent plant (1 mark).
2. Respond to 'maintain the genetic diversity'.	This ensures that in the event of a fire destroying parts of the tree, the genetic information is maintained in the dormant buds (1 mark).

KEY IDEAS

- Biodiversity is the variety of all living organisms.
- Biodiversity can refer to genetic diversity, species diversity or ecosystem diversity.
- Genetic diversity is the variation in alleles within a population.
- Increased diversity increases the chance of a species surviving.
- Chance events can reduce biodiversity in gene pools.

 Resources

 eWorkbook Worksheet 10.1 The importance of genetic diversity (ewbk-7597)

 Weblink Animals' response to bushfires

10.2 Activities

To answer questions online and to receive **immediate feedback** and **sample responses** for every question, go to your learnON title at www.jacplus.com.au. A **downloadable solutions** file is also available in the resources tab.

| 10.2 Quick quiz | 10.2 Exercise | 10.2 Exam questions |

10.2 Exercise

1. Distinguish between genetic diversity, species diversity and ecosystem diversity.
2. How does genetic diversity increase a species' chance of survival?
3. a. How do eucalypts regenerate after a bushfire?
   b. How does the regeneration of eucalypts benefit other native species?
4. Australia is moving north at a rate of 7 cm per year. Assuming this continues, describe what you believe the Victorian environment will be like in 150 years.

10.2 Exam questions

Question 1 (1 mark)
MC Species diversity refers to the
A. number of individuals within a species.
B. variety of species and the relative abundance of each species within a given habitat.
C. variety of genes within a species.
D. distribution of a species across the world.

Question 2 (1 mark)
Source: VCAA 2009 Biology Exam 2, Section A, Q11
MC A mutation is
A. a product of natural selection.
B. caused by immigration and emigration.
C. a change in an allele due to a change in DNA.
D. a random change in gene frequencies from one generation to the next.

Question 3 (1 mark)
MC In a closed population in the wild with no inward migration, new alleles can appear as a result of
A. recombination during meiosis.
B. disjunction during mitosis.
C. selection pressure.
D. gene mutation.

Question 4 (1 mark)
MC Four populations of rabbits are studied. The size of each population is 1000 rabbits. The phenotypes seen in each rabbit population and the ratios of each phenotype within each population are shown.

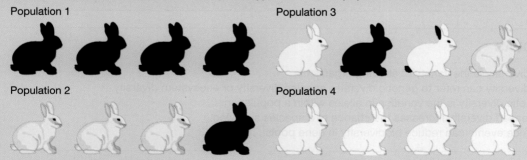

Which rabbit population is most likely to survive environmental change?
A. Population 1
B. Population 2
C. Population 3
D. Population 4

Question 5 (1 mark)
Source: VCAA 2019 Biology Exam, Section A, Q26

MC Consider the diagram below showing the gene pool of a population over 20 generations.

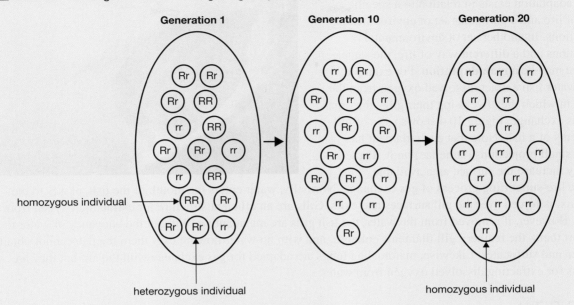

It would be correct to conclude that, over the 20 generations
A. genetic diversity is increasing in this population.
B. individuals with the genotype RR had a selective advantage in this population.
C. the frequency of each allele is equal in Generation 1 but not in other generations.
D. new advantageous alleles for this gene were introduced as individuals joined this population.

More exam questions are available in your learnON title.

10.3 Adaptations for survival

KEY KNOWLEDGE

- Structural, physiological and behavioural adaptations that enhance an organism's survival and enable life to exist in a wide range of environments

Source: VCE Biology Study Design (2022–2026) extracts © VCAA; reproduced by permission.

10.3.1 Types of adaptations

In this subtopic we will look at a range of **adaptations** that contribute to the survival of animals and plants in a variety of environments.

adaptation features or traits that appear to equip an organism for survival in a particular habitat

Adaptations

An adaptation is a structural, behavioural or physiological feature that enhances the survival of an organism in particular environmental conditions.

Adaptive features of an organism are innate — that is, they are built into its genetic make-up.

It is important to note that the value of a feature as an adaptation exists in relation to a specific way of life and a particular set of environmental conditions. In another set of environmental conditions and a different way of life, the same feature may be a **maladaptation**. For example, freshwater fish extract dissolved oxygen from the water in which they live using their gills, the organ for gas exchange. Figure 10.4 shows rows of gill filaments of a fish. Note that each filament has many small projections (lamellae) on it. These greatly increase the surface area available for the life-supporting process of gas exchange. Normally, water enters the mouth of the fish, passes to the pharynx, is forced over the gill surfaces and exits. Gills are an efficient structure for extracting oxygen from water. However, if removed from the water, the fish gills are maladaptive. Without the buoyancy of water to support them, the feathery gill filaments collapse, and with no water flowing over them the fish cannot obtain oxygen and suffocates. Likewise, mammalian lungs are adapted for gas exchange with the air, but they are useless for extracting dissolved oxygen from water.

FIGURE 10.4 Rows of gill filaments of a fish. In water, these are an effective way to extract oxygen from the water. Out of water, they collapse and are maladaptive.

10.3.2 Tolerance range

Every habitat contains a number of **abiotic factors** (environmental factors that are non-living). These make up the environmental conditions, such as temperature, **desiccation**, oxygen concentration, light intensity and ultraviolet exposure. The **tolerance range** for an organism identifies the variations in the particular environmental conditions (the abiotic factors) in which a particular species can successfully live and reproduce. The tolerance range includes an optimum range, and the extremes of the tolerance range are the **tolerance limits** for that abiotic factor. As the tolerance limits are approached, the species enter the *zone of physiological stress*. Figure 10.5 shows the tolerance range for a fish species in terms of water temperature. Notice that, as water temperature moves closer to the tolerance limits, fewer fish are found.

> **maladaptation** feature or trait of an organism that inhibits survival in a particular habitat
> **abiotic factor** nonliving factor, such as weather, that can affect population size
> **desiccation** drying out
> **tolerance range** extent of variation in an environmental factor within which a particular species can survive
> **tolerance limits** the upper and lower limits of a particular environmental condition in which a species can survive

FIGURE 10.5 Tolerance range of temperature for a fish species demonstrating the population changes between the optimum range, the zone of physiological stress and the zone of intolerance

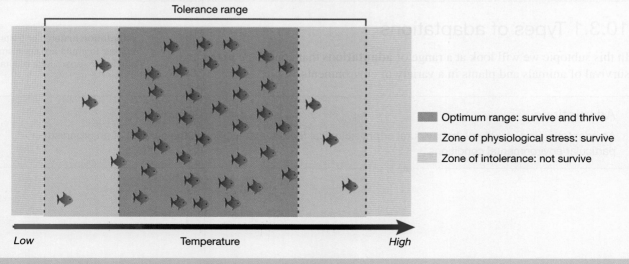

Structural, physiological and behavioural adaptations to tolerance ranges

If an abiotic factor has a value above or below the range of tolerance of an organism, that organism will not survive unless it can escape from, or somehow compensate for, the change. In some species, migration is one such escape behaviour, while others retreat underground. Tolerance ranges differ between species and are influenced by structural, physiological and behavioural features of organisms. For example, the cold tolerance of various mammals is influenced by structural features such as fur density, shape of the body and extent of insulating fat deposits; and by their behaviours, such as hibernating ('coping-with-it' strategy).

TABLE 10.1 A comparison of different types of adaptations

Type of adaptation	Definition	Example
Structural	Physical features of an organism that enable them to survive in a given environment	Blubber in seals providing a protective layer from the cold temperatures of the ocean
Physiological	Internal and/or cellular features of an organism that enable them to survive in a given environment	Vasoconstriction of blood vessels that conserves heat and increases blood pressure
Behavioural	Activities that an organism performs in response to internal and external stimuli	Huddling in penguins to stay warm, migration of birds to warmer regions over winter

In figure 10.6, can you identify which fox species is the Arctic fox (*Vulpes lagopus*) and which is the Simien fox (*Canis simensis*)? Apart from the difference in their fur thickness, these two fox species differ in their ear sizes. Smaller ears have a lower surface-area-to-volume ratio than larger ears. Which fox would be better able to conserve its body heat and tolerate lower temperatures? Why? Similarly, polar bears demonstrate a range of adaptions to survive the cold (figure 10.7).

FIGURE 10.6 Can you identify the Arctic fox (*Vulpes lagopus*) from the Simien fox (*Canis simensis*) based on their structural features?

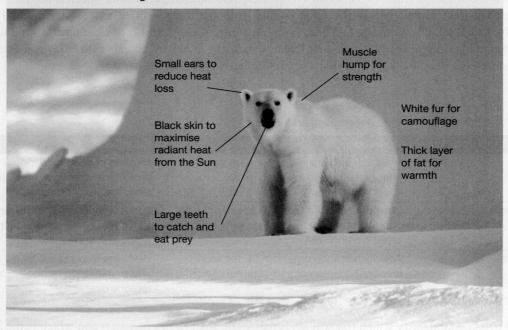

FIGURE 10.7 The polar bear has a range of adaptations to survive extreme Arctic conditions. Did you know that polar bears actually have black skin? This allows them to maximise radiant heat gain from the Sun.

Limiting factors for tolerance

Any condition that approaches or exceeds the limits of tolerance for an organism is said to be a **limiting factor** for that organism. Terrestrial and aquatic environments can differ in their limiting factors. Table 10.2 shows abiotic factors that influence which kinds of organism can survive in various habitats. Those species that can survive under certain environmental conditions have tolerance ranges that accommodate those conditions.

limiting factor environmental condition that restricts the types of organism that can survive in a given habitat

The structure and the physiology of plants and animals, and the behaviour of animals, determine their tolerance range. For each organism, the limits of its tolerance range for various abiotic factors are fixed, except for the occurrence of an enabling mutation.

TABLE 10.2 Examples of limiting abiotic factors for different habitats

Habitat	Limiting factor	Comment
Floor of tropical rainforest	Light intensity	Low light intensity limits the kinds of plants that can survive.
Desert	Water availability	Limited water supply means that only plants able to tolerate desiccation can survive.
Littoral zone (intertidal zone)	Desiccation	Exposure to air and sunlight limits the types of organism that survive.
Polar region	Temperature	Low temperatures limit the types of organism that are found.
Stagnant pond	Dissolved oxygen levels	Low dissolved oxygen levels limit the types of organism that can live there.

elog-0846

INVESTIGATION 10.1

online only

Abiotic factors and tolerance range

Aim

To measure and determine various abiotic factors in an environment and their effect on different species

CASE STUDY: Using technology to extend tolerance ranges

The human species is the only organism that makes extensive use of technology to extend the limits of its natural tolerance range. As air-breathing mammals we are not prevented from entering the watery world of the fish; technology such as scuba tanks allow us to do this. Technology enables people to survive in hostile environments on and beyond Earth, where conditions are outside the tolerance range of an unaided person. Equipment and hi-tech clothing enable a mountaineer to survive an ascent to the peak of one of the highest mountains on Earth (figure 10.8a). Extremely sophisticated transport, living quarters and spacesuits enable astronauts to live on the International Space Station and to conduct spacewalks (figure 10.8b).

FIGURE 10.8 a. Andrew Lock is Australia's most accomplished high-altitude mountaineer and the only Australian to have climbed all 14 of the world's 8000-metre-plus mountains. **b.** Expedition 61 Flight Engineers Christina Koch and Jessica Meir perform equipment lock configuration operations in the Quest Airlock in preparation for a spacewalk.

DISCUSSION

Humans use technology to enable them to climb the world's highest mountains. Do you believe that we should be able to overcome our natural limitations in this way?

 Resources

 eWorkbook Worksheet 10.2 Bills and beaks: How birds feed (ewbk-7599)
Worksheet 10.3 Types of adaptations (ewbk-7601)

10.3.3 Adaptations for survival: arid climate animals

In this section, we will look at some examples of adaptations that enable animals to survive and reproduce in desert environments/arid climates. The key environmental challenges of desert life are avoiding excessive water loss that can result in dehydration and avoiding overheating that can result in **hyperthermia**, both conditions being potentially deadly.

Humans, like most mammals, replace most of the water that they lose by drinking liquid water. This is easy when access to clean, piped water is available. In the desert, however, **free-standing water** is neither predictably nor reliably available. After an occasional heavy rain or storm, temporary creeks, transient lakes and small pools of water exist in the desert. For most of the time, however, creek beds are dry, lakes are dry saltpans, and pools and puddles of water do not exist. This lack of free-standing water in the desert can persist for years, even decades. How is survival possible for animals and plants under these conditions?

Survival without drinking: the tarrkawarra

Let's meet a water saver. The tarrkawarra, or spinifex hopping mouse (*Notomys alexis*), is a placental mammal that lives in sandy deserts in Australia (figure 10.9). Because it can survive without drinking liquid water, the tarrkawarra can endure long periods of drought. Its kidney tubules reabsorb almost all the water from the kidney filtrate so that it produces highly concentrated and almost solid urine. In fact, tarrkawarras produce the most concentrated urine of any mammal. Their kidneys can produce urine with a concentration of 9370 mOsm/L.

Table 10.3 shows a comparison of the maximum concentrating abilities of the kidneys of various mammals. This table also shows the urine-to-plasma (U/P) ratio; that is, the concentration of electrolytes — such as sodium, potassium and chloride ions — in the **plasma** relative to that in the urine.

> **hyperthermia** condition in which core body temperature exceeds the upper end of the normal range without any change in the temperature set point
>
> **free-standing water** water available for an animal to use, including to drink
>
> **plasma** the fluid portion of blood in which blood cells are suspended
>
> **water balance** the way in which an organism maintains a constant amount of internal water in hot or dry conditions

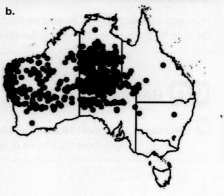

FIGURE 10.9 a. The tarrkawarra produces the most concentrated urine of any mammal. **b.** It lives in arid and semi-arid regions of Australia.

TABLE 10.3 Comparison of maximum concentration of urine in several mammals (max. mOsm/L) and maximum ratio of electrolyte concentrations in the urine versus in the blood plasma (max. U/P)

Species	Max. mOsm/L	Max. U/P
Human	1200	4
Dog	2500	7
Camel	2800	8
Rat	2900	9
Sheep	3500	11
Tarrkawarra	9000	25

How the tarrkawarra manages its **water balance** (intake and loss) to ensure its metabolic processes proceed is summarised in table 10.4 and figure 10.10.

FIGURE 10.10 An outline of how the tarrkawarra achieves water balance. For survival, water-in must balance water-out.

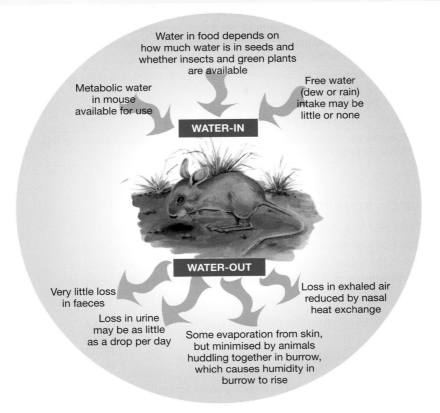

TABLE 10.4 How the tarrkawarra regulates its water balance

Source	Mode	Water gain or loss
Food	Water enters from the gut when food is digested in the small intestine. Water then travels via the bloodstream to **interstitial fluid** and cells. It enters cells via specialised channel proteins called aquaporins. Adaptation: The main food is dry seeds, with water content dependent upon the **humidity** of the air. Nocturnal habits ensure seed collection when humidity is highest, and storage in deep burrows with high humidity occurs due to animals huddling in burrows.	Gain
Metabolic water	Aerobic respiration produces water as a by-product. Rather than excreting this water, it is used throughout the body. Adaptation: Will drink free water if available but can survive on metabolic water.	Gain
Skin	There are no sweat glands, but some water is lost from diffusion through the skin. Adaptation: To minimise water loss, it burrows during the day to create a humidity chamber. The humid environment reduces the rate of evaporation.	Loss
Faeces	Faeces are very dry to reduce water loss through waste products.	Loss

(continued)

interstitial fluid fluid that fills the spaces between cells and bathes their plasma membranes

humidity measure of the amount of water vapour in the atmosphere

TABLE 10.4 How the tarrkawarra regulates its water balance (continued)

Source	Mode	Water gain or loss
Exhaled air	A special heat exchange system is used in the nostrils. Exhaled air is lower in temperature than body temperature. As the air is cooled, some of the water vapour from the lungs recondenses on the walls of the nasal passage, reducing water lost to the environment.	Loss
Urine	Nitrogenous waste must be removed from the body, and this is done via urine. Adaptation: The kidneys produce the most concentrated urine of any mammal, which conserves significant amounts of water.	Loss
Milk	The mother drinks the urine of her young, recycling the water content of this urine. This assists to replenish the water lost through milk production.	Loss

Survival by dormancy: frogs in the outback

Some frog species live in arid inland Australia. Frogs typically live in moist surroundings and need a body of water in which to reproduce. How do they survive long periods of drought in the inland? Some frog species that live near and breed in **ephemeral** waterholes respond in an amazing manner when the waterholes begin to dry out.

ephemeral short-lived or lasting a short period of time

dormancy condition of inactivity resulting from extreme lowering of metabolic rate in an organism

Frogs such as the trilling frog (*Neobatrachus centralis*) (figure 10.11) survive by doing the following:

- They burrow deeply into the soft mud at the bottom of their waterholes.
- Once underground at depths of up to 30 cm, they make a chamber that they seal with a mucous secretion.
- The frogs then go into an inactive state known as **dormancy**, in which both breathing and heart rate are minimal and energy needs are greatly reduced. Their low energy requirements are met from their fat reserves.
- They remain buried and are protected from desiccation until the next rains come; this may be a wait of one or two years.
- They come out of their dormant state only when soaking rains fall and soil moisture rises. Once activated, the frogs return to the surface to feed and breed in temporary pools.
- The completion of the life cycle is very fast. Within days of being laid, eggs undergo embryonic development, hatch, and the resulting tadpoles metamorphose to produce small frogs. These new populations of frogs feed on larvae of crustaceans and insects that have also hatched from dormant eggs.

FIGURE 10.11 a. The trilling frog (*Neobatrachus centralis*) **b.** Distribution map of the trilling frog in Australia

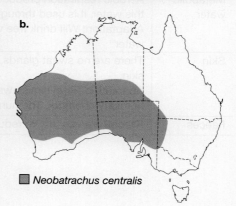

■ *Neobatrachus centralis*

Read the account written by two explorers about burrowing frogs:

> One day during the dry season we came to a small clay-pan bordered with withered shrubs … It looked about the most unlikely spot imaginable to search for frogs, as there was not a drop of surface water or anything moist within many miles …

The ground was as hard as a rock and we had to cut it away with a hatchet, but, sure enough, about a foot [30 cm] below the surface, we came upon a little spherical chamber, about three inches [76 mm] in diameter, in which lay a dirty yellow frog. Its body was shaped like an orange … with its head and legs drawn up so as to occupy as little room as possible. The walls of its burrow were moist and slimy … Since then we have found plenty of these frogs, all safely buried in hard ground.

Source: WB Spencer and FJ Gillen, *Across Australia*, Macmillan and Co., London, 1912.

Other animal species survive extended periods of drought by sealing themselves off from the drying conditions. For example, the univalve (one-shell) freshwater mollusc *Coxiella striata* seals itself inside its shell by closing the shell opening with a hard lid (**operculum**). These inland molluscs must stay sealed tightly in their shells for months or years.

> **operculum** in fish, the flaps covering the gills; in molluscs, a hard impermeable lid that closes the shell, making a watertight compartment for the animal inside

Survival by migration

Some species cope with drought by moving from affected areas to areas where conditions are more favourable. For example, banded stilts (*Cladorhynchus leucocephalus*) live near salt lakes in inland Australia and rely on these lakes for brine shrimps, which are their main food source (figure 10.12). When one salt lake dries up, these birds simply fly to another salt lake.

Another species that moves widely throughout desert areas is the budgerigar (*Melopsittacus undulatus*). Flocks of these birds move to more favourable areas in search of food and water (figure 10.13). In order to avoid the desert heat, they travel in the cooler periods of the day.

The strategy of moving quickly over large distances to seek out transient free-standing water in the arid outback of Australia is largely restricted to birds that are capable of flight. Many animal species and all plant species, however, cannot use a 'get-up-and-go' strategy in periods of drought.

FIGURE 10.12 A group of banded stilts. What strategy do they use when their salt-lake habitat dries up?

FIGURE 10.13 The Australian budgerigar lives in large flocks in the arid inland of Australia and moves to seek food and water.

Survival by reproduction

Survival can be viewed in terms of the successful survival of an individual organism that lives to reproduce on many occasions. It can also be considered in terms of survival of a species. Members of some species found in waterholes in the arid outback cannot survive long periods of drought. When the waterhole dries up, all the organisms die. Yet, these species are successful residents of the arid outback. How is this achieved?

In this case, the species survive through their offspring, as demonstrated by the crustaceans fairy shrimps and shield shrimps (figure 10.14).

When water is present in abundance, female shrimps produce eggs that are not drought resistant. Newly hatched shrimp mature and reproduce within a few days.

When water is scarce and waterholes begin to dry out, female shrimps produce drought-resistant fertilised 'eggs', which are actually cysts that each contain a fully developed embryo encased in a hard, protective shell. While the male dies after mating, the female will carry the cysts in a drought-resistant brood sac. Before the water has dried up, the female will release the cysts and then die. By the time the water has gone, all the adult shrimps are dead, but the cysts they have left behind can withstand desiccation for long periods. These cysts are in a state of dormancy and can lie in the dust of dry waterholes for more than 20 years. The next generation of shrimp will emerge only when the rains come, perhaps years later, and short-lived waterholes and pools reappear.

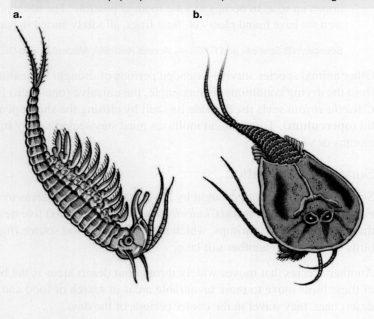

FIGURE 10.14 a. A fairy shrimp (*Branchinella* sp.) about 3 to 4 cm in length. Fairy shrimps swim with their legs uppermost. **b.** A shield shrimp (*Triops australiensis*) about 1.5 cm in length.

tlvd-1827

SAMPLE PROBLEM 2 Advantages and disadvantages of two types of egg production

To reproduce, fairy shrimp can produce eggs when water is present and cysts when water is not present. What are the advantages and disadvantages of being able to produce different types of eggs depending on the conditions? (4 marks)

THINK

1. Identify the advantages.

2. Identify the disadvantages.

WRITE

An advantage to fairy shrimp is being able to produce two types of eggs, as they can repopulate depending upon water conditions (1 mark). During dry times the gene pool can be conserved through the creation of cysts, allowing future organisms to be created (1 mark).

A disadvantage of cysts is that they may be eaten, preventing reproduction of the species and possible extinction (1 mark). There is also additional energy outlay in generating two types of eggs rather than one type of egg (1 mark).

CASE STUDY: What about the camel?

Camels, both the dromedary (*Camelus dromedarius*) and the Bactrian (*C. bactrianus*), are known as 'the ships of the desert'. Camels are large placental mammals. They are not native to Australia, but large feral herds of camels — mainly dromedaries — live in arid areas of Australia, being descendants of camels imported in the 1800s. What adaptations do camels possess that enable them to survive in a desert environment?

FIGURE 10.15 The long eyelashes and slit-shaped nostrils of the camel enable it to survive in the arid inland of Australia.

Structural features that enable camels to survive in desert conditions include:
- a double row of long eyelashes and slit-shaped nostrils that can be closed — both features protecting the camel from windborne sand particles (figure 10.15)
- bony structures in their nasal passages that enable the water vapour in their outgoing breath to be absorbed; it is then exhaled as dry air
- oval-shaped red blood cells; this enables the cells to continue circulating even when the viscosity (thickness) of the blood increases due to the camel becoming dehydrated and losing body water
- the ability to produce very dry faeces because of a long colon in their gut
- the inbuilt fat store in the hump; this is also a physiological adaption. As the fat stores are metabolised for energy production when food is scarce, the reactions also produce metabolic water for the camel.

Physiological adaptations that minimise water loss in camels include:
- the ability to produce concentrated urine because of efficient kidneys; the urine they do produce is released and runs down their legs, and its evaporation cools them
- the ability to allow their body temperature to vary over a wide range, from 34 to 42 °C, depending on the external temperature (unlike other mammals that maintain their internal body temperature within a narrow range). When it is hot during the day, the camel's body temperature rises. Sweating comes into play only when the camel's body temperature reaches 42 °C. (This is an important water conservation measure because sweating involves significant water loss.) At night, when it is cooler, this body heat is lost and the camel's body temperature falls.

Camels can lose up to 40 per cent of their body water, whereas an adult human can lose only 15 per cent. When water does become available, camels are able to drink large volumes — more than 100 litres in a day. This large intake of water, however, does not cause osmotic complications; for example, a camel's red blood cells can swell to more than double their volume before they burst, while the red blood cells of other large mammals would burst before this.

10.3.4 Adaptations for survival: arid climate plants

Vegetation types of arid Australia

As discussed in topic 4, plant systems work to maintain their water balance, to ensure water-in (by roots) = water-out (at leaves). However, in arid climates, water is scarce and the supply is often unpredictable. In this section, we will discuss a number of adaptations that allow some plants to maintain their water balance and survive in arid conditions.

Figure 10.16 shows the distribution of the major vegetation types in Australia. In terms of area, the dominant vegetation type in Australia is **hummock grassland**, which covers almost a quarter of the Australian land surface, including the sandy plains and dunes of the major deserts. The second largest vegetation type by area is shrublands.

> **hummock grassland** major vegetation type dominated by spinifex grasses and occurring over one-quarter of Australia
>
> **acacia shrubland** major vegetation type dominated by mulga, a species of *Acacia*, occurring in arid inland Australia
>
> **chenopod shrubland** major vegetation type dominated by saltbushes (*Atriplex* spp.) and bluebushes (*Maireana* spp.); occurs in arid regions with salty soils

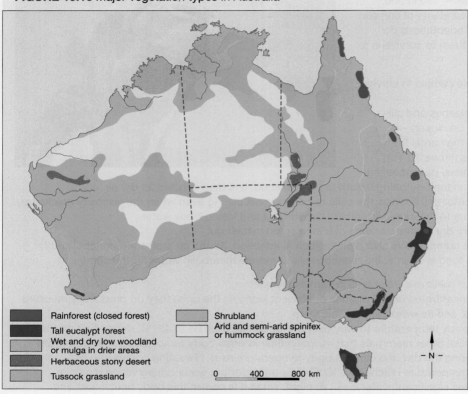

FIGURE 10.16 Major vegetation types in Australia

- Hummock grasslands are dominated by species of stiff, drought-resistant spinifex grasses (*Triodia* spp.), such as buck spinifex (*Triodia basedowii*) (figure 10.17).
- Different kinds of shrubland exist depending upon rainfall and soil type. **Acacia shrublands** occur across the arid and semi-arid areas and are dominated by mulga (*Acacia aneura*) (figure 10.18a). **Chenopod shrublands** occur in arid regions with salty soils, and are dominated by saltbushes (*Atriplex* spp.) and bluebushes (*Maireana* spp.) (figure 10.18b).

These dominant vegetation types are examined in detail in the Case study box later in this section.

Table 10.5 shows the pattern of distribution of the major vegetation types in Australia as determined by the environmental physical factors of rainfall, temperature and evaporation rate, as well as the mineral-nutrient levels (rich or poor) and salt levels of the soil (salinity).

FIGURE 10.17 Hummock grasslands are dominated by spinifex (*Triodia* spp.). Note the typical hummock shape of the spinifex plant that gives these grasslands their name.

FIGURE 10.18 Australian shrublands cover much of the continent. **a.** Acacia shrublands, dominated by mulga **b.** Chenopod shrublands, dominated by saltbushes and bluebushes

TABLE 10.5 Patterns in the distribution of various types of vegetation in Australia

Vegetation type	Climate
Hummock grasslands	Arid: lowest and erratic rainfall, high evaporation rates, high temperature Dominated by spinifex grasses (*Triodia* spp.); figure 10.17
Acacia shrublands	Arid and semi arid: low rainfall, high temperature Dominated by mulga (*Acacia aneura*); figure 10.18a
Chenopod shrublands	Arid and semi-arid: low rainfall, high temperature, salty or alkaline soils Dominated by saltbushes (*Atriplex* spp.) and bluebushes (*Maireana* spp.); figure 10.18b
Tussock grasslands	Semi-arid: annual rainfall between 200 and 500 mm, clay soils Dominated by *Astrebla* spp.
Tropical grasslands	Tropical: summer monsoons and winter drought Dominated by *Sorghum* spp.
Mallee woodlands	Temperate: intermediate rainfall, poor soil
Eucalypt forests	Temperate: high rainfall, poor soil
Rainforests	Tropical or temperate: high and reliable rainfall, rich soil

Plants that have both structural and physiological adaptations to arid conditions survive by:
- maximising water uptake
- minimising water loss
- producing drought-resistant seeds.

These adaptations result in plants that may be:

- **drought tolerant** — the plant can tolerate a period of time without water; or
- **drought resistant** — the plant can store its water and live for long periods of time without water.

Many of our Australian plants that live in water-limited environments would be classified as being drought tolerant.

drought tolerant being able to tolerate a period of time without water

drought resistant being able to store water and hence live for long periods of time without water

Adaptations: maximising water uptake

The part of a plant that takes up water is the root system. In arid areas of Australia, trees have two styles of root systems:

- **Water tappers.** Some trees growing along dry creek beds produce long, unbranched roots that penetrate to moist soil at or near the watertable. Once moisture is reached, the major root branches and forms lateral roots. The major root can grow to depths of 30 m. The part of the root that is located in the upper dry soil is covered by a corky, waterproof layer of cells that prevents water loss.
- Shallow, horizontal root systems. Other plants develop extensive root systems that spread out *horizontally*, far beyond the tree canopy but just below the soil surface. In this case, the plant takes up water from an extensive area around it.

water tappers trees that have a single main root extending to depths near the watertable before forming lateral branches

transpiration loss of water from the surfaces of a plant

stomata pores, each surrounded by two guard cells that regulate the opening and closing of the pores

cuticle waxy layer on the outer side of epidermal cells; waxy outer layer on leaves

Adaptations: minimising water loss

To minimise water loss, plants use both regulated responses and structural features.

Regulated responses to miminise water loss

Regulated responses include stomatal opening and closing and leaf rolling. As discussed in section 4.3.2 in topic 4, the major loss of water in a plant is by **transpiration**, which is controlled by the **stomata** (see figure 10.19). In a leaf, water moves from xylem into surrounding mesophyll cells. As water vapour moves out of the leaf through the stomata, water evaporates from the moist surfaces of the mesophyll cells. The stomata are typically present on the lower surface of plant leaves and this is where most water is lost, but a little also evaporates from the **cuticle**. The higher the wind speed and the higher the temperature of the leaf, the greater the rate of water loss.

For a plant to reduce its water loss, the *principal* strategy is to reduce the loss of water vapour by transpiration through closing the stomata on its leaves. Transpiration cannot be stopped permanently because it is essential for the process of moving columns of water through xylem tissue, from where water is supplied to all cells of a plant. Stomata are also the pores through which the carbon dioxide required for photosynthesis enters leaves, and they must open to allow carbon dioxide to diffuse into the leaves.

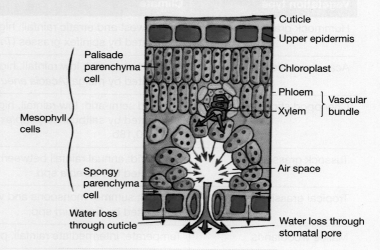

FIGURE 10.19 Water moves from xylem into surrounding moist mesophyll cells, which then evaporates out of the leaf through the stomata.

As described in topic 4, plants such as succulents can conserve water by changing when they open the stomata and undertake water-conserving photosynthesis. By opening the stomata at night, these plants take in carbon dioxide during the cool evenings and store it as an organic acid, to be released and reused for photosynthesis during the day.

Another regulated response to water deficit or heat stress is leaf rolling. Unlike stomatal regulation, the leaf rolling response is not a feature of all higher plants. It occurs in many grass species, including crops such as wheat, sorghum, rice and maize (figure 10.20). The signal for leaf rolling is a fall in soil water potential or a rise in temperature. The effector cells are a small group of large, colourless bulliform cells located on the upper epidermis.

FIGURE 10.20 Leaf rolling in maize (Zea mays)

During the daytime, as the temperature rises, maize plants lose water and their leaves begin to dehydrate. The bulliform cells quickly become flaccid. The shrinkage of these cells causes the outer edges of the leaf to be pulled inwards, causing the leaf to roll. When the temperature falls, turgor pressure is restored in the bulliform cells, they expand and the leaf unrolls. Leaf rolling reduces or prevents water loss because it creates a humidity chamber and reduces the exposure of the stomata to the wind and Sun.

Structural features to miminise water loss

Plants also have many structural features that prevent or minimise water loss. These structural features are constitutive and genetically determined traits that are present in plants regardless of whether the plant is in water deficit. However, when plants are in water deficit, these structural features reduce or prevent water loss from leaves, and are summarised in table 10.6.

TABLE 10.6 Plant adaptations to prevent water loss

Adaptation	How it prevents water loss
Presence of a thick cuticle	Water loss is minimised as the cuticle is composed of a waterproof material called cutin (figure 10.21).
Reduced number of stomata	Fewer stomata per unit area of leaves means that water loss by transpiration is reduced.
Presence of sunken stomata	Sunken stomata, located in pits (crypts) below the leaf surface, create a region of relatively higher humidity in the air space immediately surrounding the stomata. This reduces water loss by transpiration as the concentration gradient of water vapour between the air and the leaf is closer (figure 10.23).
Water storage tissue	Succulents have thick, fleshy leaves or stems that store water in their vacuoles after rainfall. Cacti (figure 10.22) stems have pleats that allow it to expand and contract quickly. Large-stem succulents may have water storage tissue in their trunks, such as the Queensland bottle tree (*Brachychiton rupestris*). Others have waster storage tissue in tubers and enlarged roots, such as magenta storksbill (*Pelargonium rodneyanum*).
No visible leaves	Succulent plants such as cacti have leaves that are reduced to spines, significantly reducing surface area (figure 10.22).
Altering leaf margins	The larger the ratio of edge length to surface area, the faster a leaf will cool. Cooler leaves have lower transpiration rates (figure 10.24a).
Altering leaf orientation and pendant branches	Leaves with a vertical orientation have less exposure to sunshine and so gain less heat and are cooler. Cooler leaves lose less water. Pendant branches may also develop, which move with the wind, creating water-catching wells in the soil under the tree (figure 10.24b).
Silver leaves	Silver or grey leaves reflect sunlight and help keep the leaf cool, which reduces water loss.
Hairs on leaf surface	When hairs are present on leaf surfaces, they slow the flow of air across the leaves, reducing the rate of water lost through transpiration.

(continued)

TABLE 10.6 Plant adaptations to prevent water loss *(continued)*

Adaptation	How it prevents water loss
Leaves that are not leaves	Members of the genus *Acacia* (wattles) have had their feathery leaves replaced by flattened leaf stalks called **phyllodes**. Figure 10.24c shows a transitional state in which the feathery true leaves of an acacia plant are starting to be replaced by leaf stalks that are gradually thickening. Phyllodes enable plants to survive in arid conditions because they provide a store of water in large parenchyma cells at their centre, and phyllodes have fewer stomata than true leaves and so lose less water by transpiration.
Shedding leaves	When plants become stressed in drought conditions they conserve water by dropping their leaves.
Drought-resistant seeds	The outer coating of the seeds contains a water-soluble chemical that inhibits germination. Dry conditions = no germination.

FIGURE 10.21 The presence of a hydrophobic waxy cuticle is revealed by the 'balling' of rainwater droplets on a leaf surface.

FIGURE 10.22 A Golden Barrel cacti (*Echinocactus grusonii*) with spines and a core of water-storing tissue

FIGURE 10.23 Light micrograph of a cross-section of the leaf of an oleander plant (*Nerium oleander*) with two crypts on the lower epidermis. Several stomata can be seen at the base of the crypt at left. Hairs lining the crypts are also visible.

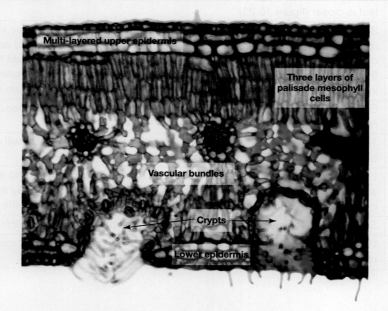

phyllodes leaf-like structures derived from a petiole (stem of a leaf)

FIGURE 10.24 a. At left is a leaf with an indented margin and at right is a leaf with an entire margin. b. Eucalypt with vertical leaves and pendant branches c. Formation of a phyllode from the leaf stalk of the original feathery leaf

elog-0848

INVESTIGATION 10.2

online only

Leaves for survival

Aim

To observe different adaptations in leaves of different plants that enhance survival

tlvd-1828

SAMPLE PROBLEM 3 Using concentration gradients to determine water loss

The graph shows the difference in concentration gradient of water between a stomata and the external environment.

a. Mark on the graph the point that water loss would be greatest and the point that water loss would be lowest. **(2 marks)**
b. The graph was based on data obtained over a 24-hour period. Suggest a position on the graph for the hottest part of the day. **(1 mark)**

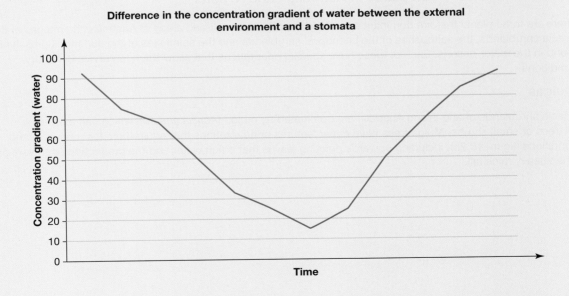

THINK

a. 1. When is water loss the greatest? How is this related to concentration gradients?

2. When is water loss the lowest?

3. Mark the points clearly on the graph and label each point where water loss is the greatest and the lowest.

WRITE

Water loss is the greatest when the difference in the concentration gradient is the greatest.

Water loss is the lowest when the difference in the concentration gradient is the lowest.

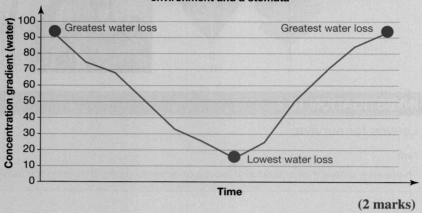

(2 marks)

b. Consider the effect of temperature on concentration gradients.

The hottest part of the day would be the time of the greatest concentration gradient and therefore greatest water loss (1 mark).

CASE STUDY: Dominant plants in Australian arid regions

There are three plants that are dominant species in the arid and semi-arid areas of Australia: the mulgas of the acacia shrublands, the saltbushes of the chenopod shrublands and the spinifexes of the hummock grasslands found in the major deserts. Each of these have specific adaptations that allow them to survive in the dry conditions.

Mulgas

Acacia shrublands of arid inland Australia are dominated by mulga (*Acacia aneura*), which can exist either as trees or small shrubs. Mulga trees have many features or adaptations that equip them for survival in arid conditions (figure 10.25), including upwards pointing leaves that catch water and the production of flowers only when heavy rains fall.

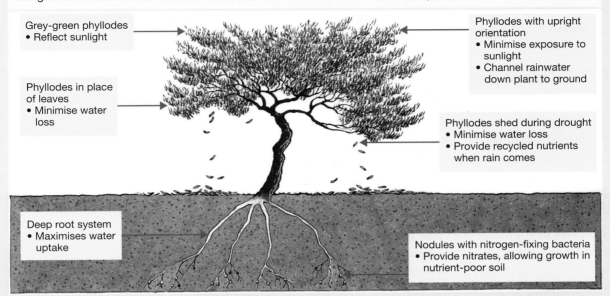

FIGURE 10.25 Some of the adaptations of the mulga tree. The vertical orientation of sparse foliage of mulga trees ensures that the little rain that falls is directed to the roots of the plant.

Saltbushes

Saltbushes have structural adaptations to conserve water, and are able to grow in regions of high salt concentration.

Their leaves have sunken stomata, are covered in hairs and are oriented so that they expose a minimal surface to the Sun's rays. Saltbushes also excrete salt from the very salty soils they grow in through their leaves. This covers the saltbush leaves in salt crystals, which reflect the Sun's heat.

The seeds have high concentrations of salt in their outer coats, and germination only occurs after the salt has been washed out after heavy rainfall. Following germination, new seedlings quickly become established. The salt inhibition of germination means the next generation of saltbush plants appears in times of good rainfall, maximising their chance of survival.

FIGURE 10.26 The saltbush (*Atriplex cinerea*) is salt tolerant and drought resistant.

Spinifexes

Spinifex grasses show physical adaptations to desert conditions. Some of these include a deep root system that extends 3 m or more into the soil, maximising water uptake to stems; rolled-up leaves that reduce heat loss; and seedlings that stay dormant in adverse conditions.

For more information about the adaptations of these dominant plants, please download the digital document.

Resources

Digital document Case study: Dominant plants (doc-36116)

Resources

- **eWorkbook** Worksheet 10.4 Adaptations in arid regions (ewbk-7603)
- **Weblink** Trees to ease drought

10.3.5 Adaptations for survival: cold climate animals

Plants and animals also show adaptations that equip them for life in cold conditions on land and in water.

Ice can damage or kill

Processes that are essential for life include chemical reactions that take place between substances that are dissolved in liquid water (i.e. in solution). These processes *cannot* take place in solid water (ice). If all the liquid water in a living organism were replaced by solid water, life would be destroyed. When ice forms, the solid water expands. If cells freeze, the expanding ice crystals rupture the cell membranes and kill the cells.

Many living things can exist on land in Antarctica or the Arctic. During winter, the air temperatures fall well below the freezing point of pure water. How do living things survive in these low temperatures?

Organisms have special features or behaviours that enable them to survive extremely low temperatures, including the following:

- Physiological adaptations include the production of **antifreeze** substances. Some insects, fishes, frogs and turtles that can survive in cold winters make antifreeze substances such as glycerol, amino acids and sugars, or mixtures of substances, at the start of the freezing season (figure 10.27). These antifreeze substances are released into their body fluids, which lowers the fluids' freezing point to well below that of the surrounding water temperatures. This means that the body fluids of these organisms stay liquid.
- Behavioural adaptations include burrowing (some frogs and toads), hibernation (see Case study box 'Survival by hibernation — pygmy possums') and huddling to avoid freezing temperatures. Emperor penguins (*Aptenodytes forsteri*) are the largest of the penguin species, with adult birds growing to more than 1 m tall and weighing 40 kg. As the only species to live on the Antarctic landmass during winter, one of their survival mechanisms is to huddle in large groups (figure 10.28).
- Structural adaptations include the growth of thick fur and layers of fat. The fat layers are also a physiological **insulating layers** of fat under the skin and thick fur, and birds have layers of feathers (figure 10.28). These insulating layers of fat are very important in cold climate marine animals (see Case study box 'Adaptations in marine mammals').

FIGURE 10.27 The Arctic woolly bear caterpillar produces antifreeze proteins to stop its cells from freezing. The long hairs covering its body provide insulation from the sub-zero temperatures.

FIGURE 10.28 Emperor penguins huddle in large groups to survive Antarctic winters.

antifreeze a chemical substance produced by an organism to prevent freezing of body fluids or tissue when in a sub-zero environment

insulating layers layers of fat under the skin of mammals that retain heat within the body

Resources

Weblinks Antarctic animals adapting to the cold

CASE STUDY: Survival by hibernation — pygmy possums

The mountain pygmy possum (*Burramys parvus*) is the only Australian mammal that lives permanently in alpine regions. Its distribution is limited to two small areas (figure 10.29): one in the Kosciuszko National Park of New South Wales and the other near Mount Hotham in Victoria.

FIGURE 10.29 **a.** The mountain pygmy possum, *Burramys parvus* **b.** Its distribution is limited to two small areas.

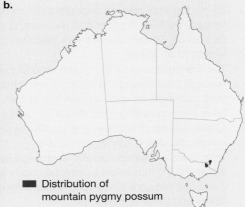

■ Distribution of mountain pygmy possum

B. parvus has both behavioural and physiological features that enable it to survive the low winter temperatures of its alpine environment:
- It collects and hides seeds and fruits for use during winter. Unlike other pygmy possums, *B. parvus* has no storage of fat in its tail.
- During winter, *B. parvus* goes into a torpor that is equivalent to hibernation. When mammals hibernate, their heartbeat slows down considerably and their breathing rate drops. Their body metabolism is significantly reduced (to between 0.6 per cent and 3.9 per cent of the normal metabolic rate) and their body temperature drops to that of the surrounding environment.
- Hibernation and the reduced metabolic rate for periods means that the amount of food required by an animal, overall, to survive in winter is reduced.

The Australian bushfires of summer 2019–20 have had a significant impact on the habitat of *B. parvus*, with much of their region impacted by the fires. With the possums listed as endangered, they are currently highly vulnerable to extinction. This is due to the geographical isolation of individual populations as a result of the mountainous region and their alpine habitat being slow to regenerate after fire.

CASE STUDY: Adaptations in marine mammals

Whales, dolphins and seals are mammals that spend their entire lives in water. Like all mammals, they are endothermic and they breathe air so must come to the surface every so often. The females give birth to young that they suckle on milk secreted by mammary glands.

Most land mammals have an insulating fur coat that assists in the regulation of body temperature. Marine mammals in cold climates rely on the following:

- They have an insulating layer of fat or blubber below the skin (figure 10.30). This layer may be up to 50 cm thick and can vary with the different seasons. They can maintain a stable body temperature of 36 to 37 °C in an environment that is usually less than 25 °C and may be as low as 10 °C. Fat may also be deposited around organs and tissues, such as the liver and muscles, and in bone in the form of oil. These deposits can make up to half of the body weight of an animal.
- The low surface area to volume ratio of these animals also reduces heat loss to the environment. Heat is readily lost from appendages such as hands and feet. Marine mammals have few protruding parts (fins and tail flukes), which means a relatively small surface area to volume ratio and heat loss across the skin is further minimised.
- They have a **countercurrent exchange system** (figure 10.31) that helps to maintain body temperature. It is a fine network of vascular tissue within the fins, tail flukes and other appendages.

In this system:
- an outgoing artery that carries warm blood from the core is paired with an incoming vein carrying cooled blood returning from the extremities. This warms the blood flowing in from the skin and prevents the venous blood from cooling the internal organs and muscles. At the same time, the blood moving out to the skin is cooled, and so the loss of heat across the skin is reduced.
- When the animal needs to conserve heat, the outermost blood vessels contract, little blood flows and heat loss from these vessels is reduced.

This countercurrent system is also present in the feet, wings and bills of penguins.

FIGURE 10.30 The crabeater seal (*Lobodon carcinophaga*) has a thick insulating layer of blubber beneath the surface.

FIGURE 10.31 A countercurrent exchange system in the skin of dolphins. Yellow arrows show direction of heat flow; direction of blood flow is shown in artery (red arrow) and in veins (blue arrows)

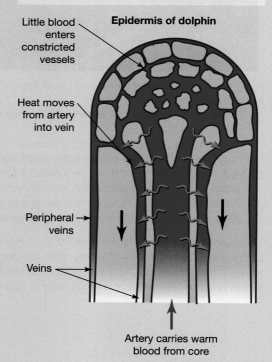

countercurrent exchange system situation in which two fluid systems flowing adjacent to each other, but in opposite directions, enables the transfer of heat or compounds from one system to the other by diffusion

10.3.6 Adaptations for survival: cold climate plants

Many plants survive in sub-zero temperatures without being damaged by these extremely low temperatures. Unlike animals, plants do not produce an 'antifreeze'. They gradually become resistant to the potential danger of ice forming in their tissues as the temperature falls below 0 °C. How does this occur? The process is described in figure 10.32.

FIGURE 10.32 Plant survival in freezing conditions occurs due to the movement of water and ions across concentration gradients.

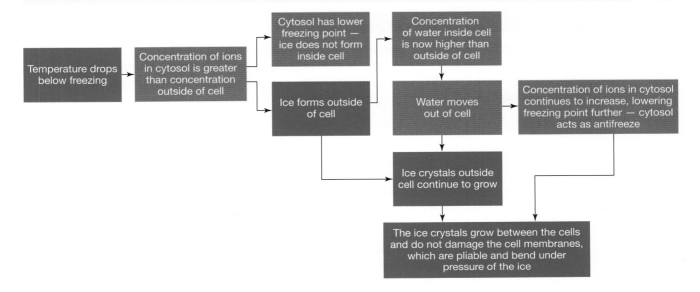

Remember that water is transported through plants in very fine xylem vessels and is subjected to a number of forces. These forces affect the way in which water behaves in plants in freezing temperatures. Remember also, that water with dissolved substances (ions) has a lower freezing point than pure water, and concentration gradients are important in how water moves through a plant.

Many species of trees are able to withstand extremely low temperatures before they are killed (table 10.7). The temperatures at which the living tissue in a tree is killed influences the latitudes at which it can grow. Which tree in table 10.7 is most likely to be found in the northern latitudes of Canada? Note that as one travels further north (or south) from the equator at sea level, the average temperature falls. The higher the latitude, the lower the temperature.

FIGURE 10.33 Although the ice punctures cell walls, the cell membranes are merely pushed inward and the cells remain intact.

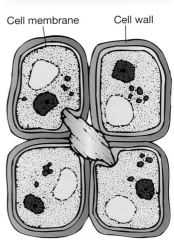

TABLE 10.7 Lethal temperatures for some trees

Species	Temperature (°C) at which it is killed
Coast redwood (*Sequoia sempervirens*)	−15
Southern magnolia (*Magnolia grandiflora*)	−15 to −20
Swamp chestnut oak (*Quercus michauxii*)	−20
American beech (*Fagus grandifolia*)	−41
Sugar maple (*Acer saccharum*)	−42 to −43
Black cottonwood (*Populus trichocarpa*)	−60
Balsam fir (*Abies balsamea*)	−80

Ultimately, if there is an excessive drop in the surrounding temperature, ice crystals form inside the cells and they die, and so the tree may die. It has been suggested that an excessive drop in temperature damages the protein molecules that form part of the cell membranes, so ions can leak out of the cell. Australia does not experience the sustained extremes of low temperatures found in many other countries and low temperature is rarely a limiting factor for plant growth. Growth of native plants in Australia is determined by whether a plant has the adaptations to survive the various altitude zones and their associated temperatures. Some plants, particularly exotic garden plants, may be killed or damaged by an unusually severe frost.

tlvd-1829

SAMPLE PROBLEM 4 How plants survive sub-zero temperatures

The woolly bear caterpillar can survive in extremely cold conditions due to the present of its thick hairlike structures and the antifreeze proteins it is able to produce within its cells. Plants are unable to produce antifreeze chemicals, yet they have adaptations that enable them to survive. Describe, with reference to osmosis, how plants stop their cells from freezing in sub-zero temperatures. (5 marks)

THINK

1. *Describe* means to provide details about the characteristics of the plant and the environment given in the question.

2. Make the link to osmosis.

3. Add a concluding statement.

WRITE

Plants transport water via the xylem (1 mark). When the temperatures drop below zero, the water outside of the cells freezes. The water inside the cell does not freeze as the ion concentration in the cytosol prevents it from freezing (1 mark).

Because ice has formed outside the cell, the concentration of water inside the cell is greater than outside (1 mark) — therefore the water leaves the cell via osmosis and the ice outside the cell continues to grow. This increases the concentration of ions within the cell, further preventing it from freezing (1 mark).

The cell membrane, although flexible, is protected from freezing by the cytosol levels in the cell. If the cell froze, the membrane would be irreparably damaged (1 mark).

 Resources

 eWorkbook Worksheet 10.5 Adaptations in cold regions (ewbk-7605)

KEY IDEAS

- The tolerance range of an organism is the range of conditions in which it can live and reproduce successfully.
- Humans use technology to extend the natural limits of their tolerance range, both on Earth and in outer space.
- Adaptations are structural, physiological or behavioural features that enable an organism to survive and reproduce in different conditions.
- Native mammals of the Australian desert show many structural, physiological and behavioural adaptations that equip them for living in arid Australia.
- A range of adaptations may be seen in desert-dwelling marsupials that enable them to minimise water loss and, if necessary, survive without drinking water.
- Other survival strategies seen in various animal species include becoming dormant, moving around and producing drought-resistant offspring.
- The arid and semi-arid areas of inland Australia are dominated by hummock grasslands, acacia shrublands and chenopod shrublands.

- Desert plants show a variety of adaptations to maximise water uptake and to reduce water loss.
- The major loss of water in plants occurs as water vapour lost by transpiration from the leaf stomata.
- Many adaptations exist in desert plants for reducing water loss from their leaves.
- Ephemeral plant species of desert regions produce drought-resistant seeds that germinate only after rain or flooding.
- Mulgas, the dominant plant of the acacia shrublands of arid regions, show various adaptations to minimise water loss and maximise water uptake.
- Saltbushes of the chenopod shrublands show adaptations for minimising water loss and also for surviving in very salty soils.
- The spinifexes of the hummock grasslands of the deserts show adaptations to maximise water uptake and minimise water loss.
- Adaptations of animals living in cold environments include the presence of insulating layers, the production of antifreeze compounds and the use of countercurrent exchange systems.
- Plants survive sub-zero conditions due to movement of water across concentration gradients and having flexible cell membranes.

10.3 Activities

learn on

To answer questions online and to receive **immediate feedback** and **sample responses** for every question, go to your learnON title at **www.jacplus.com.au**. A **downloadable solutions** file is also available in the resources tab.

| 10.3 Quick quiz on | 10.3 Exercise | 10.3 Exam questions |

10.3 Exercise

1. Provide an example of a tolerance range that may limit the distribution of fish.
2. Describe a structural, physiological and behavioural adaptation that humans use to:
   a. reduce heat loss
   b. increase heat loss.
3. Identify three adaptations of polar bears to minimise heat loss in the Arctic environment.
4. List three water-conserving adaptations that may be seen in the tarrkawarra and identify whether they are behavioural, structural and/or physiological adaptations.
5. What is an advantage of being able to produce concentrated urine?
6. Describe how an organism's ability to survive and reproduce in arid regions is enhanced by:
   a. being dormant
   b. moving around
   c. producing drought-resistant offspring.
7. What is the major avenue of water loss by plants?
8. Identify one example of an adaptation that enables a plant to conserve water by:
   a. reducing water loss from cells at the leaf surface
   b. reducing water loss from the leaf stomata
   c. reducing the absorption of heat, thus staying cooler and reducing the rate of transpiration.
9. a. What is a phyllode?
   b. How do phyllodes enable a plant to conserve water?
10. What feature prevents drought-resistant seeds from germinating before rain falls?
11. Identify an adaptation that enables:
    a. mulga shrubs to survive in nutrient-poor soil
    b. saltbushes to minimise heat absorbed
    c. spinifexes to reduce water loss.
12. What feature enables saltbushes to survive in very salty soils?
13. Predict what would happen to the cells of a plant moved from a rainforest to an arid environment.
14. Describe how countercurrent exchange is used to minimise heat loss in whales and dolphins.
15. How do mountain pygmy possums conserve energy in the winter months whilst in an alpine environment?

10.3 Exam questions

Question 1 (1 mark)

MC Echidnas have fur that is modified into spines.

This is an example of a
A. natural adaptation.
B. physiological adaptation.
C. behavioural adaptation.
D. structural adaptation.

Question 2 (1 mark)

MC Which of the following is an example of a structural adaptation that may help maintain water balance in a mammal living in a dry environment?
A. A thin covering of body hair
B. Hunting for food during the night
C. Large ears for sensing prey
D. Relatively long loops of Henle in the kidney nephrons

Question 3 (1 mark)

MC A kangaroo can be seen licking its fur on a hot day.

This is an example of a
A. structural adaptation.
B. physiological adaptation.
C. behavioural adaptation.
D. natural adaptation.

Question 4 (1 mark)

MC An example of a physiological adaptation in mice is
A. a long tail.
B. a high reproductive rate.
C. thick fur.
D. long whiskers.

Question 5 (1 mark)

MC Scientists examined the biodiversity of a marine environment. The temperature range tolerated by the cells within three fish species was determined. These are shown in the table.

Species of fish	Temperature range tolerated of cells (°C)
1	20–24
2	16–28
3	19–21
4	14–29

The current water temperatures of the marine environment range from 20 to 21 °C.

The survival of which species would most likely be affected if the temperature of their environment was to increase by 2 °C ?
A. Species 1
B. Species 2
C. Species 3
D. Species 4

More exam questions are available in your learnON title.

10.4 Survival through interdependence between species

> **KEY KNOWLEDGE**
>
> - Survival through interdependencies between species, including impact of changes to keystone species and predators and their ecological roles in structuring and maintaining the distribution, density and size of a population in an ecosystem
>
> **Source:** VCE Biology Study Design (2022–2026) extracts © VCAA; reproduced by permission.

10.4.1 What is an ecosystem?

In studying biology, it is possible to focus on different levels of organisation. Some biologists focus on the structures and functions of cells. Other biologists focus on whole organisms, others on populations. Different levels of biological organisation are shown in figure 10.34. An ecosystem is the most complex level of organisation.

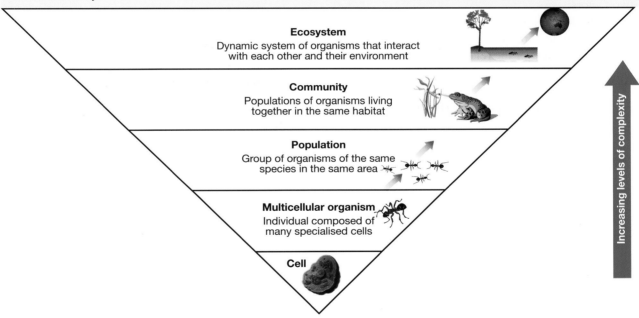

FIGURE 10.34 Different levels of biological organisation, from the most basic unit of life, the cell, to the complex unit of an ecosystem

Each ecosystem includes:
- **biotic factors** (living) — the populations of various species that live in a given region
- **abiotic factors** (non-living) — the non-living surroundings and their environmental conditions
- the interactions within and between species and their local environment.

The continuation of an ecosystem depends on the intactness of the parts and on the interactions between them. An ecosystem depends on its parts and may be destroyed if one part is removed or altered. Professor Eugene Odum, the world-famous ecologist, wrote: 'An ecosystem is greater than the sum of its parts.'

biotic factor living factor, such as predators or disease, that can affect population size

abiotic factor nonliving factor, such as weather, that can affect population size

Ecosystems can vary in size but must be large enough to allow the interactions that are necessary to maintain them. An ecosystem may be as small as a freshwater pond or a terrarium, or as large as an extensive area of mulga scrubland in inland Australia. An ecosystem may be terrestrial or marine.

The study of ecosystems is the science known as **ecology** (from the Greek, *oikos* = home or place to live; *logos* = study).

Let's now look at some living communities in their non-living physical surroundings.

10.4.2 Ecological communities

A **population** is defined as all the individuals of one particular species living in the same area at the same time.

A **community** is made up of all the populations of various organisms living in the same location at the same time.

Community 1 = population 1 + population 2 + population 3 and so on.

Different communities can be compared in terms of their **diversity**. Diversity is not simply a measure of the number of different populations (or different species) present in a community. When ecologists measure the diversity of a community, they consider two factors:
1. The richness or the number of different species present in the sample of the community (species diversity)
2. The evenness or the relative abundance of the different species in the sample.

As richness and evenness increase, the diversity of a community increases.

> **ecology** study of communities in their habitats and the interactions between them and their environment
> **population** members of one species living in one region at a particular time
> **community** biological unit consisting of all the populations living in a specific area at a specific time
> **diversity** measure of 'species richness' or the number of different species in a community

Factors affecting species diversity

Factors that affect the species diversity (or number of populations) in a community include:
1. *the physical area in which the community lives*. The number of different populations in terrestrial communities in the same region is related to the physical size of the available area. For example, the Caribbean Sea contains many islands that differ in their areas. Figure 10.35 shows the results of a study into the relationship between the sizes of the islands and the numbers of species of reptiles and amphibians (and hence the number of different populations) found on them. In general, if an island has an area ten times that of another in the same region, the larger island can be expected to have about twice the number of different species.

FIGURE 10.35 a. Relationship between the area of an island and the number of populations of different species that it contains (based on data from RH MacArthur and EO Wilson 2001, *The Theory of Island Biogeography*, Princeton University Press) **b.** Location of the islands in the Caribbean Sea

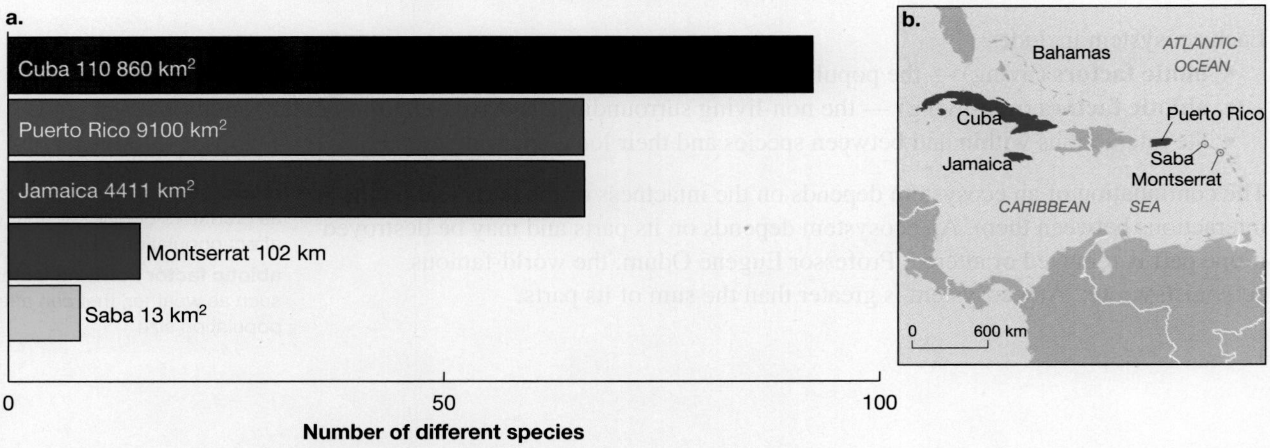

a.

Cuba 110 860 km^2
Puerto Rico 9100 km^2
Jamaica 4411 km^2
Montserrat 102 km
Saba 13 km^2

Number of different species: 0, 50, 100

b. Map showing Bahamas, Cuba, Jamaica, Puerto Rico, Saba, Montserrat in the Caribbean Sea

2. *the latitude* (or distance, north or south, from the equator). In general, as we move from the poles to the equator, species richness of terrestrial communities increases. This means that more species, and hence more populations, exist in a given area of a tropical rainforest ecosystem than in a similar area of a temperate forest ecosystem (figure 10.36). In turn, an area of temperate forest ecosystem has more populations than a similar area of a conifer (boreal) forest ecosystem in cold regions of the northern hemisphere. For example, a two-hectare area of forest in tropical Malaysia has more than 200 different tree species, while a similar area of forest 45° N of the equator contains only 15 different tree species.

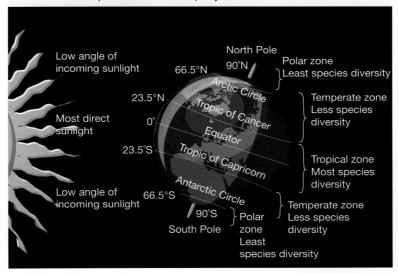

FIGURE 10.36 The equator of the Earth has the greatest species richness, with it also being the area that receives the most direct sunlight. As you move towards the poles, less direct sunlight is received and species richness rapidly declines.

CASE STUDY: Populations in Antarctic communities

Depending upon the physical area in which a community lives, different communities vary in the number of populations that they contain. The community at Cape Adare in Antarctica is dominated by one population — that of the Adélie penguins. The situation in a coral reef community is very different, with a very large number of different populations present. Since each population is made up of one discrete species, the number of populations in a community corresponds to the number of different species or the species richness of the community.

Adélie penguins are not the only species living on Antarctica. To access more information about the different plant and animal species on Antarctica, please download the digital document.

FIGURE 10.37 The Adélie penguin rookery at Cape Adare in Antarctica during summer

Resources

Digital document Case study: Populations in Antarctic communities (doc-36117)

| INVESTIGATION 10.3 | online only |

Distribution patterns of moss

Aim

To investigate the different distribution patterns of mosses in varying locations

10.4.3 Members of a community

In an ecosystem, the members of the living community can be identified as belonging to one of the following groups: **producers**, **consumers** or **decomposers**.

Producers

Producers are the members of an ecosystem community that bring energy from an external source into the ecosystem. Producers are **autotrophic** organisms that use photosynthesis to capture sunlight energy and transform it into chemical energy in the form of sugars, such as glucose, making it available within the community.

FIGURE 10.38 Cyanobacteria contain a form of chlorophyll that can carry out photosynthesis.

FIGURE 10.39 Producers from a range of ecosystems

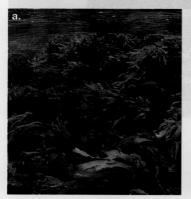

a.

Ecosystem: Temperate marine kelp forest
Producers: Algae, including the string kelp *Macrocystis angustifolia*

b.

Ecosystem: Antarctic marine ecosystem
Producers: Many species of phytoplankton

c.

Ecosystem: Temperate closed forest
Producers: Various woody flowering plants, ferns and mosses

- In aquatic ecosystems, such as seas, lakes and rivers, the producers are microscopic phytoplankton, macroscopic algae, bacteria and seagrasses (figures 10.39a and b).
- In terrestrial ecosystems, producer organisms include photosynthetic microbes and familiar green plants. These, in turn, include trees and grasses, other flowering plants, cone-bearing plants such as pines, as well as ferns and mosses (figure 10.39c).

All of these producers convert the energy of sunlight into the chemical energy of glucose through the process of photosynthesis.

producers photosynthetic organisms and chemosynthetic bacteria that, given a source of energy, can build organic matter from simple inorganic substances

consumers organisms that obtain their energy and organic matter by eating or ingesting the organic matter of other organisms; also termed heterotrophs

decomposers organisms, such as fungi, that can break down and absorb organic matter of dead organisms or their products

autotrophs organisms that, given a source of energy, can produce their own food from simple inorganic substances; also known as producers

CASE STUDY: The importance of phytoplankton

Although phytoplankton are microscopic, they are significant producers on a global scale due to their accumulated mass. Their presence is visible in colour-coded satellite images of surface chlorophyll in the Southern Ocean around Antarctica in summer (figure 10.40). Purple areas have little phytoplankton; orange areas have the highest concentration of phytoplankton. Note the high concentration of chlorophyll (and hence of phytoplankton) over the continental shelf surrounding the Antarctic continent.

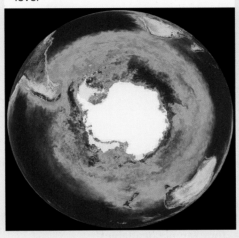

FIGURE 10.40 Colour-coded satellite image of surface chlorophyll due to phytoplankton; red indicates highest level

FIGURE 10.41 Phytoplankton are a mixture of various microscopic organisms of different types, including bacteria, protists and algae. All are autotrophic organisms that can carry out photosynthesis.

Producers

The organic compounds made by producer organisms provide the chemical energy that supports both their own needs and, either directly or indirectly, all other community members of the ecosystem.
- No ecosystem can exist without the presence of producers.
- In short, no producers equals no ecosystem.

Consumers

Consumers are **heterotrophs** that rely directly or indirectly on the chemical energy of producers (figure 10.42).

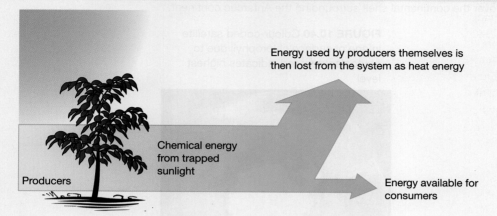

FIGURE 10.42 The chemical energy in sugars from sunlight energy trapped by producers is used mainly by the producers themselves for staying alive. A small amount of this energy is available to consumers in the ecosystem.

Consumers or heterotrophs are those members of a community that must obtain their energy by eating other organisms or parts of them. All animals are consumers.

Consumer organisms can be subdivided into the following groups:
- **herbivores**, which eat plants — for example, wallabies and algae-eating fish
- **carnivores**, which eat animals — for example, numbats, snakes and coral polyps
- **omnivores**, which eat both plants and animals — for example, humans and crows
- **detritivores**, which eat decomposing organic matter such as rotting leaves, dung or decaying animal remains — for example, earthworms, dung beetles and crabs.

FIGURE 10.43 The common dunnart (*Sminthopsis murina*) is a small Australian marsupial mammal. It is a carnivore and feeds mainly on insects.

Fragments of dead leaves and wallaby faeces on a forest floor, pieces of rotting algae and dead starfish in a rock pool are all organic matter, which contains chemical energy. Particles of organic matter like this are called **detritus**. The organisms known as detritivores use detritus as their source of chemical energy. Detritivores take in (ingest) this material and then absorb the products of digestion. Detritivores differ from decomposers (see next section) in that decomposers first break down the organic matter outside their bodies by releasing enzymes, and then they absorb some of the products.

heterotroph organism that ingests or absorbs food in the form of organic material from their environment; also known as a consumer

herbivore organism that eats living plants or parts of them

carnivore organism that kills and eats animals

omnivore organism that eat both plants and animals

detritivore organism that eats particles of organic matter found in soil or water

detritus fragments of organic material present in soil and water

Decomposers

Typical decomposer organisms in ecosystems are various species of fungi and bacteria (figures 10.44 and 10.45). Decomposers are heterotrophs, which obtain their energy and nutrients from organic matter; in their case, the 'food' is dead organic material. Decomposers are important in breaking down dead organisms and wastes from consumers, such as faeces, shed skin and the like. These all contain organic matter and it is the action of decomposers that converts this matter to simple mineral nutrients.

Decomposers differ from other consumers because, as they feed, they chemically break down organic matter into simple inorganic forms or mineral nutrients, such as nitrate and phosphate. These mineral nutrients are returned to the environment and are recycled when they are taken up by producer organisms. So, decomposers convert the organic matter of dead organisms into a simple form that can be taken up by producers.

FIGURE 10.44 The fungi *Coprinus disseminatus*, found on rotting wood that is its food

FIGURE 10.45 Fungal growth on apples. What is happening?

CASE STUDY: Considering essential groups of an ecosystem

Of the three groups described — producers, consumers and decomposers — only two are essential for the functioning of an ecosystem. Can you identify which two? One essential group comprises the organisms that capture an abiotic source of energy and transform it into organic matter that is available for the living community. Who are they? The second essential group is the one that returns organic matter to the environment in the form of mineral nutrients. Who are they?

FIGURE 10.46 The three members of a community — which two members are essential?

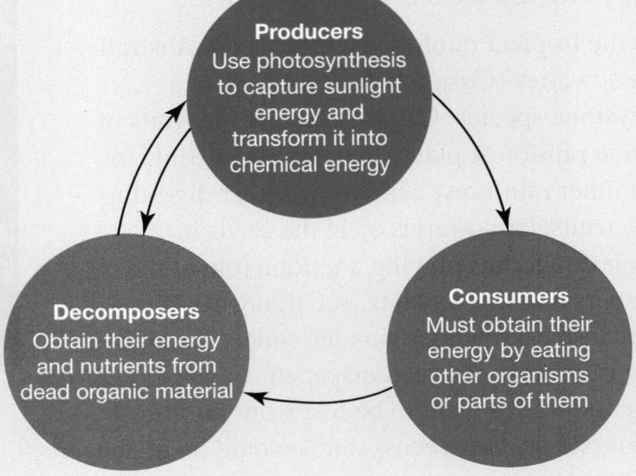

Producers: Use photosynthesis to capture sunlight energy and transform it into chemical energy

Consumers: Must obtain their energy by eating other organisms or parts of them

Decomposers: Obtain their energy and nutrients from dead organic material

10.4.4 Keystone species

An ecological community typically has many populations, each composed of a different species. Within an ecosystem, each species has a particular role — it may be as producer, consumer or decomposer; it may be as a partner or player in a particular relationship with another species. Some species have a disproportionately large impact on, or deliver a unique service to, the ecosystem in which they live. In some cases, their presence is essential for the maintenance of the ecosystem.

These species are termed **keystone species** for their particular ecosystems.

> **Keystone species**
>
> The loss of a keystone species would be expected to lead to marked and even radical changes in their ecosystems, compared with the potential impact of the loss of other species.

Keystone species: elephants

On the great grasslands of Africa, elephants (*Loxodonta* sp.) are a keystone species. Through their feeding activities, elephants consume small *Acacia* shrubs that would otherwise grow into trees. They even knock over and uproot large shrubs as they feed on their foliage. Through these activities, elephants control the populations of trees on the grassland, maintaining the ecosystem as an open grassland. The herbivores of the grasslands, including species of wildebeest, zebra and antelope, feed by grazing and depend on the existence of these grasslands. Similarly, the **predators** of the grasslands, such as lions, hyenas and painted hunting dogs, depend on the open nature of the grasslands for hunting and catching **prey**. Removal of the elephants would, over time, lead to the loss of grasslands and their conversion to woodlands or forests.

keystone species species whose presence in an ecosystem is essential for the maintenance of that ecosystem

predator an animal that actively seeks out other animals as its source of food

prey living animal that is captured and eaten by a predator

Keystone species: starfish

In some marine ecosystems, starfish are keystone species because they are the sole predator of mussels. If starfish were removed from such ecosystems, then in the absence of their only predators, mussel numbers would increase markedly and they would crowd other species. The presence of starfish in such an ecosystem maintains the species diversity of the ecosystem.

Keystone species: cassowaries

In the tropical rainforests of far north Australia, cassowaries (*Casuarius casuarius*) are a keystone species. Cassowaries eat the fruits of some rainforest plants that are indigestible to all other rainforest herbivores. After digesting the fruits, cassowaries eject the seeds in their dung, thus playing a unique role in the dispersal of these plants (see figure 10.47). Because cassowaries wander widely through the rainforest, the seed dispersal is widespread. If cassowaries were to be lost from rainforest ecosystems, these ecosystems would be in danger of extinction.

FIGURE 10.47 a. A cassowary in a rainforest in Far North Queensland **b.** Fruits of a rainforest tree species **c.** Fruits of another rainforest tree species **d.** Cassowary dung. Note the many seeds in this dung that can germinate at a distance from the tree that produced the fruit.

CASE STUDY: The wolves of Yellowstone National Park

Yellowstone National Park was the first national park in the United States, with most of its land area located in the northern state of Wyoming. Covering an area of over 9000 km^2, it is well known for its volcanic activity, with extensive geothermal activity. However, as it is a protected area, this energy is not able to be harnessed.

When Yellowstone National Park was established in 1872, wolf numbers were already declining, and by the early 1900s there were government programs designed to eliminate the wolves from the park. In 1926, the last wolf was killed.

This had a number of effects:
- Elk populations increased, as they had lost their key predator.
- As elk are a grazing species and eat willow, aspen and cottonwood trees, the increase in the elk population caused a decline in the number of these tree species.
- As the trees decreased in number due to overgrazing, the land started to erode and damage rivers. Overgrazing and river damage impacted beaver populations.
- The coyote population increased, as they had lost their main competitor.
- Increased numbers of coyotes decreased the population of antelopes in the park, as they were food for the coyotes.

FIGURE 10.48 Yellowstone National Park, Wyoming

By the 1940s, ecologists and conservationists were recommending that wolves be reintroduced into the park to restore the ecological balance; however, it was not until 1995 that eight wolves from Canada were introduced. Those with land bordering the park were concerned that the wolves would target their livestock, but their fears were unfounded.

The reintroduction of wolves to the park had the following results:
- The elk population decreased, which resulted in reduced land degradation, improved river health and the recovery of tree species.
- This in turn allowed the beaver population to increase. Prior to the reintroduction of the wolf, there was one beaver colony — today there are nine.
- As increased numbers of beavers built dams in rivers, this helped to increase the populations of amphibian and fish species.
- The structure of the rivers changed from straighter to more winding due to the increase in beaver dams and decrease in erosion on the banks.
- The coyote population decreased, which increased the populations of rabbits and mice.
- Increased populations of rabbits and mice led to increased populations of hawks and badgers.

FIGURE 10.49 A grey wolf feeding on elk in the Yellowstone National Park. The scavenging birds also benefit as they have an additional food source.

Wolves are considered a keystone species as despite their low numbers, they have a significant effect on the ecosystem in Yellowstone National Park — far greater than any other animal within the ecosystem.

SAMPLE PROBLEM 5 Keystone species

In 1995, eight wolves were reintroduced to Yellowstone National Park to try and restore its ecological balance. Wolves had since been identified as a keystone species due to the impact they have within the ecosystem. Without the wolves preying on the elk in the park, elk numbers had increased to a level that was not sustainable due to the competition for resources.

In the northern parts of Yellowstone National Park, elk numbers have decreased due to lack of resources. This can be caused by an increase in predation, extended winters and hunting by humans. Predict what is likely to happen to wolf numbers when elk numbers decline, and what effect this may have on other organisms in the ecosystem.

(3 marks)

THINK

1. What effect do elk numbers have on wolf numbers?
2. How do wolf numbers affect other species?

WRITE

Elk are a main food source for the wolf. If elk numbers decline, it is expected that wolf numbers would also decline (1 mark).

Wolves are a keystone species. Their removal from an ecosystem would have a significant effect on other organisms within the ecosystem (1 mark).

If wolves disappeared, elk numbers would increase again, habitat would be affected, and many other organisms would not be able to access sufficient resources to survive (1 mark).

 Resources

 Weblinks Role of keystone species in an ecosystem
Gray wolf: Yellowstone National Park

10.4.5 Interactions within ecosystems

Interactions are continually occurring in ecosystems as follows:
- Between the living community and its non-living surroundings through various interactions, such as plants taking up mineral nutrients from the soil and carbon dioxide from the air, and animals using rocks for shade or protection
- Within the non-living community through interactions such as heavy rain causing soil erosion, and high temperatures causing the evaporation of surface water from a shallow pool
- Within the living community through many interactions between members of the same species and between members of different species.

In the following section we will explore some of the interactions that occur within the living community of an ecosystem. These interactions and examples are summarised in table 10.8.

TABLE 10.8 Summary of interactions within ecosystems

Type of interaction	Definition	Example
Intraspecific	Competition for resources between members of the same species	Wheat crops competing for soil and nutrients
Interspecific	Competition for resources between members of different species	Leopards and lions competing for the same prey
Amensalism	One organism is inhibited or destroyed, the other is unaffected	Elephants stepping on ants — the elephant is unaffected, however the ant dies

(continued)

TABLE 10.8 Summary of interactions within ecosystems *(continued)*

Type of interaction	Definition	Example
Predator–prey	One species kills and eats the other	Polar bears (predator) eat seals (prey) in the Arctic
Herbivore–plant relationships	An animal eats a plant	Koalas eating eucalypt leaves
Parasitism	An organism living on or within another organism to derive a benefit, whilst harming the host	A flea living on a golden retriever — the flea obtains food, however the golden retriever loses nutrients
Mutualism	A beneficial relationship between two species	Hummingbird and bee balm flower — the hummingbird gets nectar, the balm flower spreads its pollen
Commensalism	One member gains a benefit from the relationship, the other is unaffected	The clown fish and the sea anemone — the clown fish obtains shelter and food scraps, and the sea anemone is unaffected

In an ecosystem, certain resources can be in limited supply, such as food, shelter, moisture, territory for hunting and sites for breeding or nesting. In situations where resources are limited, organisms compete with each other for them.

Intraspecific and interspecific competition

Competition may be between members of the same species. For example, pairs of parrots of the same species compete for suitable hollows in old trees for nesting sites. Competition between members of the same species for resources is termed **intraspecific competition** (figure 10.50).

Interspecific competition occurs between members of different species within the same population. For example, members of different plant populations in the same ecosystem compete for access to sunlight, and different animal species may compete for the same food source.

FIGURE 10.50 Intraspecific competition between southern elephant seals competing with each other for space and access to mates

Competition occurs when one organism or one species is more efficient than another in gaining access to a limited resource, such as light, water or territory. For example:

- faster growing seedlings will compete more efficiently in gaining access to limited light in a tropical rainforest than slower growing members of the same species; the faster growers will quickly shade the slower growers from the light and deprive them of that resource
- when soil water is limited, a plant species with a more extensive root system will compete more efficiently for the available water than a different plant species with shallow roots, and deprive it of that resource.

competition interaction between individuals of the same or different species that use one or more of the same resources in the same ecosystem

intraspecific competition competition for resources in an ecosystem involving members of the same species

interspecific competition competition for resources in an ecosystem involving members of one species and members of other species

TOPIC 10 Adaptations and diversity

CASE STUDY: Intraspecific competition between anenomes

Anemones compete for space and food, with behaviour that is easily observed. If one anemone encroaches too closely to another, the original occupant will inflate its tentacles and release poisoned darts from stinging cells. The intruder may retaliate and return fire, and eventually one of the anemones retires from the fight. This is also the case for other competitive interactions, such as when a number of smaller birds of one species 'mob' a larger bird of another species that enters their nesting territory.

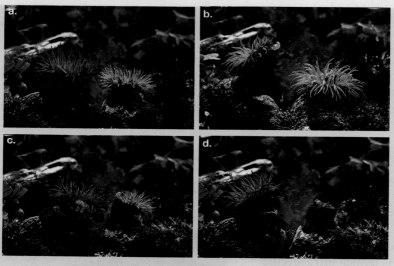

FIGURE 10.51 Anemones compete for space and food. **a.** If an anemone encroaches too closely to another, **b.** the original occupant will inflate its tentacles and **c.** release poisoned darts from stinging cells. The intruder may retaliate and return fire. **d.** Eventually one of the anemones retires from the fight.

Amensalism

Amensalism is any relationship between organisms of different species in which one organism is inhibited or destroyed, while the other organism gains no specific benefit and remains unaffected in any significant way.

Examples of amensalism include:
- the foraging or digging activities of some animals, such as wild pigs, that kills soil invertebrates or exposes them to predators. While soil invertebrates may be destroyed by the foraging of pigs, the pigs do not benefit from these deaths.
- when one species secretes a chemical that kills or inhibits another species, but the producer of the chemical is unaffected and gains no benefit from these deaths. For example, *Penicillium chrysogenum* mould produces an antibiotic that kills many other bacterial species (figure 10.52), but the mould gains no benefit from the bacterial deaths caused by the antibiotic it releases.

FIGURE 10.52 *Penicillium* mould growing on lemons

amensalism any relationship between organisms of different species in which one organism is inhibited or destroyed, while the other organism gains no specific benefit and remains unaffected in any significant way

CASE STUDY: Amensalism between hard-hooved animals and vegetation

Many grazing species were introduced species to Australia for agriculture, such as sheep, cattle, water buffalo and goats. Each of these species have hard hooves that damage the delicate native vegetation, particularly in the arid regions of Australia. In comparison, the soft feet of many native Australian species means they tread much more lightly. For example, kangaroos exert one-third of the pressure on soil compared to sheep.

This damage caused by hard-hooved grazing animals to native vegetation has a number of flow-on effects:
- Hard hooves compact the soil, which prevents native seeds from setting and germinating.
- Compacted soil allows the growth of opportunistic weeds, which reduces food sources for native species and further impacts ecosystems.
- Compaction of the soil and loss of native vegetation increases erosion, eventually leading to a reduction of moisture in the soil, causing it to crack, and soil salinity increases.

While the hard-hooved animals derive no benefit from their actions, there is a detrimental impact on the soil, native animals and ecosystems.

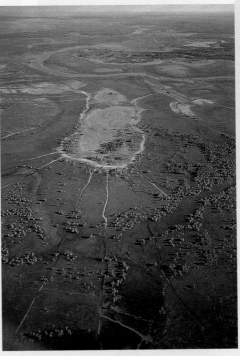

FIGURE 10.53 Wallow holes and trails made by water buffalo (*Bubalus bubalis*), which drain fresh water and cause salt build-ups that kill vegetation

Predator–prey relationships

A **predator–prey relationship** is one in which one species (the predator) kills and eats another living animal (the prey).

Predators or carnivores have structural, physiological and behavioural features that assist them to obtain food. These features include the web-building ability of spiders, claws and canine teeth of big cats, heat-sensitive pits of pythons, poison glands of snakes, visual acuity of eagles and cooperative hunting by dolphins. Some of the different ways that predators capture and eat their living prey include vacuuming, grasping, netting, ambushing, pursuing, piercing, filtering, tearing, engulfing, spearing, constricting, luring and biting.

Think about the labels 'carnivore' and 'predator'. On land, these tend to call up an image of an animal such as a lioness (*Panthera leo*) equipped with strong teeth, sharp claws and powerful muscles, which stalks its prey, pursues it over a short distance and then overpowers and kills it. However, a net-casting spider (*Deinopis* sp.) is equally a predator; it waits for its living prey to come to it to be snared on its web (figure 10.54).

If you were asked to name a predator of the seas, the powerful great white shark (*Carcharodon carcharias*), which actively hunts its prey, would probably spring to mind. However, the coral polyp (see figure 10.55b), which uses a 'sit-and-wait' strategy, is also a carnivore, equipped not with teeth but with tiny stinging cells (figure 10.55a).

> **predator–prey relationship**
> a form of interaction within a community that involves the eating of one species, the prey, by another species, the predator

FIGURE 10.54 a. A net-casting spider pulling out silk threads and crafting them into a net **b.** Spider with a completed net. In Australia, nine species of spider within the genus *Deinopis* are distributed across most states.

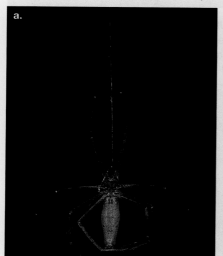
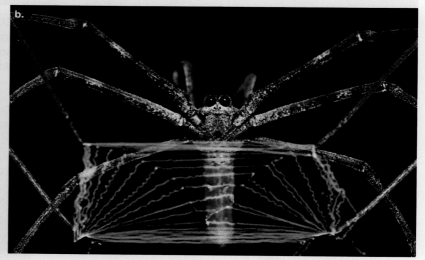

FIGURE 10.55 a. Stinging cells or nematocysts (left: charged; right: discharged). The hollow coiled thread, when discharged, penetrates the prey and injects a toxin. **b.** Coral polyps capture their prey, including fish, using the stinging cells on their 'arms'.

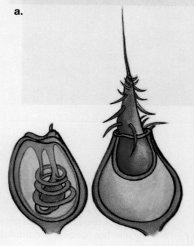

CASE STUDY: Adaptations of predators

Predators come in all shapes and sizes, and different species obtain their prey using different strategies. Let's look at three snake species. Snakes are a remarkable group of predators — legless but very efficient!
- The lowland copperhead snake (*Austrelaps superbus*) lies in wait for its prey, such as a frog or a small mammal. When the prey comes within striking distance, the copperhead strikes, injecting its toxic venom.
- The desert death adder (*Acanthophis pyrrhus*) (figure 10.56a) has a short, thick body but attracts its prey by using its thin tail tip as a lure. The death adder partly buries itself in sand or vegetation and wriggles its tail tip. When its prey is attracted by the 'grub' and comes close, the death adder strikes and injects its venom.
- The green tree python (*Chondropython viridis*) actively hunts its prey by night in trees (figure 10.56b). Its prey includes bats, birds and tree-dwelling mammals. The python locates its prey in the darkness using its heat-sensitive pits, located between their eyes and nostrils. It can detect temperature differences as small as 0.2 °C between objects and their surroundings. By moving its head and responding to information from these sense organs on either side of its head, a python is able to locate a source of relative warmth, such as a bird or a mammal. It kills its prey not by toxic venom, but by constriction.

FIGURE 10.56 a. The death adder uses its tail as a lure to attract prey. Note how the snake positions its tail tip close to its head. What advantage does this behaviour have? **b.** A green python. Like most members of the family Boidae, it has heat-sensitive pits between its eyes and nostrils.

EXTENSION: Response of prey species to predators

In the living community of an ecosystem, predators are *not* always successful in obtaining their prey. Various prey species show structural, biochemical and behavioural features that reduce their chance of becoming a meal for a potential predator. Following are examples of some features that protect prey.

Structural features of prey species:
- **Camouflage** — *look like something else!* Some insects in their natural surroundings look like green leaves, dead leaves or twigs; for example, the stick insect.
- **Mimicry** — *look like something distasteful!* Viceroy butterflies (*Limentitis archippus*) mimic or copy the colour and pattern of monarch butterflies (*Danaus plexippus*), which are distasteful to birds that prey on butterflies.

Behavioural features of prey species:
- *Stay still!* Prey animals such as some rodents and birds reduce their chance of being eaten by staying still in the presence of predators.
- *Keep a lookout!* Meerkats gain protection from predators by having one member of their group act as a sentry or lookout when the rest of the group is feeding (figure 10.57). The lookout signals the approach of a predator, such as an eagle, and the group immediately flees to shelter.
- *Schooling — safety in numbers!* Individual organisms in a large group, such as a school of fish, have a higher chance of not being eaten than one organism that is separated from the group.

Biochemical features of prey species:
- *Produce repellent or distasteful chemicals!* Larvae of the monarch butterfly (figure 10.58) feed on milkweeds, which contain certain chemicals that cause particular predator birds to become sick. The larvae store these chemicals in their outer tissues and the chemicals are also present in adult butterflies. Predator birds that are affected by this chemical rapidly learn that the monarch butterflies are not good to eat. In fact, monarch butterflies advertise that they are distasteful with bright **warning colouration**.

FIGURE 10.57 Meerkats on sentry duty while a pup is feeding

camouflage an adaptation that allows organisms to blend into their environment

mimicry a situation in which one species has an appearance similar to that of a different but distasteful species, where that similarity apparently gives protection against predators

warning colouration a conspicuous colouring that warns a predator that an organism is toxic or distasteful

FIGURE 10.58 A monarch butterfly in its caterpillar stage

Herbivore–plant relationships

One of the most common relationships seen in living communities is the **herbivore–plant relationship**. Herbivores are organisms that obtain their nutrients by eating plants. They include many mammals, such as kangaroos, koalas and cattle, but the most numerous herbivores are insects, such as butterfly larvae (caterpillars) (figure 10.58), bugs, locusts, aphids and many species of beetle. Plants under attack from herbivores cannot run, hide or physically push them away. What can plants do?

herbivore–plant relationship a form of interaction within a community between plants and the animals that eat them

Plants can protect themselves from damage by herbivores by physical means, such as thorns and spines, as seen in cacti, and also by means of stinging hairs, as in nettles.

CASE STUDY: Pandas and bamboo — an unusual herbivore–plant relationship

Giant pandas (*Ailuropoda melanoleuca*) are native to south central China. While their diet is made up almost exclusively of bamboo shoots and leaves, they are classified in the order Carnivora (meat eaters). They will occasionally eat grasses, tubers or meat. They have the teeth and the digestive system of a carnivore.

FIGURE 10.59 The giant panda's diet is largely comprised of bamboo.

Bamboo has a hard, outer layer that requires a significant investment from the panda to access the nutrients. While its leaves have a higher protein content than the stems, it is not a nutritionally dense food, which acts as a deterrent for other organism that require a more nutrient-dense meal. Pandas have a short intestinal tract, which allows a large amount of indigestible material to pass through quickly but also inhibits microbial digestion. Although they gain little energy or protein from eating bamboo, they have few competitors for their main food source.

They have adapted to their bamboo diet with a low surface area to body volume ratio, resulting in a low metabolic rate and therefore, a sedentary lifestyle. The panda has a strong jaw and modified bone on its paws that acts like a thumb — both adaptations to allow them to eat bamboo. Genome sequencing of ancient pandas has also suggested that mutations resulted in the loss of the umami taste receptor, which may have made eating meat less palatable and encouraged the change to a herbivorous diet.

Parasitism in animals

Parasitism occurs when one kind of organism (the **parasite**) lives on or in another kind (the **host**) and feeds on it —typically without killing it, but the host suffers various negative effects in this relationship and only the parasite benefits. A variety of organisms are parasitic, including insects, worms, crustaceans, plants, fungi and microbes. It is estimated that parasites outnumber free-living species by about four to one.

Parasites can be grouped into:
- **exoparasites**, which live *on* their host. Fleas live all of their lives on their host, while ticks and leeches feed from their hosts at specific times only; at other times, they are off the host. Other external parasites are fungi, such as *Trichophyton rubrum*, a fungus that feeds on moist human skin, causing tinea and athlete's foot.
- **endoparasites**, which live *inside* their host.

> **parasitism** a form of interaction within a community that involves one species, the parasite, living on or in another species, the host, typically without killing the host
>
> **parasite** organism that lives on or in another organism and feeds from it, usually without killing it
>
> **host** organism on or in which a specific parasite lives
>
> **exoparasite** parasite that lives on its host
>
> **endoparasite** parasite that lives inside its host

Parasites are also found in freshwater and marine ecosystems. Parasitic lampreys have round sucking mouths with teeth arranged in circular rows (figure 10.60a). Lampreys attach themselves to their host's body using their sucker mouth and rasp the skin of the host fish, then feed on its blood and tissues (figure 10.60b). Other examples of parasite–host relationships include roundworms and tapeworms that are endoparasites in the guts of mammals, such as the beef tapeworm (*Taenia saginata*) and the pork tapeworm (*T. solium*).

FIGURE 10.60 a. Mouth of a wide-mouthed lamprey (*Geotria australis*) showing large teeth above the mouth and radiating plates with smaller teeth around the mouth **b.** Lamprey attached to its host

CASE STUDY: The life cycle of a paralysis tick

In a temperate forest, a wallaby hops through the undergrowth, carrying some 'passengers' that are attached to the animal's face. The passengers in this case are adult female paralysis ticks (*Ixodes holocyclus*), which are native to Australia. Female ticks must attach to and feed on a host, engorging in size after feeding on their host's blood, increasing in size from about 3 mm in length to about 12 mm.

To access more information on this case study and learn about the life cycle of paralysis ticks, please download the digital document.

FIGURE 10.61 Life-size representations of a tick

Unfed Half-fed Fully fed (engorged)

on Resources

Digital document Case study: The life cycle of a paralysis tick (doc-36119)

Parasitism in plants

Parasite–host relationships also exist in the Plant kingdom. The most frequent type of plant parasitism is **hemiparasitism** (from the Greek, *hemi* = half).

Mistletoe plants are the best-known example of a hemiparasite. Australia has many species of mistletoe that are parasitic on different host plants. Mistletoes form connections, known as **haustoria**, with their host plants and the parasites obtain water and mineral nutrients from their hosts through the haustoria (figure 10.62) (see the section on Mutualism).

Another type of plant parasitism is **holoparasitism**, in which the parasite is totally dependent on the host plant for all its nutrients. Although holoparasitism is rare in plants, an example is the *Rafflesia* genus of plants found in the rainforests of Borneo and Sumatra — they have no leaves, stems or roots and, most of the time, they grow as parasites hidden inside the tissues of one specific vine (*Tetrastigma* sp.). A *Rafflesia* parasite forms a bud on the roots of its host vine and, over a 12-month period, the bud swells until finally it opens out into a single giant flower (figure 10.63) that has the smell of rotting meat. The pollinators for *Rafflesia* flowers are flies and beetles that usually feed on dead animals (carrion).

FIGURE 10.62 A mistletoe stem (right) and its host plant (left). Note the connections between the parasite and its host. These connections (haustoria) are modified roots.

FIGURE 10.63 *Rafflesia* flowers are holoparasites.

hemiparasitism form of parasitism in which a plant parasite obtains some nutrients and water from its host plant but also makes some of its own food through photosynthesis

haustoria thin strand of tissue through which a plant parasite makes connection with its host; singular = haustorium

holoparasitism form of parasitism in which a plant parasite depends completely on its host for nutrients and water

Mutualism

Mutualism is a prolonged association of two different species in which both partners gain some benefit. Examples of mutualism include:

- *mistletoebirds (Dicaeum hirudinaceum) and mistletoe plants*. The birds depend on mistletoe fruits for food and, in turn, act as the dispersal agents for these plants. The birds eat the fruit but the sticky seed is not digested. It passes out in their excreta onto tree branches where it germinates.
- *fungi and algae that form lichens* (figure 10.64). The fungus species of the lichen takes up nutrients made by the alga, and the alga appears to be protected from drying out within the dense fungal hyphae.
- *fungi and certain plants*. A dense network of fungal threads (hyphae) becomes associated with the fine roots of certain plants to form a structure known as a **mycorrhiza** (figure 10.65). Plants with mycorrhizae are more efficient in the uptake of minerals, such as phosphate, from the soil than plants that lack mycorrhizae. This is because the mycorrhiza increases the surface area of root systems. The fungal partner gains nutrients from the plant.

FIGURE 10.64 Lichen on a tree trunk

FIGURE 10.65 a. Transverse section through a plant root showing thin threads (hyphae) of the associated fungus b. Longitudinal section through root showing fungal hyphae

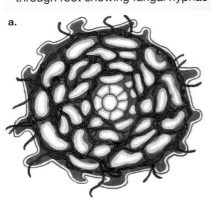
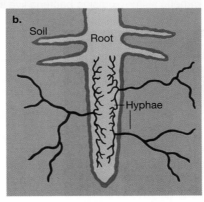

mutualism an association between two different species in a community in which both gain some benefit

mycorrhiza fine threads formed by a fungus that form a large surface area for uptake of nutrients

nitrogen-fixing bacteria bacteria able to convert nitrogen from the atmosphere into ammonium ions

- **nitrogen-fixing bacteria** *and certain plants*. Plants require a source of nitrogen to build into compounds such as proteins and nucleic acids. Plants can use compounds such as ammonium ions (NH^{4+}) and nitrates (NO^{3-}) but cannot use nitrogen from the air. However, bacteria known as nitrogen-fixing bacteria can convert nitrogen from the air into usable nitrogen compounds. Several kinds of plants, including legumes (peas and beans), and trees and shrubs such as wattles, develop permanent associations with nitrogen-fixing bacteria. These bacteria enter the roots and cause local swellings called nodules (figure 10.66). Inside the nodules, the bacteria multiply. Because of the presence of the bacteria in their root nodules, these plants can grow in nitrogen-deficient soils.

FIGURE 10.66 Root nodules contain large numbers of nitrogen-fixing bacteria.

CASE STUDY: Mutualism in marine environments

One stunning example of mutualism on the Great Barrier Reef involves small crabs of the genus *Trapezia* and a specific species of coral, *Pocillopora damicornis*. The crab gains protection and small food particles from the coral polyps. The coral also receives a benefit from the crab. Look at figure 10.67 and you will see a tiny *Trapezia* crab defending the coral polyps from being eaten by a crown-of-thorns starfish (*Acanthaster planci*). The crab repels the starfish by breaking its thorns.

FIGURE 10.67 *Trapezia* crab repelling a crown-of-thorns starfish from eating coral polyps

To date, we have dealt with sunlight-powered ecosystems. However, a striking example of mutualism exists in deep-ocean hydrothermal vent ecosystems. **Sulfur-oxidising producer bacteria** bring energy into these ecosystems through chemosynthesis, using energy released from the oxidation of hydrogen sulfide to build glucose from carbon dioxide. These producer bacteria form microbial mats around the vents that release hydrogen sulfide. In addition, some producer bacteria form relationships with other organisms in the ecosystem, including mussels, clams and giant tube worms (figure 10.68a). Giant tube worms (*Riftia pachyptila*) have no mouth, no digestive system and no anus. They have plumes that are richly supplied with blood and they possess an organ known as a **trophosome** (from the Greek, *trophe* = food; *soma* = body). Chemosynthetic bacteria live inside the cells of the trophosome (figure 10.68b). The tube worms absorb hydrogen sulfide, carbon dioxide and oxygen into blood vessel in their plumes; from there, it is transported via the bloodstream to trophosome cells, where it is taken up by the bacteria living inside those cells.

FIGURE 10.68 a. Giant tube worms form part of the community of a deep-ocean hydrothermal vent. Note their red plumes. **b.** Chemosynthetic sulfur-oxidising bacteria live inside cells of a specialised organ called the trophosome.

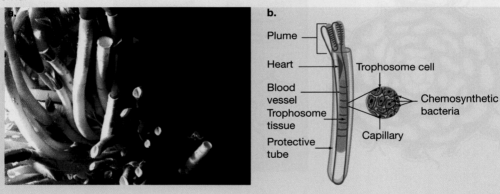

Commensalism

Commensalism ('at the same table') occurs when one member gains benefit and the other member neither suffers harm nor gains apparent benefit. An example of commensalism is seen with clownfish and sea anemones (figure 10.69). The clownfish (*Amphiprion ocellaris*) lives among the tentacles of the sea anemone and is unaffected by their stinging cells. The clownfish benefits by obtaining shelter and food scraps left by the anemone. The anemone appears to gain no benefit from the presence of the fish.

sulfur-oxidising producer bacteria one group of bacteria that gain their energy by oxidising sulfur compounds

trophosome an organ found within deep sea worms that is colonised by bacteria that supply the host worm with food and energy

commensalism association between two different species in a community in which one benefits and the second apparently neither gains nor is harmed

FIGURE 10.69 The clownfish (*Amphiprion ocellaris*) lives among the tentacles of a sea anemone in tropical seas and is unaffected by the anemone's stinging cells.

Symbiosis

Symbiosis ('living together') is the general term for a prolonged association in which there is benefit to at least one partner. It includes interactions such as parasitism, mutualism and commensalism.

symbiosis prolonged association between different species in a community in which at least one partner benefits; includes parasitism, mutualism and commensalism

tlvd-1831

SAMPLE PROBLEM 6 Amensalism and commensalism

With the aid of an example, compare and contrast amensalism and commensalism. **(4 marks)**

THINK	WRITE
1. *Compare* means to find the similarities.	Both amensalism and commensalism have one species unaffected by the actions of the other (1 mark).
2. *Contrast* means to find the differences.	Commensalism provides a benefit to one species, whereas in amensalism, one species is harmed (1 mark).
3. Provide an example.	An example of commensalism is the clown fish benefitting with food and habitat, while the sea anemone is unaffected (1 mark). An example of amensalism is wild pigs killing invertebrates found in the soil; the pigs derive no benefit from the death of the invertebrates (1 mark).

INVESTIGATION 10.4

Interactions in ecosystems

Aim

To explore different interactions within ecosystems, including the presence of keystone species

TABLE 10.9 Summary of relationships between organisms

Interaction	Species 1	Species 2
Amensalism	Harmed	No effect
Predator–prey	Benefits	Harmed
Herbivore–plant	Benefits	Harmed and benefits
Parasitism	Parasite: benefits	Host: harmed
Mutualism	Benefits	Benefits
Commensalism	Benefits	No harm done or benefit gained

Resources

eWorkbook Worksheet 10.6 Relationships in ecosystems (ewbk-7613)
Worksheet 10.7 Members of a community — case study for Lamington National Park (ewbk-7615)

Digital document Case study: Ancient commensalism (doc-36120)

Weblink Symbiosis: The Art of Living Together

10.4.6 Distribution of populations

Distribution refers to the *spread* of members of a population over space. Populations may have identical densities but their distributions can differ. Figure 10.70 shows three populations with identical densities but different horizontal distributions. Clumped (population C) and uniform (population A) distributions are both non-random patterns, with the most common pattern observed in populations being a clumped distribution.

FIGURE 10.70 Three populations, A, B and C, with different distributions. What might cause a clumped distribution?

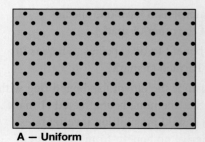
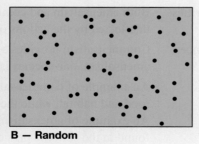
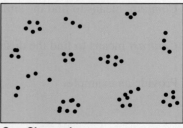

A — Uniform   B — Random   C — Clumped

Changes in the distribution of populations can occur over time. Animal populations that have a random distribution at one period, such as the non-breeding season, may show a different distribution during the breeding season.

Clumped distributions

Clumped distributions occur in:
- plant populations when only some areas within a sample area are suitable for germination and survival of a plant species; the areas without plants are unsuitable for survival because of the pH of the soil, a lack of water or the ambient temperature
- plant populations that reproduce asexually by runners or rhizomes, with new plants appearing very close to the parental plant
- moss populations in open forests, which are confined to damp, sheltered areas; a distribution map of mosses in a forest corresponds to the distribution of damp, sheltered areas
- animal populations that aggregate in the more favourable parts (for example, more protected or closer to water) (figure 10.71)
- populations of mammals that form herds or schools as a strategy for reducing predation.

FIGURE 10.71 A group of feral goats in central Australia — an example of clumped distribution

Uniform distributions

A uniform distribution may indicate a high level of intraspecific competition so that members of a population avoid each other by being equidistant from each other. Uniform spacing is seen in plants when members of a population repel each other by the release of chemicals (figure 10.72). In animals, uniform distribution occurs when members of a population defend territories.

FIGURE 10.72 Spinifex covers the level ground and hills of this area of Western Australia — an example of uniform distribution

Random distributions

A random distribution is expected when:
- the environmental conditions within the sample area are equivalent throughout the entire area
- the presence of one member of a population has no effect on the location of another member of the population.

Both of these conditions rarely occur and, as a result, a random distribution pattern of members of a population is rare in nature.

10.4.7 Variables affecting population size

The size of the population of a particular species in a given area is not always stable. Fluctuations can occur — a population may decline or it may suddenly explode, such as has occurred from time to time with the crown-of-thorns starfish population in the Great Barrier Reef. What determines the size of a population?

FIGURE 10.73 Migration of a wildebeest population

Primary ecological events

The four **primary ecological events** that determine population size are:
1. births
2. deaths
3. immigration (movement of individuals into the population)
4. emigration (movement of individuals out of the population).

Measuring changes in population sizes

A population will increase in size if the sum (births + immigration) is greater than the sum (deaths + emigration). A population will decrease in size if the sum (births + immigration) is less than the sum (deaths + emigration). The growth rate of the population is the change in population size over a set period of time.

To understand population **growth rates**, the following terms may be useful:

- **Positive growth rates** occur when population size increases over a stated period; for example, 200 organisms per year.
- **Negative growth rates** occur when the population size decreases over a stated period.
- **Zero population growth** occurs when growth by births and immigration matches losses by deaths and emigration over a stated period.
- **Open populations** experience migration of individuals into the population (immigration) or out of the population (emigration). For example, wildebeest live in the open woodlands and grassy plains of Africa. Each year they migrate north in May/June in search of food. One and a half million wildebeest make this trek, along with zebras, gazelles and other migratory species. As they are not confined to one location, they live in an open population (figure 10.73).
- **Closed populations** do not experience migration (immigration or emigration) as they are isolated from other populations of the same species; for example, a lizard population on an isolated island. Other closed populations include monkey populations in closed forests on various mountains, where the mountains are separated by open grassland and desert that the monkeys cannot cross. Closed populations are less common among bird species.

primary ecological events in population dynamics, the factors that contribute directly to population density, such as births and deaths

growth rate the change in population size over a set period of time

positive growth rates population increases over a stated period of time

negative growth rates population decreases over a stated period of time

zero population growth occurs when growth by births and immigration matches losses by deaths and emigration over a stated period

open populations refers to populations that experience migration of individuals into the population (immigration) or out of the population (emigration)

closed populations refers to populations that do not experience migration (immigration or emigration) as they are isolated from other populations of the same species

Secondary ecological events

Apart from the four primary ecological events discussed previously, many other factors can affect population size. These include both biotic factors, such as predators or disease, and abiotic factors, such as weather. These factors are called **secondary ecological events** because they influence one or more of the primary events of birth, death, immigration and emigration. For example, events such as droughts, cyclones, bushfires and outbreaks of disease increase deaths in a population. In contrast, events such as favourable weather conditions, removal of predators and increased food supply would be expected to increase births in a population.

One striking example of the impact of a secondary event on population size can be seen with the red kangaroo (*Macropus rufus*). From 1978 to 2004, populations of red kangaroos were surveyed over a large area of South Australia using aerial belt transects. Over that time, the population size varied from a high of 2 175 200 in 1981 to a low of 739 700 in 2003. The major factor affecting the numbers of red kangaroos was drought.

Density-independent or density-dependent?

Some of these secondary events, such as weather events, are said to be **density-independent factors**. This means that they affect all individuals in a population, regardless of the size of the population. So, a sudden frost will kill a high percentage of members of a population of frost-sensitive insects. It does not matter if the population size is small or large. Both a small and a large population would experience the same mortality (death) rate. Likewise, a population of plants in a forest, whether large or small, will all be affected when a bushfire races through their habitat (figure 10.74). Other density-independent events include cyclones, flash floods and heatwaves.

FIGURE 10.74 Bushfires are density-independent secondary events.

In contrast, other secondary events are said to be **density-dependent factors**. These are events that change in their severity as the size of a population changes. The impact of density-dependent factors varies according to the size of a population.

Examples of density-dependent factors include:
- *the outbreak of a contagious disease*. The spread of this disease will be faster in a large, dense population than in a small, sparsely distributed population. As a result, the impact of the disease outbreak is greater in the large population as compared with the small population. The primary means of controlling pandemics, such as COVID-19, is social isolation, which reduces the population exposed to the virus.
- *predation*. Predators are more likely to hunt the most abundant prey species, rather than seek out prey from a small population.
- *competition for resources*. Members of a population need access to particular resources. For plants, these resources include space, sunlight, water and mineral nutrients; for animals, necessary resources include food, water and space for shelter and breeding. These resources are limited in supply.

secondary ecological event in population dynamics, abiotic or biotic factors that influence changes in population density, such as temperature

density-independent factor factor whose impact on members of a population is not affected by the size of the population

density-dependent factor factor whose impact on members of a population is dependent on the size of the population

As a population increases in size, the pressure on these resources increases because of competition between members of the same population for the resources, as well as competition from members of other populations that live in the same habitat and compete for the same resources figure 10.75. The impact of competition on individuals in a population depends on the population size. When a population is small, the impact of competition is low or absent. However, when a population becomes large, competition has a major impact on each member of a population, and survival and reproductive success are threatened.

FIGURE 10.75 Interspecific competition for food. The lioness is trying to defend her prey from her successful hunt from a pack of hyenas.

Variables affecting population size

- The primary ecological events affecting population size are births, deaths, immigration and emigration.
- Secondary ecological events can be due to both biotic and abiotic factors.
- Density-independent factors affect all individuals in a population, regardless of the size of the population; for example, tsunamis or bushfires.
- Density-dependent factors change in their severity depending on the size of the population; for example, a disease spreads quicker through a population with greater numbers as there are more available hosts.

 Resources

 eWorkbook Worksheet 10.8 Variables affecting population size and density (ewbk-7617)

10.4.8 Models of population growth

A new species is introduced to an island where it has no predators, no diseases that might affect it, and food and other resources are in plentiful supply. What will happen to the size of the population?

Two models of population growth in a closed population can be identified:
- the exponential or unlimited growth model (J curve)
- the logistic or density-dependent model (S curve).

Exponential growth: the J curve

Exponential growth is seen in the growth of bacteria over a limited period of time. For example, consider a bacterial species in which each cell divides by asexual binary fission to give two cells every 20 minutes. Figure 10.76 shows the theoretical outcome of this pattern of growth starting with a single bacterial cell. This pattern is known as the J curve.

exponential growth population growth that follows a J-shaped curve but cannot continue indefinitely

Exponential growth can also potentially occur in other species, such as the Australian bush fly (*Musca vetustissima*). If a female fly lays 100 eggs and 50 develop into females, then if exponentiatial growth occurred, there would be 31 billion flies in just over one year. However, conditions required for exponential growth — unlimited resources such as food and space — can last for only a few generations. Every habitat has limited resources and can support populations of only a limited size. So, let's look at another model of population growth that has a better fit with reality.

Density-dependent growth: the S curve

A pair of rabbits in a suitable habitat with abundant food and space initially multiplies and, over several generations, the population grows faster and faster; this is a period of exponential growth. However, this rate of growth cannot continue. As the population increases in size, the pressure on resources increases, competition grows, and the population growth slows and then stops. At this point, the so-called **carrying capacity** of the habitat is reached. The carrying capacity is the maximum population size that a habitat can support in a sustained manner.

The growth of a population under the density-dependent condition is shown in figure 10.77. The growth is at first like the exponential growth pattern, but, as the population grows, the rate of growth slows and finally stabilises at the carrying capacity. This pattern is known as an S curve.

FIGURE 10.77 An S-shaped curve that is typical of the growth of most populations. The upper limit of this curve is determined by the carrying capacity of the habitat. The arrow marks the point of maximum growth of the population.

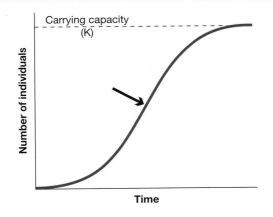

FIGURE 10.76 Exponential growth of bacteria over a seven-hour period, starting with a single cell

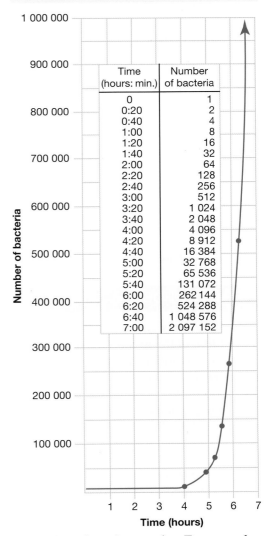

Time (hours: min.)	Number of bacteria
0	1
0:20	2
0:40	4
1:00	8
1:20	16
1:40	32
2:00	64
2:20	128
2:40	256
3:00	512
3:20	1 024
3:40	2 048
4:00	4 096
4:20	8 912
4:40	16 384
5:00	32 768
5:20	65 536
5:40	131 072
6:00	262 144
6:20	524 288
6:40	1 048 576
7:00	2 097 152

Populations affect other populations

The population size of one species can be affected by the size of the population of another species. For example, the size of a plant population is affected by the sizes of the populations of herbivores that feed on that plant.

Other density-dependent factors that influence the size of one population include the sizes of populations of its parasites and its predators. Let's look at how predator and prey populations interact and the impacts on their population sizes.

Predator and prey population numbers

The population size of a prey species can be affected by the size of the population of a predator species that feeds on it. Over time, several outcomes are possible:
- If the predators are absent, the prey population will increase exponentially but will eventually 'crash' when its numbers become too high to be supported by the food resources in the habitat.
- If the prey population is too small, the predator population will starve and die.

carrying capacity the maximum population size that a habitat can support in a sustained manner

In some cases, cycles of 'boom-and-bust' can be seen in both populations, with the peak in the predator population occurring after the peak in the prey population. Why? Figure 10.78a shows the theoretical expectation of these boom–bust cycles while figure 10.78b shows the result obtained in an actual experimental study.

FIGURE 10.78 a. Fluctuations in population size in a prey population and in the predator population that feeds on it. Which population peaks first in each cycle: predator or prey? **b.** Results from a study of boom–bust cycles in predator and prey populations

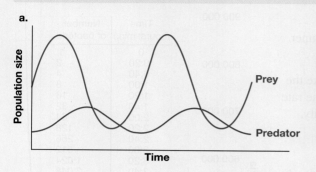
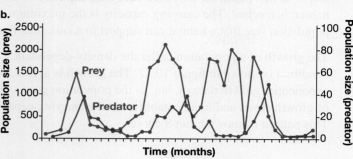

In the next section we will see how the different intrinsic rates of growth of populations can affect their ability to colonise new habitats and their ability to rebuild numbers after a population crash.

tlvd-1832

SAMPLE PROBLEM 7 Predator–prey relationships

Predator–prey relationships are sometimes referred to as an arms race, with prey driving the evolution of the predator. Predators must select for traits that enable them to catch prey, such as being able to camouflage to ambush a future meal or find another food source.

Provide an explanation for why:
a. a rise in prey population is followed by a rise in the predator population and then a decrease in the prey population (2 marks)
b. a crash in the prey population causes a delayed crash in the predator population. (2 marks)

THINK

a. 1. Why is an increase in the prey population followed by an increase in the predator population?

2. What effect does this increase in predator population have on the prey population?

b. If the prey population crashed, what effect would this have on the predator population?

WRITE

When more prey are available there is more food available for predators. This means the predators have more energy to mate, resulting in an increase in the predator population (1 mark).

As the predator population increases, more prey are eaten, and so their population will decrease (1 mark).

If food (prey) has been in abundance, but the population crashes, predators will have energy stores to survive in the short term. However, in the longer term, this decrease in the predators' food source will see a reduction in the predator population (1 mark). Fewer predators will enable the prey population to increase again (1 mark).

elog-0854

INVESTIGATION 10.5 — online only

A population study — modelling breeding in long-nosed bandicoots

Aim
To model the breeding of long-nosed bandicoots

elog-0856

INVESTIGATION 10.6 — online only

Carrying capacity limits and population growth

Aim
To investigate population growth in plants and determine the carrying capacity limit

10.4.9 Populations of a species: intrinsic growth rates

In the previous section we looked at growth of populations in general. Populations of different species vary in their intrinsic rates of increase, typically denoted by the symbol 'r'.

r-selected populations have a 'quick and many' reproduction strategy. Features of r-selected populations include the following:
- They are short-lived (short generations) populations that produce very large numbers of offspring.
- Their population sizes fluctuate widely over time.
  - When resources are plentiful and other environmental conditions are favourable, the numbers of a species can increase very rapidly because of their short generation times and the large numbers of offspring.
  - When conditions are unfavourable, the very low survival rates of offspring mean that population numbers drop sharply. However, populations can recover quickly because of their high growth rates.
- They are adapted for life in 'high risk' and unstable environments (for example, flood and drought-prone areas), and in 'new' habitats such as on fresh lava flows, new land elevated from the sea as a result of earthquake or volcanic activity, or even a cleared patch of land in the suburbs (most weeds are r-selected).
- Examples include bacteria, oysters, cane toads, crown-of-thorns starfish, many species of reef fish, clams, coral polyps, many weed species, rabbits and mice.

In general, r-selection is directed to *quantity* of offspring.

K-selected populations have a 'slower and fewer' reproduction strategy. Features of K-selected populations include the following:
- They are longer-lived populations producing fewer offspring at less frequent intervals.
- Species put their energy into the care, survival and development of their offspring.
- They are adapted to cope with strong competition for resources.
- Population sizes cannot increase rapidly.
- If population size drops sharply as a result of fire, habitat loss or overharvesting, the population will not recover quickly because of their long generation times and their low rates of increase.
- K-selected species are at great risk of extinction if their population numbers fall because their initial rate of replacement is very slow.
- Examples include humans, gorillas, whales, elephants, albatrosses, penguins, southern bluefin tuna and many shark species.

In general, K-selection is directed to *quality* of offspring.

r-selected species organisms that live in unstable environments and produce many offspring with the likelihood that few will survive to adulthood

K-selected species organisms that live in stable environments and produce few offspring in each litter with a greater chance of survival to adulthood

Table 10.10 identifies some of the differences between the extremes of r-selected and K-selected species. Not every species displays all the features of one strategy and many have intermediate strategies.

TABLE 10.10 Extremes of reproductive strategies compared

Feature	r-selected strategy	K-selected strategy
General occurrence	Commonly seen in oysters, clams, scallops, bony fish, amphibians, some birds and some mammals, such as mice and rabbits	Commonly seen in sharks, some birds such as penguins, and some mammals such as whales and gorillas
Lifespan	Shorter	Longer
Number of offspring	Many	Few
Energy needed to make an organism	Smaller	Greater
Survivorship	Very low in young	High in young
Growth rate	Faster	Slower
Age at sexual maturity	Early in life	Late in life
Parental care	Little, if any	Extensive
Adapted for	Rapid population expansion	Living in densities at or near carrying capacity
Relative energy investments	Higher investment into numbers of offspring; lower into rearing offspring	Lower investment into numbers of offspring; higher into rearing offspring

CASE STUDY: K- and r-selected strategies in aquatic species

Bodies of water, such as oceans and rivers, are vast regions with many different species in various niches. Aquatic species use a variety of different reproductive strategies to suit their requirements. This includes both r-selected and K-selected strategies.

Aquatic species with r-selected growth strategies include cane toads (*Rhinella marina*) and crown-of-thorns starfish (*Acanthaster planci*), both of which are significant pests to native populations. Aquatic animals with K-selected strategies include humpback whales (*Megaptera novaeangliae*) and orange roughies (*Hoplostethus atlanticus*).

To access more information on this case study on K-selected and r-selected aquatic species, please download the digital document.

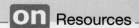

Digital document Case study: K- and r-selected strategies in aquatic species (doc-36121)

FIGURE 10.79 The spawn of cane toads can be readily distinguished from those of native frogs by their appearance as black eggs embedded in long, jelly-like strings. How many eggs are produced on average in a cane toad spawning?

Not all species dovetail into r-selected or K-selected species. For example, green sea turtles (*Chelonia mydas*) exhibit r-selection in terms of the numbers of eggs produced and the lack of parental care, but they also show some K-selection through features such as their slow growth rates, the period required for sexual maturity (estimated at 40 to 50 years) and their long lifespan (estimated at 70 years).

Green sea turtles are common in the waters of the Great Barrier Reef (figure 10.80a). A female lays a clutch of about 100 eggs on a sandy beach at night, covers them with sand and returns to the water (figure 10.80b). During the breeding season, she returns to the same beach approximately every two weeks and may lay 3 to 9 clutches of about 100 eggs per clutch. About two months later, baby turtles hatch from the eggs, dig their way out of the nest and make their way to the sea (figure 10.80c).

Sea turtles use environmental sex determination, where the temperature at which the eggs develop determines the sex of the turtles; lower temperatures produce males, while higher temperatures produce females. As a result of climate change, 65–69% of turtles on the southern end of the Great Barrier Reef and 99% of turtles on the northern end of the reef hatching are female.

FIGURE 10.80 a. Green sea turtles are common in the waters of the Great Barrier Reef. This species has a worldwide range in tropical and semitropical waters. **b.** Female green turtle laying eggs **c.** Turtle hatchlings making their way to the sea. If this occurs during the day, predators such as sea birds and crabs take many of the hatchlings before they reach the sea.

Resources

eWorkbook Worksheet 10.9 How fast are they growing? (ewbk-7621)

KEY IDEAS

- An ecosystem consists of a living community, its non-living physical surroundings, and the interactions both within the community and between the community and its physical surroundings.
- Ecosystems are the most complex level of biological organisation.
- A community is composed of several populations.
- Each ecosystem has a living community composed of several populations.
- A keystone species is one that has a disproportionately larger effect on the ecosystem in which it lives relative to other species.
- Every ecosystem must have a continual input of energy from an external source.
- The organisms in an ecosystem can be grouped into the major categories of producers, consumers and decomposers.
- Producers use the energy of sunlight to build organic compounds from simple inorganic materials.
- Consumers obtain their energy and nutrients from the organic matter of living or dead organisms.
- Consumers can be subdivided into various groups.
- Decomposers break down organic matter to simple mineral nutrients.
- Interactions occur continuously between and within the various components of an ecosystem.
- Relationships between different species in a living community of an ecosystem can be grouped into different kinds, with effects on species involved being beneficial, harmful or benign.

- Population size is determined by four primary events: birth, death, immigration and emigration.
- Population size is also affected by secondary events that impact on the rate of births and deaths.
- The impact of some secondary events depends on the size of a population and these are said to be density-dependent.
- Exponential population growth follows a J-shaped curve but cannot continue indefinitely.
- Logistic population growth follows an S-shaped curve that levels off at the carrying capacity of the ecosystem concerned.
- The population of one species may be affected by the population size of another species in the community.
- Populations differ in their intrinsic rates of growth.
- Species can be identified as being r-selected or K-selected.
- r-selection and K-selection are the extremes of a range.
- r-selected species are adapted for living in newly created and in unstable habitats.
- K-selected species are adapted for living in stable habitats and at densities at or near the carrying capacity of a habitat.

10.4 Activities

learn on

To answer questions online and to receive **immediate feedback** and **sample responses** for every question go to your learnON title at **www.jacplus.com.au**. A **downloadable solutions** file is also available in the resources tab.

| 10.4 Quick quiz on | 10.4 Exercise | 10.4 Exam questions |

10.4 Exercise

1. When measuring diversity of a community, what two factors need to be considered?
2. Suggest a reason for why species richness varies with latitude.
3. Why is a keystone species so important in an ecosystem?
4. a. Outline the role that the cassowary has in the rainforest of Far North Queensland.
   b. Suggest what would happen to native fruits that grow in the rainforest should the cassowary become extinct.
5. Green plants are often identified as the main producer in an ecosystem. Provide two other examples of producers and explain how they convert energy into a form that can be used within an ecosystem.
6. a. What is the role of detritivores in an ecosystem?
   b. Describe how decomposers differ from other consumers within an ecosystem.
7. What would happen to the green python if it was unable to detect heat from its prey?
8. Meerkats always have a member of their group acting as a lookout. What advantage does this social behaviour have for the overall survival of the species?
9. Describe a way that plants defend themselves from being eaten by herbivores.
10. Describe how the formation of lichen is beneficial to both fungi and algae.
11. Explain why some asexually producing organisms are more likely to have clumped distributions.
12. Growth rate of a population is linked to the number of births and immigration being greater than the number of deaths and emigration. The outbreak of a disease impacts the growth rate of a population. Would the outbreak of a disease have a greater impact on an open or closed population? Explain your answer.
13. Describe the effect of carrying capacity on a population.
14. What can cause a crash in a prey population?
15. K-selected growth is often described as slower and fewer. Compare the energy invested in offspring from the K-selected growth strategy with the energy invested with offspring from the r-selected growth strategy. To populate a disturbed area, which strategy is most effective?

10.4 Exam questions

Question 1 (1 mark)
MC In a tropical rainforest, plants of faster-growing species will shade plants of slower-growing species. The plants of slower-growing species will not receive enough sunlight and will die.

This is an example of
A. mutualism.
B. amensalism.
C. predation.
D. parasitism.

Question 2 (1 mark)
MC Whales are often seen with barnacles attached to their bodies. The tail of the whale in the photograph has barnacles attached to it.

The barnacles are in a position to obtain food and are transported through the ocean by the whale. The barnacles neither harm nor benefit the whale.

This is an example of
A. predation.
B. commensalism.
C. mutualism.
D. amensalism.

Question 3 (2 marks)
There is a lake that contains 10 different species of fish. In one season there is a lot of rain in the area surrounding the lake and the lake floods. The growths of the fish populations are affected.

State whether this would be a density-dependent or density-independent effect on the populations and give a reason for your choice.

Question 4 (4 marks)
Consider the diagram of the tapeworm *Echinococcus granulosis*. The tapeworm lives in the digestive system of dogs.

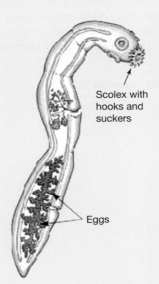

a. What is the function of the scolex? **1 mark**

A single *E. granulosis* contains both male and female sex organs.
b. Of what benefit is this to the organism? **1 mark**
c. Describe the relationship between *E. granulosis* and a dog. **2 marks**

▶ Question 5 (6 marks)

The graph displays the growth of a population over time.

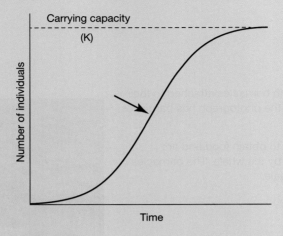

a. **MC** Consider the position of the graph marked by the arrow.
At this point you would expect
A. the amount of food available for individuals in the population to be limited.
B. the population to be in the lag phase of growth.
C. the population to be growing at maximum growth rate.
D. population growth to be influenced by a density-independent factor. **1 mark**
b. After a period of time, the carrying capacity of the habitat is reached, as marked by the dashed line on the graph.
  i. What does the term *carrying capacity* mean? **1 mark**
  ii. What type of growth does this curve represent? Provide an example to illustrate your response. **2 marks**
c. Dingoes are a keystone species in many arid Australian habitats that have a similar population growth rate similar to data shown in the graph. The presence of dingoes protects mammal biodiversity in these areas and the impact of introduced species. Dingoes have been introduced to areas with large feral pig populations.
  i. Suggest why dingoes have been introduced to areas with large feral pig populations. **1 mark**
  ii. Predict the changes on the population size of other native mammals after dingoes are introduced to areas with feral pig populations. **1 mark**

More exam questions are available in your learnON title.

10.5 Australia's First Peoples

> **KEY KNOWLEDGE**
>
> - The contribution of Aboriginal and Torres Strait Islander peoples' knowledge and perspectives in understanding adaptations of, and interdependencies between, species in Australian ecosystems
>
> *Source:* VCE Biology Study Design (2022–2026) extracts © VCAA; reproduced by permission.

10.5.1 First human settlement of Australia

Archaeological evidence shows that Aboriginal peoples have lived in Australia for over 60 000 years, with Torres Strait Islanders being present for at least the last 2500 years. The Aboriginal and Torres Strait Islander peoples, whilst both being acknowledged as the First Peoples of Australia, have very diverse cultures and identities. Aboriginal Australians are considered to have a land-based culture, whereas Torres Strait Islanders are considered to have a sea-based culture. There are over 500 Aboriginal Australian groups on mainland Australia, and 5 island clusters of Torres Strait Islanders encompassing 274 small islands off the far northern coast of Queensland.

FIGURE 10.81 An Aboriginal Australian cave painting in Kakadu National Park in the Northern Territory. This cave art is culturally important in communicating stories to future generations.

Europeans arrived in Australia less than 250 years ago and hence, the knowledge and perspectives of the original custodians of the land and surrounding seas are recognised as essential in maintaining viable ecosystems moving forward.

10.5.2 Seasons and the use of fire

In the Western tradition, four seasons are recognised — winter, spring, summer and autumn — and these are determined by the Gregorian calendar. Whether you live in the tropical north of Queensland, or the sub-temperate climate of Tasmania, you are in one of these four seasons. Seasons in the northern hemisphere are opposite to the seasons in the southern hemisphere.

FIGURE 10.82 The position of the dark emu constellation signals stages of the emu's life cycle including nesting times.

Aboriginal and Torres Strait Islander peoples recognise different numbers of seasons depending where in Australia their land is. The recognition of these different seasons is the result of Aboriginal and Torres Strait Islander peoples' complex and subtle knowledge of the stars, as well as the life cycles of plants and animals and changing weather conditions over thousands of years of observation (figure 10.82). Thus, these seasons are localised to specific areas. This wholistic understanding of individual ecosystems is fundamental to successful survival and land management. For example, in Melbourne there are seven annual seasons, as well as non-annual fire and flood seasons. Flood season is generally every 28 years, with fire being every 7. However, in the Grampians to the west of Melbourne, there are six annual seasons. (See the Weblinks in this subtopic for more information).

Burning techniques

Aboriginal Australians have used fire for many centuries, and it is an important part of both culture and survival. In the tropical north of Australia, cool burning has been used to manage the land.

Cool burning has many features that contribute to land management:
- It occurs early in the dry season, before temperatures reach their annual peak.
- Reducing the fuel load of the area ensures that if a fire comes through later in the season, the ferocity of the fire is reduced, resulting in less damage to the ecosystem.
- It stimulates new growth, as some plants, such as eucalypts, rely on fire to germinate their seeds (see topic 9).
- Less carbon dioxide is released into the atmosphere when compared to hot burning. With rising carbon dioxide levels contributing to global warming, this is essential to reducing the carbon footprint of Australia.
- It reduces the number of hot burns that can occur. This is beneficial to native wildlife, who continue to have food sources and habitats to live in. Cool burns rarely burn logs that have fallen, which are essential for small animals as hiding spots and burrows for protection from predators and/or the environmental conditions. This supports biodiversity, as species do not lose all food and habitat resources.

FIGURE 10.83 Cool burning in Kakadu National Park. This technique burns the forest floor and undergrowth, with the canopy being unaffected. This ensures that habitat is still protected for organisms that rely on the flora for shelter and food.

Furthermore, cool burning is also used as a means of attracting organisms for hunting purposes, to provide food. Creating open grassland expanses that are inviting to kangaroos provides a food source for the human inhabitants.

The grass plains apparent in southern Australia before the arrival of Europeans have since become regions of scrub and bushland as the use of fire to manage the land declined. These areas are now more prone to hot burns. Hot burns decrease biodiversity, as they do not enable animals to escape as readily, and the habitat and food sources of organisms that do manage to survive are severely impacted.

10.5.3 Maintenance of biodiversity

Aboriginal and Torres Strait Islander peoples have a strong connection to the land. In the Arrernte language, spoken in central Australia, *anpernirrentye* means 'kin relationships among all things'. Traditional Aboriginal Australian life relies on the environment and all things in it for food, medicine and other materials commonly acquired at a supermarket or hardware store. Tea tree oil, an antibacterial product now commonly found in many Australian homes, was first used by Aboriginal Australian peoples. They also use a particular desert mushroom to cure a sore mouth. This specialised knowledge has been passed down through generations of Australian Aboriginal peoples. Table 10.11 provides some examples of bush medicine utilised by Aboriginal and Torres Strait Islander peoples. In contrast to most non-Indigenous Australians, Aboriginal and Torres Strait Islander peoples see themselves as part of an ecosystem, not the managers of the ecosystem. Topic 11 includes a discussion of biopiracy, which is the theft of indigenous knowledge of biological resources without permission.

TABLE 10.11 Indigenous bush medicine

Traditional medicine	Properties and effects
Tea tree oil (*Melaleuca alternifolia*)	Antiseptic and antifungal properties; leaves can be crushed and applied as a paste to wounds
Eucalyptus oil (*Eucalyptus* sp.)	Leaves can be infused for pain relief, fever and chills
Kakadu plum (*Terminalia ferdinandiana*)	Richest source of vitamin C in the world; the plums are a major source of food
Desert mushrooms (*Pycnoporus* sp.)	Cures a sore mouth or lips; the mushrooms can also be used as an infant teething ring
Emu bush (*Eremophila* sp.)	Used to wash sores and cuts; research is currently being undertaken to see whether the emu bush can be used for sterilising implants such as artificial hips
Witchetty (witjuti) grub	May be crushed into a paste to help heal burns and soothe skin
Snake vine (*Tinospora smilacina*)	Crushed snake vine can be used to treat headaches; it also is an effective anti-inflammatory.
Sandpaper fig (*Ficus opposita*) and stinking passionflower (*Passiflora foetida*)	Treatment of itchy skin involves crushing the leaves of the sandpaper fig, soaking them in water and placing them on the itchy area until bleeding; the pulped fruit of the stinking passion flower is then smeared over the itchy area.
Kangaroo apple (*Solanum laciniatum* and *S. aviculare*)	The fruit contains a steroid that is useful to relieve swollen joints.
Goat's foot (*Ipomoea pes-caprae*)	Provides pain relief from stingray stings; the leaves are crushed and heated, then applied directly to the skin

Aboriginal and Torres Strait Islander peoples' knowledge of the balance within ecosystems is the result of observations of interactions between species. This knowledge has been passed on to subsequent generations through story (see the Case study box later in this section). These observations are not entirely altruistic; rather, Aboriginal and Torres Strait Islander peoples recognise that a healthy ecosystem also provides benefits to their populations through increased food or other resources.

For Aboriginal Australian peoples, 'Country' refers to more than the physical land on which they were born. It includes all living things, seasons, stories and creation spirits. One issue that faces many Aboriginal Australian communities trying to maintain Country is the introduction of invasive species. For example, the cane toad has decimated the goanna population. To ensure the longevity of the species, goannas are not hunted as frequently for food in order to enable them to repopulate, as Aboriginal Australians seek to increase their numbers.

Inspired by nature

This connection to the land and the organisms that are part of it has enabled Aboriginal and Torres Strait Islander peoples to live within Australian ecosystems for millenia. Their knowledge has also been used to solve problems, akin to **biomimicry**.

FIGURE 10.84 Witchetty grubs, a nutritious bush snack that is sweet-tasting and high in protein, with a liquid centre. They are a useful treatment for burns when ground into a paste.

biomimicry the imitation of designs found in nature to solve human problems using technology

The kangaroo pouch provides a safe place for a joey to continue to develop prior to becoming independent of its mother. Aboriginal Australians designed a similar 'pouch' to carry their babies. The benefit of this is that their hands were free to carry other things, and the energy expenditure to carry their child was reduced.

The barb of a stingray, perhaps made most famous in the death of environmentalist Steve Irwin, is a structural feature observed in an organism that was then modified for use. The top of a stingray's stinger has a sharp point to easily pierce the skin, and backward serrations that often contain venom. The serrations act like a fishhook, making it difficult to remove and immobilising the prey or threat. Weapons and tools made by Aboriginal and Torres Strait Islander peoples used the stingray barb as a spear tip.

Maintenance of biodiversity in the Torres Strait Islands

Aboriginal and Torres Strait Islander peoples have been stewards of the Torres Strait, and the region has significant cultural significance, in addition to its significance in providing resources for survival. The Torres Strait Protected Zone, in accordance with the Torres Strait Treaty signed in 1978 and enacted in 1985, allows Torres Strait Islanders and those from coastal Papua New Guinea to move within the region without passports in order to enable them to maintain traditions and cultural practices.

In acknowledgment of the economic, cultural and spiritual connection that indigenous groups have with the region, Torres Strait Islanders are governed by *Ailan Kastom* (Island Custom), with Aboriginal Australians in the region governed by Aboriginal Lore (Law).

Part of the Torres Strait Treaty enables those who are indigenous to the region to fish. With their intricate knowledge of the land and sea, from centuries of observations passed down through generations, they have an awareness of when species are in abundance. For example, in the wet season there are ample shellfish, so this is the time to hunt and eat them. When milkwood blooms, it signals the arrival of the stingrays (figure 10.84). This local knowledge demonstrates the intricate understanding that Torres Strait Islanders have of the interrelationships between species. It is this respect for the sea and other ecosystems that allows those within the region to maintain the biodiversity by not overfishing or seeking to catch particular organisms when they are not in high numbers. This ensures that organisms can repopulate during the breeding seasons, aiding in the long-term survival of the species and providing a long-term food source. Whilst traditionally this food was caught using a canoe, now you are more likely to see the use of outboard motors or dinghies.

FIGURE 10.85 The grey milkwood, native to Northern Queensland, Papua New Guinea and the Torres Strait. When it begins to bloom, Torres Strait Islanders identify this as an indication that stingrays are arriving in the waters.

SAMPLE PROBLEM 8 Identifying relationships within ecosystems

tlvd-1833

Through knowledge passed down through generations, Torres Strait Islanders use the blooms of milkwood plants to signal the arrival of the stingray hunting season. If the milkwood plants were to become extinct in Far North Queensland, Papua New Guinea and the Torres Strait, what impact could this have on the stingray population in the first year, and subsequent years? **(2 marks)**

THINK

1. What effect would the extinction of the milkwood plant have?

2. How is this linked to the stingray population?

WRITE

If the milkwood plant became extinct, the blooms would not appear. This is one of the seasonal indicators used to identify when other organisms, such as stingrays, are ready to be hunted.

If the indicator is not present that it is time to hunt the stingrays (the milkwood blooms), the stingray population may increase in the short term as they would not be hunted (1 mark).

In subsequent years you could expect that an alternative marker would be used to indicate the stingray hunting season can commence, and their population would decline (1 mark).

CASE STUDY: Using storytelling to protect ecosystems

Within Australia there are hundreds of different Aboriginal Australian cultures. Each group has their own language and their own creation story. One way that these stories and culture are passed down to the next generation is through storytelling. These stories may be collective histories, spiritual narratives, life histories or cultural practices. Similarly, Torres Strait Islanders also use storytelling as a means to pass their cultural knowledge to subsequent generations.

Many of these stories describe significant events in the changing landscape of Australia, and it is important that they are transmitted accurately as they contain important information about finding water or navigating different areas.

The Dreaming is the environment that Aboriginal and Torres Strait Islander peoples once lived in; it is the connection between the spiritual world of the past and today. One example the Gunai/Kurnai Traditional Custodians' story of Jiddelek the frog. He selfishly drank all the water in their environment, which left all the animals thirsty. However, by working together, the other animals found a solution to their problem. This story demonstrates the importance of working as a community to protect the environment and to ensure everyone's survival.

 Resources

 Weblinks Indigenous weather knowledge
Indigenous astronomy and seasonal calendars
Creating a seasonal calendar
Cultural burning
Seasonal calendar for the Melbourne area
The story of Jiddelek

KEY IDEAS

- Cool burning reduces the fuel load in the warmer months and helps to maintain biodiversity.
- Aboriginal and Torres Strait Islander peoples identify themselves as part of an ecosystem, rather than managers of the ecosystem.
- Aboriginal and Torres Strait Islander peoples' observations of nature enabled them to create objects to enhance their survival.
- Aboriginal and Torres Strait Islander peoples identify the change in seasons by the changes in the environment.
- Aboriginal and Torres Strait Islander peoples' management of resources is based on observations of the natural world and understanding of the balance in ecosystems.

10.5 Activities

To answer questions online and to receive **immediate feedback** and **sample responses** for every question go to your learnON title at **www.jacplus.com.au**. A **downloadable solutions** file is also available in the resources tab.

| 10.5 Quick quiz | 10.5 Exercise | 10.5 Exam questions |

10.5 Exercise

1. Describe the difference between the Gregorian calendar and the Aboriginal and Torres Strait Islander calendars.
2. Describe three advantages of cool burning to an ecosystem.
3. What effect do hot fires have on the long-term health of an ecosystem?
4. Describe why it is important to hunt organisms that are in abundance at a particular time of year, rather than hunt year-round.
5. Do you believe that we should incorporate more Aboriginal and Torres Strait Islander peoples' knowledge into our treatment of illness? Justify your response.

10.5 Exam questions

Question 1 (1 mark)

MC Aboriginal and Torres Strait Islanders peoples have a rich understanding of the ecosystems around them and the flora and fauna within these. Much of the native flora is used for various medicinal purposes. One such plant used is *Melaleuca alternifolia*, which is used to extract tea tree oil.

The properties and effects of tea tree oil include
A. treatment of burns.
B. soothing of itchy skin.
C. antiseptic and antifungal properties.
D. being a rich source of vitamin C.

Question 2 (3 marks)

The Torres Strait is a region north of Queensland that has significant cultural significance. The Torres Strait Protected Zone sits within the Torres Strait. It is an area that is fundamentally important in regards to resources and culture. Within this region, Aboriginal and Torres Strait Islander peoples and those from coastal Papua New Guinea can freely move without passports in order to carry out traditional activities. These activities are carried out in a way that uses local knowledge of the awareness of the land, sea and surrounding ecosystems. Regulations around this were part of the Torres Strait Island Treaty.

a. Describe one example of a traditional activity undertaken by Aboriginal and Torres Strait Islander peoples that allows for the preservation of ecosystems and biodiversity in the Torres Strait. **2 marks**
b. Explain why it is important that Aboriginal and Torres Strait Islander peoples can freely travel through the Protected Zone. **1 mark**

Question 3 (4 marks)
Cool burning is used as a land management technique across Australia.
a. Why is cool burning undertaken early in the dry season? **1 mark**
b. How does cool burning help maintain Australia's biodiversity? **1 mark**
c. Suggest why cool burning can be used as part of Australia's plan to reduce carbon emissions. **2 marks**

Question 4 (4 marks)
Biomimicry is the development of tools by humans based on, or inspired by, designs in nature. Identify two examples of biomimicry in Aboriginal and Torres Strait Islander peoples' tools, providing a reason for their usefulness.

More exam questions are available in your learnON title.

10.6 Review
10.6.1 Topic summary

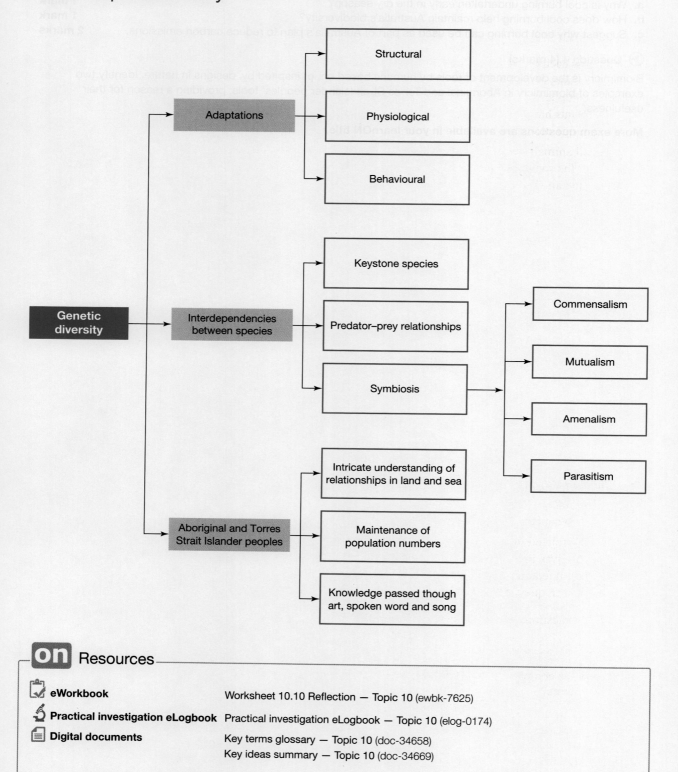

Resources

eWorkbook — Worksheet 10.10 Reflection — Topic 10 (ewbk-7625)

Practical investigation eLogbook — Practical investigation eLogbook — Topic 10 (elog-0174)

Digital documents — Key terms glossary — Topic 10 (doc-34658)
Key ideas summary — Topic 10 (doc-34669)

10.6 Exercises

To answer questions online and to receive **immediate feedback** and **sample responses** for every question go to your learnON title at **www.jacplus.com.au**. A **downloadable solutions** file is also available in the resources tab.

10.6 Exercise 1: Review questions

1. Explain how each of the following features assists a plant to survive in a very hot environment:
   a. Desert plants generally have deeply penetrating root systems.
   b. Succulent plants (that store water) have stomata that open only at night.
   c. Some plants have special cells, called hinge cells, on the surface of their leaves that also have stomata. When hinge cells lose water, the leaf rolls up with the hinge cells on the inside of the rolled leaf.

2. Many small animals that are solitary over summer tend to become social during winter and often construct nests under the snow. The temperature inside a communal nest of beavers was compared with the temperature of the outside air. The results are shown in the figure.

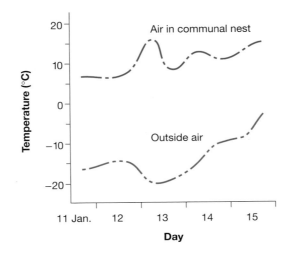

   a. What is the maximum difference between the temperature of the beaver nest and the temperature of the outside air?
   b. Suggest what causes this difference in temperature.

3. Two closely related mammalian species differ in their range. One species, A, lives in a sandy desert, while the other, B, lives in a cool temperate grassland. Devise four relevant questions about structural, physiological or behavioural features of these mammals that will show which species lives in which habitat, and give the answers to each question.

   Show your questions and your answers by constructing a table similar to the one shown.

Question	Species A	Species B
1. What is the comparative ear length of each species?	Longer	Shorter

4. The figure shows a morning-glory vine (*Ipomoea purpurea*) that has overgrown plants, some of which can still be seen in the lower right-hand corner of the image.
   a. What will happen to the underlying plants?
   b. What name might be given to the relationship between the vine and the plants below it?

5. Consider the following relationships between organisms in a community:
   i. Termites ingest the cellulose of wood but they cannot digest it. Protozoan organisms living in the termite gut secrete cellulases, which are enzymes that digest cellulose, releasing nutrients that can be used by the termites.
   ii. Giant tube worms (*Riftia pachyptila*) living in deep-ocean hydrothermal vents have no mouth or digestive system. Within their cells live chemosynthetic bacteria.
   iii. Adult barnacles are sessile animals that are filter feeders. Barnacles are generally found attached to rocks. Some barnacles, however, attach to the surface of whales.
   iv. Bacteria of the genus *Vampiro coccus* attach to the surface of other species of bacteria, where they grow and divide, consuming the other bacteria.

In each case:
   a. identify the type of relationship that exists between the organisms
   b. briefly describe the outcome of the relationship for each member of the pair of interacting organisms.

6. a. Explain why most stems and leaves on plants have a waterproof cuticle and yet roots do not.
   b. When cuttings of plants are first potted, they have no roots and may wilt. Wilting is prevented if the pot is enclosed in a plastic bag and shaded from sunlight. Explain why this treatment prevents wilting.

7. Suggest an explanation for, or comment on, each of the following observations:
   a. The young of tiny bats such as the little bent-wing bat (*Miniopterus australis*) huddle very tightly together in large groups on the walls of caves where they live, rather than being widely separated.
   b. One plant species grows well in soils with a high salt content but a second plant species dies if the salt concentration in the soil exceeds a low value.
   c. Brown trout (*Salmo trutta*) are found in cold, fast-flowing mountain streams, but are absent from warm, sluggish waters.
   d. If placed in seawater, goldfish (*Carassius auratus*) will die.
   e. Desert mammals are typically active at night.
   f. A person who suffers from obesity has a lower body water content than a lean person.

8. Elk in the Yellowstone National Park were said to be at carrying capacity prior to the reintroduction of the wolves in 1995. Explain what factors would have limited an increase in their population size.

10.6 Exercise 2: Exam questions

Resources

Teacher-led videos Teacher-led videos for every exam question

Section A — Multiple choice questions

All correct answers are worth 1 mark each; an incorrect answer is worth 0.

Question 1
What advantage does the presence of a lignotuber have for the survival of eucalypts following a fire?

A. It has stored chlorophyll for the plant to be able to photosynthesise.
B. It stores starch for energy use when the plant is regenerating and unable to photosynthesise.
C. It provides an extensive root system for the eucalypt to obtain nutrients from the soil.
D. It protects the phloem, enabling transport of nutrients through the soil.

Question 2
Which of the following is an example of a tolerance range that may affect the survival chances of a marine species?

A. Water temperature
B. pH of the water
C. Dissolved oxygen concentration
D. All of the above

Question 3
What is the advantage of having sunken stomata in an arid environment?

A. Sunken stomata can create a humidity chamber to reduce water loss.
B. Sunken stomata only open at night when it is cool.
C. Sunken stomata are smaller in order to reduce water loss.
D. Sunken stomata create a dry environment to reduce water loss.

Question 4
An example of intraspecific competition is

A. two albatross competing for the same nesting site.
B. a lion and a tiger competing to catch a gazelle.
C. a brown bear trying to catch a migrating salmon.
D. a plant competing with a weed for sunlight.

Question 5
Monarch butterflies are distasteful to birds that prey upon them. With their distinctive markings, birds that are hunting for food will avoid them and seek other sources of prey. The viceroy butterfly does not produce the same chemical that is distasteful to birds; however, they mimic the appearance of the monarch butterfly to confuse potential predators.

This is an example of

A. camouflage.
B. mimicry.
C. commensalism.
D. schooling.

▶ **Question 6**

The clownfish, made famous by the film *Finding Nemo*, lives among the tentacles of the sea anemone. It obtains food and shelter and appears unaffected by the sea anemones stinging cells. The sea anemone derives no benefit from this relationship, but also does not suffer any harm.

What term best describes the relationship between the sea anemone and the clownfish?

A. Predator–prey
B. Parasitism
C. Mutualism
D. Commensalism

▶ **Question 7**

Bacteria replicate through binary fission and can increase their population sizes rapidly, as is evident by their exponential growth. Exponential growth, represented with the J curve, is not sustainable over an extended period of time.

The reason for this is

A. bacteria die quickly, so exponential growth cannot occur.
B. a disease may kill a significant number of the population.
C. every habitat has limited resources, such as food and space.
D. bacteria would convert to mitosis to replicate, which is a slower process.

▶ **Question 8**

Populations that operate under the 'slower and fewer' mode of reproduction are referred to as

A. S-selected.
B. K-selected.
C. r-selected.
D. F-selected.

▶ **Question 9**

When is the traditional time of year that Aboriginal Australians undertake cool burning?

A. Early in the dry season
B. Early in the wet season
C. Late in the dry season
D. Late in the wet season

▶ **Question 10**

What advantage did Aboriginal Australians derive from creating open grasslands using cool burning?

A. More space was created to build areas for communities to live and sleep in.
B. They attracted native animals that were traditionally hunted.
C. Clear areas were created that could be used for agricultural and horticultural purposes.
D. They provided a region that was easier for rain to run off in heavy downpours.

Section B — Short answer questions

Question 11 (4 marks)

In nature, two variants of the peppered moth exist: white and black. Prior to the Industrial Revolution of the 1800s, the white moth was favoured and more prevalent in moth populations. Following the Industrial Revolution, buildings began to become covered in black soot and the black peppered moth became more prevalent in populations.

a. With reference to genetic variation, explain what would happen if a natural disaster occurred and the environment of the moth became lighter. **2 marks**

b. Explain what would happen to bird species that rely on the black peppered moth as their primary food source. **1 mark**

c. How does genetic variation increase the survival chances of the peppered moth? **1 mark**

Question 12 (5 marks)

The figure shows the great crested newt (*Triturus cristatus*), a forest-dwelling amphibian found in the United Kingdom. Female great crested newts lay two to three eggs each day from March to July, with about 200 eggs being sealed within aquatic leaves.

One great crested newt found that her habitat had been tainted by the presence of plastic bags. She sealed one of her eggs within this bag in place of a leaf.

a. i. Compare and contrast the structural properties of a plastic bag and leaves. **2 marks**

ii. How will the egg being laid in a plastic bag affect the larvae's ability to hatch after the two to three-week incubation period? **1 mark**

b. What structural features enable the newt to seal eggs within a leaf? **1 mark**

c. What type of growth strategy does the great crested newt employ? **1 mark**

Question 13 (4 marks)

Kelp forests in marine ecosystems are valuable due to their ability to absorb CO_2 from the atmosphere and convert it into usable energy and O_2. Within kelp forests live organisms such as sea otters, sea urchins and many small invertebrates. Many of the small invertebrates rely upon the kelp for both habitat and food resources.

Within the kelp forests, sea otters feed on sea urchins. If the population of sea urchins increases, the amount of kelp decreases as the sea urchins feed upon it.

a. What term is used to describe the role of sea otters within the ecosystem? **1 mark**

b. What effect would an increase in sea urchins have on the population of small invertebrates? **1 mark**

c. Unlike most marine species, sea otters lack the layer of blubber that acts as insulation. What adaptations does the sea otter have that enables it to maintain its temperature in the cold marine environment? **2 marks**

Question 14 (8 marks)

The kangaroo rat is a small nocturnal organism that is found in North America. It lives in incredibly dry environments and thus does not consume any water.

a. If the kangaroo rat does not drink water, from what sources can it obtain water? **1 mark**

b. Kangaroo rats produce extremely concentrated urine. Explain how extremely concentrated urine would be produced and why this is an important adaptations in arid regions. **3 marks**

c. Kangaroo rats are exposed to extremely hot conditions. Outline two physiological and two behavioural responses that would help them reduce their core temperature. **4 marks**

Question 15 (6 marks)

Dromedary (*Camelus dromedarius*) and Bactrian (*Camelus bactrianus*) camels are both well-adapted to survive arid conditions. They employ both physiological and structural adaptations.

a. A camel's body temperature can vary over a wide range (34 to 42 °C). How does this help them survive in extreme heat conditions? **1 mark**

b. Camels can tolerate dehydration, and when water is available, can drink large quantities of water (more than 100 litres in one day). What adaptations of the camel's red blood cells would allow for:

   i. tolerance of dehydration **1 mark**

   ii. maintenance of osmotic balance after drinking large quantities of water? **1 mark**

c. A biologist wanted to test whether the hair of camels had insulating properties. They formulated the hypothesis that 'removal of the hair from a camel leads to an increase in the rate of evaporation from the camel's skin'. The results of the biologist's investigation are summarised in the graphs. The two camels were in the same environment. The hair on camel X was removed; camel Y was left with its natural hair. Graph P represents the results at the end of one day.

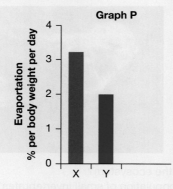
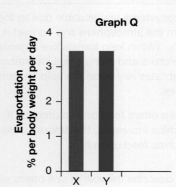

   i. Identify the control group for the experiment summarised in graph P. **1 mark**

   ii. Identify the independent variable for the experiment summarised in graph P. **1 mark**

   iii. Later, camels X and Y were both shorn and the evaporation rate for each camel measured again. The results are shown in graph Q. Suggest a reason why the second stage of the experiment was most likely carried out. **1 mark**

10.6 Exercise 3: Biochallenge online only

Resources

eWorkbook Biochallenge — Topic 10 (ewbk-8080)

Solutions Solutions — Topic 10 (sol-0655)

teachon

Test maker
Create unique tests and exams from our extensive range of questions, including practice exam questions. Access the assignments section in learnON to begin creating and assigning assessments to students.

Online Resources

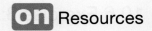

Below is a full list of **rich resources** available online for this topic. These resources are designed to bring ideas to life, to promote deep and lasting learning and to support the different learning needs of each individual.

eWorkbook

- 10.1 eWorkbook — Topic 10 (ewbk-3172)
- 10.2 Worksheet 10.1 The importance of genetic diversity (ewbk-7597)
- 10.3 Worksheet 10.2 Bills and beaks: How birds feed (ewbk-7599)
  - Worksheet 10.3 Types of adaptations (ewbk-7601)
  - Worksheet 10.4 Adaptations in arid regions (ewbk-7603)
  - Worksheet 10.5 Adaptations in cold regions (ewbk-7605)
- 10.4 Worksheet 10.6 Relationships in ecosystems (ewbk-7613)
  - Worksheet 10.7 Members of a community — case study for Lamington National Park (ewbk-7615)
  - Worksheet 10.8 Variables affecting population size and density (ewbk-7617)
  - Worksheet 10.9 How fast are they growing? (ewbk-7621)
- 10.6 Worksheet 10.10 Reflection — Topic 10 (ewbk-7625)
  - Biochallenge — Topic 10 (ewbk-8080)

Solutions

- 10.6 Solutions — Topic 10 (sol-0655)

Practical investigation eLogbook

- 10.1 Practical investigation eLogbook — Topic 10 (elog-0174)
- 10.3 Investigation 10.1 Abiotic factors and tolerance range (elog-0846)
  - Investigation 10.2 Leaves for survival (elog-0848)
- 10.4 Investigation 10.3 Distribution patterns of moss (elog-0850)
  - Investigation 10.4 Interactions in ecosystems (elog-0852)
  - Investigation 10.5 A population study — modelling breeding in long-nosed bandicoots (elog-0854)
  - Investigation 10.6 Carrying capacity limits and population growth (elog-0856)

Digital documents

- 10.1 Key science skills — VCE Biology Units 1–4 (doc-34648)
  - Key terms glossary — Topic 10 (doc-34658)
  - Key ideas summary — Topic 10 (doc-34669)
- 10.3 Case study: Dominant plants (doc-36116)
- 10.4 Case study: Populations in Antarctic communities (doc-36117)
  - Case study: The life cycle of a paralysis tick (doc-36119)
  - Case study: Ancient commensalism (doc-36120)
  - Case study: K- and r-selected strategies in aquatic species (doc-36121)

Teacher-led videos

- Exam questions — Topic 10
- 10.2 Sample problem 1 The importance of lignotubers to maintain genetic diversity of a species (tlvd-1826)
- 10.3 Sample problem 2 Advantages and disadvantages of two types of egg production (tlvd-1827)
  - Sample problem 3 Using concentration gradients to determine water loss (tlvd-1828)
  - Sample problem 4 How plants survive sub-zero temperatures (tlvd-1829)
- 10.4 Sample problem 5 Keystone species (tlvd-1830)
  - Sample problem 6 Amensalism and commensalism (tlvd-1831)
  - Sample problem 7 Predator–prey relationships (tlvd-1832)
- 10.5 Sample problem 8 Identifying relationships within ecosystems (tlvd-1833)

Weblinks

- 10.2 Animals' response to bushfires
- 10.3 Trees to ease drought
  - Antarctic animals adapting to the cold
- 10.4 Role of keystone species in an ecosystem
  - Gray wolf: Yellowstone National Park
  - Symbiosis: The Art of Living Together
- 10.5 The story of Jiddelek
  - Indigenous weather knowledge
  - Indigenous astronomy and seasonal calendars
  - Creating a seasonal calendar
  - Cultural burning
  - Seasonal calendar for the Melbourne area

Teacher resources

There are many resources available exclusively for teachers online

To access these online resources, log on to **www.jacplus.com.au**

UNIT 2 | AREA OF STUDY 2 REVIEW

AREA OF STUDY 2 How do inherited adaptations impact on diversity?

OUTCOME 2
Analyse advantages and disadvantages of reproductive strategies, and evaluate how adaptations and interdependencies enhance survival of species within an ecosystem.

PRACTICE EXAMINATION

STRUCTURE OF PRACTICE EXAMINATION		
Section	Number of questions	Number of marks
A	20	20
B	8	30
	Total	50

Duration: 50 minutes
Information:
- This practice examination consists of two parts. You must answer all question sections.
- Pens, pencils, highlighters, erasers, rulers and a scientific calculator are permitted.

SECTION A
All correct answers are worth 1 mark each; an incorrect answer is worth 0.

1. Amoebae are tiny shapeless, unicellular organisms. They replicate their genetic material through miotic division and produce two daughter cells of the same size in the process.
   The method of reproduction that amoebae use is best described as
   A. fragmentation.
   B. binary fission.
   C. budding.
   D. spore formation.

2. Which of the following is **not** an advantage of asexual reproduction?
   A. The ability to repopulate areas quickly
   B. Energy is not expended on finding a mate.
   C. Offspring are well suited to the current environment.
   D. There is little genetic diversity in the population.

3. Sexual reproduction provides variation in a population through the fusion of a sperm and egg with varying genomes. Aside from independent assortment of chromosomes that occurs during meiosis, what other event occurs during meiosis that increases genetic variation in the offspring?
   A. Crossing over of chromosomes
   B. Unwinding of sister chromatids
   C. Removal of the chiasma
   D. Replication of the plasmid

4. Zhong Zhong and Hua Hua are macaque monkeys created using somatic nuclear cell transfer technology. With reference to their genome, they would be
   A. identical to the somatic donor.
   B. identical to the sperm donor.
   C. identical to the surrogate.
   D. a combination of alleles from both the egg and sperm donors.

5. Genetic variation within a population is advantageous to the survival of the species as it
   A. reduces intra-specific competition.
   B. allows faster rates of reproduction.
   C. makes it easier to identify who belongs to a given population.
   D. provides favourable alleles to select for future generations.

6. The mountain pygmy possum, *Barramys parvus,* is found in the alpine regions of Victoria and New South Wales. To survive the cold winter months when food resources are scarce, the mountain pygmy possum hibernates under an insulating layer of snow. Using snow to provide insulation during hibernation is an example of a
   A. physiological adaptation.
   B. structural adaptation.
   C. behavioural adaptation.
   D. symbiotic relationship.

7. Plants have many structural adaptations to enable them to survive in dry and arid environments. One such adaptation is the ability to roll their leaves to reduce water loss. This reduces water loss by
   A. creating a dry air chamber, which reduces the rate of transpiration.
   B. creating a humidity chamber, which reduces the rate of transpiration.
   C. forcing the closure of the stomata.
   D. blocking the xylem, preventing transpiration from occurring.

8. The wildflowers, shown in the image, compete for access to soil nutrients and sunlight.

   This type of competition is referred to as
   A. intraspecific competition.
   B. interspecific competition.
   C. resource guarding.
   D. monopolistic competition.

9. Which of the following graphs best represents the relationship between predator and prey populations?

A.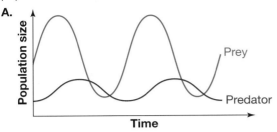

B.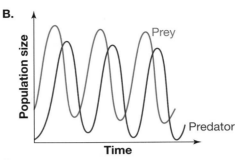

C.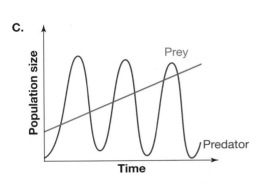

D.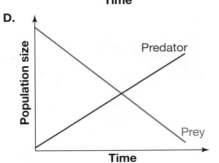

10. In some marine ecosystems, sea urchins have no natural predators. Sea urchins feed on algae, which provides oxygen for the ecosystem. In other ecosystems, sea otters feed on sea urchins, which keeps the sea urchin numbers in check. Depending upon their location, sea otters are prey for killer whales or great white sharks. What term best describes the role of the sea otter in the marine ecosystem?

   A. Keystone species
   B. Symbiotic organism
   C. Apex predator
   D. Top order consumer

11. Selective breeding of animals, such as the golden retriever with her pups shown in the photograph, reduces genetic diversity in the gene pool of the offspring.

   This is because
   A. all offspring are clones of the parent.
   B. of a random assortment of alleles that produces a desired phenotype for a buyer.
   C. of the selection of identical desirable traits in both parents.
   D. only one parent contributes to the genome of the daughter offspring.

12. Which of the following is an advantage of using tissue culture cloning to propagate plants?
    A. Virus-free tissue can be used to limit the spread of disease.
    B. Sterile growing conditions reduce the need for quarantine when exporting plants.
    C. Slow growing plants can be created in large numbers.
    D. All of the above

13. The pistol shrimp is a marine organism that creates burrows in the sea bed with its sharp claws. They have poor eyesight, which places them at risk of predation. Another marine species, the goby fish, is unable to create burrows as it lacks claws. It resides with the pistol shrimp within the burrow and spends much of its time on the lookout for predators. When a threat is detected, the goby fish darts into the burrow, which also alerts the shrimp to the danger.

    The relationship between these two organisms is best described as
    A. parasitic.
    B. amensalism.
    C. mutualism.
    D. commensalism.

14. Some sexually reproducing organisms, such as crayfish and Komodo dragons, can reproduce asexually if no mate is available. In this process, an egg is able to develop into an embryo without fertilisation occurring. This process is known as
    A. heterogametic sex.
    B. homogametic sex.
    C. parthenogenesis.
    D. somatic cell nuclear transfer.

15. As populations increase in size, competition for resources, including food, habitat and space, increases. As the population increases, its growth rate slows before it finally stops. This is known as the carrying capacity of a population.
Which of the following graphs best represents the carrying capacity of a population after an initial increase in size?

A.
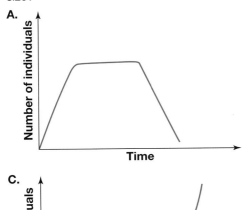

B.
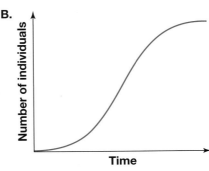

C.
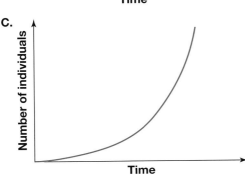

D.
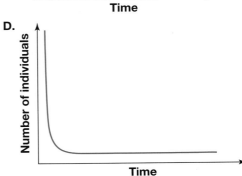

16. Organisms have different reproductive strategies to repopulate. For some, such as mammals, there is a long gestation period and a significant investment in parental care over a period of time. This is referred to as a K-selected strategy. For others, the 'quick and many' strategy is used, with potentially thousands of offspring created from a single mating event, with no parental care following birth. This is referred to as an r-selected growth strategy. Which of the following organisms is likely to reproduce using the r-selected strategy?

   A. Emperor penguin
   B. Southern right whale dolphin
   C. Cane toad
   D. Orangutans

17. Which of the following statements regarding population size is correct?
   A. If births and immigration is greater than deaths and emigration, the population will increase in size.
   B. If births and emigration is greater than deaths and immigration, the population will increase in size.
   C. If births and immigration is less than deaths and emigration, the population will increase in size.
   D. If births and emigration is less than deaths and immigration, the population will increase in size.

18. All organisms have a tolerance range in which they can exist. Within this range, there is an optimal region, a zone of physiological stress and a zone of intolerance, where the organism can no longer survive.

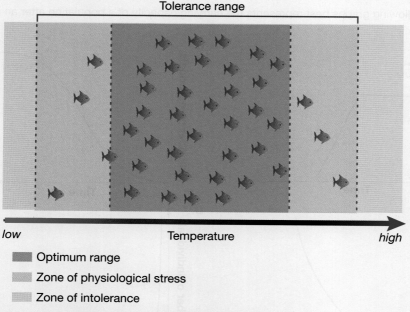

What is the tolerance range for a particular organism dependent upon?
A. Limiting factors
B. Abiotic factors
C. Biotic factors
D. Structural adaptations

19. It is widely appreciated that Aboriginal and Torres Strait Islander peoples have a strong connection to the land and have been stewards of the Australian continent for over 60 000 years. During this time, they have observed features of organisms and adapted these to improve their own practices. Which of the following is an example of imitating nature to solve a problem?
A. Burning during the hot months
B. Storytelling to pass cultural knowledge to future generations
C. Creating a pouch to carry young
D. Cave painting to maintain a historical record of key events and beliefs

20. Cool burning is a practice used by Aboriginal and Torres Strait Islander peoples to reduce leaf litter and maintain ecosystems. It is beneficial to the environment because
A. hollow logs are burnt, making hunting easier.
B. forest canopies are burnt to allow for more sunlight to reach the forest floor.
C. it can reduce the ferocity of the fires in peak fire season.
D. it encourages thick scrub growth.

SECTION B — Short answer questions

Question 21 (4 marks)

The two naturally occurring types of reproduction are asexual and sexual. List two advantages and two disadvantages for asexual and sexual reproduction.

	Advantages	Disadvantages
Asexual reproduction		
Sexual reproduction		

Question 22 (2 marks)

An organism similar in size to a chloroplast is found to show exponential growth in population size, with daughter cells the same size as the parent cells. What type of organism and what method of reproduction could this refer to?

Question 23 (2 marks)

The black-flanked rock-wallaby lives in small isolated populations in Western Australia, South Australia and the Northern Territory. They are dusk feeders, sheltering in rock piles and caves during the day. Their numbers have declined due to habitat loss, introduction of pest species and changes to fire patterns.

a. What effect would isolation of the populations have on the genetic diversity of the species? **1 mark**
b. What fire strategy would you recommend to reduce the risk of the black-flanked rock-wallaby being affected by fire? **1 mark**

Question 24 (6 marks)

The emu, found in grasslands and forests, is a native Australian bird renowned for its inability to fly. Emus are prey for dingoes and wedge-tailed eagles, with their eggs being eaten by reptiles such as goannas. Their high protein eggs are an important seasonal food source for Aboriginal peoples across Australia. Despite being ground dwellers, emus can move at nearly 50 kilometres per hour. They have a sharp beak and feed on a diet of leaves, grass shoots, larvae and beetles. Emus have a double set of eyelids, one to lubricate the eye and one to act as a translucent layer to protect the eye from dust when they are running through the dirt. Their long, double-quilled feathers allow them to maintain a constant temperature of about 40 °C. A nasal turbinate allows them to reabsorb water into the body from the nasal passages when water is scarce. Unlike many species, after the female lays her eggs she leaves the nest never to return, while the male incubates the eggs. The male continues to offer protection to the young for up to two years.

a. What is the difference between a structural and a behavioural adaptation? **2 marks**
b. Identify two structural adaptations of emus and explain how they contribute to their survival. **2 marks**
c. What type of reproductive strategy does the emu employ? Explain. **2 marks**

Question 25 (4 marks)

Compare predator–prey relationships with parasite–host relationships, providing an example of each. **4 marks**

Question 26 (5 marks)

Beavers have been referred to as 'nature's engineers' due to their natural tendency to build dams, which are essential for maintaining wetlands and the species that they support, such as young salmon and frogs. Dams slow water movement, which prevents erosion as the slower-moving water does less damage to the banks of rivers and streams. Beavers have webbed back feet, a broad oar-like tail and valves in their mouth that enable them to gnaw while underwater.

If beavers were to be removed from an ecosystem, there would be a significant effect.

a. What term is used to describe the role of a beaver in an ecosystem? **1 mark**
b. Predict what would happen to salmon numbers should beavers be removed from the ecosystem. **2 marks**
c. Name and describe one adaptation that enables beavers to construct dams. **2 marks**

Question 27 (1 mark)

Cloned organisms can be created through both somatic cell nuclear transfer and embryo splitting. Compare the two processes.

Question 28 (6 marks)

A scientist wanted to test the effectiveness of different antibiotics against the bacterium *Staphylococcus aureus*. He prepared agar plates, and, in a sterile environment, covered these with a strain of *Staphylococcus aureus*. He then placed an antibiotic ring on the plate, sealed it and incubated the plate for 48 hours at 40 °C. Following the 48-hour incubation period, he analysed his results. The greater the ring around each antibiotic (zone of inhibition), the greater its effectiveness.

His results are summarised below:

Antibiotic	Zone of inhibition
Te 30	6
S 10	3
N 30	2
L 2	1.5
MY 100	0

a. Write a hypothesis for his experiment. **1 mark**
b. Why did he incubate the bacteria at 40 °C? **1 mark**
c. Suggest an improvement for his experiment. **1 mark**
d. If bacteria reproduced sexually, would you expect the same results? Explain. **3 marks**

END OF EXAMINATION

UNIT 2 | AREA OF STUDY 2

PRACTICE SCHOOL-ASSESSED COURSEWORK

ASSESSMENT TASK — A report of a laboratory activity investigating the tolerance range of temperature for bacterial growth

In this task you will complete a report of a laboratory activity, including the generation of primary data.
- Practical activities must be completed over two days together. Day 1 will be the setting up of the experiment and incubation overnight. Day 2 will be the recording of the results and the SAC task.
- The SAC task will follow the collection of results.
- The SAC task is a scientific report based on your experimental results.

Total time
- 50 minutes to set up agar plates
- 20 minutes to collect data
- 50 minutes to complete practical report

Total marks

Practical report: 30 marks

EXPERIMENT: The zone of tolerance for asexual reproduction

Bacteria reproduce asexually through the process of binary fission. *E. coli* is an example of a bacteria that is part of the natural flora of the human gut, and some strains have a beneficial relationship with humans. Some strains of *E. coli* can be harmful and are a common cause of food poisoning. As with all organisms, bacteria have a tolerance range in which they can exist. In this experiment, the zone of tolerance for temperature will be investigated.

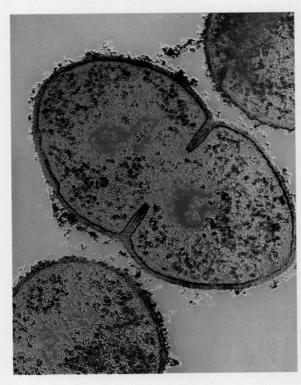

Required materials

- 5 agar plates with lids
- 4 mL of *E. coli* culture K-12 strain live broth
- Bunsen burner
- Heat mat
- 2 Incubators
- Parafilm
- 1 L-shaped cell spreader
- Micropipette (1 mL capacity) and tips or disposable 1 mL pipette

Method

Before you begin: Write a clear aim and hypothesis in your logbook.

1. After putting on a pair of disposable gloves, label the plate around the edge of the base with your name, the date and *E. coli*.
2. Set up and light a Bunsen burner. Make sure it is on a heatproof mat. All your work should now be done near the flame to prevent contamination.
3. Using a micro pipette, add 1 mL of *E. coli* to the agar plate. Using the L-shaped cell spreader, swab each of your plates with the *E. coli* bacterial culture. Make sure the whole plate is covered with the bacterial culture.
4. Seal the 4 experimental plates with parafilm and incubate the plates overnight at:
   - 40 °C
   - 60 °C
   - 4 °C
   - 20 °C.
5. Seal the final agar plate with no treatment with *E. coli*.
6. The following day examine the plates. Do not open the seal on the agar plates. Determine the percentage cover of the agar plate by using a marker to separate the plate into quarters and estimating the coverage of each quarter. Add these percentages and divide by 4 to determine the percentage cover of the plate.

$$\frac{Q1 + Q2 + Q3 + Q4}{4} = \text{total percentage cover of } E.coli$$

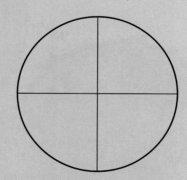

Results

After washing your hands thoroughly and disposing of your agar plates as directed by your teacher, copy and complete the following table in your logbook.

TABLE 1 Percentage cover of *E. coli* growth at experimental temperatures

Temperature	Quarter 1	Quarter 2	Quarter 3	Quarter 4	Total % cover
4 °C					
20 °C					
40 °C					
60 °C					

ASSESSMENT TASK — Scientific report on results

In this task you will write a scientific report based on the results of your experiment on the zone of tolerance for asexual reproduction.
- You may use your practical logbook, scientific calculator and general stationery to complete this task.

Total time: 50 minutes writing time

Total marks: 30 marks

With reference to your logbook, write a scientific report based upon your results. Your report must address the following:
- Aim
- Hypothesis
- Reference to the method (including any modifications, if any, that you did)
- Results: this should include a graph
- Discussion which, as a minimum, should respond to the following:
  - Identification of the dependent and independent variables
  - The optimal temperature for growth of *E.coli* and implications for optimal conditions for growth, making reference to the method of reproduction
  - Limitations of your experiment
  - Suggestions to improve accuracy
  - Suggestions for further experimentation
  - Other factors that would impact *E.coli* growth and how these could be tested
  - Considerations for minimising food poisoning in food outlets
- Conclusion
- Any references and acknowledgements

See rubric provided for marking criteria.

Resources

Digital documents Unit 2 Area of Study 2 School-assessed coursework (doc-35101)
Unit 2 Area of Study 2 School-assessed coursework rubric (doc-35102)

AREA OF STUDY 3 HOW DO HUMANS USE SCIENCE TO EXPLORE AND COMMUNICATE CONTEMPORARY BIOETHICAL ISSUES?

11 Exploring and communicating bioethical issues

KEY KNOWLEDGE

online only

In this topic you will investigate:

Scientific evidence
- the distinction between primary and secondary data
- the nature of evidence and information: distinction between opinion, anecdote and evidence, and scientific and non-scientific ideas
- the quality of evidence, including validity and authority of data and sources of possible errors or bias
- methods of organising, analysing and evaluating secondary data
- the use of a logbook to authenticate collated secondary data

Scientific communication
- biological concepts specific to the investigation: definitions of key terms; use of appropriate biological terminology, conventions and representations
- characteristics of effective science communication: accuracy of biological information; clarity of explanation of biological concepts, ideas and models; contextual clarity with reference to importance and implications of findings; conciseness and coherence; and appropriateness for purpose and audience
- the use of data representations, models and theories in organising and explaining observed phenomena and biological concepts, and their limitations
- the influence of social, economic, legal and political factors relevant to the selected research question
- conventions for referencing and acknowledging sources of information

Analysis and evaluation of bioethical issues
- ways of identifying bioethical issues
- characteristics of effective analysis of bioethical issues
- approaches to bioethics and ethical concepts as they apply to the bioethical issue being investigated

Source: VCE Biology Study Design (2022–2026) extracts © VCAA; reproduced by permission.

This topic is available online at **www.jacplus.com.au**

Online Resources

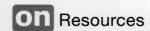

Below is a full list of **rich resources** available online for this topic. These resources are designed to bring ideas to life, to promote deep and lasting learning and to support the different learning needs of each individual.

eWorkbook

- 11.1 eWorkbook — Topic 11 (ewbk-3174)
- 11.2 Worksheet 11.1 Analysing bioethical issues (ewbk-7629)
- 11.3 Worksheet 11.2 Evaluating secondary sources (ewbk-7631)
- 11.4 Worksheet 11.3 Communicating effectively in science (ewbk-7633)
- 11.6 Worksheet 11.4 Acknowledgements and referencing (ewbk-7635)
- 11.7 Worksheet 11.5 Reflection — Topic 11 (ewbk-7639)
- Biochallenge — Topic 11 (ewbk-8081)

Solutions

- 11.7 Solutions — Topic 11 (sol-0656)

Digital documents

- 11.1 Key science skills — VCE Biology Units 1–4 (doc-34648)
- Key terms glossary — Topic 11 (doc-34659)
- Key ideas summary — Topic 11 (doc-34670)

Teacher-led videos

- Exam questions — Topic 11

Weblinks

- 11.2 Human Research Ethics Committees
- Principles of Biomedical Ethics
- 11.3 The mind of a scientist, *The Conversation*
- 11.5 Theory, fact and the origin of life: Stephen Jay Gould
- 11.6 Creative commons for teachers and students
- Copyright — The Australian Government
- Citing and referencing guides
- Online citation generator

Teacher resources

There are many resources available exclusively for teachers online

To access these online resources, log on to **www.jacplus.com.au**

GLOSSARY

abiotic factor nonliving factor, such as weather, that can affect population size

abomasum final part of the stomach of a ruminant where enzymes are secreted and used in digestion; food then passes to the small intestine.

acacia shrubland major vegetation type dominated by mulga, a species of *Acacia*, occurring in arid inland Australia

accuracy how close an experimental measurement is to a known value

acinar cells cells of the pancreas that produce and transport enzymes that are passed into the duodenum where they assist in the digestion of food

active boundary a barrier that is constantly changing and responsive to the environment

acute hypothermia occurs when a person is suddenly exposed to extreme cold

adaptation features or traits that appear to equip an organism for survival in a particular habitat

adenine (A) one of the purine bases found in the nucleotides that are the building blocks of DNA (and RNA)

adenosine triphosphate (ATP) the common source of chemical energy for cells

adhesion attraction of water molecules to other kinds of charged molecules, such as the walls of xylem tubes

adipocytes cells with large deposits of stored fat

adrenal glands small endocrine glands located on top of the kidneys that produce various hormones

adult somatic cells body cells that have differentiated into their cell type

adult stem cells undifferentiated cells obtained from various sources and capable of differentiating into related cell types; also known as somatic stem cells

aim a statement outlining the purpose of an investigation, linking the dependent and independent variables

albinism inherited condition in which pigment production does not occur normally

aldosterone a steroid hormone produced by the adrenal cortex that regulates levels of salt and water and so controls blood pressure

alimentary canal the whole passage from mouth to anus; see gastrointestinal tract

alleles the different forms of a particular gene

amensalism any relationship between organisms of different species in which one organism is inhibited or destroyed, while the other organism gains no specific benefit and remains unaffected in any significant way

anaphase stage of mitosis during which sister chromatids separate and move to opposite poles of the spindle fibre within a cell

anecdote an individual's story based on personal experience rather than strong evidence

Angelman syndrome a genetic disorder resulting from the loss of function of a gene on chromosome-15 inherited from the mother; it primarily affects the central nervous system

Animalia a group of eukaryotic multicellular organisms whose cells lack a cell wall

anterior pituitary anterior lobe of the pituitary gland; it is made of glandular tissue that synthesises and secretes several releasing hormones that activate other endocrine glands.

antidiuretic hormone hormone produced by neurosecretory cells in the hypothalamus; increases reabsorption of water into the blood from distal tubules and collecting ducts of nephrons in the kidney

antifreeze a chemical substance produced by an organism to prevent freezing of body fluids or tissue when in a sub-zero environment

antigens proteins on the plasma membrane of the red blood cells, the presence or absence of which determines phenotype

apoptosis the programmed death of cells that occurs as a normal and controlled part of an organism's growth or development

apoptosome a large protein formed during apoptosis; its formation triggers a series of events that leads to apoptosis

apoptotic body vesicle containing parts of a dying cell

aquaporins protein channels in the plasma membrane that allow the rapid flow of water into and out of cells

archaea a group of prokaryotes that live in extreme environments; also known as extremophiles

areolar loose, irregularly arranged connective tissue

asexual reproduction that only requires one parent, leading to the production of a clone

asexual reproduction type of reproduction that does not require the fusion of gametes, where offspring arise from a single parent and are genetically identical to that parent

autophagy breakdown by lysosomes of non-functioning cell organelles that are old and/or damaged and in need of turnover

autosomal genes have two copies of each gene located on an autosome

autosome any one of a pair of homologous chromosomes that are identical in appearance in both males and females

autotrophs organisms that, given a source of energy, can produce their own food from simple inorganic substances; also known as producers

bacteria a group of prokaryotes that can reproduce by binary fission

bar graph a graph in which data is represented by a series of bars. They are usually used when one variable is quantitative and the other is qualitative.

basal metabolic rate the minimum amount of heat generated in the body by metabolic processes

bias the intentional or unintentional influence on a research investigation

binary fission process of cell multiplication in bacteria and other unicellular organisms in which there is no formation of spindle fibres and no chromosomal condensation

biodiversity total variety of life forms, their genes and the ecosystems of which they are a part; biodiversity can relate to planet Earth or to a given region

bioethics the study of ethics, or the moral principles which guide good behaviour, in the fields of biology, medicine, research and biotechnology

biomacromolecules large biological polymers such as nucleic acids, proteins and carbohydrates

biomimicry the imitation of designs found in nature to solve human problems using technology

biopiracy the use and appropriation of knowledge of farming and indigenous communities by others for economic gain and control of intellectual property

biotic factor living factor, such as predators or disease, that can affect population size

bivalent a pair of homologous chromosomes that are held together by at least one crossover

blebs bulges of the cell membrane created as the cytoskeleton of the cell breaks down; these break off to form apoptotic bodies

budding asexual reproduction where a new organism develops from cell division at an outgrowth from the parent

bulb an underground storage organ with short stems, a central bud and many closely packed, fleshy leaves

bulk transport the movement of material into a cell (endocytosis) or out of a cell (exocytosis)

camouflage an adaptation that allows organisms to blend into their environment

cancer a disease in which cells divide in an uncontrolled manner, forming an abnormal mass of cells called a tumour

capsule polysaccharide layer outside the cell membrane for protection

carnassial teeth paired upper and lower premolars and molars which do not meet, and hence allow a shearing action to tear food

carnivore organism that kills and eats animals

carrier an organism that has inherited an allele for a recessive genetic trait but usually does not display that trait or show symptoms of the disease

carrier protein protein that binds to a specific substance and facilitates its movement through the membrane

Casparian strip a ring of waterproof material, composed of lignin, deposited on the walls of endodermal cells that forces all soil water to move into the cells of the endoderm before it can reach the xylem

caspase cascade a group of proteins that are sequentially activated to bring about apoptosis

caspases protease enzymes that break down proteins during apoptosis

causation when one factor or variable directly influences the results of another factor or variable

cell basic functional unit of all organisms

cell-based therapies the use of stem cells in the treatment of human disorders or conditions to repair the mechanisms of disease initiation or progression

cell cycle the series of events of cell growth and reproduction that results in two daughter cells

cell division division of a cell into two genetically identical daughter cells
cell elongation any permanent increase in size of a cell
cell (plasma) membrane partially permeable boundary of a cell separating it from its physical surroundings; boundary controlling entry to and exit of substances from a cell
cell theory theory that all living things are made of cells
cell wall semi-rigid structure located outside the plasma membrane in the cells of plants, algae, fungi and bacteria
cellular respiration process of converting the chemical energy of food into a form usable by cells, typically ATP
cellulose complex carbohydrate composed of chains of glucose molecules; the main component of plant cell walls
central nervous system the part of the nervous system composed of the brain and spinal cord
centrioles a pair of small cylindrical organelles, used in spindle development in animal cells during cell division
centromere the position where the chromatids are held together in a chromosome
channel proteins trans-membrane proteins involved in the transport of specific substances across a plasma membrane by facilitated diffusion
Chargaff's rules the relative proportions of bases A and T are equal and C and G are equal
chemical digestion the chemical reactions changing food into simpler substances that are absorbed into the bloodstream for use in other parts of the body
chenopod shrubland major vegetation type dominated by saltbushes (*Atriplex* spp.) and bluebushes (*Maireana* spp.); occurs in arid regions with salty soils
chiasma a point at which paired chromosomes remain in contact during the first metaphase of meiosis, and at which crossing over and exchange of genetic material occur between the strands
chief cells in the parathyroid gland, secretory cells that produce parathyroid hormone (PTH) (Note: different cells with same name are present in the gastric pits of stomach.)
chitin a fibrous substance, mainly composed of polysaccharides, used in the cells walls of fungi
chlorophyll green pigment required for photosynthesis that traps the radiant energy of sunlight
chloroplast chlorophyll-containing organelle that occurs in the cytosol of cells of specific plant tissues
cholesterol sterol compound important in the composition of cell membranes
chromatid one of two identical threads in a replicated DNA molecule
chromatin a mass of genetic material composed of DNA and proteins that condense to form chromosomes during eukaryotic cell division
chromosome a thread-like structure composed of DNA and protein
chyme slurry of partially digested food produced in the stomach which passes into small intestine
cilia in eukaryote cells, whip-like structures formed by extensions of the plasma membrane involved in synchronised movement; singular = cilium
circular folds permanent macroscopic folds (ridges) in the mucosa of the small intestine
clones groups of cells, organisms or genes with identical genetic make-up
closed populations refers to populations that do not experience migration (immigration or emigration) as they are isolated from other populations of the same species
codominance the relationship between two alleles of a gene such that a heterozygous organism shows the expression of both alleles in its phenotype
cohesion tendency of water molecules to stick together through the formation of hydrogen bonds between one another
collenchyma thick, flexible walled cells in plants; the main supporting tissue of stems
colloid gel-like material inside follicles that is the source of thyroglobulin, a prohormone of the T3 and T4 thyroid hormones
colony several individuals living together in close association
commensalism association between two different species in a community in which one benefits and the second apparently neither gains nor is harmed
community biological unit consisting of all the populations living in a specific area at a specific time

competition interaction between individuals of the same or different species that use one or more of the same resources in the same ecosystem

concentration gradient occurs when there is a difference in solute concentration from one area to another

conclusion a section at the end of the report that relates back to the question, summarises key findings and states whether the hypothesis was supported or rejected

connective tissue diverse solid tissues that connect and support other tissues and organs, or store fat deposits, and fluid tissues that transport materials such as nutrients, wastes and hormones

consumers organisms that obtain their energy and organic matter by eating or ingesting the organic matter of other organisms; also termed heterotrophs

continuous data quantitative data that can take any continuous value

control group a group that is not affected by the independent variable and is used as a baseline for comparison

controlled variable variable that is kept constant across different experimental groups

core body temperature temperature of internal cells of the body; in humans, core temperature is around 37 °C

corm a rounded underground storage organ present in plants, consisting of a swollen stem base covered with scaly leaves

correlation measure of a relationship between two or more variables

corticoid hormones steroid hormones, including aldosterone and cortisol, produced by cells of the cortex of adrenal glands

cortisol a steroid hormone produced by the adrenal cortex that has many roles, including control of blood glucose levels during stress and the body's recovery from the stress response

countercurrent exchange system situation in which two fluid systems flowing adjacent to each other, but in opposite directions, enables the transfer of heat or compounds from one system to the other by diffusion

covalent bonds a type of bond between atoms where electrons are shared. The bond is very strong.

crenation shrinking of cell due to water loss

crossing over an event that occurs during meiosis, involving the exchange of corresponding segments of non-sister chromatids of homologous chromosomes

crypts tube-like depressions of the mucosa located in the intestine and the site of glandular cells

cuticle waxy layer on the outer side of epidermal cells; waxy outer layer on leaves

cuttings a type of vegetative propagation that involves taking pieces of shoots, roots or leaves and planting them

cytochrome c a protein that has a role in the formation of ATP in mitochondria; its leakage from the mitochondria leads to apoptosis

cytogeneticist a scientist who specialises in the study of human karyotypes

cytokinesis division of the cytoplasm occurring after mitosis

cytoplasm formed by cell organelles, excluding the nucleus, and the cytosol

cytosine (C) one of the pyrimidine bases found in the nucleotides that are the building blocks of DNA (and RNA)

cytoskeleton network of filaments within a cell

cytosol the aqueous part of the cell

death receptor receptors on the surface of the cell that, when activated, lead to apoptosis of the cell

decomposers organisms, such as fungi, that can break down and absorb organic matter of dead organisms or their products

deletion type of chromosome change in which part of a chromosome is lost

density-dependent factor factor whose impact on members of a population is dependent on the size of the population

density-independent factor factor whose impact on members of a population is not affected by the size of the population

deoxyribonucleic acid (DNA) nucleic acid containing the four bases — adenine, guanine, cytosine and thymine — which forms the major component of chromosomes and contains coded genetic instructions

dependent variable the variable that is influenced by the independent variable. It is the variable that is measured.

dermal tissue tissue that protects the soft tissues of plants and controls water balance

dermis underlying part of the skin

desiccation drying out

detritivore organism that eats particles of organic matter found in soil or water

detritus fragments of organic material present in soil and water

differentiation the process by which cells, tissues and organs acquire specialised features

dihybrid cross a cross in which alleles of two different genes are involved

diploid ($2n$) having two copies of each specific chromosome in each set

discrete data quantitative data that can only take on set values

discussion a detailed area of a report in which results are discussed, analysed and evaluated; relationships to concepts are made; errors, limitations and uncertainties are assessed; and suggestions for future improvements are discussed

diversity measure of 'species richness' or the number of different species in a community

dominant a trait that is expressed in the heterozygous condition; also a trait that requires only a single copy of the responsible allele for its phenotypic expression

dormancy condition of inactivity resulting from extreme lowering of metabolic rate in an organism

double-blind type of trial in which neither the participant nor the researcher is aware of which participants are in a control or experimental group

Down syndrome (DS) a chromosomal disorder due to the presence of an additional number-21 chromosome, either as a separate chromosome (trisomy or triplo-DS) or attached to another chromosome (translocation DS)

drought resistant being able to store water and hence live for long periods of time without water

drought tolerant being able to tolerate a period of time without water

duplication type of chromosome change in which part of a chromosome is repeated

ecdysone steroid hormone which controls moulting in insects

ecology study of communities in their habitats and the interactions between them and their environment

economic factors factors that are related to the financial implications of an action or decision; these may be on an individual level or on a larger community or national level

ecosystem diversity the variety of physical environments in which organisms live and with which they interact

ecosystems biological units comprising the community living in a discrete region, the nonliving surroundings and the interactions occurring within the community and between the community and its surroundings

ectoderm the most external primary germ or cell layer that differentiates into epithelial tissue, which covers the outer surfaces of the body; epithelial tissue that covers the outer surface of the body

embryo early stage of a developing organism; in humans this includes the first eight weeks of development

embryo splitting process of separating the totipotent cells of a very early embryo, so that the resultant cells are each able to form a complete embryo

embryonic stem cell (ESC) an undifferentiated cell obtained from early embryonic tissue that is capable of differentiating into many cell types

endocrine glands ductless glands that distribute hormones via the bloodstream

endocrine system system of ductless glands that produce hormones and release them directly into the bloodstream

endocytosis bulk movement of solids or liquids into a cell by engulfment

endoderm the innermost primary germ layer that differentiates into digestive lining and organs like the lungs; a layer of cells that forms the innermost part of the cortex and encircles the vascular bundle that includes the xylem

endoparasite parasite that lives inside its host

endoplasmic reticulum cell organelle consisting of a system of membrane-bound channels that transport substances within the cell

endosymbiosis a special case of symbiosis where one of the organisms lives inside the other

endosymbiotic theory see endosymbiosis; a theory proposed by Lynn Margulis

enucleated cell a cell from which the nucleus has been removed

environment the external conditions (both biotic and abiotic factors) that surround and affect an organism

environmental sex determination (ESD) the sex of the offspring is established by environmental conditions rather than genetic factors

ephemeral short-lived or lasting a short period of time

epicormic shoot growth occurring from dormant buds under the bark after crown foliage is destroyed
epidermis the outer layer of cells; in human skin it consists of three layers (outer region of dead cells, layers of living keratinocytes, and a basal layer of melanocytes and constantly dividing stem cells)
epigenetic factors external factors which change genes, but not the base DNA sequence
epigenetic inheritance the inheritance of epigenetic tags across generations
epigenetics the study of changes in organisms caused by modifications of gene expression rather than alteration of the genetic code itself
epithelial tissue sheets of cells that cover the external surface of the body and also line internal surfaces that connect to the external environment
error differences between a measurement taken and the true value that is expected; errors lead to a reduction in the accuracy of the investigation.
ethical issue a problem or situation that requires a person or organisation to choose between alternatives that must be evaluated as right (ethical) or wrong (unethical)
ethics acceptable and moral conduct determining what is 'right' and what is 'wrong'
eukaryote any cell or organism with a membrane-bound nucleus
evidence reliable and valid data used to support or refute a hypothesis, model or theory
excretion process of removal from the body of various types of waste material arising from its metabolic activities
exhaustion hypothermia occurs when a person is exposed to a cold environment and cannot generate sufficient metabolic heat to maintain their core body temperature due to exhaustion or lack of food
exocytosis movement of material out of cells via vesicles in the cytoplasm
exoparasite parasite that lives on its host
experimental group test group that is exposed to the independent variable
exponential growth population growth that follows a J-shaped curve but cannot continue indefinitely
external fertilisation union of sperm and egg occurring outside the body of the female parent
extremophile microbe that lives in extreme environmental conditions, such as high temperature and low pH
extrinsic coming from outside
facultative parthenogenesis a type of parthenogenesis in which females can reproduce via both sexual and asexual reproduction
falsifiable able to be proven false using evidence
feedback loop a process in which the response (output) in a stimulus–response action affects the stimulus (input), either increasing or decreasing it
fertilisation the union of egg and sperm to form a zygote
filtrate fluid composed of blood plasma minus large proteins that is filtered into Bowman's capsule from the glomerulus
flagella whip-like cell organelles involved in movement; singular = flagellum
flame cells excretory cells with flagella present in members of phylum *Platyhelminthes*
fluid mosiac model a model proposing that the plasma membrane and other intracellular membranes should be considered as two-dimensional fluids in which proteins are embedded
follicles secretory cells that enclose the thyroid follicles and that synthesise and release the hormones, T3 and T4
free-standing water water available for an animal to use, including to drink
fungal spore microscopic biological particle that allows fungi to reproduce asexually
Fungi any of a wide variety of organisms that have a cell wall made of chitin and reproduce by spores, including the mushrooms, moulds, yeasts, and mildews
G1 checkpoint a check that occurs during G1 of interphase that makes sure the DNA is not damaged and is ready to undergo replication
G1 stage of interphase the first stage of interphase in the cell cycle where the cell grows, increasing the amount of cell cytosol
G2 checkpoint a check that occurs during G2 of interphase where the replicated DNA of the cell is checked for completeness and lack of damage; if the cell passes this checkpoint, it can then advance to mitosis

G2 stage of interphase the third stage of interphase where proteins are synthesised and the cell continues to grow in preparation for division
gall bladder sac shaped organ which stores the bile after it has been secreted by the liver
gamete an egg (ovum) or a sperm cell
gastric glands glands of the stomach that contain various epithelial secretory cells, producing neutral mucus, stomach acid and enzymes
gastric juice acid fluid secreted by the stomach glands for digestion in the stomach
gastrodermal tissue epithelial tissue that lines the inner cavity of the body
gastrointestinal tract the whole passage from mouth to anus; see alimentary canal
gene a section of a chromosome that codes for a protein through the order of the nucleotide base sequence it possesses
genetic diversity the variety of genes or the number of different inherited characteristics present in a species
genetic screening testing of persons to detect those with the allele responsible for a particular genetic disorder
genetic variation variation exhibited among members of a population owing to the action of genes
genome the sum total of an organism's DNA
genomics the study of the entire genetic make-up or genome of a species
genotype both the double set of genetic instructions present in a diploid organism and the genetic make-up of an organism at one particular gene locus
glucagon hormone produced by alpha cells of the pancreas that acts on liver cells resulting in increased release of glucose from the liver cells into the bloodstream
gluconeogenesis cellular production of glucose using non-carbohydrate precursors; occurs mainly in liver cells
glycogen a polysaccharide storage carbohydrate built from glucose; found mainly in liver and muscle tissue
glycolipids small chains of carbohydrates (sugars) attached to the phospholipids and proteins of the plasma membrane; aid in cell recognition
glycoprotein combination formed when a carbohydrate group becomes attached to the exposed part of a trans-membrane protein
Golgi apparatus organelle that packages material into vesicles for export from a cell (also known as Golgi complex or Golgi body)
grana stacks of membranes on which chlorophyll is located in chloroplasts; singular = granum
ground tissue all plant tissues that are not dermal or vascular
growth the process of increasing in size
growth rate the change in population size over a set period of time
guanine (G) one of the purine bases found in the nucleotides that are the building blocks of DNA (and RNA)
gut microbiota the population of organisms which live in the gut and play a crucial role in maintaining immune and metabolic homeostasis and protecting against pathogens
halophile an organism that grows in or can tolerate saline conditions
haploid (*n*) having one copy of each specific chromosome in each set
haustoria thin strand of tissue through which a plant parasite makes connection with its host; singular = haustorium
heat exhaustion an increase in core body temperature; symptoms include poor coordination, slower pulse and excessive sweating; may develop into heat stroke
heat stroke a critical and life-threatening condition where brain function is affected; symptoms include high core body temperature in excess of 40 °C, slurred speech, hallucinations and multiple organ damage
hemiparasitism form of parasitism in which a plant parasite obtains some nutrients and water from its host plant but also makes some of its own food through photosynthesis
hemizygous the genotype with respect to any gene carried on either the X or the Y chromosome, which comprises just a single allele for each gene
hemolymph the internal fluid of insects, analogous to blood in vertebrates; mostly water; it also contains ions, carbohydrates, lipids, glycerol, amino acids, hormones and some cells.
herbivore organism that eats living plants or parts of them
herbivore–plant relationship a form of interaction within a community between plants and the animals that eat them

hermaphrodite organism with both male and female reproductive organs
heterogametic organisms with two different sex chromosomes; the prefix hetero, from the Greek term *heteros*, means to be different
heterotroph organism that ingests or absorbs food in the form of organic material from their environment; also known as a consumer
heterozygous a genotype at a particular gene comprising of two different alleles; for example, *Aa*
histogram a graph in which data is sorted in intervals and frequency is examined. This is used when both pieces are data are quantitative. All columns are connected in a histogram.
histone a protein found in eukaryotic chromosomes that assists in packaging the DNA
holoparasitism form of parasitism in which a plant parasite depends completely on its host for nutrients and water
homeostasis condition of a relatively stable internal environment maintained within narrow limits
homogametic organisms with two of the same sex chromosomes; the prefix homo, from the Greek term *homos*, means same
homologous matching pairs of chromosomes that have the same genes at the same positions
homozygous a genotype at a particular gene comprising of two identical alleles; for example, *AA* or *aa*
hormones chemical messengers, released by endocrine glands, that regulate the function of distant organs, each with a specific receptor for its hormone
host organism on or in which a specific parasite lives
humidity measure of the amount of water vapour in the atmosphere
hummock grassland major vegetation type dominated by spinifex grasses and occurring over one-quarter of Australia
hydrophilic substances that dissolve easily in water; also called polar
hydrophobic substances that tend to be insoluble in water; also called non-polar
hyperglycaemia a condition where glucose levels in the blood rise above normal
hyperthermia condition in which core body temperature exceeds the upper end of the normal range without any change in the temperature set point
hyperthyroidism condition in which there is an overabundance in thyroid hormone production
hypertonic having a higher concentration of dissolved substances than the solution to which it is compared
hyphae long, branching, filamentous structures of a fungus that make up mycelium; singular = hypha
hypoglycaemia glucose levels in the blood drop below normal
hypothermia condition in which an individual has an extremely low body temperature and is at risk of death
hypothesis a tentative, testable and falsifiable statement for an observed phenomenon that acts as a prediction for the investigation
hypotonic having a lower concentration of dissolved substances than the solution to which it is compared
ideogram a stylised representation of a haploid set of chromosomes arranged by decreasing size
incomplete dominance the appearance in a heterozygote of a trait that is intermediate between either of the trait's homozygous phenotypes
independent assortment the formation of random chromosome combinations during meiosis that contributes towards producing variation
independent variable the variable that is changed or manipulated by an investigator
induced pluripotent stem cell (IPSC) a stem cell that has been genetically reprogrammed to return to an undifferentiated embryonic state
insulating layers layers of fat under the skin of mammals that retain heat within the body
insulin hormone produced by beta cells of the pancreas that acts to increase the uptake of glucose from the blood by body cells
integral proteins fundamental components of the plasma membrane that are embedded in the phospholipid bilayer
intermediate filaments one of the components of the cytoskeleton of a cell, composed of protein; they form a rope-like arrangement and give mechanical support to cells
internal fertilisation union of sperm and egg occurring inside the body of the female parent
interphase a stage in the cell cycle that is a period of cell growth and DNA synthesis

interspecific competition competition for resources in an ecosystem involving members of one species and members of other species

interstitial fluid fluid that fills the spaces between cells and bathes their plasma membranes

intraspecific competition competition for resources in an ecosystem involving members of the same species

intrinsic coming from inside

isotonic having the same concentration of dissolved substances as the solution to which it is compared

juvenile hormone insect hormone in the nymphal and larval stage of all insects; it is a lipid, derived from a fatty acid.

K-selected species organisms that live in stable environments and produce few offspring in each litter with a greater chance of survival to adulthood

karyotype an image of chromosomes from a cell arranged in an organised manner

ketones acid; breakdown product of fat metabolism

keystone species species whose presence in an ecosystem is essential for the maintenance of that ecosystem

kinetochore a special attachment site of a chromatid by which it links to a spindle fibre

lacteals the vessels of the lymphatic system which absorb digested fats

laws universal ideas often accepted as facts that describe what is expected to occur

legal factors factors that are related to the legality of an action

ligand a substance that forms a complex with a biomolecule to serve a biological purpose, such as the production of a signal upon binding to a signal

lignin a complex, insoluble cross-linked polymer

limitations factors that have affected the interpretation and/or collection of findings in a practical investigation

limiting factor environmental condition that restricts the types of organism that can survive in a given habitat

line graph graph in which points of data are joined by a connecting line. These are used when both pieces of data are quantitative (numerical).

line of best fit a trend line that is added to a scatterplot to best express the data shown. These are straight lines, and are not required to pass through all points.

linkage group genes that are physically close to each other on a chromosome and are likely to be inherited together as a single unit

linked genes describes genes whose loci are located on a given chromosome

lipophilic 'lipid-loving' molecules that dissolve in lipids (hydrophobic)

lipophobic 'lipid-fearing' molecules that do not dissolve in lipids (hydrophilic)

locus the position of a gene on a chromosome; plural = loci

logbook a record containing all the details of progress through the steps of a scientific investigation

lumen the inside space of a tubular structure

lysis bursting of a cell

lysosome vesicle filled with digestive enzymes

M checkpoint a check that occurs during mitosis where the connection between chromatid and spindle fibres is checked and corrected

maladaptation feature or trait of an organism that inhibits survival in a particular habitat

marsupials the order of non-placental mammals that are born at a very early stage of development and then grow inside their mother's pouch

maternal inheritance inherited traits that only occur through the mother's egg to her offspring; only her daughters can pass the trait on

measurement bias a type of influence on results in which an experiment manipulates their results to get a desired outcome. This may be unintentional (i.e. through the placebo effect) or intentional.

mechanical digestion digestion that uses physical factors such as chewing with the teeth

meiosis a type of cell division that produces haploid gametes

melanin the pigment produced by melanocytes; in humans it gives skin, hair and eyes their colour

melanocytes the pigment-producing cells in the basal layer of the epidermis

melanoma cancer derived from the pigment-producing cells (melanocytes)

membrane-bound organelle an organelle that has a membrane surrounding it

meristematic tissues plant tissue found in tips of roots and shoots that is made of unspecialised cells that can reproduce by mitosis

mesoderm the middle primary germ layer that differentiates into various tissues and organs, including the heart

messenger RNA (mRNA) single-stranded RNA formed by transcription of a DNA template strand in the nucleus; mRNA carries a copy of the genetic information into the cytoplasm

metamorphosis process of transformation — in insects (holometabolous) and amphibians — from an immature form to an adult that involves major changes in body structure and physiology

metaphase stage of mitosis during which chromosomes align around the equator of a cell

metastasis a process where malignant tumours spread throughout the body

methyl group a group of one carbon atom bonded to three hydrogen atoms ($-CH_3$)

methylation the addition of a methyl group ($-CH_3$) to a cytosine base of DNA, usually to repress gene transcription

microfilaments one of the components of the cytoskeleton of a cell; very thin threads composed of the protein actin; they allow cells to move and change in shape

microscopic anything that is not visible to the naked eye and requires a microscope to view it

microtubules part of the supporting structure or cytoskeleton of a cell, made of sub-units of the protein tubulin

microvilli sub-microscopic outfoldings of the plasma membrane of enterocytes of the small intestine that form the so-called 'brush border' of these cells

migrate to move from one part of something to another

mimicry a situation in which one species has an appearance similar to that of a different but distasteful species, where that similarity apparently gives protection against predators

mitochondria in eukaryotic cells, organelles that are the major site of ATP production; singular = mitochondrion

mitochondrial DNA (mtDNA) the DNA within mitochondria

mitosis process involved in the production of new cells genetically identical with the original cell; an essential process in asexual reproduction

models representations of ideas, phenomena or scientific processes; can be physical models, mathematical models or conceptual models.

monohybrid cross a cross in which alleles of only one gene are involved

monosomy a condition in which a cell or organism has only one copy of a particular chromosome that is normally present as a homologous pair

monotremes the order of non-placental mammals that lay leathery-shelled eggs and secrete milk through pores in the skin

moult in insects, the periodic event during development of larvae or nymphs which involves growth in epidermal cell numbers, shedding of the noncellular outermost cuticle and production of a new expanded cuticle

mucosa the innermost lining of the digestive system

multiple fission process of division in which multiple cells are produced from a single starting cell

multipotent a cell that can differentiate into a number of closely related cell types

muscularis muscle tissue of the gut

mutualism an association between two different species in a community in which both gain some benefit

mycelium the vegetative part of a fungus, consisting of a network of fine, white filaments

mycorrhiza fine threads formed by a fungus that form a large surface area for uptake of nutrients

negative feedback a regulatory mechanism in which a stimulus results in a response that is opposite in direction to that of the stimulus

negative growth rates population decreases over a stated period of time

nephron functional unit of the kidney

neurosecretory cells cells that receive nerve impulses and respond by a chemical stimulus

nitrogen-fixing bacteria bacteria able to convert nitrogen from the atmosphere into ammonium ions

nominal data qualitative data that has no logical sequence

nondisjunction failure of normal chromosome separation during cell division

non-homologous non-matching chromosomes

nuclear envelope membrane surrounding the nucleus of a eukaryotic cell

nuclear pore complex (NPC) protein-lined channel that perforates the nuclear envelope

nucleolus a small, dense, spherical structure in a nucleus that is composed of RNA and produces rRNA

nucleosome a section of supercoiled DNA around histones

nucleotide a sub-unit of DNA consisting of a phosphate group, a nitrogenous base and a sugar

nucleus in eukaryotic cells, membrane-bound organelle containing the genetic material DNA

obligate parthenogenesis a type of parthenogenesis in which a species can reproduce only through asexual reproduction

oligopotent a cell that has the ability to differentiate into a few different cell types

omasum third part of the stomach of a ruminant which acts as a filter and where water is absorbed

omnivore organism that eat both plants and animals

oncogene a gene that signals cells to continue dividing

open populations refers to populations that experience migration of individuals into the population (immigration) or out of the population (emigration)

operculum in fish, the flaps covering the gills; in molluscs, a hard impermeable lid that closes the shell, making a watertight compartment for the animal inside

opinions statements based on an individual's personal belief and experience

ordinal data qualitative data that can be ordered or ranked

organelle any specialised structure that performs a specific function

organism any living creature

osmoregulation process by which the volume of body fluids and their solutes concentrations are controlled

osmosis net movement of water across a partially permeable membrane without an input of energy and down a concentration gradient

osmotic adjustment a control mechanism for preventing water loss due to reverse osmosis in plants under water or salt stress; involves root cells accumulating additional solutes that lower the cell water potential to below that of the soil water

osmotic flow the net movement of water molecules from a solution of high water concentration to lower water concentration (or alternatively, from a region of low to high solute concentration)

outlier result that is a long way from other results and seen as unusual

oxytocin a hormone that induces labour and milk release from mammary glands in females

p53 a protein that is coded for by a gene of the same name and regulates the cell cycle, hence functioning as a tumour suppressor

palliative care care to improve the quality of life of patients with life-limiting illness and their families through prevention and relief of suffering

pancreas organ that secretes digestive enzymes into the duodenum and hormones into the bloodstream

panting an increase in breathing rate that acts to reduce body temperature through evaporation

parafollicular cells secretory cells, located in small areas between follicles, that produce the thyroid hormone, calcitonin

parasite organism that lives on or in another organism and feeds from it, usually without killing it

parasitism a form of interaction within a community that involves one species, the parasite, living on or in another species, the host, typically without killing the host

parathyroid hormone a protein hormone synthesised and secreted by the parathyroid glands in response to a falling blood calcium levels

parenchyma thin-walled living plant cells; in leaves the site of photosynthesis and in roots, tubers and seeds the site of stored starch and oils

parental (noncrossover) gametes sex cells that contain the same combination of alleles along their chromosomes as were present in the parental cells

parthenogenesis one form of asexual reproduction in which new individuals are produced from unfertilised eggs

parthenote potential source of embryonic stem cells, derived from unfertilised human eggs that are artificially stimulated to begin development

paternal inheritance inherited traits that only occur through the transmission of Y-linked genes from the father's sperm to his sons

pedigree a graphic representation, using standard symbols, showing the pattern of occurrence of an inherited trait in a family

peripheral proteins either anchored to the exterior of the plasma membrane through bonding with lipids or indirectly associated through interactions with integral protein

peripheral nervous system the part of the nervous system containing nerves that connect to the central nervous system

peripheral surface temperatures temperature of cells on the outside of the body; may be many degrees cooler than core temperature

peristalsis involuntary constriction and relaxation of muscles in the alimentary canal to push food to the stomach

permanent tissues plant cells that can no longer divide and includes dermal tissue, ground tissue (parenchyma, collenchyma and sclerenchyma) and vascular tissue (xylem and phloem)

peroxisome small membrane-bound organelle rich in the enzymes that detoxify various toxic materials that enter the bloodstream

personal errors human errors or mistakes that can impact results, but should not be included in analysis

phagocytosis bulk movement of solid material into cells

phenotype an observable or measurable characteristic of an organism that is the product of genetic and environmental factors

phloem complex plant tissue that transports sugars and other organic compounds

phospholipid major type of lipid found in plasma membranes

photosynthesis process by which plants use the radiant energy of sunlight trapped by chlorophyll to build carbohydrates from carbon dioxide and water

phyllodes leaf-like structures derived from a petiole (stem of a leaf)

pinocytosis bulk movement of material that is in solution being transported into cells

placebo substance or treatment that is not designed to affect an individual (used as a control)

plant tissue culture technique used to clone plants in large numbers

Plantae a group of organisms that include land plants and algae. Cells have a cell wall of cellulose, a large permanent vacuole and some contain chloroplasts.

plantlets tiny young plants that develop from the meristem tissue along plant margins

plasma the fluid portion of blood in which blood cells are suspended

plasmid small ring of DNA found in prokaryotes

plasmolysis shrinking of the cytoplasm away from the wall due to water loss

political factors factors that are related to the actions of government

polysaccharides long chains of sugars joined together to form large molecules

population members of one species living in one region at a particular time

positive feedback a process in which the response to a stimulus causes an amplification of the initial change, increasing its level further away from the starting value

positive growth rates population increases over a stated period of time

posterior pituitary posterior lobe of the pituitary gland; it is made of neural tissue that stores and releases hormones sent from the hypothalamus.

precision how close multiple measurements of the same investigation are to each other

predator an animal that actively seeks out other animals as its source of food

predator–prey relationship a form of interaction within a community that involves the eating of one species, the prey, by another species, the predator

prey living animal that is captured and eaten by a predator

primary data direct or firsthand evidence about some phenomenon

primary ecological events in population dynamics, the factors that contribute directly to population density, such as births and deaths

primary source a document which is a record of direct or firsthand evidence about some phenomenon

procreation the production of offspring

producers photosynthetic organisms and chemosynthetic bacteria that, given a source of energy, can build organic matter from simple inorganic substances

proenzyme a precursor of an enzyme that must be activated to form the functional enzyme

prokaryote any cell or organism without a membrane-bound nucleus

prophase stage of mitosis in which the chromosomes contract and become visible, the nuclear membrane begins to disintegrate and the spindle forms

proteins macromolecules built of amino acid sub-units and linked by peptide bonds to form a polypeptide chain

Protista (protists) a group of organisms similar in the fact that they do not fall into any other kingdom; generally unicellular; can have aspects of both plant and animal cells

proto-oncogene a gene that leads to the production of proteins which initiate the cell cycle

psoriasis chronic autoimmune condition in which skin cells are overproduced, resulting in raised patches of red, inflamed skin, often covered in a crust of small silvery scales

pumps special transport proteins embedded across the plasma membrane that carry out the process of active transport

Punnett square a diagram used to determine the probability of an offspring having a particular genotype

pyloric sphincter a sphincter at the join of the stomach and the duodenum of the small intestine; it controls the flow of acidic chyme into the alkaline duodenum.

qualitative data data with labels or categories rather than a range of numerical quantities; also known as categorical data

quantitative data numerical data that examines the quantity of something (e.g. length, time)

r-selected species organisms that live in unstable environments and produce many offspring with the likelihood that few will survive to adulthood

random errors chance variations in measurements

randomised the assignment of individuals to an experimental or control group that is done in a deliberately random way

recessive a trait that is not expressed but remains hidden or masked in a heterozygous organism

recombinant (crossover) gametes sex cells that contain new combinations of alleles along their chromosomes than were present in the parental cells

reduction division production of gametes through cellular division in meiosis I that reduces the chromosome number from diploid to haploid

regenerative medicine an experimental field of research involving stem cells in medicine that raises promise for the treatment of degenerative conditions and severe trauma injuries

reliability the consistency of a measurement across multiple trials

repair to restore something damaged or faulty to a good condition

repeatability how close the results of successive measurements are to each other in the exact same conditions

replication copying or reproducing something

reproducibility how close results are when the same variable is being measured but under different conditions

response bias a type of influence on results in which only certain members of the target population respond to an invitation to participate in the clinical trial, resulting in an unrepresentative sample of the larger population

results a section in a report in which all data obtained is recorded. This is usually in the form of tables and graphs.

reticulum second part of the stomach of a ruminant that is the smallest of the four chambers; collects any heavy objects

reverse osmosis a process in which water moves by osmosis from root cells to the soil; occurs when an increase in the solute concentration in the soil water lowers its water potential to below that of the cell water

rhizoids fine, root-like structures present in some plants, such as mosses

rhizomes horizontal underground stems

ribonucleic acid (RNA) type of nucleic acid consisting of a single chain of nucleotide sub-units that contain the sugar ribose and the bases A, U, C and G

ribosomal RNA (rRNA) ribosomal ribonucleic acid; synthesises proteins and is the primary component of ribosomes

ribosome organelle containing RNA that is the major site of protein production in cells

risk assessment a document which examines the different hazards in an investigation and suggested safety precautions

root hairs extensions of cells of the epidermal tissue that forms the outer cellular covering of the root, responsible for absorption and uptake of liquid water

root system the below-ground system of plants which anchors the plant in the soil, is responsible for the absorption and conduction of water and minerals, and the storage of excess sugars (starch)

rough endoplasmic reticulum endoplasmic reticulum with ribosomes attached

rumen first part of the stomach of a ruminant where digestion occurs with the aid of gut microbiota

ruminants animals that absorb nutrients by fermenting food in a specialised stomach prior to digestion

runners stem-like growths extending from a mother plant's growing point

S stage of interphase the stage where the parent cell replicates its DNA; at the end of the S stage the parent cell contains two identical copies of its original DNA

sample size the number of trials or individuals being tested in an investigation

sampling bias a type of influence on results in which participants chosen for a study are not representative of the target population

scatterplot a graph in which two quantitative variables are plotted as a series of dots

scientific investigation methodology the principles of research based on the scientific method

scientific method the procedure that must be followed in scientific investigations that consists of questioning, researching, predicting, observing, experimenting and analysing; also called the scientific process

sclerenchyma dead cells with thickened walls for strength and rigidity

secondary data comments on or summaries and interpretations of primary data

secondary ecological event in population dynamics, abiotic or biotic factors that influence changes in population density, such as temperature

secondary source a document that comments on, summarises or interprets primary data

selection bias a type of influence on results in which test subjects are not equally and randomly assigned to experimental and control groups

selectively permeable (semipermeable) allows some substances to cross but precludes the passage of others

septum a wall, dividing a cavity or structure into smaller ones

serosa outer connective tissue which encloses the gut

set point midpoint of a narrow range of values around which a physiological variable fluctuates in a healthy person

sex chromosomes a pair of chromosomes that differ in males and females of a species; allosomes

shoot system the above-ground system of plants, the site of photosynthesis, transport of sugars and the site of reproductive organs

shunt vessels direct connections between arterioles and venules that bypass capillary beds

sister chromatids identical copies of DNA formed by the replication of a chromosome

smooth endoplasmic reticulum endoplasmic reticulum without ribosomes

social factors factors that influence both a whole society and the individuals within the society

socioeconomic the interaction of social and economic factors

sodium–potassium pump protein that transports sodium and potassium ions against their concentration gradients to maintain the differences in their concentrations inside and outside cells

solute substance that is dissolved

solution liquid mixture of the solute in the solvent

solvent liquid in which a solute dissolves

somatic cell nuclear transfer a cloning technique that involves the nucleus of a somatic cell being transferred into the cytoplasm of an enucleated cell that is then stimulated to divide

somatic stem cells undifferentiated cells obtained from various sources and capable of differentiating into related cell types; also known as adult stem cells

specialisation the adaptation of something for a specific function

species diversity the number of different species — that is, different populations — in a community

sphincters thickened rings of muscle which control the opening and closing of a tube

spindle fine protein fibres that form between the poles of a cell during mitosis and to which chromosomes become attached

spindle fibres clusters of microtubules, composed of the contractile protein actin, that grow out from the centrioles at opposite ends of a spindle

sporangium a plane aerial structure where spores are formed by mitosis

spore in bacteria, a reproductive structure that is resistant to heat and desiccation; also formed by fungi and some plants

stimulus–response model a representation of an action that starts with a stimulus and ends with the response to that stimulus

stomata pores, each surrounded by two guard cells that regulate the opening and closing of the pores

stress response combined physiological reactions to stress, also known as the 'fight or flight' response, that results from release of the hormones, adrenaline and noradrenaline from the adrenal medulla

stroma in chloroplasts, the semi-fluid substance between the grana, which contains enzymes for some of the reactions of photosynthesis

suberin a non-cellular waxy substance that forms a waterproof barrier found on the cell walls of some plant cells, including root epidermal cells (except for the root tip) and root endodermal cells

sub-mucosa connective tissue forming the second layer of the gut lining

sulfur-oxidising producer bacteria one group of bacteria that gain their energy by oxidising sulfur compounds

surface area to volume ratio a measure that identifies the number of units of surface area available to 'serve' each unit of internal volume of a cell, tissues or organism

symbiosis prolonged association between different species in a community in which at least one partner benefits; includes parasitism, mutualism and commensalism

systematic errors errors that affect the accuracy of a measurement that cannot be improved by repeating an experiment; usually due to equipment or system errors

telophase stage of mitosis in which new nuclear membranes form around the separated groups of chromosomes

tentative not fixed or certain, may be changed with new information

test cross a cross used to determine the genotype of an individual with an unknown genotype by crossing it to an individual with a homozygous recessive genotype

testable able to be supported or proven false through the use of observations and investigation

theory a well-supported explanation of a phenomena, based on facts that have been obtained through investigations, research and observations

therapeutic cloning cloning carried out to create an embryo from which stem cells can be harvested

thermograph an instrument that shows the surface temperature or heat signature of an object

thermoregulation the maintenance of core body temperature

thymine (T) one of the pyrimidine bases found in the nucleotides that are the building blocks of DNA

thyroid gland endocrine gland located in throat that produces and secretes the hormones including T3 and T4

tolerance limits the upper and lower limits of a particular environmental condition in which a species can survive

tolerance range extent of variation in an environmental factor within which a particular species can survive

tonoplast in plant cells, the membrane of the plant vacuole; separates it from the rest of the cytosol

totipotent a cell that is able to give rise to all different cell types

tracheids major water conducting cells in the xylem of all vascular plants

transitional epithelium type of stratified epithelium present only in the hollow organs of the excretory system

translocation type of chromosome change in which a chromosome breaks and a portion of it reattaches to a different chromosome

trans-membrane proteins spanning the plasma membrane but have parts exposed to the exterior and interior of the cells

transpiration loss of water from the surfaces of a plant

trisomy a condition in which a cell or organism has three copies of a particular chromosome that is normally present as a homologous pair

trophosome an organ found within deep sea worms that is colonised by bacteria that supply the host worm with food and energy

tuber a thickened underground part of a stem or rhizome
tumour-suppressor gene a type of gene that produces a protein that signals for cells to stop dividing
turgid swollen and distended
turgor pressure force or pressure potential within a fully hydrated plant cell that pushes the plasma membrane against the cell wall
type 1 diabetes a condition that results when the homeostatic mechanisms that regulate blood glucose levels fail when insulin production fails, characterised by a blood glucose level that is higher than normal
uncertainty limit to the precision of equipment; it is a range within which a measurement lies
unipotent a cell that has the ability to produce only cells of their own type
urinary tract a series of hollow organs comprising ureters, bladder and urethra that transport urine to the outside of the body
vacuole structure within plant cells that is filled with fluid-containing materials in solution, including plant pigments
validity credibility of the research results from experiments or from observations; a measure of how accurately and precisely results measure what they are intending to through fair testing
vascular plants plants with xylem and phloem tissue — the majority being flowering plants and conifers
vascular tissue plant tissue composed of xylem and phloem
vasoconstriction narrowing of the diameter of blood vessels
vasodilation widening of blood vessels to increase blood flow
vegetative propagation asexual reproduction in plants
vessels major water conducting cells in the xylem of angiosperms
villi outfoldings or projections of the mucosa of the small intestine
warning colouration a conspicuous colouring that warns a predator that an organism is toxic or distasteful
water balance the way in which an organism maintains a constant amount of internal water in hot or dry conditions
water tappers trees that have a single main root extending to depths near the watertable before forming lateral branches
X-linked gene a gene located on the X chromosome; also refers to a trait that is controlled by such a gene
xylem the part of vascular tissue that transports water and minerals throughout a plant and provides a plant with support
Y-linked gene a gene located on the Y chromosome; also refers to a trait that is controlled by such a gene
zero population growth occurs when growth by births and immigration matches losses by deaths and emigration over a stated period

INDEX

A

abiotic factors 568, 617
ABO blood type 462
ABO gene in humans 463–4
abomasum 198
abscisic acid (ABA) hormone 260
acacia shrubland 578
acinar cells 187
active boundary 60
acute hypothermia 282
adaptations 567
   arid climate animals 572–7
   arid climate plants 577–86
   cold climate animals 586–9
   cold climate plants 589–92
   marine mammals 587
   predators 606
   tolerance range 568–72
   types of 567–8
adenine (A) 357
adenosine triphosphate (ATP) 29
adipocytes 214
adrenal glands 212–3
   adrenal cortex 212–3
   adrenal medulla 213
adult somatic cells 546
adult stem cells 121
albinism 449
aldosterone 213, 216
alimentary canal 175
alleles 360
   carrier 422–3
   codominance 423–4
   complete dominance 421–2
   dominant 416
   gene 360–3
   incomplete dominance 424–9
   multiple alleles 361–2
   recessive 416
   TYR gene 460–1
   variation in plants 362–3
amensalism 604
amniotic egg 533
amoebas 13
amphibians thyroid hormones 224
anaphase 95
anenomes 604
angelman syndrome 388
animal cells 42–3
   cycle 107
   organisation, cellular level 164–5
   organs in 169–71
   systems in 171–4
   tissues 165–9
animalia 40
animals regulate blood glucose
   blood glucose levels 296
   feedback loops 302–5
   glucose for body cells 295–6
   glucose-releasing hormone 298–9
   glucose-storing hormone 298
   glycogen 299
   releasing glucose 299
   too high 301–5
   too low 300–1
anterior pituitary 209
antidiuretic hormone (ADH) 308
antifreeze 586
antigens 462
apoptosis 110
   mechanisms of 112
apoptosome 112
apoptotic bodies 111
aquaporins 256
aquatic species 622
archaea 7
archaea vs. bacteria 9
arid climate animals
   frogs, dormancy 574–5
   migration 575
   reproduction 575–7
   tarrkawarra, without drinking 572–3
arid climate plants
   arid Australia, vegetation types 577–9
   dominant plants 584
   maximising water uptake 580
   minimising water loss 580–6
artificial cloning of mammals
   downside 547–8
   embryo splitting 544–5
   reproductive cloning 548–52
   somatic cell nuclear transfer 545–7
artificial insemination 539
asexual reproduction 91, 511
   advantages of 522–4
   best of both worlds 523–4
   binary fission in eukaryotes 513–5
   binary fission in prokaryotes 512–3
   disadvantages of 524–9
   mitosis in eukaryotes 515–22
   vs. sexual reproduction 536
Australian bushfire season 108
Australia's first peoples
   biodiversity maintenance 628–33
   burning techniques 628
   first human settlement 627
   seasons, use of fire 627–8
autophagy 36
autosomal dominant traits 447–9
autosomal genes 414
autosomal recessive traits 449
autosomes 370, 388
autotrophic organisms 596

B

bacteria 7
bacteria vs. archaea 9
bacterium cells 9
bamboo 608
basal metabolic rate 285
basal stem cells 122–3
binary fission 91–4
   eukaryotes 513–5
   prokaryotes 512–3
biodiversity 563
   maintenance 628–33
   nature 629–30
   Torres Strait Islands 630–3
bioethics 105
biomacromolecules 56
biomimicry 629
biotic factors 617
birds, digestive system
   crop 202
   large intestine 202–5
   mouth 201
   oesophagus 201
   small intestine 202
   stomach 202
bivalent 394
blebs 111
blood glucose in animals
   glucose for body cells 295–6
   regulate blood glucose levels 296
   too high 301–5
   too low 300–1
body temperature in animals
   behavioural changes 286–7, 291–2
   core body temperature 278–9

regulate body temperature 279–82
   too high 287–95
   too low 282–7
   vasoconstriction 283–4
breast showing normal tissues 114
brown adipose tissue (BAT) 285–6
budding 520–1
bulb 516
bulk transport 64, 72–8

C

camel 577
cancer 105, 113–8
cancer cells 13
capsule 8
carbohydrates 60
cardiac muscle tissue 167
carnassial teeth 194
carnivores 598
carrier proteins 66–8
carrying capacity 619
casparian strip 256
caspase cascade 112
caspases 112
cassowaries 600–2
cell-based therapies 121
cell cycle 94–103
   animals 107
   fungi 109–10
   humans 106–7
   plants 107
   programmed cell death 110–3
   regulation of 103–6
cell differentiation 118, 432
   stem cells 119–20
cell division 94
cell elongation 91
cell (plasma) membrane 8
cell replication 91
   drug effect of 115
   eukaryotes 93–103
   prokaryotes 91–3
cells 6
   basic structure 7
   liverworts 109
   mitochondria 29–32, 355
   multiple nuclei 28
   overview 4–5
   prokaryotic cells and eukaryotic cells 7–10
   shapes 12–17
   size 10–2
   theory 4
   types of 11
cell size

substance movement limits 17–24
surface area to volume ratio 17–18
V ratios limit 18
cell theory 6
cellular respiration 29
cellulose 46
cell wall 8, 45–48
centrioles 42–3, 105
centromere 96, 370, 376
channel proteins 66
Chargaff's rules 357
chemical action 433–7
chemical digestion 176
chemoreceptors 266
chenopod shrubland 578
chiasma 535
chief cells 185, 214
chitin 43
chloroplasts 43–5, 106
chlorophyll 44
cholera 72
cholesterol 59–60
chromatids 95
chromatin 96, 354
chromosomes 91, 353
   abnormal 384–92
   errors in sex 387–8
   eukaryotic 353–5
   important 353
   look the same 375–7
   number 379–82
   number of 384–7
   organising 382–3
   parts of 388
   prokaryotic 355–7
   rearrangements 388–92
   size differences 377–9
chyme 183
cilia 14, 39–40
circular folds 188
clones 511
closed populations 616
clumped distributions 615
codominance 423–4
codominant alleles 468
cohesion 162
cold climate animals
   ice can damage or kill 586–9
cold climate plants 589–92
collenchyma 155
colonies 40
colour blindness in humans 465
commensalism 612–4
community 594

competition 603
concentration gradient 64
connective tissues 166
consumers 598
coral spawning 529
core body temperature 278–9
corm 516
corticoid hormones 212
cortisol 216
countercurrent exchange system 588
covalent bonding, water 64
crenation 69
crossing over chromosome 394
crypts 119, 123, 187
cuticle 580
cuttings 516
cystic fibrosis (CF) 71–2, 416
cytochrome c 112
cytogeneticists 383
cytokinesis 94, 97–103
cytoplasm 17
cytosine (C) 357
cytoskeleton 38–9
cytosol 8

D

death receptor 112
decomposers 599
deletion 383, 388
deletion chromosome 388
density-dependent factors 617
density-dependent growth 619
density-independent factors 617
deoxyribonucleic acid (DNA) 7, 353
   bases 357–8
   genes 358–60
   mitochondrial 355
   protein 359
dermal tissue 154
dermis 123
desiccation 568
detritivores 598
detritus 598
deviant cell behaviour
   cancer 113–8
   psoriasis 113
different animals, digestive system
   alimentary canal 197
   eats 192–3
   hollow organs 196–201
   in birds 201–5
   microbial fermentation 197–201
   stomach 197
   teeth size, continuously growth 195–6
   teeth, mammals 193–6
different animals, excretory system

diffusion of ammonia   237–42
    invertebrates   237
differentiation   118–9
digestive system, animal   174–7
    connective tissue   181–2
    cow   197
    different animals   192–205
    epithelial tissues   179–80
    large intestine   191
    liver, bile   186–7
    mechanical and enzymic digestion
        mouth   184–5
        stomach   185–6
    muscle tissue   182–3
    oesophagus   185
    organs   183–92
    pancreas   187
    small intestine   187–9
    tissues in   177–83
    variation   178–9
dihybrid cross   413, 461
    ABO and *TYR* genes in humans   475–7
    fur colour   477–8
    two genes in action   475–8
diploid (2*n*)   363
distribution of populations
    clumped distributions   615
    random distributions   615–6
    uniform distributions   615
diversity   594
DNA replication   93
dominant plants
    Australian arid regions   584
    hummock grasslands   585
    saltbushes   585
    spinifexes   585
dormancy   574–5
down syndrome (DS)   383, 385
drought resistant   579
drought tolerant   579
duplication chromosome   388

E

ecdysone   221
ecological communities
    consumers   598
    decomposers   599
    producers   596–8
    species diversity   594–6
ecology   594
ecosystems   563, 593–4
    commensalism   612–4
    diversity   563
    herbivore-plant relationships   608–9
    interactions   602–14
    intraspecific and interspecific competition   603–4
    parasitism in animals   609–10
    predator-prey relationships   605–8
    storytelling   631
ectoderm   120, 165
ectotherms regulate temperature   279
elephants   600
embryo   120
embryo splitting   544–5
embryonic stem cells (ESCs)   121
endocrine glands   205
endocrine system   265
    negative feedback loops   273
endocrine system in animals   205
    different animals   220–7
    insects   221–7
    organs   215–20
    tissues   209–15
endocytosis   73–4
endoderm   120, 256
endoparasites   609
endoplasmic reticulum (ER)   25   32–4
endosymbiosis   48
endosymbiotic theory   48–53
energy source   5
enucleated cell   545
environment   419
environmental sex determination (ESD)   372–5
ephemeral   574
epicormic shoots, bushfire   107–9
epidermis   106
epidermis basal stem cells   122–3
epigenetics   430–7
    factors   431–2
    inheritance   430
epithelial tissues   165, 166, 179–81
ethical issues   126
eucalypt regeneration, bushfire   565
eukaryotes   7, 363
    cell replication   91–103
    cells types   40–2
    organelles for   27–40
eukaryotes, binary fission   513–5
    multicellular eukaryotes   514–5
    unicellular eukaryotes   513–4
eukaryotic animal cell   25
eukaryotic cells   8–10
eukaryotic chromosome   353–5
excretion   227
excretory system in animals   227
    different animals   236–42
    organs   230–6
    tissues   228–30

exhaustion hypothermia   282
exocytosis   74–8
exoparasites   609
exponential growth   92, 618–9
external fertilisation   529–30
    fish   530
    frogs   530
extremophiles   9
extrinsic   111

F

facultative parthenogenesis   522
feedback loops   271
fertilisation   363
flagella   14, 39–40
flame cells, planarians   238
fluid mosaic model   53–4
follicles   213
fragmentation   520
free-standing water   572
fungal spore   109
fungi   40
    cell cycle   109–10

G

G1 checkpoint   104
G1 stage of interphase   94
G2 checkpoint   104
G2 stage of interphase   94
gall bladder   186
gametes   363
gastric glands   179
gastric juices   185
gastrodermal tissue   165
gastrointestinal tract (GT)   175
genes   353
    alleles   360–3
    DNA   358–60
    multiple alleles   361–2
genetic diversity   534–5, 563
    biodiversity versus   563
    importance of   563–7
    lignotubers   565
genetic screening   484
genetic variation   524
genome   358, 363–9
    eukaryotes   363
    prokaryotes   364–9
genomics   363
genotypes   360
    Mendel's contribution   413
    pedigrees   454–60
    sex chromosomes   415
    writing   415–9
glucagon   297–9
gluconeogenesis   295
glucose-releasing hormone   298–9

glucose-storing hormone 298
glycogen 295, 299
glycolipids 60
glycoprotein 56
Golgi apparatus 34–5
grana 44
ground tissue 155
growth 91
growth rates 616
guanine (G) 357
gut microbiota 192

H

haematopoietic stem cells 124
halophiles 513
haploid (n) 363
haploid gametes
    chromosomes 392–3
    crossing over 394–6
    meiosis 393–4
hard-hoofed animals 605
haustoria 610
heat-conserving mechanism
    piloerection 284
    vasoconstriction 283–4
heat exhaustion 287
heat loss mechanism
    sweating 289–91
    vasodilation 289
heat-producing response
    brown adipose tissue 285–6
    shivering 284–5
heat stroke 287
HeLa cells 105
hemiparasitism 610
hemizygous genotypes 415
hemolymph 238
herbivore-plant relationships 608–9
herbivores 598
hermaphrodites 529
heterogametic 522
heterotrophs 598
heterozygous 414
histones 353
hollow organs, excretory system
    ureter 234
    urethra 235–6
    urinary bladder 234
homeostasis 263–6
    body systems 265–6
    control loops 270–1
    loops in action 269–70
    negative feedback loops 272–4
    stimulus-response loop model 268–9
homeostatic mechanisms
    hyperthyroidism 322–6

hypoglycaemia 320–2
    type 1 diabetes 315–20
homogametic 522
homologous chromosomes 369–70
homozygous 414
hormones 206
    moulting in insects 221
horticultural practice, cloning 542–3
host 609
human genome project 364
human karyotypes 389
human male sex chromosomes 359
humans
    ABO gene 463–4
    colour blindness 465
    TYR gene 461–3
humidity 573
hummock grassland 578
hydrophilic 55, 206
hydrophobic 55, 206
hyperglycaemia 316
hyperthermia 287, 572
    heat stroke 290
hyperthyroidism 322–6
hypertonic 68–70
hyphae 109
hypoglycaemia 319–22
hypothermia 282
hypotonic 68

I

identical twins 399
ideogram 369
imprinted genes 432–3
incomplete dominance 424–9
independent assortment 394, 396
induced pluripotent stem cells (iPSCs) 121, 125
inheritance patterns
    autosomal dominant traits 447–9
    autosomal recessive traits 449
    maternal pattern 451
    pedigree chart 447
    pedigrees, genotypes 454–60
    traits 452–3
    X-linked dominant traits 449–50
    X-linked recessive traits 450–2
    Y-linked traits 452
insects
    endocrine system 221–7
    malpighian tubules of 238
insulating layers 586
insulin 297–8
integral proteins 56
intermediate filaments 39
internal fertilisation 531
    birds 532–3

    insects 531–2
    mammals 533–4
    marine animals 531
    reptiles 532
interphase 94
interspecific competition 603
interstitial fluid 573
intestinal stem cells 123–4
intraspecific competition 603
intrinsic 111
isotonic 68

J

jellyfish, tissues 165
juvenile hormone (JH) 221

K

karyotypes 382
    analyse 383–4
    organising chromosomes 382–3
ketones 318
keystone species 600
    cassowaries 600–2
    elephants 600
    starfish 600
kidney
    disease in humans 235
    excretion 233
    filtrate 231–2
    tubular reabsorption 232–3
    tubular secretion 233
kinetochore 105
komodo dragons 522
K-selected populations 621

L

leaf epidermal layer 160
Legionnaire's disease 11
lethal genes 468–9
ligand 112
lignin 46
limiting factors, tolerance 570–2
linkage group 478
linked genes 479
    behaviour of 479
    in blood 479–80
    detecting linkage 480
    estimating distance between 481–2
    outcomes of crosses 482–7
lipophilic 62
lipophobic 62
liquid water 5
liver cells 28
living sponge 165
loci 369, 462
lumen 177

lysis   68
lysosomes   35–6

M

major histocompatibility complex   61
maladaptation   568
malpighian tubules of insects   238
mammalian tissues   166–9
mammals
   organs in   169–71
   temperature-sensitive colouration   430
marsupials   534
maternal inheritance   451
M checkpoint   104
mechanical digestion   176
mechanoreceptors   266
meiosis   371, 393–4, 516
   stages of   398
melanin pigment   460
melanocytes   113, 460
melanomas   113
membrane-bound organelles   7
membrane transport method
   active transport   70–2
   facilitated diffusion   66–8
   osmosis   68–70
   passive vs. active transport   64–5
   simple diffusion   65–6
Mendel's contribution   413
meristem   108
meristematic tissue   107, 154, 516
mesoderm   120
messenger RNA (mRNA)   25, 359
metabolic rate   285
metacentric chromosomes   377
metamorphosis   222
metaphase   95
metastasis   114
methyl groups   431
methylation   431
microbial cells   11
microfilaments   39
microscopic   10
microtubules   39
microvilli   187
migrate   91
mitochondria   29–32, 106
mitochondrial DNA (mtDNA)   355–451
mitosis   94–511
   drug effect of   115
mitosis in eukaryotes
   body cells, replacement   515
   budding   520–1
   fragmentation   520
   parthenogenesis   521–2
   spore formation in fungi   515–6
   vegetative propagation   516–20
mitotic spindle   104–6
monohybrid crosses   413, 461
   ABO gene in humans   463–4
   codominant alleles   468
   lethal genes   468–9
   plants   464
   test   469–74
   TYR gene in humans   461–3
   variations to   467–9
   X-linked *CBD* colour-vision gene   465–7
   X-linked genes   464–7
monosomy   384
monotremes   533
moults   222
mucosa   177
multicellular eukaryotes   514–5
multiple fission   514
multipotent   120
muscle tissues   166
muscularis   177
mutualism   612
mycelium   516
mycorrhiza   611

N

negative feedback   272
   blood calcium level   273–4
   thyroid hormone secretion   273
negative growth rates   616
nephridia of earthworm   240
nephron   228
nervous tissues   166
neurosecretory cells   211
newborn screening   484
nitrogen-fixing bacteria   611
nitrogenous wastes (N-wastes)   237
nondisjunction   386
non-fever hyperthermia   287–9
non-homologous   369
nuclear envelope   27
nuclear pore complexes (NPCs)   27
nucleolus   28–9
nucleosome   353
nucleotides   357
nucleus   7
nucleus control centre   27–9

O

obligate   521
oesophagus   185
oligopotent   120
omasum   198
omnivores   598
oncogenes   106
open populations   616
operculum   575
organelles   7
   animal cells   42–3
   for eukaryotes   27–40
   inside the cell   24
   plants   43–8
   ribosomes   25–7
organisms   5
   classification of   7
   V ratio   20
organs, endocrine system
   adrenal glands   216
   parathyroid gland   218–20
   pituitary gland   215–6
   thyroid gland   216–8
organs, excretory system
   hollow organs   234–6
   kidneys   230–3
osmoreceptors   266
osmoregulation   307
   antidiuretic hormone   309
   brain   308
   kidneys   308
osmosis   68–70
osmotic adjustment   257
osmotic flow   68
oxytocin   274

P

p53   104
pancreas   297
pandas   608
panting   290
parafollicular cells   214
paralysis tick   609
parasite   609
parasitism in animals   609–10
parathyroid glands   214–5
parathyroid hormone (PTH)   214
parenchyma   155
parental (noncrossover) gametes   479
parthenogenesis   522
parthenotes   121
passive vs. active transport   64–5
paternal inheritance   452
pedigrees
   chart   447
   genotypes   454–60
   pets   538
peripheral proteins   56
peripheral surface temperatures   279
permanent tissues   154
peroxisomes   36–8

phagocytosis 73
phenotypes 413, 419–21
   alleles 421–9
   environment 429–30
   epigenetics 430–7
   prokaryotes 420
phenylketonuria (PKU) 429
phloem tissue 155
phospholipids 55–6
photoreceptors 266
photosynthesis 43, 45–565
phyllodes 582
phytoplankton 597
piloerection 284
pinocytosis 73
pituitary gland 209–12, 285
plant cells
   getting to the top 161–4
   intake of water, tissues 156–7
   loss of water, tissues 159–64
   movement of water, tissues 157–9
   organs and tissues 153–6
   without soil 159
plant tissue culture 542
plantae 40
plants
   acid-sensitive colouration 429
   cell cycle 107
plantlets 516
plasma 572
plasma membrane
   components of 54–60
   crossing 62–4
   fluid mosaic model 53–4
   functions of 60–2
   proteins in 57–9
plasmids 8
plasmolysis 69
pluripotent 120
polysaccharides 35
population 594
population growth model
   density-dependent growth 619
   exponential growth 618–9
   populations affect other populations 619
   prey population numbers 619–21
population size
   density-independent or density-dependent 617–8
   primary ecological events 616–7
   secondary ecological events 617
populations 563
populations of species 621–6
positive feedback 272
   childbirth 274–8

positive growth rates 616
posterior pituitary 209, 211–2
Prader-Willi syndrome 433
predator-prey relationships 605–8, 620
predators 600
prey 600
primary ecological events 616–7
procreation 91
producers 596–8
proenzyme 185
programmed cell death 110–3
prokaryotes 7, 364–9
   binary fission 512–3
   cell replication 91–3
   phenotype 420
prokaryotic beginnings 48–53
prokaryotic cells 8
prokaryotic chromosome 355–7
prophase 95
protein 25
   importance of 26–7
   production 26
proteins 25, 56–9
prothoracicotropic hormone (PTTH) 221
protista 40
proto-oncogenes 106
psoriasis 113
pumps 70
Punnett squares 461, 471
pygmy possums 587
pyloric sphincter 183

R
random distributions 615–6
receptors 266
recombinant (crossover) gametes 479
reduction division 393
regenerative medicine 124–6
regulate body temperature 279–82
releasing glucose 299
repair 91
reproductive cloning technologies
   artificial cloning of mammals 543–52
   horticultural practice 542–3
reticulum 198
reverse osmosis 257
rhizoids 109
rhizomes 516
ribonucleic acid (RNA) 28
ribosomal RNA (rRNA) 25
ribosomes 8, 25–7
root hairs 156, 255
root systems 153

absorbing water 255–6
finding water 254–5
grow towards moist soil 255
water entry, xylem 256–7
water stress 257
waterlogging 257–8
rough endoplasmic reticulum 32–3
royal X-linked recessive disorder 466
r-selected populations 621
rumen 198
ruminants 195, 197
runners 519

S
saving burns victims 106
scanning electron micrograph (SEM) 255
sclerenchyma 155
secondary ecological events 617
selectively permeable 54
semipermeable 54
septum 91
serosa 177
set point 264
sex chromosomes 370–1, 388
sex chromosomes, genes 415
sex determination
   birds and reptiles 372
   environmental sex determination 372–5
   mammals 371–2
sexual reproduction
   advantages and disadvantages 536–42
   external fertilisation 529–30
   fusion of gametes 529
   genetic diversity 534–5
   internal fertilisation 531–4
   vs. asexual reproduction 534
shivering 284–5
shoot system 153
shunt vessels 284
sister chromatids 375
small intestine, digestion 187–9
   digestion of carbohydrates 189
   digestion of fats 189
   digestion of proteins 189–90
   functions of 189–90
smooth endoplasmic reticulum 32–4
sneaker fish 537
sodium-potassium pump 70
solutes 65
solution 68
solvents 68
somatic cell nuclear transfer (SCNT) 544–7

somatic stem cells   121–4
specialisation   118–9
species diversity   563
sphincters   183
spindle   94
spindle fibres   96
sporangia   516
spore   514
S stage of interphase   94
stable environmental conditions   5
starfish   600
stem cells   119–20
   haematopoietic   124
   intestinal   123–4
   medicine   124–9
   regenerative medicine   124–6
   somatic   122–4
   sources of   120–2
   therapeutic cloning   126–9
stimulus-response models   266–71
   homeostatic   268–9
   open   267–8
stomata   159, 258–63, 580
   action of   160–1
   guard cells   261
   pore closing   260
   to minimise water loss   259–61
stress response   213
stroma   44
suberin   255
submetacentric chromosomes   377
sub-mucosa   177
sulfur-oxidising producer bacteria   612
surface area to volume ratio   17
sweating   289–91
symbiosis   613

T

tarrkawarra   572–3
telophase   95
test cross, monohybrid   469–74
therapeutic cloning   126–9
thermoreceptors   266
thermoregulation   280
thymine (T)   357
thyroid gland   213–4, 285
thyroid hormone secretion   273
thyroid-stimulating hormone (TSH)   210

thyrotropin-releasing hormone (TRH)   210
thyroxine   285
tissues in endocrine system
   adrenal glands   212–3
   parathyroid glands   214–5
   pituitary gland   209–12
   thyroid gland   213–4
tissues in excretory systems
   connective tissue   230
   epithelial tissues   228–9
   muscle tissue   230
tolerance limits   568
tolerance range   568–72
   limiting factors   570–2
   physiological and behavioural adaptations to   569
   technology   571
tonoplast   45
totipotent   120, 546
tracheids   157
trans-membrane proteins   56
transitional epithelium   228
translocation   386
transmission electron microscope (TEM)   11
transpiration   159, 580
   water loss   258
trilling frog   574–5
trisomy   384
trophosome   612
tuber   516
tumour-suppressor genes   106
turgid   68
turgor pressure   253
type 1 diabetes   315–20
   diagnosis of   318
   glucose levels   316–8
   symptoms of   318
   treatment of   318–20
TYR gene   461–3

U

unicellular eukaryotes   513–4
uniform distributions   615
unipotent   120
ureter   234
urethra   235–6
urinary bladder   234
urinary tract   227

V

vacuole   43
vascular plants   153
   major types of tissue   154
vascular tissue   155
vasoconstriction   283–4
vasodilation   289
vegetative propagation   516–20
vesicles   35–8
   lysosomes   35–6
   peroxisomes   36–8
vessels   157
villi   187

W

water balance   572
water balance in animal
   amphibians   312–4
   birds   312
   different animals   312–4
   fish   312
   regulate water   305–9
   too high   310–2
   too low   309–10
water balance in vascular plants   253
   water loss control   258–63
   water uptake control   253–8
water loss control   253–8
water tappers   580
white blood cells   13
WZ/ZZ system   372

X

X-inactivation   432
X-linked *CBD* colour-vision gene   465–7
X-linked dominant traits   449–50
X-linked genes   464–7
X-linked recessive traits   450–2
XX/XY system   371–2
xylem   253
xylem tissue   155

Y

Y-linked traits   452
Yellowstone National Park   601

Z

zero population growth   616